25-9	1·00	1·[illegible]	[illegible]
1616	1·26	1·8[illegible]	[illegible]
3240	1·82	4·38	3·17
4300	2·29	8·27	4·83

Use of correct cascade theory instead of Serber's solution would not change general conclusion, though the actual numerical rates of decay to primary soft component.

Johnson & Shutt '42 P. R. 61. 380-1.
Cloud Chamber Track of a Mesotron stopped by Gas

—ve meson track ends in gas after passing through 1 cm Pb plate. Track above plate thin, below, heavily ionizing. No ~~de~~ decay electron visible from end of track

Origin of the soft component of cosmic rays.

No: D_2 of cascade electrons due to meson decay calculated (using Serbers' ~~method~~ cascade calculations) Those due to fast mesons D_1 (penetrate 15 cm Pb, momentum $> 3 \times 10^8$ e) and slow mesons D_1 separated ~~(D_1)~~.

No: of fast mesons, N, slow mesons N_1, no: of electrons due to meson collision, C

Depth in rad. units	N_1/N	γ/N	D_1/N	C/N	$\frac{N_1 + D + D_1 + C}{N}$	Observed
23.41	·07	·22	·010	·10	·40	·50
19.93	·09	·31	·014	"	·51	·74
16.47	·13	·39	·025	"	·65	1·20
14.33	·27	·47	·060	"	·90	1.81.

Shows that actual no: of soft component increase much more rapidly than no: ~~of~~ if produced by mesons ~~comp.~~ through decay + collision. At S.L. two numbers comparable, but at 4300 m. some half at least must be due to direct cascades produced by electrons at the top

...ce not all soft-comp: above S.L. can be

HOMI J.
BHABHA

HOMI J. BHABHA

A LIFE

BAKHTIAR K. DADABHOY

RUPA

Published by
Rupa Publications India Pvt. Ltd 2023
7/16, Ansari Road, Daryaganj
New Delhi 110002

Sales Centres:
Bengaluru Chennai
Hyderabad Jaipur Kathmandu
Kolkata Mumbai Prayagraj

P-ISBN: 978-93-5520-818-7
E-ISBN: 978-93-5520-820-0

Second impression 2024

10 9 8 7 6 5 4 3 2

Printed in India

In memory of R.M. Lala and S.S. Tarapore

The heights by great men reached and kept
Were not attained by sudden flight,
But they, while their companions slept,
Were toiling upward in the night.

—'The Ladder of St. Augustine'
Henry Wadsworth Longfellow

CONTENTS

INTRODUCTION

The origins of this book lie in a detailed profile I wrote on Homi J. Bhabha for my book *Sugar in Milk: Lives of Eminent Parsis* over a decade ago. Until then, I had only known what everybody knows about Bhabha: that he was a distinguished physicist who pioneered fundamental research and atomic energy in India. At the time, I was only dimly aware of what a fascinatingly diverse man he was: a lover of music, an accomplished painter (and, later, an avid collector), a violinist and a budding poet—truly a Renaissance man. I also did not know much about Bhabha the man—his loves, his likes, his bond with his parents, his close relationship with a lady companion and more. He drew upon a profound knowledge of science, history and English literature, and his assimilation into western, particularly British life, was so complete that his real life-space was little understood by his European colleagues. No one expected to find a scientist from a British colony who was indistinguishable from his western counterparts. Bhabha cast serious doubts on the dichotomy of metropolis and colony inherent in the institutions of science at the time.

Learning all this, I was surprised to discover that no full-length biography of him had been written until then. Bhabha was a man of many talents and well worth being the subject of a popular biography—one that would attempt to do justice to his important role in Indian science, to the singularity of his mind, and to the depth and diversity of his intellectual interests. The more I looked into his life, the more remarkable it seemed; I am sure that the reader will agree that his is a story very much worth telling. Since writing that profile, a few books on Bhabha have been written, all with undeniable merit, but none that could qualify as a full-fledged biography.

People who feel that it is possible to understand a person's work without knowing anything about their lives overlook the fact that understanding individuals is a worthwhile and interesting pursuit in itself. We should aspire to understand Bhabha not merely to learn about his work but also because he was such an interesting man—a 'humanist–scientist' with diverse talents, a physicist admired for his contributions to elucidating the fundamental features of the constitution of matter and an artist deeply interested in all aspects of human culture. There is no doubt that Bhabha's cultivation was genuine and has been much-commented upon—his philosophical sensibilities and his cultural cosmopolitanism counted for much. All great human endeavours consume and transform their doers, who devote time and energy in the service of these goals. Those who undertake such endeavours are, in turn, rewarded by becoming most truly themselves. Nowhere does this apply more than to Bhabha, whose uncommon mind was unafraid of its own capacity for living fully.

The stature of a man can be judged by the breadth of his mental horizon. Bhabha was both a creator and a product of his times. He was a man who bridged the old and the new, the East and the West. Equally at home in the world of science and of the arts, he was both a visionary and a doer—a man of thought as well as of action. A great deal is known of his astonishing intellect but less about his human side. I have tried to combine both aspects of him, hoping to present a more complete picture. It is important to remember that one cannot treat science as a realm of pure objectivity, forgetting that it is a pursuit of humans with all their hopes, prejudices, arrogance and other foibles. The methods employed, conclusions reached and goals achieved by scientists rely heavily on their own background and the culture to which they belong.

This, of course, means that I am very interested in the social and political background to Bhabha's life and, indeed, devote the second chapter to the Parsi community into which he was born. The legacy of that community is important in understanding the mainspring of Bhabha's later endeavours. Recognizing that Bhabha's intellectual and personal development can only be reflected clearly in the mirror of his own times, I have tried to situate him firmly in the culture, society and politics of the 1930s, 1940s and 1950s India. In doing so, I hope to not only avoid the trap of a one-dimensional portrait but also to illuminate more brightly the roots of his thoughts and the reasons for his actions.

To write the biography of a man who, thanks to his extraordinary talents, managed to bridge what the English novelist C.P. Snow called the gap between science and the humanities is an arduous and presumptuous task. Bhabha, with a mastery over English that rivalled his expertise in physics and mathematics, would have been the best person to write about himself. Perhaps, at a later stage in his life, he may have been tempted into confidential expansion, speculation which must remain in the realm of conjectures since his life ended at the relatively young age of 56. Bhabha's place in history, his impact on Indian society and its impact on him are all interesting topics, and ones that a biographer cannot ignore. However, since Bhabha the person is also of great interest, it was important for me to write a biography that aims to not only describe the main events of Bhabha's life and work but also to understand the internal workings of Bhabha as a person—his ideas about science, the complexity of his influences, and the full and vivid sweep of his life.

There is nothing personal about facts; what is personal is their choice and grouping. Events may not be new but their grouping is akin to turning a kaleidoscope—the pieces remain the same but they fall differently each time, forming ever-changing patterns. A biography is a form of creative non-fiction that allows the writer the freedom to select the stories and anecdotes to be highlighted, and set the pace and the tone of the narrative. I have attempted to draw as sharply defined a picture as I could of Bhabha's work and character as well as the personalities and events that decisively influenced it.

This biography aims to provide as detailed, accessible and comprehensive a portrait of Bhabha's life and times as possible. As the endnotes testify, I owe a considerable debt to the labours of those who have preceded me. However, I have not hesitated to suggest new interpretations where appropriate and to shine a light into those corners of Bhabha's life that have hitherto remained in the shadows, attempting to make his life as vivid and immediate as the evidence allows.

This account of Bhabha's life is patchy since there are many gaps in the record. But evidence of his participation in the developments of his time survives in his epistolary legacy, official documents and even in works of art, all of which are vivid testimonies of his impact on the epoch. I have given ample consideration to these, and hope to make apparent the full richness of his personality as well as allow for a more realistic assessment of not only his successes, but also of his failures and shortcomings, which have often been minimized or completely overlooked in more deferential (read hagiographic) accounts of his life.

A challenge biographers face is the distribution of periods and phases—the right proportion among varied fields of public and personal activities, and the right scale for the many episodes of a subject's life. This decision is a highly subjective one and depends on the author's agenda. In this case, however, I have no agenda except to portray Bhabha in as complete and well-rounded a manner as my abilities allow, setting the facts before the readers and letting them form their own opinions. Another challenge is to remain unbiased given the fact that it is admiration, at least initially, that makes a biographer invest his time and energy in writing someone else's life story. I have tried my best to be impartial and it is for the reader to decide how far I have succeeded.

Ralph Waldo Emerson observed that an institution is but the lengthened shadow of one man. Both the Tata Institute of Fundamental Research (TIFR) and the Atomic Energy Establishment at Trombay (AEET)—institutions that Bhabha founded—bore his unmistakeable imprimatur: there was no activity that he did not personally oversee and contribute to. However, his work was by no means confined to fundamental research and atomic energy, but ranged outward towards electronics and computers, the setting of architectural standards and the establishment of forward-looking administrative structures around the rising industrial complexes. More important was the building up of a pool of scientists and engineers—many of whom were trained abroad—who, inspired by Bhabha, embraced his vision of a new India. Equally important was the confidence in their own abilities, which contact with Bhabha engendered. Whatever Bhabha did, he pursued with rare passion, intensity and an almost obsessive eye for detail. He was not just interested in scientific research but had also expressed the view that someday TIFR might research the history of Indian science along the lines of the Institute for Advanced Study (IAS) at Princeton, which had a faculty of humanistic studies.

Bhabha epitomized the nuclear scientists' role as the avant-garde of India's scientific elite, leading the country on its path to modernity. This perception of him left no room for critical self-reflection on his work. It is said that the principal task of a historical biography is to use the life as a window to the times. Bhabha's story is compelling in itself and, thus, it never lacks a strong narrative thrust. The story of his life also provides a good vehicle for telling the story of Indian science and of the foundations of India's atomic energy programme. Although he is remembered for his work on cosmic rays which led to his nominations for the Nobel Prize, he is better remembered today as the founder of TIFR and the architect of India's atomic energy programme. Thus, this tome is as much an early history of TIFR and India's atomic energy programme as it is an account of Bhabha's life and work. Little wonder then that concision has eluded me: this book has ballooned to twice the size I had originally intended it to be.

Bhabha's life was a true embodiment of the trajectory of India in the twentieth century. Only a few can claim to be the carriers of the spirit of their times: Bhabha was, in fact, not only the carrier but also the creator of the spirit of his times. The impact of his work was as broad as it was deep. Thus, his constituency at a macro level was India, since he helped shape national policy and founded institutions of national importance. At a more micro level, his constituency can be said to be TIFR and the Department of Atomic Energy (DAE). Bhabha wanted an India with its face turned optimistically to the world, not as a former colony but as a significant contributor to a more equal, inclusive and collaborative global scientific order. To this end, he continuously promoted research and pedagogic innovations in scientific thinking and cultural expressions in research and learning. He emphasized excellence, modern infrastructure, modern non-hierarchical management, an international presence and aesthetics in the workplace.

Bhabha's vision envisaged science and technology as playing a key role in development and nation-building. His ideal of science was profoundly intertwined with the ideal of a better world based on his conviction that science has the potential to transform lives. He believed that the pursuit of science and its practical application are not subsidiary activities but form the basis of the whole social structure without which life as we know it today would be inconceivable. Science provided freedom from long hours of manual drudgery and the world was at the cusp of an age in which every person would have the opportunity to develop himself spiritually to his fullest stature.

Is it worth spending so many pages on things like electrons, protons, neutrons and mesons? Would anyone seriously interested in Bhabha seek out a biography for this? To answer this, it is important to understand just how captivating science is to a true scientist. The American physicist, J. Robert Oppenheimer, once reportedly said, 'I need physics more than friends.' The same could be said of Bhabha who is not an easy biographical subject: the complexity and cultural diversity of his life provoke interest about him but for the biographer they constitute practical problems. Does he cut out some aspects and only address what remains? Bhabha as physicist? Bhabha as institution builder? Bhabha as scientist–administrator? I can only hope that I have achieved an appropriate blend of all the above in this narrative, and indeed the book is structured in a manner that addresses these three aspects section-wise.

It is difficult to fashion a narrative out of the facts of someone's life. It is even more difficult in the case of someone like Bhabha, whose life spanned separate themes and different worlds, making a traditional sequential chronicle impossible. Thus, this book does not read like a diary; it is divided into four sections and 36 chapters, and the narrative unfolds as a theme-based story. Each section has its own introductory passage that sets the background.

The first section 'Scientist as Researcher' deals with Bhabha's background, early life, education and initial research. The first chapter introduces the reader to C.V. Raman, Meghnad Saha and Shanti Swarup Bhatnagar, three eminent scientists with whom Bhabha was closely associated with in his career. It also provides a brief history of physics in India, with particular reference to Raman and Saha in the period before Bhabha arrived on the scene and touches upon the rivalry and ego clash between Raman and Saha. The second chapter gives the reader a background on the Parsi community and Bhabha's roots, which played an important role in shaping his future endeavours. I have tried to explain his contribution to theoretical physics in the earlier chapters as simply as the subject allows. However, since the subject is both abstract and abstruse, and I am no scientist, I crave the indulgence of the general readers if the recondite areas of physics and the arcana of concepts like spin, time dilation, parity and conservation of energy escape them. They are unavoidable (even if they make the narrative less readable), since this is, after all, the biography of a scientist.

Those trained in science, especially physics, will face no problems. That said, given the limitations of the author, there is no mathematics involved, and certainly no scholarly rigour in the treatment of the scientific details. There is also an entire chapter devoted to the quantum revolution, which, though not directly connected to Bhabha, was of such moment and the happenings of such interest that I have taken the liberty of concluding that this digression is not unjustifiably tangential to the main theme. Similarly, the introduction to the third section 'Scientist as Administrator' gives details about the discovery of nuclear fission and the Manhattan Project, which made the first atomic bombs that were used on Japan during the Second World War.

The second section 'Scientist as Institution Builder' traces the history of the TIFR and how Bhabha nurtured it into a world-class institution. The research carried out at the institute, including details of the School of Mathematics and the pioneering efforts at developing a computer are covered in some detail. His emphasis on combining science with art and appropriate aesthetics, including his own skill as an artist, form the subject of separate chapters.

The third section 'Scientist as Administrator' traces the arc of the atomic energy programme and institutions, like the AEET, and reactors, like Apsara and CIRUS (Canada India Reactor Utility Services), are discussed in some detail. Bhabha's sterling role at the Geneva Conference in 1955 forms the subject of a different chapter. Bhabha's views on a nuclear weapon are discussed in the concluding chapters. The fourth section 'Coda' contains only one chapter and deals with Bhabha's premature demise. A brief epilogue rounds off the book.

I have chosen to write for a general readership, and it is hoped that this book will appeal to a larger audience. For those interested in both Bhabha, the man, and his contribution to fundamental research and the nuclear energy programme, this book will definitely be of interest. For those exploring atomic energy as it is now, an understanding of the people who made it what it is today will certainly enhance their experience. It will also appeal to all those who simply like a good story.

This biography is intended to be a popular account and does not purport to be a scholarly work. However, I hope I am not wrong in assuming that I have not compromised on the rigour of scholarship. The writing is based on well-researched history and the wide scope of the work means that not all the primary research is mine; much of it is based on the works of others available to me through published writings. The sources of my information are clearly set out in the notes and bibliography, and give the interested reader an opportunity to follow up with further reading. They are, however, not essential to enjoying the book.

Before I conclude, a word about the names of places in the book: Bombay is now Mumbai, Bangalore is now Bengaluru, Madras has been renamed Chennai, Calcutta has become Kolkata and Poona is now Pune, but the older names have been retained, since these were used during the period covered in the book. Thus, throughout the book, the names of places, roads, activities and more, refer to the time period of the events being discussed.

This biography largely addresses readers who already know something about Bhabha and his work, but it can also be read with profit and pleasure by those who are making Bhabha's acquaintance for the first time. No biography, however, is perfect. It is possible that the present offering is more imperfect than others. There are inevitable gaps in knowledge and lacunae in the evidential record, and my reconstruction of Bhabha's life, thoughts and motivations is based on the facts at my disposal. Everyone may not agree with everything I have written in this book. This offering will have more than fulfilled its purpose if it imparts to the reader a small fraction of the pleasure I have derived from writing it and inspires others to delve more deeply into the life of this remarkable man. About biographies, Lytton Strachey once wrote, 'We do not reflect that it is perhaps as difficult to write a good life as it is to live one.' Bhabha undoubtedly lived a good life. How far I have succeeded in portraying that life is for the reader to decide.

SCIENTIST AS RESEARCHER

When Homi J. Bhabha returned to India from the University of Cambridge for a holiday in the summer of 1939, he would never have guessed that his days of residence abroad had ended. The onset of the Second World War led him to permanently relocate to India. He returned to Bangalore, and C.V. Raman, the head of the physics department at the Indian Institute of Science (IISc), took Bhabha under his wing. The other two scientists that Bhabha would be closely associated with in his career were the physicist Meghnad Saha and the physical chemist Shanti Swarup Bhatnagar.

Bhabha's relationship with Saha started off cordially, but later soured. Saha became a rival and his most trenchant critic. Bhatnagar, on the other hand, remained a source of constant support since he, too, was intimately associated with science administration in India. Relations between Raman and Saha were less than cordial, if not downright acrimonious and antagonistic, something that finds mention in the first chapter.

On his return from England, Bhabha brought to India his appreciation for that great, occidental invention: modernity. During this time, he also underwent his own personal coming of age: the Indian-born Anglophile who was forced to return from Cambridge to India during the Second World War, transformed, by the end of his life, into the very archetype of the patriotic Indian scientist. The three stages of Bhabha's life—student, scientist and science administrator—represent not only the trajectory of his personal journey but also the arc of the nuclear programme. An in-depth investigation of Bhabha's trajectory is essential for a more comprehensive understanding of the institutions he set up.

Theoretical physics was Bhabha's forte, but he could have risen to the top in about half a dozen other fields. He would have made a brilliant architect, and the TIFR building, with its splendid proportions and open spaces, is a testament to his genius for design.

The buildings are also filled with one of the finest collections of modern art in the country. Bhabha was, above all, a scientist, but he was also interested in many other pursuits that make up a complete life: music, painting and sculpture; landscapes, gardens, trees and flowers; international friendships; and the application of science for the benefit of society. Bhabha's interests were not superficial in nature: he had a deep knowledge of everything he was interested in and also succeeded in giving practical shape to such interests.

Bhabha, who was an artist of rare skill and feeling, could have been a successful painter had he chosen to be one. His paintings adorned the walls of his home—Mehrangir. Apart from being a scientific leader and administrator, he was an intensely sensitive human being. He was as versed in western culture as he was in Indian culture, and derived joy and guidance from both. His own talents as an artist made him especially sensitive to the beauty of his surroundings, and the two institutions that he established and guided till his death have a rare beauty that reflect this facet of his life. The flower beds, landscaping and the architecture at TIFR and AEET, now renamed as the Bhabha Atomic Research Centre (BARC) in his memory, bear witness to his keen perception of colour, form and design.

The Indian nuclear programme was established by Bhabha and his political patron, Prime Minister Jawaharlal Nehru. It was their combined vision of a new India at the vanguard of the modern world that inspired its creation. Bhabha was a non-government man by outlook and wanted room to manoeuvre. Nehru understood Bhabha's need for independence and gave it to him. Bhabha's programmes were officially sanctioned and supported by Nehru, and he enjoyed great autonomy. Notwithstanding his declamation against hierarchies, eventually, Bhabha became the State with respect to nuclear energy.

Bhabha's institutionalizing work developed in a new world order, where science was controlled by the perceived security needs of the

Indian nation state as well as by the constraints of the Cold War international order. During this period, the use of science changed—the idea of the scientist as entrepreneur, pursued by Raman and Saha before the Second World War, transformed into the scientist as administrator, something that Bhabha epitomized in his person. His role demonstrated not just the degree of autonomy that scientists and technocrats could negotiate vis-à-vis the Indian state in the period after Independence but also, significantly, how much the political leadership relied on such experts.

Chapter 1

BEFORE HOMI BHABHA

THE PIONEERS

The institutionalization of physics research in India is said to have started with the establishment of the Indian Association for the Cultivation of Science (IACS) by Mahendra Lal Sircar in 1876 in Calcutta. Sircar, the second MD to graduate from the Calcutta Medical College, was also a social reformer and, more importantly, a propagator of scientific studies in nineteenth-century India. Aware that English education in India emphasized western philosophy and literature, which were in consonance with the colonial agenda of 'civilizing' India in Lord Macaulay's sense of the word, Sircar had often spoken about the necessity of introducing more science courses in the curriculum. The IACS came into being on 15 January 1876 with Sircar as the secretary and had as its aim the dissemination of scientific knowledge. The name of the association left no one in doubt of its objective: science would be cultivated here and this was its sole objective.[1]

The Thursday evening lectures by Father Eugene Lafont, a Jesuit missionary who was the co-founder of the institution, and other eminent scientists were one of the association's main activities. Apart from Sircar and Lafont, Jagadish Chandra Bose (physics), Ram Chandra Dutta (chemistry), Pramatha Nath Bose (geology) and

Sir Asutosh Mookerjee (mathematics) delivered lectures there every week, which attracted many listeners. Sircar himself delivered lectures on physics, and in his lectures and practical demonstrations, he was forever performing experiments, repeating works done by great scientists in the field of electricity and magnetism.

J.C. Bose was perhaps the first Indian of international scientific repute. A biologist, physicist and botanist who pioneered the investigation of radio and microwave optics, he laid the foundation of experimental science in the Indian subcontinent. He received his DSc degree from the University of London in 1896 and, upon returning to India, became the first Indian to occupy the post of professor of physics at Presidency College. Considered one of the fathers of radio science, his work on electromagnetic waves continues to be recognized as the inspiration for the work of, among others, Guglielmo Marconi. He invented the crystal semiconductor diode, the most crucial part of a radio at the time, and held the patent for it. In fact, it was Bose who should have received the Nobel Prize in 1919 for his research on radio waves. He was knighted in 1917 and elected to the Royal Society in 1920. It is believed that J.C. Bose set the physics question paper of the Indian Financial Service examination, which C.V. Raman topped.

The IACS was more like a scientific gentleman's club for dilettantes than an association for serious research. The lack of activities in the association made Sircar despondent and he died in February 1904, thinking that he had toiled in vain. However, times changed three years after his death. In August 1907, the ghost kingdom, which his son Amritalal had inherited, received a visitor. A lanky young man who wore a turban in the South Indian style knocked on the doors of the association at 210, Bow Bazar Street, and introduced himself to Asutosh Dey, the laboratory assistant and assistant secretary of the IACS. He said that he worked in a government office all day but was an amateur scientist who needed a place to pursue his scientific work in the evenings and on weekends. Ashu Babu, as the assistant was called, took him to Amritalal who was overjoyed and welcomed the young man warmly. Raman, for that was the young man's name, soon became a regular visitor to the IACS.[2]

Raman had obtained a BA in 1904 from the University of Madras, winning a gold medal in physics. In 1907, he completed an MA from the University of Madras with the highest distinction. Aware of his talent for physics, his teacher wanted him to go to England, but a medical examination revealed that Raman's lungs would not be able to withstand the British weather. Unable to study abroad, Raman followed in his elder brother C.S. Ayyar's footsteps and qualified for the Indian Finance Service (the precursor of what is known as the Indian Audit and Accounts Service today), standing first in the entrance examination in February 1907. Never known for his modesty, he is reported to have told his brother, 'I took one look at all the candidates who had assembled, and I knew I was going to stand first.'[3] An early example of the famous Raman ego! He was sent to Calcutta as assistant accountant general in June 1907.

Though the IACS eventually received government support, it was private and nationalist in intent and most of its early members were Bengalis. Raman was the first person to regularly use the facilities there. He would arrive early in the morning, go to work and return for another round of experimentation and research in the evening. This is exactly what Sircar had in mind when he had set up the association. Raman proved to be the perfect example of the intelligent, curious Indian, who wanted to engage in experimental science. This pattern continued, save for a short break, when Raman was posted out of Calcutta, until 1917, when he resigned from his government employment to take up Sir Asutosh's offer of the Palit Professorship at the University College of Science and Technology.

Sir Asutosh was a towering figure in the educational system of the time. Like his friend, Sircar, he too was a visionary. An educationist, jurist, barrister and mathematician, he was the first student to be awarded a dual degree (MA in mathematics and physics) from the University of Calcutta. He also received a degree in law and was a judge of the Calcutta High Court for nearly two decades. He was the president of the IACS at the time of his death in 1924. Calcutta in the 1920s saw the emergence of Raman, Satyendra Nath Bose and Saha, all of whose contributions were important enough to win a Nobel Prize, but fortune favoured only Raman. Of all the factors that contributed to such a flourishing environment, the most important was Sir Asutosh's stewardship of Calcutta University and the scientific climate created by scientists like J.C. Bose and Prafulla Chandra Ray. Major scientific contributions—like Saha's thermal ionization equation (1920), Bose–Einstein (now Bose) statistics (1924) and the Raman Effect (1928)—must be seen within this historical context.

Sir Asutosh was a man of great administrative ability. The second Indian vice chancellor of Calcutta University, he continued in that position for four consecutive two-year terms from 1906 to 1914 and a fifth one from 1921 to 1923. He founded the Bengal Technical Institute in 1906 and the University College of Science and Technology in 1914, among other institutions. He was also the president of the inaugural session of the Indian Science Congress (ISC) in 1914. Sir Asutosh believed that India should take its place in the world of science. To this end, he asked his rich and powerful contemporaries to donate money to start the University College of Science and Technology and to establish various departments that would be chaired by distinguished Indian scientists. Among those who gave freely of their largesse, was Sir Taraknath Palit, lawyer and philanthropist, who donated ₹15 lakhs for this purpose.

For the Palit Professorship in physics, Sir Asutosh had his eye on Raman. But could the latter be persuaded to give up a well-paying government job and sacrifice what would certainly have been a very successful career? There was no harm in trying, and an offer was made to the young assistant accountant general. To his great surprise, Raman accepted the offer with alacrity.[4] He had done most of his important work on vibrations and sound before he became the Palit Professor. While his many contributions in this area have been overshadowed by the one great discovery he made later, his earlier work was equally important.[5]

Under the terms of his appointment, Raman was not required to do any routine teaching. But he loved teaching and lecturing, and it was impossible to keep him away from the lecture room. He had lectured extensively at the IACS and had built a formidable reputation for himself as a lecturer in popular science. His booming voice, superb diction, lively demonstrations and sense of humour, all served to keep his audience spellbound. Physicist G. Venkataraman, who wrote

Raman's biography observed, 'He was perhaps the greatest salesman science has ever had in this country.'[6] His students often recalled how inspired they felt after attending his lectures.

Raman continued to use the IACS laboratory and he began operating on both fronts. The laboratory soon became the research arm of the University of Calcutta.[7] Slowly, Raman gained control over the association, becoming the honorary secretary after Amritalal's demise and bringing about changes that he thought would be beneficial. These changes, however, marked the beginning of tensions between him and the older members of the association, which subsequently became a Bengali vs South Indian conflict, with Saha and Shyama Prasad Mookerjee, the son of Sir Asutosh and the future founder of the Bharatiya Jana Sangh, becoming involved.

Raman attempted to limit new membership, admitting only those who met with his approval. He was accused of favouring South Indians, of whom there were only a small number, ignoring the fact that the association invited Bengalis along with people from across India. His main assistant was Kariamanikkam Srinivasa Krishnan, better known as K.S. Krishnan, from Madras. Not only people from Madras but from all the other states, such as Mysore and Kerala, were lumped together to form the amorphous group called 'South Indian'. Before we discuss this conflict, it would do well to take a look at Raman's discovery of the effect named after him.

THE RAMAN EFFECT

In 1921, C.V. Raman went abroad for the first time when he represented the University of Calcutta at the Congress of Universities of the British Empire at Oxford. While returning, he observed the blue opalescence of the Mediterranean Sea and wondered about the origin of this beautiful phenomenon. Raman did not accept Lord Rayleigh's explanation that the colour of the sea came from the reflection of the blue sky in it. He used a simple Nicol prism to examine the colour of the sea during the voyage.

By the time he got back to Calcutta, he was quite convinced that the blue colour of the sea was due to the scattering of the incident sunlight by the molecules of liquid water and that Lord Rayleigh's earlier suggestion was erroneous. Using a diffraction grating, he showed that the maximum spectral intensity was different for the blue sky and the blue sea. During this voyage, Raman sent two papers to the journal *Nature* positing that the colour of the sea came from light being scattered by the water molecules—a phenomenon he called 'molecular diffraction'. Thus began Raman's new research obsession: the molecular basis of light scattering.[8]

From 1921 onwards, the subject of light scattering became the main theme of the Calcutta group, led by Raman, and it culminated in the discovery of the Raman Effect in 1928. By 1924, Raman's outstanding researches had won him international recognition and he was elected a Fellow of the Royal Society of London (FRS) when he was only 36. In 1923, a detailed study of the scattering of light in water was taken up by Kalpathi Ramakrishna Ramanathan, one of Raman's research scholars. Raman thought that the 'weak fluorescence' that they encountered might be something analogous to the recently discovered Compton Effect in X-ray scattering.[9]

In 1927, Arthur Compton won the Nobel Prize in Physics for his eponymous effect and Krishnan rushed to give Raman the news. The astrophysicist Subrahmanyan Chandrasekhar's father, and Raman's elder brother, C.S. Ayyar, was also with Raman at the time. On hearing the news, Raman enthusiastically said, 'Excellent news…very nice indeed. But look here, Krishnan. If this is true of X-Rays, it must be true of light too. I have always thought so. There must be an optical analogue to the Compton Effect. We must pursue it and we are on the right lines. It *must* and shall be found. The Nobel Prize must be won.'[10] Later, Krishnan observed the same phenomenon while studying the scattering of light in 65 different carefully purified liquids. Raman suspected that it was an entirely new type of secondary radiation, quite distinct from the usual fluorescence.

The Raman Effect was finally discovered on the afternoon of 28 February 1928. Raman described the discovery at a meeting of the South Indian Science Association in Bangalore on 16 March 1928. Upon returning to Calcutta, Raman's address was written up immediately, printed overnight and copies posted to scientists all over the world.

Soon, the scientific world got to know of the discovery, and the Raman Effect became an active field of research. National Science Day is now celebrated on 28 February in India.

After the publication of 'A New Radiation', on the Raman Effect, in the 16 March 1928 issue of the *Indian Journal of Physics*, Raman mailed hundreds of reprints to scientists all over the world. He knew the importance of establishing priority and had been sending letters to *Nature*, informing them of the work he was doing. This was just as well because almost at the same time (on 21 February) G.S. Landsberg and L.I. Mandelstam of the Union of Soviet Socialist Republics (USSR) had studied the scattering in crystalline quartz and succeeded in seeing wavelength-shifted lines. They, however, did not publish their findings immediately. This delay cost them the Nobel Prize.

If Raman was a prolific investigator, he was also a very skilled and aggressive communicator. He aggressively courted the media, reaching out to journalists with press reports for publication. It was this relationship that helped Raman publicize his work and made him famous. He was knighted in 1929 and won the Nobel Prize a year later for his investigations on the scattering of light and the effect named after him. Never short on confidence, he knew the importance of his discovery and had expected to win the Nobel Prize. He also actively canvassed for it. Raman was so confident of winning that he booked tickets to Sweden in July that year. Had he waited for the announcement in November, he may not have been able to make it in time for the ceremony in December. His nominators for the prize included big names like Ernest Rutherford, Niels Bohr, Eugène Bloch and Charles Thomson Rees Wilson.

The effect would be known, at least in the English-speaking world, as the Raman Effect. Many European physicists felt that credit should have been shared but the Nobel Committee decided that the award should go to Raman alone. Unfortunately, after 1953, Raman and Krishnan's relationship deteriorated and broke down completely. There were rumours that the latter felt deprived of the credit due to him. There was also talk that he had played a significant role in the discovery and had deserved more credit than had been given to him. Krishnan believed that some jealous person had poisoned Raman's mind against him and that the latter, in his childlike innocence, had believed the scurrilous statements.

In his Nobel lecture, Raman had given due credit to all his research assistants, including Krishnan. In public lectures, Raman would repeatedly acknowledge his role. He said that Raman had not deprived him of any credit and acknowledged that the most important decisions in the sequence of events that led to the discovery had been taken by Raman and not by him.[11] Krishnan never uttered a word against Raman. He had kept a diary of the events leading up to the discovery. The diary has a curious history and the only reference to it by Krishnan was indirect. He told Chandrasekhar about it and the latter referred to it in a conversation with his biographer Kameshwar Wali.

Chandrasekhar said that Krishnan had told him that he had kept a diary for the period from January to April 1928 and that he had intended to deposit it with the Royal Society. The existence of the diary was known only to members of his family and his close associates in Calcutta. Strangely, after Raman's death, extracts from the diary started appearing in books and articles whenever the Raman Effect was discussed. For 40 years, there had been no reference to the diary and most did not even know that it existed. What happened to the diary? According to Krishnan's family, it was taken away by K.R. Ramanathan a few days after Krishnan died in 1961, and it was returned to the family many years later.[12]

Chandrasekhar discovered that it had been with K.R. Ramanathan in 1971. He recalled his conversation with Krishnan in 1951, and on realizing that the latter had not deposited it with the Royal Society, he made a copy of it and sent it to the Royal Society with his own recollections. Chandrasekhar placed a time embargo on them, so these papers are not currently available. When the diary came back to the family, it was found to have only 16 pages, covering the period from 5 to 28 February. The entry on the last day ended abruptly with an unfinished sentence, which had started to describe the actual discovery. A substantial part of the diary was missing, including the record of events on the day of the discovery and the subsequent work done until April 1928.[13]

Chandrasekhar, who was in Calcutta almost immediately after the discovery, thought that both Raman and Krishnan deserved credit. He believed that the Raman Effect had been discovered because two absolutely original scientists had complemented each other.[14] The common picture of science based on the premise of individual genius is fallacious. Science does not issue from singular minds—it is a community effort that thrives on fellowship and collective skill. The Nobel Prize does not reflect how science is done; it merely reinforces the fiction of individual genius. The Nobel is an individual prize but this no longer conforms to the multidisciplinary and collective nature of contemporary research.

During his stint as the director of the National Physical Laboratory (NPL) in New Delhi, Krishnan greatly upset Raman when he offered financial assistance from the government to the Raman Research Institute on the condition that Raman should submit a financial statement. Krishnan had intended to announce this during Raman's 60th birthday celebrations but Raman flew into a rage, expressing outrage at being asked for a report. Even after Krishnan's demise in 1961, Raman's bitterness had not run its course. In a statement to a newspaper reporter, he called Krishnan the greatest charlatan he had ever known and someone who had masqueraded all his life in the cloak of another man's discovery.[15] Chandrasekhar observed: 'Nothing I have known justified such a statement. Besides Raman's published Nobel lecture includes ample reference to Krishnan and is fair.'[16]

MEGHNAD SAHA

Another major player on the scientific stage of the time, Meghnad Saha, was born on 6 October 1893, the fifth of eight children of a poor, lower caste shopkeeper, Jagannath Saha, and his wife, Bhubaneswari Devi, in the town of Seoratali in East Bengal (present-day Bangladesh).

In 1905, having finished middle school, ranking first in class at the age of 12, he was awarded a scholarship to study at the Dacca Collegiate School. However, his stay there proved to be short-lived. He was expelled soon after his admission because of his participation in a protest against the partition of Bengal earlier that year. Together with some other senior students—including Nil Ratan Dhar, who went on to become a close friend and, later, a colleague at the University of Allahabad—Saha staged a boycott of the school visit by the British lieutenant governor of East Bengal and Assam.[17] This cost him his scholarship and with it the resources to pay for his tuition. Subsequently, Saha went to a small private school and passed the entrance examination for Calcutta University in 1909, standing first among thousands of students from the schools of East Bengal.

His scientific education began at the age of 16 in Dacca College (it was not yet a university). He also started learning German from the German botanist Paul Johannes Brühl, who taught at the Bengal Engineering College. In order to become more proficient in English, Saha attended Bible classes at the Dacca Baptist Mission. He proved his mettle there as well, winning a deluxe copy of the Bible in a Bible competition.

Saha joined Presidency College for a BSc in Mixed Mathematics in 1911. While studying at Presidency College from 1911 to 1913, Saha stayed in a hostel where he faced discrimination because he was a Shudra, a member of the lowest caste. Students objected to eating in the same dining hall with him and he was also prevented from making a religious offering to the goddess of learning, Saraswati. For the next two years, Saha ate at another students' mess nearby.

It was at Presidency College that Saha earned the nickname 'Eigenschaften' for his 'invincibility' and his knowledge of German. Saha's classmates included S.N. Bose, Jnan Chandra Ghosh, Nikhil Ranjan Sen and Jnanendra Nath Mukherji, all of whom became famous scientists in their own right. Dhar, who studied chemistry and, later, invited Saha to come to Allahabad University, was two years his senior. The statistician Prasanta Chandra Mahalanobis, who later set up the Indian Statistical Institute (ISI), was one year his senior, and the famous Subhash Chandra Bose was three years his junior. At that time, Prafulla Chandra Ray, J.C. Bose and the famous mathematician D.N. Mallick, all of whom used science as a tool for promoting the spirit of nationalism, were among the teaching faculty.

For Saha, the pursuit of science and aspirations for national independence were linked together even more closely. While in college, he encountered the militant nationalists of Bengal, including the famous revolutionary 'Bagha' Jatin Mukherjee and Pulin Das. The extent of Saha's involvement with the Bengal revolutionaries is not yet fully known because of the clandestine nature of their activities.[18] In 1915, when Saha wanted to sit for the Indian Financial Service examination, (the same exam Raman had topped in 1907) his past caught up with him. He was refused permission to write the examination because of his association with the revolutionaries. Left with no choice, he continued his studies in applied mathematics and physics at the newly established University College of Science and Technology. There was also the matter of earning a living, and Saha did this by giving private tuitions.

Along with playing an important role in Raman's career, Sir Asutosh was also crucial to Saha's transition from a graduate student to a professional physicist. He offered the post of lecturer to both Saha and S.N. Bose in the department of mathematics. Later, Sir Asutosh transferred them to the physics department. Sir Asutosh had heard that the younger generation wanted more modern subjects to be included in the university curriculum and asked the duo what subjects they were competent to teach. At that time they had only heard of the developments in Europe and the contributions of famous scientists, such as Max Planck, Albert Einstein and Bohr.

The supply of foreign academic journals had stopped because of the First World War and there seemed to be no way of getting them. Advanced books on science and European scientific literature were available in the libraries of Calcutta only to a very limited extent. Sir Asutosh put his faith in the greenhorns and gave them a special allowance of ₹125 per month. S.N. Bose recalled that Saha had to study quantum theory while he had to learn Einstein's Theory of Relativity. They came away committing themselves to being able to teach these subjects within a year. But where were they to get the required books, most of which were German? Professor Brühl saved the day by providing them with many of the books they needed.[19]

Saha was awarded a DSc degree from the Calcutta University in 1918 on the basis of his work in the field of electromagnetic theory and radiation pressure, and for the many other papers he had published. Saha's greatest contribution has undoubtedly been the theory of high-temperature ionization and its application to stellar atmospheres. The Saha ionization equation was first mentioned in the paper 'On Ionization in the Solar Chromosphere', published in the *Philosophical Magazine* in October 1920.[20] It is pertinent to observe that the ionization theory was formulated by Saha working all on his own in Calcutta. People were often incredulous that his earliest and most pioneering work had been done in India and not abroad.

The theory of thermal ionization introduced a new epoch in astrophysics and the impetus given to it by Saha's work can scarcely be overestimated. Using mathematical tools, Saha developed his now famous ionization formula, which gave new insights into the interior of the sun and stars, their chemical compositions, densities of various elements and gravitational forces on their surfaces. His theorizing transformed spectroscopy into a precise quantitative method, and nearly all later progress in this field has been influenced by it. Much of the subsequent work has simply refined Saha's ideas. Saha was elected an FRS in 1927 for this work. Chandrasekhar wrote, 'Meghnad Saha's place in the history of astrophysics and in the history of modern science in India is unique.' His theory of thermal ionization, which explained the origin of stellar spectra, was one of India's most important contributions to world science during the twentieth century and was an epoch-making discovery.

Saha sailed for England in September 1919 on the back of two scholarships. He worked with Professor Alfred Fowler in his spectroscopy laboratory in the Imperial College of Science and Technology, London. After spending about five months in Fowler's laboratory, Saha went to German chemist Walther Nernst's laboratory in Berlin and worked on the conductivity of heated caesium vapour to seek an experimental verification of the theory of thermal ionization. The results were inconclusive. Some years later, Saha returned to this problem when he was at Allahabad, and he jointly published a paper on the subject with K. Majumdar and N.K. Sur in 1926.

During his time in Berlin, Saha attended the colloquium at the University of Berlin every week and got to meet stalwarts, like Planck, Einstein and Max von Laue. In November 1921, he returned from

Berlin and joined Calcutta University as the Khaira Professor of Physics, a new chair created from the endowment of Kumar Gurprasad Singh of Khaira. Sir Asutosh had written to Saha in Berlin, asking him to join quickly. At that time, Calcutta University was going through a deep financial crisis. Saha soon realized that given the financial crunch, he would be able to achieve very little there. There was the laboratory of the IACS but that was Raman's kingdom, and Saha and Raman did not see eye-to-eye on a few issues—disagreements that would turn into almost full-blown enmity later. Saha loved Calcutta but seeing little opportunity there, decided to leave in 1923.

Saha had been looking for professional opportunities outside Calcutta, especially in North India, where he wanted to create a school of physics. He received offers from Banaras Hindu University (BHU) and Aligarh Muslim University (AMU) but rejected both in 1923 to accept the position at Allahabad University, which Dhar, his college friend and a professor of chemistry, arranged for him. Dhar had joined Allahabad University in 1919 and had been actively engaged in research activities there ever since.

In Allahabad, Saha found that there were virtually no facilities and everything had to be built from scratch. There was a lack of funds, and, more importantly, a lack of understanding on the part of those who controlled them. Saha had to contend with the tyranny of petty officials throughout, but things changed for the better when he was elected an FRS. Saha stayed in Allahabad as a professor and the head of the physics department until 1938. The higher teaching load and the lack of staff in the physics department left him with little time for research. He spent a large part of his time teaching undergraduate and postgraduate classes. His reputation attracted a group of young and talented scientists, who made a mark for themselves later in life, and included Daulat Singh Kothari, Bishwambhar Nath Srivastava and Ramesh Chandra Majumdar. His house in Allahabad, called 'Science Villa', was always open to students.

◆

What of Saha's classmate S.N. Bose, who had topped both his graduation and postgraduation degrees? S.N. Bose, who was essentially a theoretical mathematical physicist moved to the new University of Dacca (now Dhaka, Bangladesh) as a reader in 1921. With Saha, he co-authored the first English translation of the original papers of Einstein and Hermann Minkowski that the University of Calcutta Press published in 1920 as a book titled *Principles of Relativity*, with an introduction by Mahalanobis.

S.N. Bose wrote two papers while working in Dacca in 1924. The most important one was titled 'Planck's Law and the Hypothesis of Light Quanta', which was ignored by the *Philosophical Magazine* of the Royal Society. Then, in what must now rank as one of the seminal moments in the history of Indian science, he decided to write to Einstein. It was a bold decision and became one of the great moments in modern science. S.N. Bose sent his article with a handwritten letter to Einstein, who immediately appreciated the significance of the paper he had received.

He translated it into German and sent it for publication to the prestigious journal *Zeitschrift für Physik* with a note that read, 'Bose's derivation of Planck's law, signifies in my opinion, an important step forward. The method used here gives also the quantum theory of an ideal gas, as I shall show elsewhere.'[21] Bose's novel derivation of Planck's radiation formula, his only significant contribution to physics, came at a turning point between the old quantum theory of Planck, Einstein, Bohr and Arnold Sommerfeld and the new quantum mechanics of Werner Heisenberg, Paul Dirac and Erwin Schrödinger.[22]

This became the basis of Bose statistics. The mathematical derivation by S.N. Bose was an original one, based on the concept of the 'indistinguishability' of photons—the light quanta. Einstein also translated Bose's second paper and got it published, but with a note of dissent. It is generally thought that these two papers, published before the birth of quantum mechanics as we know it today, laid the foundations of a quantum theory of the electromagnetic field.

S.N. Bose also sent a third paper to Einstein in 1925, which was never published and remains untraceable. Since S.N. Bose did not have a PhD, Einstein's recommendation helped him become a professor and head of the department at the University of Dacca. Earlier, he had used Einstein's letter to take leave to travel to Europe.

S.N. Bose returned to Calcutta University in 1945. The particles that were discovered based on Bose statistics were christened 'bosons' by Dirac. Similarly, the particles that were discovered based on the Fermi-Dirac statistics are known as 'fermions'. Einstein developed S.N. Bose's concept further and predicted what is known as the Bose-Einstein Condensation.

Sadly, it is said that the relationship between the two friends, Saha and S.N. Bose, deteriorated to the point that the two avoided being in the same room together. As we shall see in the following section, Raman and Saha couldn't stand each other either. According to Chandrasekhar, the primary motivation for the flowering of Indian science in the first three decades of the twentieth century was the desire to show that Indians were as good as the British. Such a motivation was acceptable in the early stages but 'when recognition, glamour, reputation continued to dominate as the primary motives in individual scientists lives, deleterious effects took hold.'[23]

RAMAN–SAHA FACE-OFF

Even though Saha was in Allahabad, he kept abreast of the happenings in Calcutta, particularly at the IACS. Raman's attempts to professionalize the IACS had threatened Bengali prominence, and Saha became a leader of the opposition to Raman and his plans. The confrontation between the two slowly grew over the years. Raman and Saha both had huge egos. There was also an element of jealousy involved: Saha had struggled to make ends meet, whereas Raman had a cushy, well-paid job and good career prospects.

Things started to deteriorate in 1928, after Raman's famous discovery. Arnold Sommerfeld, the famous German physicist, visited India that year, and Raman delivered a lecture on the Raman Effect. After the lecture, which Sommerfeld also attended, Saha said that the discovery was no more than a confirmation of the German physicist Adolf Smekal's surmise. Saha also sent a letter to *Nature*, belittling Raman's work and suggesting it was wrong. While Saha was not wrong in his reference to Smekal, Raman had made no secret of the fact that he had, in fact, been testing and proving 'Smekal's surmise'.

The major flashpoint came in 1934, but there had been simmering discontent between the two for years before the IACS episode brought it to a boil. During these intervening years, Raman had earned his knighthood in 1929, and won the Nobel Prize in 1930. Saha's own scientific stature too had changed with his becoming an FRS in 1927.

Raman's Exit from the IACS

One of the charges against Raman was that he had gathered a coterie of South Indians around himself at the IACS. There was an unfortunate tendency (prevalent even today to some extent) to group people hailing from diverse regions into one category. Thus, people from southern states like Kerala, Karnataka and Andhra Pradesh, were often referred to simply as 'Madrasis'! K.R. Ramanathan, for example, was from Kerala, Suri Bhagavantam was from Andhra Pradesh, and Seshagiri Rao was from Mysore, but to Raman's opponents, such differences mattered little. In reality, the IACS had students from across India, with the possible exception of the Bombay Presidency.

Raman had numerous Bengali colleagues and associates, and his great regard for Sir Asutosh was well known. Since Raman was in charge of the association's facilities, it is unlikely that these people could have used them without Raman's approval. Students arrived at Calcutta attracted by Raman's fame. When Raman was offered the directorship of the IISc, he wanted to ensure a fallback in case things didn't work out in Bangalore. To this end, Raman persuaded the Sircar family to establish a new chair in physics at Calcutta University in memory of M.L. Sircar. This chair was to be attached to the IACS, thus ensuring that Raman could return to Calcutta, if necessary. Saha was also interested in this position but Raman believed that the IACS needed a younger researcher, who had not plateaued, as he thought Saha had.[24]

In a bid to have complete control over the association, Raman had put both his wife and his elder brother, C.S. Ayyar, on the management committee. The president of the IACS, Sir R.N. Mukherjee had sent his resignation and it was accepted in a meeting of the association's management committee. Ayyar proposed that Raman should take over the vacant post of president. This, too, was accepted, although, the names of S.P. Mookerjee, R.P. Mookerjee and J.N. Mukherjee had also been proposed for election but the proposal was defeated. This was to have serious repercussions in the following year when S.P. Mookerjee wrested control from Raman in an unruly general body meeting.[25]

When the M.L. Sircar chair was created in May 1933, Raman proposed Krishnan's name for it. Furthermore, Raman installed the latter as the association's secretary. Thus, Raman's control over the IACS, even in absentia, was complete. But his arrogant and authoritarian ways greatly upset the scientific circles in Calcutta. It rankled that he was in command even when he was away in Bangalore. On 19 June 1934, S.P. Mookerjee, with the help of Saha, turned the tables on Raman. It proved to be an acrimonious exit. Raman resigned as president, and the committee was reorganized with only five members, which included Krishnan from the old committee. Krishnan continued to be the M.L. Sircar Professor but without any administrative responsibilities. The next day, Raman resigned from the Palit Professorship, ending his long scientific relationship with Calcutta.[26] Over the years, Saha's influence over the IACS increased. From 1943 onwards, he had sole control over the destiny of the association. He became its president in 1946 and was responsible for building a new laboratory in South Calcutta, near the new Jadavpur University. After Saha left Calcutta University in 1952, he became the director of the IACS laboratory.

The Battle of the Academies

The Saha–Raman conflict in the IACS also showed up in the form of rival scientific academies and journals. This served to strengthen the regional tendencies in Indian science and delayed the growth of a national consensus over the organization of science in India. Until the 1930s, the ISC had served as both the forum for technical exchange and for political negotiations. It was founded in 1911 and first met in Calcutta in 1914 in the rooms of the Asiatic Society, Calcutta, with Sir Asutosh as its first president. But the ISC met only once a year, creating little opportunity for fruitful intercourse. Something more was needed and many voiced the need for the creation of an Indian Academy of Sciences.

In May 1933, an editorial appeared in *Current Science* (of which Raman was the editor) about the need for a national academy of science. It was later revealed that Raman himself had written the piece.[27] It was decided that the question of forming an academy should be discussed at a session of the ISC. It met in Bombay in January 1934 and Saha, in his presidential address, proposed the formation of an Indian Academy of Science, with the Royal Society of London as a model. Raman was the co-chairman, with Saha, of the physics and meteorology section.

It was decided that a special academy committee would be formed, which would look into the matter in detail. But there were disagreements from the start. Raman resigned from the academy committee and, on 27 April 1934, announced the establishment of an Indian Academy of Science in Bangalore, pre-empting the academy being planned by the ISC. He registered it in Bangalore, under the Societies Registration Act, appointing himself president. He reorganized his journal *Current Science* to serve the new academy in Bangalore.

Later, a mollified Raman suggested a federation of the various existing academies in India. This federation was to be an alternative to the academy originally proposed under the auspices of the ISC. After some negotiations, a deal was struck and Raman withdrew his resignation from the academy committee. According to the new arrangement, a National Institute of Sciences of India (NISI) was to be formed.

The first meetings of NISI were held on 3 January 1935 in the Senate Hall of Calcutta University and the rooms of the Asiatic Society. It functioned in Calcutta until 1946, when it was moved to Delhi. (In 1957, Homi Bhabha became its president). In 1970, it was renamed as the Indian National Science Academy (INSA). In 1935, under P.C. Ray's patronage, Saha started the popular monthly science journal *Science and Culture*. He used the journal frequently to express his views in the form of unsigned editorials. *Science and Culture* and Raman's *Current Science* were in direct competition. Now there were two popular science journals, catering mostly to scientists, and three science academies, one in Allahabad, the second in Calcutta and the third in Bangalore.[28]

The Conflict in Bangalore

This was not the first time that Saha and Raman had disagreed: they had also clashed over the affairs at the IISc. In the history of science in India, two events stand out: the first, as discussed, was the founding of the IACS. The second was the conception of the IISc by Jamsetji Tata in 1898. Jamsetji's philanthropic principles were rooted in the belief that for India to climb out of poverty, its finest minds would have to be harnessed. He believed that western education and training in the sciences would be extremely useful in the development and material progress of the country. His interest in education for scientific research was fuelled by his frequent travels abroad. His inspiration came from the Governor of Bombay Lord Reay's convocation address in 1889, in which he stated that it was imperative to evolve a teaching university. Jamsetji decided to address the deficiency.

This idea assumed a tangible form in September 1898, when Jamsetji announced that he had decided to set aside 14 buildings and four landed properties in Bombay for an endowment to establish a university of science. The properties were supposed to contribute £8,000, with the balance coming from the government. But it was hardly that simple because the viceroy, Lord Curzon, was initially not in favour of the idea. Jamsetji died in May 1904, uncertain of the future of his scheme, and it was left to his son Dorabji Tata to carry forward the scheme that had been so dear to his father. Hormusji Bhabha, Homi Bhabha's grandfather, also played a role in the initial stages of this project. What is important to note here is that both the IISc and TIFR (which Homi Bhabha would establish in 1945), evolved as a result of private philanthropy and public support working in concert for the benefit of institutional growth.

Raman left Calcutta to become the director of the IISc in April 1933. A department of physics was created to enable him to continue his research. In the 22 years before Raman was appointed as its director, the IISc had done nothing to suggest that it was dramatically impacting the industrial development of India. A popular saying then was that the institute was a sanatorium to which a few laboratories were attached!

Raman tried to get things moving and it is no surprise that he encountered stiff opposition. Saha, who was on the council, refused his request for a scientific assistant and wanted Raman to agree to a reduction in pay by half, something which greatly annoyed the Nobel laureate. Upon assuming office, Raman made three major changes, all of which caused problems for him. He started a new physics department, restructured some of the existing departments and reorganized the workshop. A new department needed money and he reappropriated certain sums to aid the new physics department, something that invited charges of defalcation! The reorganization antagonized both the professors of chemistry and electrical engineering, who resigned.

Aware of the fact that excellence flows from having the best personnel, Raman strongly felt that the institute needed fresh talent. At that time many eminent scientists were fleeing Hitler's Germany, and Raman thought it would be a good idea to bring them to Bangalore. One particular appointment, which he pursued with enthusiasm, was that of the distinguished German émigré physicist, Max Born, and it created much controversy. When Born arrived in IISc in September 1935, he was already a distinguished physicist. In the 1920s, along with the likes of Schrödinger, James Franck and Heisenberg, he had been instrumental in developing the foundations of quantum mechanics.

Born studied mathematics at Breslau, Heidelberg and Zurich before getting his doctorate from the University of Göttingen. He became professor of theoretical physics at the University of Berlin in 1915, where he met Einstein, and they became close friends. In 1919, he moved to Frankfurt, before becoming director of the newly established Institute for Theoretical Physics in Göttingen. The institute became an important centre for theoretical physics and attracted bright students from around the world. Raman was very keen that Born should become a permanent member of IISc. He proposed to the senate of the institute that it recommend the appointment of Max Born as professor of mathematical physics, but it was met with acrimonious rejection. Regardless, Raman continued to campaign for Born's appointment, and, in November 1935, the council accepted his suggestion for the creation of such a chair.

In 1935, the viceroy nominated Sir James Irvine, an organic chemist and vice chancellor of St Andrews University to head a review committee that included Bhatnagar, who was the professor of chemistry at Panjab University at that time. The first review committee meeting was held in Bangalore on 24 February 1936. The committee was very clear in interpreting the founder Jamsetji's intention as being one of material gain flowing from scientific activity. It also came to the conclusion (without a shred of evidence) that applied research did not receive Raman's support and that the resources of the institute were now being concentrated on research in pure physics. If the committee was unimpressed with Raman's research, it was even less enamoured by Born's work, which lacked direct connection with industry and was, therefore, unlikely to be of any practical service to India. According to the committee, Raman was guilty not only of administrative irregularities but also of financial lapses. Furthermore, it was critical of Raman's efforts in implementing economy measures.

Thanks to the Irvine Committee, Born's appointment as the professor of mathematical physics was rescinded. The accusation against Raman of nurturing a South Indian coterie reared its ugly head again, even though the records did not show any statistically significant change in the demographic composition of the student population after Raman had come to Bangalore. There can be little doubt that the Irvine Committee had been tasked with getting rid of Raman, and it did what it was meant to do.

The Saha–Raman conflict had been extended to Bangalore via IISc. Dr Guha, an assistant professor of chemistry, provided S.P. Mookerjee, P.C. Ray and others with all the local information they needed. Raman considered Guha an inferior scientist and wanted to get rid of him. Raman's own character defects did nothing to endear him to people. Born, who corresponded with Rutherford on the subject, felt that Raman did not respect other people's feelings. He wrote: 'He is not a bad character; he is rather an eternal and badly educated child.'[29]

The Irvine Committee submitted its report to the viceroy in May 1936. At a council meeting in August, S.P. Mookerjee attacked

Raman on various counts and moved a resolution that the council declare its lack of confidence in the director on the matter of preparing the council minutes. The council recorded its 'sense of strong disapproval' and appointed a special sub-committee to enquire into the charges contained in the memorandum.

Things were spiralling out of control. Raman was offered two choices: either continue as the professor of physics or resign with effect from 1 April 1938. He was warned that if he declined both options, he would be suspended. With no room to manoeuvre, Raman reluctantly agreed to continue as the professor of physics. He still appeared to retain his old enthusiasm and spirit but deep inside, he was much shaken. He was never the same man again, increasingly prone to cloudy judgement, where both persons and scientific matters were concerned.

SHANTI SWARUP BHATNAGAR

The third of the troika, Shanti Swarup Bhatnagar, was educated at Dayal Singh College and the Forman Christian College in Lahore, from where he got a BSc in physics in 1916, having failed—difficult as it may be to believe—the chemistry exam! Bhatnagar prepared himself for the MSc exams as a private student, and with coaching from diligent and dedicated British and Indian science teachers, particularly in chemistry, he prepared his MSc thesis on the effect of absorbed gases on the surface tension of water. He ranked first in the physical and inorganic chemistry exams, and his thesis won the Dayal Singh Five-Year Travelling Scholarship to study in America. He used this scholarship, however, to pursue his postgraduate research in London under Professor Frederick G. Donnan of the University College, London.

Professor Donnan agreed to take Bhatnagar under his wing for the DSc degree of the London University. After two years of work in the Ramsay Memorial Laboratory at the University College, London, he submitted a thesis on the 'Solubilities of bi-and trivalent salts of higher fatty acids in oils and their effect on the surface tension of oils' and received his degree in 1921.[30]

Like Saha, Bhatnagar arrived in London in 1920 where the two met for the first time. Saha had also arrived in England on the back of two scholarships, but since he still could not afford to attend Cambridge or Oxford as he wished, he took admission in Imperial College, London.

Saha nicknamed his friend 'Steamship Bhatnagar' because of his energy and determination and the name seems to have stuck. Bhatnagar accompanied Saha when the latter went to Berlin. They were part of a group of 14 students visiting from the University College, London. Bhatnagar carried the letter of introduction to Nernst from Donnan. Nernst had just won the Nobel Prize for Chemistry a few months before, in late 1920, and held numerous patents, including one for the nerve gas used during the First World War. Nernst, at first, refused to invite any of the students from London into his laboratories but later relented because he believed that the last blow to the British Empire would come from India.

Through his visit to Nernst in Berlin, Bhatnagar connected with German scientific world, particularly with Fritz Haber, who, along with Nernst, invented nerve gas and won the Nobel Prize for Chemistry in 1918 for making ammonia from nitrogen in the air. All this gave Bhatnagar an early entry into the field of colloid chemistry. The stimulus of the war resulted in the formation or expansion of large American, British, French and German international chemical corporations, all of which had big offices in London. There was now a huge global market for chemicals.

On his return to India in August 1921, Bhatnagar was appointed as the professor of chemistry at BHU, where he worked for three years. He then migrated to Lahore on his appointment as professor of physical chemistry and director of the University of Punjab's chemical laboratories. During this period of active teaching and research in the two universities, he had a large number of able students and collaborators.

In 1926, Compton, of the University of Chicago, came to Lahore to deliver a series of lectures at the University of Punjab. On landing in Calcutta, he found that he was supposed to lead an expedition to Gulmarg, Kashmir, to study cosmic rays but no equipment had been provided for it. In an early example of Indian jugaad, Compton, with Raman's help, fashioned an electroscope from the bowl of a hookah and other parts that were fabricated by Raman's mechanic. Part of the trip was on horseback, and Compton's wife, Betty, not only proved to be a good rider but also an able assistant.[31] Bhatnagar himself made new equipment in his laboratories for Compton's experiments. The equipment he became familiar with in this project led him to think about the invention of other research equipment for physical chemists. Bhatnagar had never visited America, but this interaction with Compton helped him establish contacts there.

While working in Compton's group in Kashmir, Bhatnagar met Nazir Ahmed, a physicist who knew a lot about instrumentation because he was completing his PhD in physics under Rutherford at the Cavendish Laboratory at Cambridge University. At the time, the Cavendish was famous as a place where students built their own instruments for experiments. Ahmed was so impressed by Compton's wife that he decided to get Muslim girls educated.

In 1928, with his old friend from Benares, K.N. Mathur, he produced the Bhatnagar–Mathur magnetic interference balance, which was an extremely sensitive instrument for measuring very small variations in magnetic properties. This instrument was manufactured by a British firm and exhibited in London at the Royal Society in 1931. Bhatnagar benefitted from the royalties for years.

He completed his PhD in 1927 and returned to Lahore to teach physics. Ahmed moved from Lahore to Bombay to work in cotton research and was a member of the exclusive team of scientists that toured Allied nuclear research installations in 1944–1945 with Bhatnagar and Saha. He was also involved in planning the CSIR and was, thus, an important figure in Indian science planning and politics in the mid-1940s. After Partition, in 1947, he became the first chairman of the Pakistan Atomic Energy Commission.

Lahore was an important educational and cultural centre at the time, even more than Delhi. Bhatnagar set about building academic and political connections in the scientific community, a good example of which was his organizing the meetings of the ISC in 1927 in Lahore, in direct consultation with its president J.C. Bose. The 16 years (1924–1940) he spent as a professor at the Punjab University constituted the most active period of his life for original scientific work. Bhatnagar enjoyed teaching and earned a formidable reputation as a very inspiring teacher. His major fields of study were colloid chemistry and magneto-chemistry. He did considerable work in applied and industrial chemistry—areas in which he had a natural interest from his early student days. Bhatnagar was also very interested in solving problems brought to him by people outside the university. He developed a process for the industrialist Sir Ganga Ram, which allowed the conversion of bagasse into cattle fodder.

One of his most striking successes came in 1933, when he was working on what was called the 'mud problem' for Messrs Steel Brothers & Co., London. While drilling for oil, the Attock Oil Company at Rawalpindi had been using a drilling mud, which set into a solid mass when it came into contact with saline water, thereby rendering all drilling impossible. Bhatnagar solved the problem by adding an Indian gum that had the property of lowering the viscosity of the mud suspension.

Pleased with Bhatnagar's elegant solution, the petroleum subsidiary of Rangoon-based steel brothers, Attock Oil Company, decided to offer him a grant of ₹150,000 over five years for research work on any subject related to petroleum in 1934. This was a huge sum of money for Indian scientists, particularly so during the post-Depression years. Bhatnagar decided that he would not personally accept this money but, instead, channel it through his university. This grant led to the foundation of a new department of petroleum research. Bhatnagar was to also receive an honorarium of ₹6,000, a sum that amounted to almost half his annual salary. In addition, he was to be assisted by a fully qualified British petroleum research chemist who would not only be his personal assistant but also look into the minutiae of daily administration.[32] Bhatnagar had also planned six research scholarships, which were named after five eminent men of Punjab and Professor Donnan, his teacher.

In January 1939, the ISC met in Lahore again, and Bhatnagar was in charge of all local arrangements, under the presidency of Calcutta-based chemist, J.C. Ghosh. It was a great success. Bhatnagar's fame and the goodwill he earned led to a visit from Sir Ramaswamy Mudaliar, the commerce member of the viceroy's executive council, in December of the same year. Mudaliar was impressed enough to write to the governor of the Punjab to ask for Bhatnagar's services as an advisor in scientific and industrial research. Bhatnagar set conditions on his acceptance of the offer to move to Delhi, including the relocation of six Steel scholars and the provision of laboratories for them wherever he went.

In April 1940, soon after the outbreak of the Second World War, the Government of India established the Board of Scientific and Industrial Research (BSIR) with Bhatnagar as director. The laboratories of the board were first located in Calcutta but threats of an imminent Japanese invasion resulted in them being shifted to Delhi. In March 1942, the BSIR became the Council of Scientific and Industrial Research (CSIR) based in New Delhi. As director of the CSIR, Bhatnagar served under several members of the viceroy's executive council, such as Mudaliar, Sir Azizul Haque and Sir Ardeshir Dalal. Later, when the Congress took over the government, he served under C. Rajagopalachari and, ultimately, under Nehru, who was equally enthusiastic about the development of science and of national laboratories.[33]

ENTER HOMI BHABHA

In 1939, the year Homi Bhabha returned to India, Bhatnagar and Saha were poised to undertake new ventures in institution building. Saha was beginning to build his new institute within Calcutta University, and, a little later, Bhatnagar commenced his role at the BSIR. Both Bhatnagar and Saha had taken part in the Raman affair at Bangalore and had, since then, developed a very friendly relationship with the new director of IISc, J.C. Ghosh. Saha was already an FRS; Bhatnagar became one in 1943. They were now ready to plan scientific institutions in collaboration with industrialists, politicians and government officials.

It was this milieu that the young Homi Bhabha stepped into when he returned to India. Within the span of a decade-and-a-half, he was monopolizing 'big science' in India. Bhatnagar became his supporter and partner, and Saha an acerbic critic. Bhatnagar and Bhabha who occupied positions of scientific eminence went on to occupy positions of power, providing scientific leadership and policy advice to Nehru. The consensus between the scientific and political elite was both strong and unprecedented, and it was this rapport with the highest political power in the land that allowed Bhabha to be as effective as he was. Nationalism and government investment in science contributed to the belief in the power of science to liberate India from poverty and ignorance. With Bhabha at the helm, nuclear energy started being seen as the way to speedy development: the authority of science became unequivocally established among the educated elite and the illiterate poor.

As we shall see, Bhabha was unique in his ability to successfully combine science with politics and private capital with government. It has been argued that Bhabha did not have to convince people about science having the solution to both India's economic underdevelopment and political subjugation. What he had to convince people about was the fact that the specific scientific solutions he had to offer were what the country needed, and that it was only he who could provide those solutions. In other words, only Bhabha could split the atom for India.[34] By virtue of being a practitioner of the branch of science that was most in vogue in the mid-1940s, and with the impeccable credentials acquired in the West, Bhabha's path to power was an easy one.[35] How his outstanding talents and abilities put him in an almost unassailable position, and what he did with that power to benefit India's scientific development, forms the subject of this book.

Chapter 2

THE HERITAGE

The Parsi community, to which Homi Bhabha belonged, is a distinct ethnic and religious minority in Indian society. Its members are descendants of the Iranian Zoroastrians, who migrated to India over a thousand years ago. They first landed on the island of Diu, in southern Saurashtra, where they stayed for 19 years. From Diu, they sailed along the coast and finally landed in Sanjan in Gujarat.

The *Kisseh-i-Sanjan* (the story of Sanjan), an epic poem written in 1599, recounts the early history of the Zoroastrian settlers in India. It elaborates on their travails and persecution after the fall of the Sassanid Empire in AD 651. It is the only record of the history of the Parsis from the time they reportedly migrated from Iran to escape religious persecution. Written by Bahman Kaikobad, a Parsi priest from Sanjan, the *Kisseh* is more a story than history—in fact, it is a multifaceted religious myth. There are virtually no details about the number of people who migrated from Iran, the number of ships they arrived on or even the time when they reached India, with the dates differing from the eighth to the tenth century. The Parsis have engaged in bitter disputes over whether the poem is a work of fiction or history. Perhaps, it would be best to treat it as an oral account passed down through generations of priests. In the words of D.F. Karaka, the author of an authoritative history of the Parsis in the mid-nineteenth century, '[v]arious meagre and unsatisfactory traditions exist concerning the tide of emigration, the manner in which it was effected, and the total number of those who left the shores of the Gulf'.[1]

Even today, most Parsis believe that their forefathers arrived in one boat, braving storms and other hardships before landing on Indian shores. However, the truth is quite different. The earliest Persians came to India as traders by land before the Sanjan migration. That they were essentially a mercantile group is borne out by records as well as by excavated material. The idea that a shipload of migrants arrived after a hazardous journey is little more than a myth but there is little doubt that religious pressures played a part in the migration. The Abbasid Caliph, Al-Mansur, ordered an oath of allegiance in AD 766 which seems to have been a signal for the first exodus of the refugees who came to India, seeking, in Henry Lord's words, 'liberty of conscience'.[2] The *Kisseh* speaks of an interaction between a group of Zoroastrian priests and Jadi Rana, the Hindu king who ruled over Gujarat when they arrived. A Parsi priest, who had learnt Sanskrit in Diu, is said to have explained the tenets of the Zoroastrian religion in the form of 16 Sanskrit slokas. The ruler was impressed and granted them sanctuary after imposing a few preconditions.

However, an oral account is the more popular narrative of the Parsi settlers being accepted in India. According to a popular legend, Jadi Rana, during his interaction with the priests, presented them with a bowl of milk filled to the brim, symbolically denoting that his kingdom had no place for them. A wise Parsi priest added some sugar to the milk, suggesting the adaptive and accommodating attitude of the Parsis. They would be like sugar in milk, enhancing the population without overwhelming it.[3] Jadi Rana was impressed by this symbolism, and the refugees were accepted. There are slight variations in how the story is told, but the essence always remains the same. Although only a fable, it is a significant marker for the Parsi community as a whole through its relevance to their attitude to their host countries—historically, India, but now globally.

To understand Bhabha, it is essential to have some knowledge about the community to which he belonged and how it shaped him and his endeavours. The qualities of hard work, thrift, honesty, integrity and charity that the Parsi community displayed would see it emerge from the obscurity of small-time merchants and agriculturalists to a dominant position in Bombay in the eighteenth century. In fact, the development of Bombay in the eighteenth and nineteenth centuries is indistinguishably linked with the prosperity of the Parsis.

Their progress was assisted in no small measure by the security and encouragement afforded to them by the colonial British rulers. The Parsis have excelled in almost all spheres: industry, public life, scientific endeavour and the professions. The community also enjoys a prominence far out of proportion with what could have been expected from such a minuscule minority. These contributions explain the justifiably high regard that the community is held in at the national level. Through their actions, they have fulfilled their legendary pledge to be the sugar in the milk of the local population.

One of the defining elements of the refugee experience is negotiating old and new identities. All refugees face the dilemma of preserving their religion, language and traditions in their adoptive home. Though the Parsis have assimilated into Indian society, they have remained an endogamous group, preserving their religious identity while becoming an integral part of Indian society.

In many ways, the turning point for the Parsis in India was the arrival of the British. From the eighteenth century, the adaptive Parsis started establishing economic and sociopolitical contacts with British interests and emerged as a prosperous and influential community under British colonialism. By the nineteenth century, they had become a highly westernized community with a pluralistic outlook.

Thus, the Parsis are, by definition, a diasporic people that have flourished outside their Persian homeland. They migrated to the edges of Britain's imperial reach and beyond for trade and, later, for education and to take up professional and governmental positions. They balanced a strong sense of Persianness with close ties to the imperial British metropole.[4]

PARSI IDENTITY AND THE URBAN SETTING

In the early seventeenth century, Surat was essentially a Parsi base, and the British chose to establish their first factory there. The Parsis were quick to learn English and became translators for the British. They initially came in contact with the British as interpreters, hawkers, traders and contractors. By the eighteenth century, they had become commercial agents who handled a range of activities including extending credit and buying ships. Such commercial success is best exemplified by Rustum Maneck of Surat, who was the chief broker to both the Portuguese and the British. His activities went far beyond what the word broker signifies today. He was virtually an ambassador to the Europeans and even the Nawab of Surat sought his help on matters of state.

However, it was in Bombay that the Parsis set the foundations for the accomplishments of the community in India and abroad. Bombay was gifted to Britain by the Portuguese in 1661 as part of the dowry of Catherine of Braganza who married Charles II of England. When it became a colonial possession, Parsi merchants were among the first to settle there and take advantage of the new centre of commercial activity. It also attracted other communities in search of similar opportunities, something which would give Bombay its cosmopolitan social culture. Parsis became *dubashs* (translators) to the East India Company. They were also contractors, building the Fort to protect against the invasions of the Marathas and Siddis.

In 1669, Gerald Aungier, the governor of Bombay, realizing that Bombay would be a more viable port than Surat, issued an open invitation to the weavers, carpenters and shipbuilders of Gujarat to settle in Bombay. Parsis formed the majority of these pioneers and most of them settled in the enclosed area of Bombay known as Fort. As Bombay developed into a major commercial centre, the Parsis started to prosper. Bombay became their new home and its development became an expression of their influence in the new setting.

The Wadia family, beginning with Lovji Nusserwanjee Wadia in 1730, led the migration of the influential shipbuilders from Surat to Bombay. Invited by the British to establish a dry dock, he built ships which were of the highest class. The Parsis went from being mere artisans to owners. They bought ships and started trading, first along the Malabar Coast, and then engaged in the very lucrative trade with China. The lack of caste prejudice and interdictions on crossing the seas, a facility with languages and a reputation for honesty helped them participate in the lucrative overseas trade with China and, eventually amass much wealth. The Camas, Banajis, Readymoneys and Sir Jamsetjee Jejeebhoy were all linked to the China trade.

Jejeebhoy, who went on to become the first knight and baronet in India, was the best known among the early nineteenth-century traders. A substantial part of this trade included the traffic in opium. The British were getting millions of people addicted to opium both in China and India, 'enabling the longest-running drug deal in the history of the world'.[5] The East India Company has been described as, 'a drug cartel masquerading as a joint stock corporation masquerading as a government'.[6] Jejeebhoy became fabulously wealthy due to his partnership with one of the largest opium traders of the time—Jardine, Matheson & Co.

Several Parsi families made their fortunes; many more were obsessed with the potential benefits to be gained from the China trade since it was the most valuable branch of the commerce in Bombay. The Parsis were involved in the opium trade from the beginning. As early as 1756, Hirji Jivanji Readymoney was trading with China. Jejeebhoy not only became extremely wealthy but later in his life also became a legend for his charitable endeavours that changed the face of Bombay forever. He set up more than 125 charitable institutions and donated about £245,000 in his lifetime, a figure which, if inflated to its present value, would be more than ₹100 crore.

The Parsis traded cloth and opium for silver, spices and tea from China and Southeast Asia. Huge fortunes were made from the opium trade, and a look at Parsi philanthropy shows that the greatest traders of opium were also the greatest benefactors of Bombay. Many of the fine public buildings, colleges and hospitals owe their origin to the liberality and public spirit of the Parsis. It was opium that linked Bombay to the international capitalist economy and the western Indian hinterland. It was also crucial to the emergence of Bombay as a metropolis. There are close linkages between the opium enterprise of western India and the creation of early Victorian Bombay.[7] Amar Farooqui has observed that Bombay's destiny as a great commercial centre was 'born of its becoming an accomplice in the drugging of countless Chinese with opium, a venture in which the Indian business class showed great zeal alongside the East India Company'.[8] This zealous Indian business class included a number of prominent Parsi families. A fact that most discussions on Parsi philanthropy tend to ignore is the role played by profits from the opium trade in the Parsi charity of the times.

After the First Opium War (1839–1842), Parsi merchants began to look for new areas for business. Jeejebhoy Dadabhoy, an agent, broker banker and merchant introduced steam navigation for commercial and passenger traffic along the west coast of India. He helped set up the Bombay Chamber of Commerce and also played a role in establishing the Great Indian Peninsula Railway (now Central Railway). The cotton industry, too, owed much to the Parsis. In 1854, India's first spinning mill, the Bombay Spinning and Weaving Company, was set up by Cowasji Nanabhai Davar. A few years later, it became the first cotton mill to be run on steam. By 1858, Maneckjee Petit established the first composite spinning and weaving mill. The Wadias, Tatas and the Petits dominated cotton manufacturing in Bombay.

The incorporation of the Parsis into the economic and political world of British power from the late eighteenth century reflected their successful transformation from being a minority community in a provincial setting to an influential colonial elite of Bombay. The Parsis' cooperation with the British was based on a keen appreciation of the opportunities they provided, both to individuals directly in contact with them and to the community as a whole. The assimilation of British values was based on the personal affinity the Parsis developed in contact with all things British. However, this assimilation functioned alongside the Parsis' commitment to their traditional identity. The ability to actively assimilate and subsume British values and influences to support their identity distinguishes the cultural history of the community from the mid-nineteenth century.[9]

William Thomas Fee, the US consul general in Bombay, perceptively observed in 1905, 'There is now a marked desire on the part of the Parsees to adapt themselves to the manners and customs of the Europeans [...] A Greek historian has remarked that of all nations the ancient Persians were most distinguished by their readiness in imitating foreign manners and customs. This peculiarity their descendants have retained to the present day.'[10]

PROGRESS IN EDUCATION

The Parsis also pioneered educational advancement in Bombay. The social and political connection between the British and the Parsis led to cooperation in the establishment of educational institutions. This progress resulted in the rise of a new generation of Parsis who were well prepared to lead the advancement of the community. Education became an integral characteristic of the Parsi identity and both social advantage and cultural identification followed from it.[11]

In the colonial age, the Parsis mastered the attributes of intermediacy, urbanity, mobility and flexibility that were considered central to the 'modern' condition.[12] The incorporation of western values into the Parsi value system enabled their assimilation into modernity and the Parsi intelligentsia played a major role in this transmission of western values. The Parsis, now regarded as the most influential, literate and progressive of communities, were able to exploit opportunity and direct change that would benefit them. The quest for economic advantage can be seen as an attempt to preserve identity and remain economically, socially and politically relevant as a community.[13]

Around 1850, a Parsi renaissance seemed to have begun, which resulted in the flowering of their creativity and genius on all fronts. By 1853, Dadabhai Naoroji, who was a trained mathematician, was teaching at the Elphinstone Institution, which later became a college. Parsi reformers started a society to provide school facilities, and they were the first to pioneer education for women. Naoroji and his friends went from house-to-house to persuade parents to allow their daughters to go to school. Along with educational awakening came social reform. The Parsis took the lead and urged the British government to set up a commission to look into Parsi marriage, divorce and succession. While Parsis allowed remarriage of widows, bigamy was accepted. In 1865, with the passing of the Parsee Marriage and Divorce Act, bigamy was made illegal. K.R. Cama, a layman, was one of the first Parsis to be sent to study the Avestan language through western methodology and training. Later, the K.R. Cama Institute was established in Bombay as a repository of books and a centre for learning.[14]

The generation after the Shetias (the business magnates) and their great wealth was that of the Parsi professional. Following the economic upsurge brought by the trade with China, the Parsis were also leaders in the professions of law, medicine and engineering. They played a major role as lawyers and reformers, greatly impacting both the social and political fields. The Parsis also added the political dimension to the convergence of educational, social and economic forces. Naoroji and Pherozeshah Mehta were among those present at the launch of the Indian National Congress in 1885. Jamsetji—who almost single-handedly launched India on the path to modern science and industry, risking his personal fortune in the process—was also present at the launch in his own unobtrusive way.[15] It is pertinent to note that of all the Indian nationalist political leaders, Naoroji was one of the few who opposed the opium trade, refusing to touch the opium profits of the firms he had invested in. He observed, 'This opium trade is a sin on England's head, and a curse on India for her share in being the instrument.'[16]

A NATIONAL OUTLOOK AND POLITICAL NATIONALISM

The same forces that led to the educational and economic progress of the Parsis also led them to move from a parochial to a national outlook: they were Indians first and Parsis later. The Shetias, who were early Parsi leaders, had cooperated with the British, resulting in them being recognized as the leaders of the Parsi community. If the Parsis were considered among the most loyal colonial subjects, it was also from their ranks that individuals who shaped the foundations for the self-government of India emerged.[17]

The advancement of English education had predisposed a new class of Indians to be loyal in its allegiance to the British while desiring the rights and opportunities of western culture and colonial rule. The early Indian nationalists, such as Naoroji, Mehta and Dinshaw Wacha, were perhaps the best examples of the benefits of an English education. Their outlook and identities had been shaped by their educational background, and while they remained loyal to British rule, they sought a greater role for Indians in public affairs. Naoroji and others aimed at actively guiding the British rulers to better understand Indians and their needs, and eventually they sought political responsibility from the British rulers.[18] The rise of the Parsis within the nationalist movement was in consonance with the intellectual maturation and politicization of educated Indians. The careers of these nationalists exemplify the transition of some Parsis to a sense of national identity.

A 'DOUBLE CONSCIOUSNESS'

This, however, was not the full story. As the Parsis of Bombay prospered, they became increasingly westernized, identifying more closely with the British. A new consciousness of a larger cosmopolitan identity emerged among the Parsi community, and as the economy of Bombay grew, so did their loyalty to the colonizers. It was perhaps this loyalty that led many affluent Parsis in Bombay to become Anglophiles, more or less separated from Indian culture and society, and from their roots in Gujarat.

The rapid Anglicization of Parsi manners, particularly among the educated and prosperous classes reflected the new cultural imprint that marked Parsi identity. Their efforts at self-fashioning included negotiating change in a number of personal aspects like dress, deportment, dining and domesticity. The highly urbanized nature of the community accelerated the process of acculturation of Occidental practices. The attraction for the British was reflected through a host of cultural forms—the Parsis adopted British taste and style in music, sport, literature and the desire for western consumer goods. The love of western classical music was reflected in the growing number of Parsis who took violin and piano lessons, their patronage of music, and the largely Parsi audiences at western classical music concerts in Bombay.[19] Bhabha and his family's love for western classical music can be easily understood given the general predilection of the community for such pursuits.

Parsi gymkhanas opened and the community took to sports like swimming, wrestling, boxing, fencing and marksmanship during the last quarter of the nineteenth century. They also took to tennis, cycling, motor sports and aeronautics. The Parsi enthusiasm for sport was best symbolized through their commitment to cricket. Two Parsi cricket tours went to England in 1886 and 1888 and though the team did not fare well, the aim of learning and playing the game alongside the British was a major accomplishment of the tours.[20] Dorabji was passionate about cricket and his wife, Meherbai Tata, Bhabha's paternal aunt, participated in tennis competitions in Britain and Europe.

In many ways, Parsi modernity also rested in the city of Bombay. Colonial Bombay was not only the setting for Parsi self-fashioning but it also became part of their new identity. Parsis in the city referred to themselves as the 'Parsis of Bombay', and they developed discourses of othering that were framed in terms of the relative disadvantage of residing in the 'country' or the 'mofussil' as compared to Bombay. Bombay was the seat rather than the setting of Parsi modernity.[21] By the last quarter of the nineteenth century, Parsi businessmen had made a steady profit by importing and distributing British and western-style goods in India, which included furniture, clothing, personal items and motor cars.[22]

The Parsis had prospered due to their contact with the British and it was, therefore, no surprise that the British cultural model became a standard and exemplar for the Parsi desire for business. To be modern was to be western: the Bombay Parsis now considered themselves superior to their non-westernized brethren in Gujarat. Religion, too, assumed a secondary role; their Savile Row suits, Royal Worcester porcelain and expensive crystal no longer fitted in with their earlier outlook and way of life.

This unquestioning absorption of English cultural and societal mores marginalized their own identity. Robert E. Park and V. Stonequist, who formulated the concept of the marginal man, defined such a person as a cultural hybrid, a man living and sharing intimately in the cultural life and traditions of two distinct peoples, which never completely interpenetrate or fuse.[23] This leads to a dual personality, a 'double consciousness', and is a characteristic of racial and cultural hybrids. The westernized Parsis like Bhabha faced this dilemma, and as we shall see later, this dual identity presented a problem to Bhabha's European colleagues who did not know quite what to make of him. While this was true of the elite South Bombay Parsis, equally, a large population continued to be rooted in their traditional culture and values. They appreciated the importance of western education and the opportunities it afforded them, but continued to be conscious of their roots in Surat, Navsari and other parts of Gujarat.

It is said that the elite minority tends to fall prey to the culture of imitation. The reason is its proximity to power as well as the desire to mimic that power. Postcolonial theorist Homi K. Bhabha calls mimicry 'the most elusive and effective strategies of colonial power and knowledge'.[24] He says 'colonial mimicry is the desire for a reformed, recognizable Other, as a subject of a difference that is almost the same, but not quite'.[25] The ambivalent figure of the colonized appears as neither this nor that—a doubly inauthentic person.

CHARITY AND PARSI IDENTITY

An important feature of the Parsi religion and community is its emphasis on charity—philanthropy is a defining feature of the community's conscience. This emphasis on charity greatly influenced the development of the Parsi communities in Gujarat and Bombay, particularly during the nineteenth and twentieth centuries, providing the financial resources to establish their infrastructure, and on a bigger stage, the infrastructure of Bombay. This included not only fire temples but hospitals, libraries, schools, technical colleges, housing colonies, rest homes for the elderly, endowment policies for the disadvantaged and a myriad charitable trusts.

The nature of Parsi charity draws its inspiration from the life-affirming Zoroastrian philosophy to contribute to the welfare of the world. *Humata, hukhta, hvarshta* (good thoughts, good words and good deeds) are key maxims of the faith. While Zoroastrianism is not alone among religions in emphasizing these values, what makes Parsi charity unique has been the setting up of very large and powerful charitable trusts. Charitable gifting to these trusts, and the use made of the trusts by individuals has in the past, had a major influence on Parsi identity, both individually and collectively. To be a good Parsi, one had to give away a portion of one's wealth to benefit others, not just within the Parsi community but to anyone in need.[26]

This secular nature of Parsi charity distinguished it from the charity practised by other communities, at least in the years gone by. Charity was never confined to helping only the Parsi community. Mahatma Gandhi observed, 'I am proud of my country, India, for having produced the splendid Zoroastrian stock, in numbers beneath contempt, but in charity and philanthropy perhaps unequalled and certainly unsurpassed.'[27] Stories abound regarding donations by Parsi people to relieve suffering, not only across India but also abroad. Records show that the Tatas and Wadias donated large sums of money to the Gandhi Memorial Fund and that the Tatas and Godrejs donated generously to the Indian Defence Fund at the time of the Indo-Chinese War of 1962.

During the nineteenth century, charity became strongly linked to not only acquiring power, particularly political power within the Parsi community but also as a means of gaining favour with the British rulers in India. Within the community, this link caused its leadership to pass from the priests to the leading mercantile houses such as the Wadias, Readymoneys, Jeejeebhoys, Petits and Tatas. Many of the heads of these houses acquired British knighthoods or even baronetcies in recognition of their charitable deeds.

Philanthropy has been a vital mechanism through which Parsi individuals and groups have sought to acquire and maintain their social status and leadership. According to Jesse S. Palsetia, professor of history at the University of Guelph, Jejeebhoy's charity has often been viewed as solely the altruistic impulse of an enlightened mind seeking to benefit society. However, it had multiple motivations, including political and social recognition, leadership within the Parsi community and a desire to satisfy his pious impulses. If his main aim had been to raise his status in British colonial culture, he certainly succeeded. The accolades he received included a knighthood, which culminated in the bestowal of a hereditary baronetcy.[28] Charity provided the framework around which a new set of relations with the colonial order could be designed.

Parsi charity continues to be a hallmark of Parsi identity in Bombay and India. In Bombay, close to a thousand Parsi charities have been listed in the Directory of Public Trusts under the Public Trusts Act since 1950. Though declining in importance in present times, Parsi charitable trusts remain a crucial aspect of Parsi culture. It was one such trust, the Sir Dorabji Tata Trust, which played an important role in the life of Bhabha, and through him, of India's scientific and technological development.

Chapter 3

A SCIENTIST IN PUPA

> The Cathedral High School did much to foster a love of science in me. One's school years are often the most formative ones and I remember enjoying my school days very much.
>
> —Letter to C.H. Hammond, former headmaster of Cathedral and John Connon Boys' School, Bombay, 14 October 1955[1]

The child who would become famous as Homi Jehangir Bhabha was born into privilege, culture and entitlement on Saturday, 30 October 1909, in Bombay. He was born in a bungalow named Kenilworth, located on 53 Pedder Road in South Bombay. Kenilworth belonged to Bhabha's maternal aunts, Bachubai and Hirabai and, later, his other aunt, Cooverbai Panday. In a common naming custom among Hindus, the first sound of the newborn's name is astrologically determined, since it is supposed to bring the child luck. Many Parsis follow this custom but it is not known if Jehangir and Meherbai Bhabha, the child's highly westernized parents believed in this. He was called Hormusji, after his grandfather, a name that was shortened to Homi by all those who knew him, and it was as Homi Bhabha that he courted national and international fame. Almost five years later, on 21 August 1914, another son, Jamshed, was born to the family, who who made a substantial contribution to art and culture in Bombay.

Bhabha's family was very well off but by no means extraordinarily wealthy. However, they had a long tradition of learning and service in the field of education. His paternal grandfather, Hormusji Bhabha's association with Mysore began in 1876 when he was appointed vice principal of Central College in Bangalore. (More than six decades later, his grandson Homi would be a guest lecturer in physics in the same college). Eight years later, Hormusji became the headmaster of Maharaja's College in Mysore. In 1890, he was appointed secretary in the Education Department, and then from 1895 to 1909, he served as the first Indian inspector general of education in the state of Mysore.

During his years in the department, Hormusji helped implement the Maharaja's vision for educational reform. In recognition of his services, he was given the title Munir-ul-Taleem in 1909 by the Mysore government.[2] Hormusji was also a leader in his own community, being president of the Iranian Association and chairman of the standing committee of the Zoroastrian Conference. He was also a member of the Syndicate of Bombay University, along with being on the boards and committees of various other charitable and educational institutions.

Mysore was one of the few progressive princely states in India at the time and Hormusji substantially contributed to its development. He was a forward-thinking individual who eschewed religious orthodoxy. He deposed in support of Bella in the famous *Saklat vs Bella*[3] case, which ended in 1925. Bella was a girl from Rangoon, Burma, born out of the union of a Parsi mother and non-Parsi father, and the issues to be decided upon were whether she could be considered a Parsi after her navjote[4] had been performed and if she could enter the fire temple. After victories in the lower courts, the matter went up to the Privy Council, which, citing the judgement in another very famous case, *Petit vs Jeejeebhoy* (1908)[5], ruled against her. The question of whether the child of a Parsi Zoroastrian mother and a non-Parsi father should be admitted to the Zoroastrian faith remains a vexed question to this day.

AN IMPRESSIVE PEDIGREE

There is no doubt that the advantage of being born a Bhabha gave Homi a leg up in the world, but the fires burned in him and you could always see the glow—he was a product of two moulding forces: the external conditions and his hidden germinal energies. Intellectually, life at home was as stimulating as any gifted child could wish for. He was encouraged from an early age to cultivate an aristocratic appreciation for the fine arts. This passion for art encompassed all that the young Bhabha did, and his thirst for western music and culture only grew as he aged. When he was in his early twenties he would spend a lot of time at concerts in England, listening to his favourite composers again and again.

Bhabha had an ear for music from a very early age: his parents only had to play some music to stop him from crying. The family owned a large collection of western classical music records and from his childhood, Bhabha was exposed to the works of leading European musicians. His interest in western classical music was nurtured by his maternal aunt Cooverbai, who also stayed with the Bhabhas at Kenilworth. Handicapped and unable to move independently, she had an enviable collection of 78 RPM gramophone records that Bhabha, his younger brother and his cousin, Dinshaw Panday, listened to with rapt attention. Bhabha also played the violin and the piano.

The young men soaked up all the music on offer—Beethoven, Mozart, Bach, Haydn, Verdi, Wagner and Schubert, to name just a few. They would take turns to wind the gramophone and turn over the record. They were familiar with Beethoven's symphonies since, as Jamshed put it, they were 'tiny'.[6] By the time he was eight, young Bhabha was familiar with the concertos and symphonies of not only Beethoven but also Mozart, Verdi and Wagner. Thus, by the time he was a teenager, a foundation had been laid for his lifelong appreciation of symphonies and operatic music. He frequented concerts in Vienna, Boston and Bombay. Jamshed, for his part, was a lifelong aficionado of Beethoven and even got the Indian postal authorities to issue a stamp in his honour.

Bhabha also played with Meccano sets which his father who had hoped to make him an engineer bought him. His creative mind was on display there too—he showed a marked preference for trying to make models that were not part of the accompanying booklet.[7] Naturally curious about things, he was ready to try them out as well. One particular experiment would have ended in tragedy had it not been for the timely intervention of his older cousin, Dinshaw. Bhabha first heard of parachutes around the time of the First World War and decided that there was no harm in attempting a practical demonstration. He and his second cousin, Rustom Vatcha, who was a few months younger, stood on the ledge of a first-floor balcony at Kenilworth holding two umbrellas, ready to 'parachute' down. Dinshaw noticed the young 'paratroopers' in the nick of time and a great tragedy was averted.[8] The tenor of both anecdotes suggests the potential for an overbearing precocity.

We do not know if Bhabha experienced the simple childhood pleasures of rambunctiousness and mischief that arise from playing with young companions. Perhaps he did, if we take a clue from this 'parachute' experiment. Unlike Oppenheimer (who Bhabha befriended later in life), whose intellectual interests and abilities developed well beyond his years but whose social skills remained stunted, Bhabha does not seem to have created the same sense of separation between himself and other people. Still, there were not many people who could claim to have known his warmth later in his life: friendly, but never familiar, he kept his distance from all but a very few. Perhaps, like the mathematician John von Neumann (also a friend of Bhabha's), who was ill at ease with peers from more modest backgrounds, Bhabha, too, found it difficult to bridge the gap with such people. While there was no hint of decadence or overindulgence in the boys' upbringing, there was no doubt that it was a luxurious life and a very sheltered one too.

Bhabha blossomed into an artist of great skill and, as a boy, regularly won prizes for his drawings at annual exhibitions organized by the Bombay Art Society. He took painting lessons from the artist, Jehangir Lalkaka, and at the age of 17 his self-portrait won a prize at an exhibition of the Bombay Art Society.[9] His sketches were so good that later Air India designed a calendar based on them. M.F. Husain compared his lines to those of the great Leonardo da Vinci![10] In the arts, Bhabha's knowledge of classical European painters was such that his understanding could have made him an amateur art critic in his own right. It was not for nothing that Raman introduced Bhabha as the 'modern equivalent of Leonardo da Vinci' to an audience of Indian scientists at Nagpur in 1941.[11]

Noticeably missing from Bhabha's rich artistic repertoire at this stage was any mention of Indian art. Growing up in Bombay in the early twentieth century, he certainly must have had access to Indian art and its omission suggests a high level of Anglophilia. Indeed, for the young Bhabha, it is clear that art, music and poetry primarily meant those of the western kind. It was the great classics of the West that inspired him, and figures like Beethoven, Mozart and Shakespeare were his role models. However, he seems to have been well acquainted with Indian history, writing to his friend Homi Seervai, as we shall see, about how history in India should be written.

Bhabha's familiarity with western art came from his father's excellent collection of books on the subject, but he was always quick to point out the impact of seeing the great works of art in the museums and galleries of Europe. While on vacations, he developed an interest in the great sculpture and architecture at places like Ajanta, Ellora and Sanchi. However, his appreciation for Indian art and artists blossomed only when he returned from England. During his stay in Bangalore, he attended concerts of Hindustani and Carnatic music and developed an admiration for eminent exponents of Carnatic music, like A.R. Iyengar and M.S. Subbulakshmi. According to M.G.K. Menon, Bhabha's protégé and successor as director of TIFR, the background already existed for the great feeling of identity that he established with his country. 'He was deeply conscious of the great cultural heritage and traditions of India, of which he was proud and from which he derived great pleasure; and he began to be conscious of the possibility of economic prosperity and social change, based on science and technology, which had been so remarkably demonstrated in the West.'[12]

THE UPBRINGING

Bhabha's father, the Oxford educated Jehangir, also started his career in Mysore after joining the judicial service of the state. Later, he was a legal adviser to the House of Tatas, serving on the board of directors of many Tata companies. He was a connoisseur of western classical music and painting and had an abiding interest in flowers, trees and gardens, an interest which the young Bhabha inherited. Bhabha had all the advantages of a privileged childhood. It is perhaps no exaggeration to say that the Bhabha boys studied in school but received their real education at home. It was an impressive pedigree, signalling elegance, wealth and membership in India's academic and artistic elite. There is little doubt that it played an important role in Bhabha becoming a cosmopolitan savant later in his life.

Jehangir Bhabha married Meherbai Panday, the daughter of Ruttonbai and Framji Dorabji Panday of Bombay. Ruttonbai was the daughter of the famous Sir Dinshaw Petit, the first baronet, who was widely respected in Bombay for his philanthropic endowments. She died prematurely in November 1885, when she was only 31. Her father, Sir Dinshaw, laid the foundation stone of the Bai Ruttonbai Panday Girls' High School in June 1893 in her memory. The school building was declared open five years later in June 1898. In later life, Bhabha was very proud of the school, and called it 'my grandmother's school'.[13]

The Bhabhas were also connected to the Tatas through marriage—a connection that was to have far-reaching consequences for Bhabha in his adult life. Bhabha's paternal aunt, Meherbai Bhabha, or Mehri, as she was generally called, was married to Sir Dorabji Tata, Jamsetji Tata's elder son. Jamsetji seems to have had a hand in the selection of his daughter-in-law. He used to visit Mysore often from 1890 onwards since he thought the state had industrial potential. During one of these visits, he met and befriended Hormusji Bhabha.

Jamsetji was charmed by the young Mehri, Hormusji's 18-year-old daughter. He felt that she would be a good match for his son Dorabji, who was approaching 40 at the time. But he was wise enough not to make it obvious to his son, merely hinting that he should have a cup of tea with the Bhabha family the next time he was in Bangalore. Dorabji visited the Bhabhas and sure enough fell under the spell of the attractive young Mehri, who served the refreshments. He proposed to her in 1897, and it is said that a very pleased Jamsetji wrote to Meherbai Bhabha and warmly expressed his great admiration for her. They were married on 14 February 1898. Tata historian and J.R.D. Tata's biographer, Russi M. Lala, said that this incident was narrated to him by Jamshed.[14]

Bhabha's family connections with the leading industrial houses in Bombay, and his familiarity with the ethos and culture of Parsi philanthropy were key factors that enabled him to garner support for his institution building later in life. This support, too, came from the Tatas—Parsi philanthropy is a story that is woven into the collective biography of this illustrious family.

After their marriage, the young couple, Jehangir and Meherbai Bhabha, relocated to Bombay from Mysore. Their house contained a large number of books, the joint collection of Hormusji, Jehangir and Meherbai. Hormusji had collected a large number of books on literature, education and the arts, to say nothing of scientific and technical subjects. To this rich collection, Jehangir added books on painting, which he had collected during his student days. Later, books on flowers, trees and gardens were also added. When Bhabha's talents in science and mathematics became evident, books on these subjects formed part of the collection. The house also had a garden and dogs. Bhabha's precocious intelligence manifested from a very early age, and his parents did everything they could to stimulate his intellectual and artistic interests.

As a child, Bhabha used to sleep very little and the family doctor's prescription didn't help. Scylla Vatcha, whose mother-in-law was a first cousin of Bhabha's mother, Meherbai, narrated that when the Bhabha family visited Europe in 1913, they were told of a famous paediatrician in Paris. They took an appointment with the child specialist, who started talking to the young Bhabha. After some time, the nurse informed the other patients that their appointments had been cancelled, as the doctor wanted to spend more time with the young Bhabha. The doctor told a concerned Jehangir and Meherbai that their son fascinated him. He told them that the child had an extraordinarily active brain and this was the reason that he could not sleep. With uncanny prescience, he told the parents that all they needed to do was to provide the right environment and he would grow up to be a genius.[15]

The Bhabhas provided a most congenial environment to the young Homi, not only in terms of education and culture but also through the closeness of their bond as husband and wife. Jamshed recounted that their parents had been very happily married to each other. The whole environment had been very happy, which is crucial when one is young. He felt that it is very important to be brought up in an atmosphere free of anxiety, strain and hostility. He considered himself very fortunate to have had such a wonderful environment. They also had dogs who were lovely.[16]

The two brothers also shared a very close bond and, if Jamshed is to be believed, Homi was like a second father to him. M.G.K. Menon became quite close to Jamshed after Bhabha's death in January 1966. Menon observed, 'He spoke to me of the wonderful background at home in which they grew up, and the opportunity for both brothers to develop, particularly in their shared interest of western music, poetry, literature and drawings/paintings. Homi was regarded as a star by all. Jamshed never resented this or felt any insecurity. He truly loved Homi and, in many ways, revered him.'[17]

Journalist and author Shyam Bhatia, writing in *The Tribune*, gave a somewhat different twist to the relationship between the two brothers. He said that Jamshed remained jealous of his elder brother's success and the high regard he was held in by Nehru and other members of the ruling establishment.[18] One can only speculate about the extent of resentment, if any, but Tata old-timers, who worked under Jamshed, testify to the fact that he rarely spoke of his elder brother, something which, if true, does not fit in with the image of a doting younger brother.

Bhabha himself noted the deep bonds of affection he shared with his parents. After his father died in 1942, he wrote to the physicist Wolfgang Pauli and the palaeobotanist Birbal Sahni about his love for them. However, it was his mother Meherbai who remained his emotional fulcrum throughout his life. After his father's death, he took her on a vacation every summer. Even in her later years, Meherbai used to be impeccably dressed in a white silk sari and lacy Victorian blouse, with pearls and diamonds around her neck and fashionable pointy-toed fine leather shoes bought from abroad.[19]

Bhabha spoke to her in English and Parsi-Gujarati, a kind of Gujarati that often causes great amusement among those who speak a more authentic version of the language. Homi Sethna, who was chairman of the Atomic Energy Commission (AEC) from 1972 to 1983, and knew the Bhabha family personally, recounted that Meherbai was an extremely gracious lady who epitomized all that was noble. Bhabha made it a point to see her before he left for office every day.[20] He also gave her a nickname: Twit. What led him to give her this name is not known, since Meherbai was anything but stupid.

SCHOOL AND COLLEGE

Bhabha attended the Cathedral and John Connon Boys' School[21] in Bombay. Esplanade House, Jamsetji Tata's grand home, was a stone's throw from Bhabha's school, and he used to have his lunch there apart from spending many fruitful hours in the well-stocked library. Thus, in addition to the many books at home, Sir Dorabji Tata's rich library was also available for his reference. His aunt, Meherbai Tata, was childless, and the Bhabha brothers received the affection reserved for sons. Bhabha, who was a voracious reader, was fascinated by science. He also read widely on other subjects, laying the foundation for his multifaceted interests and achievements. That he was an avid reader is evident from an early photograph in which an adolescent Bhabha is dressed in a dark suit and tie, reclining in a formal pose in an armchair with a book on El Greco on his lap.

Meherbai Tata was a lady of remarkable talents and the perfect foil for Dorabji. A champion tennis player, she won many tournaments in a sari, something which drew much comment from spectators. She represented India at the Paris Olympics in 1924 and was a familiar face at Wimbledon. A good rider, she also drove a motor car and harboured hopes of being the first Indian woman to fly. Musically inclined, she played the piano well enough to be in demand for public concerts in Mysore. She was also the owner of the famous Jubilee diamond, a gift from her husband, which, at 245.35 carats, is the sixth largest diamond in the world. Meherbai played an active part in raising contributions during the First World War and was an active member of the Indian

Red Cross Society. She was made a Commander of the British Empire by King George V in recognition of her services.

Concerned about the condition of women, she was actively involved in the enactment of the Child Marriage Restraint Act, 1929, popularly known as the Sarda Act. She was not only consulted on the Sarda Act but also campaigned for it in India and abroad. Her advocacy of the Sarda Act was only one aspect of her vast contribution to women's emancipation in India. She was committed to women's suffrage, girls' education and the removal of the purdah system. She championed these causes on global platforms. Unfortunately, Meherbai died of leukaemia in 1931 in Wales and is buried at the Brookwood cemetery in Surrey.

While Bhabha was fortunate in being born into a family that recognized his talents and had the wherewithal to nurture them, he was also lucky to be so closely associated with the Tatas. He witnessed the emerging world of Indian industry and saw and heard the nationalist politicians of the day—Gandhi, Vallabhbhai Patel, Motilal Nehru and his son Jawaharlal all used to visit the Tata home while they were in Bombay. Gandhi stayed with the Tatas during the launch of the first Civil Disobedience movement.

British physicist, Sir John Cockcroft, observed: 'Homi Bhabha's family on both sides was strongly nationalist and, as a young boy, Homi used to listen to the conversation of Mahatma Gandhi when he visited his maternal aunt at the time of the launching of India's first civil disobedience movement. Friendly ties also existed between the Nehru family and the families of Tata and Bhabha.'[22] Dorabji's younger brother, Ratan Tata, had come to Gandhi's aid in 1909 with a donation of ₹25,000 to be used for his struggle in South Africa. Ratan spent a great deal of time on the Continent and in England—he had a grand home in Twickenham—but was very interested in Indian politics. He was a good friend of the Indian liberal politician and social reformer, Gopal Krishna Gokhale, and contributed ₹6,000 a year to his Servants of India Society.

Notwithstanding the thesis of the marginal man, many Parsis learnt to master and manage two cultures and were not simply stuck between the two. By the time he went abroad at the age of 18, Bhabha understood a way of life that allowed his community to prosper in a tumultuous political world. As Robert S. Anderson has observed, 'In a colonial situation, where collaboration with foreign firms was necessary for industry's survival, Bhabha met foreign leaders along with famous nationalist politicians as houseguests and learned that nationalists too were adaptive and strategic in their relationships.'[23]

According to Cockcroft, being privy to discussions on national issues, particularly regarding economic development and the setting up of industries with India's top leaders, was a very important influence on the young Bhabha. This exposure resulted in not only a broad vision but also brought home the importance of having the ability to plan and organize within national and international contexts. Such exposure is uncommon among scientists who tend to be ivory tower academics or researchers with a very limited focus.

By the time he took his Senior Cambridge examination, Bhabha had read Einstein's Special Theory of Relativity and gained some understanding of it. Such precocity was in itself remarkable given the fact that Einstein's theory was understood by very few people at the time and certainly not by boys of Bhabha's age. Given his ability to think in abstract terms, he was aware of the limitations of observing the natural world through ordinary sense perceptions. In a speech many years later, he said that science expands the mental horizon '[...] by demonstrating the limitations of commonsense ideas based upon the world immediately perceived by our senses [...]' since '[...] our senses perceive only a very small fraction of the phenomena of nature around us.'[24]

Bhabha passed the Cambridge Local Examination (Juniors) in 1924 with honours. He was awarded the Headmaster's Prize for English Essay and the Hudson Prize for Mathematics. After he became a national and international figure, his old headmaster, C.H. Hammond, wrote to him congratulating him 'for the eminent position you have reached in the world'[25] saying that he was proud to have had some role, however small, in his education. Bhabha responded warmly by acknowledging the part played by the school in his success.[26] Whatever the relative contributions of upbringing, schooling and Bombay society, in Bhabha's case everything aligned propitiously to produce not only a scientific mind of rare ability but also the sensibilities of a true aesthete.

He was only 15 when he finished school and since universities abroad had a minimum age restriction of 18, Bhabha had to wait to go to Cambridge. He spent the academic year 1925–1926 at the Elphinstone College, studying Arts, and 1926–1927 at the Royal Institute of Science in the first year of the BSc class. Many years later, Bhabha reminisced about his time at the Royal Institute of Science and of the time he first heard of cosmic rays from Compton, a subject that later became his special field of study. He said that the year he spent studying at the Royal Institute of Science before he left for Cambridge was a very happy one.

The next year, in 1927, Compton won the Nobel Prize in physics for his 1923 discovery of the effect named after him—the Compton Effect, which demonstrated the particle nature of electromagnetic radiation. It was a sensational discovery at the time: the wave nature of light had been well demonstrated, but the idea that light had both wave and particle properties had not been demonstrated conclusively. In the same year, Bhabha set sail for Cambridge for his Mechanical Sciences[28] Tripos from Gonville and Caius College. This was to be the start of a fascinating journey, beginning with mechanical engineering and leading to mathematics, theoretical physics and much else.

Chapter 4

CAMBRIDGE

> Physics is my line [...] I am burning with a desire to do physics [...] I have no desire to be a 'successful' man or the head of a big firm.
>
> —Letter to his father, Jehangir, 8 August 1928[1]

Cambridge and its great rival, Oxford, which have become the embodiment of elitism, are not merely educational standard-bearers but capital cities of knowledge and all forms of intellectual life.[2] They are talked of in the same breath and, indeed, the same word, 'Oxbridge', a synthetic product that was first given currency by the author William Makepeace Thackeray. No other private institutions have had greater influence on England's history than these two universities, providing church and state with teachers, priests and officials who not only spread their language and culture to all parts of the land but also to the whole world. It has often been said that Oxford is for the arts and Cambridge is for science, but this is not true. Oxford has a scientific tradition that goes back all the way to the natural philosopher Roger Bacon, and the most imaginative of all mathematicians, Lewis Carroll was also from Oxford.[3]

Cambridge is a network of some 30 colleges scattered about the town, each self-governing under a master and staffed by fellows, who could also be university lecturers or professors. Each college selects its own undergraduates and oversees all their needs, providing a social as well as academic focus. An important aspect of the college education is the tutors' close, almost parental supervision of the students, covering their general welfare as well as academic progress. However, programmes of study, lectures and examinations are organized by the university.

Cambridge has produced more Nobel laureates, but Oxford has seen more prime ministers. Isaac Newton taught at Cambridge, and Charles Darwin began there as a theologian but ended up becoming a naturalist. The pioneers of computers, Charles Babbage and Alan Turing, along with the nuclear physicists Ernest Rutherford and James Chadwick all played their part in the Cambridge phenomenon, to say nothing of the discoverers of the structure of DNA, Francis Crick and James Watson. Stephen Hawking, who sat in Newton's professorial chair, held forth on black holes in the universe. Of all the towns in Europe, Bill Gates chose Cambridge as the place for his first Microsoft branch outside the US.

GONVILLE AND CAIUS

Gonville and Caius, the college to which Bhabha was admitted, not only has a double name but was also founded twice: first, in 1384, by Edmund Gonville, a parson from Norfolk and, then, in 1557, by John Keys, a wealthy doctor from Norwich who was physician to Edward VI and Mary Tudor. Keys, who Latinized his name to Caius (also pronounced 'keys'), had studied at Gonville Hall and used his wealth to revitalize his old college. The college has three gates through which students must pass: Humilitatis, Virtutis and the Gate of Honour. The last gate is central to an academic rite of passage that is still practiced ceremonially in the college. It is through this gate that students pass to collect their degrees, one of the two occasions on which the gate is opened, the other being when a fellow dies.[4]

Bhabha's choice of college was no doubt influenced by the fact that his uncle, Sir Dorabji Tata, was an alumnus of Gonville and Caius, and, later, a generous patron who donated £25,000 to the college in

1920. Dorabji, however, did not complete his graduation because he was recalled to India in 1879 after just two years of beginning his studies because his grandfather, Nusserwanji Tata, could not bear to be separated from him any longer. When Bhabha arrived at Cambridge in 1927, he enrolled for a BA in mechanical engineering. Both his father and his uncle hoped that he would return to India after completing his degree and join the House of Tatas.

It was by no means an unrealistic expectation, since a career in the pure sciences held little promise, at least in the colonies. Engineering was the more pragmatic choice, and the steel works at Jamshedpur beckoned, but Bhabha had other plans. Even though he had a talent for mechanical drawing and had little trouble in understanding his subject, engineering did not inspire him. If his correspondence with his namesake, Homi Seervai, is any indication, there were times when the dislike verged on contempt, even hatred. Seervai courted fame later in life as a noted constitutional jurist and as the advocate general of Maharashtra. He is best remembered for his monumental multivolume *Constitutional Law of India* and for his role in India's most famous constitutional case, *Kesavananda Bharti vs State of Kerala*, popularly known as the Fundamental Rights case.

In a letter dated 2 August 1928, Bhabha expressed his dislike for his current syllabus and stated with some vehemence that eventually he would like to pursue physics and astrophysics and nothing could stop him from doing so.

> I am doing engineering now, but it is not very interesting [...] However, I am determined to do physics and astrophysics eventually. Nothing will stop me [...] The mere circumstances that there is an opening in steel or hydro will not change my course [...] I do not care two hoots for anybody. I am going to go my own way against the world if it is in my way [...] How can a person like me be ever expected to work harmoniously in a steel work, when at any instant I might explode against it.[5]

After these strong words, he asked his friend to intervene with his father on his behalf. 'If you see my father, I earnestly ask you to impress this view upon him and convince him that scientific research is my only line. Leave no stone unturned. By the way, could you find out and let me know what openings there are in India for the same.'[6] Another scientist, arguably the greatest of all time, had expressed similar sentiments to a friend about how he had rejected the idea of pursuing engineering in the mid-1890s. Albert Einstein's father, Hermann, owned an electrical business and hoped that his son would join his business one day. 'I was originally supposed to become an engineer,' Einstein later wrote to a friend, 'but the thought of having to expend my creative energy on things that make practical everyday life even more refined, with a bleak capital gain as the goal, was unbearable to me'.[7] Bhabha seems to have had similar feelings.

Bhabha and Seervai shared a love of philosophy and their letters reflect this interest. They also discussed books, logic, forms of government and historical figures. In a letter written on 2 January 1929 Bhabha observed: 'I do not think it worth writing to you unless I can write an intelligent letter, because I don't think it can be of great interest to you that I had Christmas cake on Christmas unless it causes some reflections. Anybody can eat Christmas cake. After all, we are what we think.'[8]

In the same letter, he put forth his views on democracy.

> I agree that for a really good and efficient Democracy there ought to be a widely diffused culture throughout the masses. You will be surprised how uncultured the average Englishman is... He has only acquired a certain sense of responsibility, and a certain sense and forbearance in political matters through long experience in this form of government, which makes the government more or less stable. This sense can only be formed by experience, and therefore, it is unreasonable to expect a democracy in India which shall not make some blunders in the beginning, some even of a serious nature.[9]

Bhabha continued with his prescription for India.

> Unity of all mankind cannot be achieved by blotting out one civilization and implanting another. As a preliminary step, a new and unbiased history of India ought to be written which ought to be taught in schools. The past that ought to be more studied

is not the Mogul period, but some of the more ancient periods, such as those of Chandragupta and later—I mean the real Hindu empires.[10]

'SCIENCE SHALL BE MY WIFE, ART MY MISTRESS'

Apart from the fact that Bhabha disliked mechanical engineering and, therefore, did not invest too much of himself in it, there were many distractions for a young man in Cambridge. He plunged into extracurricular activities and sports with a vengeance. Soaked in history and tradition, replete with chapels, gothic spires and medieval arches, carefully tended lawns and flowing through it all, the gentle Cam, Cambridge held Bhabha in its thrall. He also enjoyed some of Cambridge's pleasures: punting on the Cam, Sunday picnics on the riverbank and afternoon teas. The days were busy with supervisions (the Cambridge equivalent of Oxford tutorials), where students were taught in groups of two to four. The evenings, too, were full, and Bhabha must have gone walking along the Cam or for a run on the university cricket ground, popularly known as Fenners. And, of course, there were plenty of parties to attend.

On his arrival at Cambridge, he took to coxing[11] in the boat races. Bhabha's skill at this activity seems to have left a lasting impression because more than three decades later, he would be remembered as the 'cox of the fifth boat'. Wilfrid Bennett Lewis, a fellow student who later headed the atomic energy programme in Canada (and had much to do with Bhabha in that capacity), recounted:

> We first met as freshmen at Cambridge in the same college, Gonville and Caius, in 1927, and in its boat club [...] Being at first light in weight, Homi Bhabha was appreciated as a coxswain in the rowing eights; he was, however, growing and putting on weight so he transferred his athletic activities to track running at Fenners [...] and to tennis.[12]

The motto of the boat club was, *labor ipse voluptas*, which is Latin for 'the pleasure is in the work itself'. This motto could not have been more apposite because later in their lives both Lewis and Bhabha would find great pleasure in their work. The two competed for a research fellowship at the college, which Lewis won. After Bhabha's death, he reminisced, 'Perhaps my need was greater for it fell to me. We remained good friends and later the balance was corrected. We both became Fellows of the Royal Society, he first; and then he was made an honorary fellow of Caius College, a very high distinction.'[13]

Bhabha also indulged his passion for western classical music and painting while at Cambridge. He listened to gramophone records, bought sheet music to understand the composition and structure, and even attempted to write a few compositions himself. He wrote to Seervai about how he was studying counterpoint for an hour every day and wrote to his brother, Jamshed, about a performance of Beethoven's Ninth Symphony which he had witnessed in London in November 1927. He wrote that it was almost 1.00 a.m. and he could not help writing to Jamshed. He said that ever since he had come back from the concert at about 11.00 p.m. or earlier, he had been constantly thinking of the Ninth Symphony. He was so overwhelmed by it that he could neither talk nor think of anything else. He told his brother that he had walked back in a dream and was surprised that he had got back at all. It had been the sublimest and happiest time he had ever had in his life.[14]

Bhabha also appreciated music in a wider historical context.

> Mozart came at the end of an epoch, Beethoven at the beginning of another. It is Beethoven's birth after the French Revolution that explains the new force in his music and the new spirit of faith in human progress [...] In Beethoven's time, the French Revolution had shown that inequalities of opportunity and class could be overcome [...] This is one main reason for the great power of the music of Beethoven and his heroic themes.[15]

Jamshed, who was five years younger than Bhabha, credited himself for his brother's love of poetry. He used to give Bhabha books by poets like P.B. Shelley and John Keats to read. As a result, Bhabha could quote freely from Shelley, Keats, Byron and Milton. Jamshed joined his brother in Cambridge a few years later to do his Historical Tripos there.

Bhabha was passionate about music and, if his letter to Seervai is any indication, it ranked above his love for painting and mathematics.

> I am manifestly efficient (I don't mean in any superlative degree) at mathematics and painting, and I like music more than I like either of those, in fact more than anything else—it would therefore be natural to suppose that in the end I will be able to compose as well as I paint or do maths. I don't know if the reasoning is correct. I often wonder what it is that makes us like music as much. It is not as if I possessed a ear that could detect subtleties of harmony etc. which others can't. I would like your views on the subject. I would value your suggestions and opinions.[16]

Bhabha made it a point to visit art galleries wherever he went. His observations to Seervai show his deep insight and understanding of the art he saw.

> It is interesting to note that the works, in art or music, produced just before the war all manifest violence and force. Frank Rutter observes that in the abstract paintings of Wassily Kandinsky painted just before the war, the pictures contain guns, falling houses etc. He pointed this out to the artist, who admitted that when he painted them, he was not even aware that a war was imminent; nor was he consciously trying to predict. I suppose the violence and conflict of forces in the intellectual atmosphere affect artists more than others, and it unconsciously finds its way into art.[17]

Frank Rutter was the art critic for *The Sunday Times* from 1903 until his death in 1937. He was an early champion of modern art in England, founding the French Impressionist Fund in 1905 to buy work for the national collection. Bhabha had bought one of his books, *Evolution in Modern Art*, from his prize money at the Bombay Art Exhibition. In a letter to Seervai dated September 1929, he quoted Rutter's views on Jacob Epstein's sculpture depicting industrialization and the dangerous mania for machinery.[18] Many years later, Bhabha would buy Epstein's bust of Einstein for TIFR. In an earlier letter, in March 1928, he had written: 'Ask my brother to take you to listen to our new loudspeaker.'[19]

Many years later, when M.G.K. Menon asked Bhabha why he had chosen science as a career instead of the arts, the latter replied that while the arts made life worth living, he wanted to pursue science as a profession. The art critic, Rudolf von Leyden, a German émigré who became a close associate of Bhabha in Bombay in the 1940s and 1950s, and author of the introduction that accompanied a portfolio of Bhabha's drawings, observed: 'Dr Bhabha felt very strongly that the cerebral activity of the scientist had to find its counterpoint in the activity of the senses, in art.'[20] His activities at Cambridge certainly provided strong evidence for this idea. Bhabha painted and sketched, participated in exhibitions of the Art Society and designed the sets and decor for plays and operas. Rudolf von Leyden, in an article in *Marg*, commented on Bhabha's style: 'While a student in Cambridge, he drew and painted from models whenever he found time. He painted in a romantic symbolist style and experimented intensively in abstract visual interpretations of musical themes, based as they were, on his knowledge of the mathematical structure of music.'[21]

Bhabha also corresponded with a friend, Jessie Maver, whom he had met in Vienna in 1933 and with whom he kept in touch all his life. They shared an interest in music and painting and his very first letter written to her in September 1933 (on air-mail paper in pencil) is full of references to musicians, painters, places of artistic interest and places to eat (the last proving—not that there was ever any doubt—that he was a true Parsi!). He wrote that he had also spent some time at the Mozarteum reading Mozart's letters. He believed that those who did not know these letters did not know Mozart—they were charming, bubbling with humour and the goodness of his nature. Mozart was, Bhabha wrote, perhaps the greatest artist that lived, and in music his only rivals were Bach or Beethoven. He added that Beethoven never equalled the best of Mozart until the works of his last period, particularly the late quartets and sonatas, and that Beethoven was unknown to the public.[22]

He displays a withering contempt for Richard Wagner and Johannes Brahms, a dislike that seems to stem from their personal characteristics rather than their music. 'And this naturalness of Mozart is one whole better than the behaviour of that fat and sensual swine

Wagner with his mutton chop whiskers, and that bloated pig Brahms with his flowing beard—who was disgusting enough to eat a tin of sardines and then pour the oil down his throat.'[23]

After discussing Vienna, which he called a dead city, and expressing astonishment that it produced Haydn, Mozart, Beethoven and Schubert, he advised Maver on a restaurant in Florence and a gallery in Milan. He said that if she should ever go to Florence, there was a restaurant where one could eat really well. He said he had forgotten its name but it usually had a small boy in a red uniform standing just inside the door. In Milan, he advised her to go to the Brera Gallery where he wanted her to see a painting by El Greco. Bhabha praised it as the best, saying that the exquisite painting of the face and the dress made even early Raphael appear gross in his handling.[24]

Was there also an element of romance involved? We will never know, but the use of 'darling' as a term of endearment used in the letter points in that direction. He wrote 'Darling, I could go on for ever,' adding that he had so much to say but didn't have the time to do so. Maver was supposed to sail on the 25th, but Bhabha wanted her to stay in London till the end of the month so that they could meet. He informed her that he would be in London on the 22nd or so. He wanted her to sail a month or so later, perhaps even two. That way, they could meet in Paris or Rome. Bhabha added that if there was no hurry and money wasn't a concern, she should stay on because there was lots to see in Europe, and he volunteered to be her guide. He asked her to consider this seriously and write to him by airmail when she was in Europe.[25]

In her response to Bhabha's letter, Maver had thanked him for giving her a spark. In response to this, with the ebullience of a self-confident young man, Bhabha compared himself to a fire, claiming that it always threw off sparks. One did not have to thank it for that. A fire is only glad when it finds something it can set alight. Most people, he said, had such damp souls that even the fires of hell could not make them burn. They would merely emit grey and disgusting smoke.[26]

Maver, in an interview for a documentary made by the Films Division of India in 1973, six years after Bhabha's death, narrated how he had decided to live life to its fullest. He had written to her saying that since the duration of life could not be controlled, he would increase it by living more intensely. He seemed determined to fill Rudyard Kipling's 'unforgiving minute with sixty seconds' worth of distance run.'[27] When asked about his failures, if any, she said that Bhabha was never known to be in the state of 'being enough', clearly indicating that anything even remotely approaching placid contentment was made impossible by the frenzied activity and the high pitch at which he lived his life.[28] Sabyasachi Bhattacharya, a former director of the TIFR, echoed a similar view. Bhabha always seemed to be in a hurry—a kind of creative or purposeful impatience that was unique to him. Bhattacharya described Bhabha's creative energies to be those of an active dreamer, one who not only dreamed big but also wanted to achieve those dreams very quickly.[29]

Bhabha honed his considerable artistic skills, sketching and painting frequently. He designed an 'artistic and symbolic' new cover for his college magazine, *Caian*. It depicted a head with long flowing hair with the college's 'Gate of Virtue' towering above it in the background. It was, according to Cockcroft, too radical a design for the students of the college at that time.[30] Bhabha drew inspiration from music for his paintings. He designed the sets for a student performance of the seventeenth century Spanish dramatist Pedro Calderon de la Barca's play *Life is a Dream*. Later, in 1939, he also designed the sets for the Cambridge Musical Society's performance of Mozart's *Idomeneo*.

According to his brother, Jamshed, Bhabha was also at one time the conductor of the Cambridge University Orchestra. Centenarian Mithoo Coorlawala, a social worker and historical researcher was also at Cambridge at the same time as Bhabha but never met him there. She recalls that he was the first violin at a concert in Cambridge. Bhabha also tried his hand at writing blank verse in the style of Shakespeare, in a skit called *Shakespeare Orders Lunch*, which had the Bard and a serving man as characters.

Bhabha was photographed by Lettice Ramsey, the widow of the mathematician and philosopher Frank Plumpton Ramsey, who was famous for her portraits of scientists and intellectuals. There are photographs of Bhabha standing in front of one of his paintings, along with one of him working on his painting inspired by 'The

Countess' Aria' from Mozart's *The Marriage of Figaro*. He not only had himself photographed but later also acquired Lettice's portraits of Lord Rutherford, Paul Dirac and J.J. Thomson for the TIFR in 1950. Her studio, Ramsey & Muspratt, which she shared with the Madras-born Helen Muspratt, was the venue for social gatherings for young intellectuals in Cambridge.

In one of his letters to Seervai, Bhabha wrote,

> Yesterday, for the first time in Rugby I went out sketching, drawing a railway bridge at dusk. I shall paint it tomorrow. I have been reading a technical book on oil painting by S.J. Solomon. One thing appears here. That he is a very good technician, but a man without an original mind, in fact not a great artist. All academicians are like this, or they would not be academicians.[31]

He put his old teacher in Bombay, Jehangir Lalkaka, in the same category. 'Lalkaka, like the Academicians here is merely a painter, not a creative artist. The creative artists are those who are so often stigmatized as cranky and appreciated only in small circles.'[32] The English-trained Lalkaka developed a high-profile practice as a portraitist with his subjects being drawn from a wide spectrum of late-colonial public life, including Dadabhai Naoroji, Sir Pherozeshah Mehta, Lord Brabourne and even King George V. Later, he became the deputy director of the Sir J.J. School of Art. The impact of galleries and museums Bhabha visited in Europe was strong. His brother Jamshed remarked: 'Seeing the originals of Impressionist and post-impressionist masters stimulated him to change his drawing and painting from the dull academic style learnt in Bombay to a more vigorous individual style.'[33]

Praise for Bhabha's sketches came from an authoritative source—Roger Fry, the English artist and critic credited with coining the term 'post-impressionism' and of bringing artists like Paul Cézanne, Henri Matisse and Vincent van Gogh to the public in England. Fry, a member of the famous Bloomsbury Group who became the Slade Professor of Fine Art at Cambridge in 1933, thought that Bhabha had great talent and encouraged him to create works on a large scale, including possibly a fresco at New Delhi. But the praise was not without qualification. He thought Bhabha lacked a 'certain sensitiveness'. Bhabha, whose style was still evolving, was a trifle disappointed upon hearing this. He conveyed his feelings to his parents a few years later, after an exhibition of his works at Cambridge. He wrote, 'I wish Roger Fry could have seen the latest stuff.'[34]

It is possible that such praise made Bhabha consider becoming a professional artist, but in the end physics won. As he wrote to Seervai in July 1930, 'He who would pursue Science or Art must give up thoughts of wife or home, country or race, and do whatever he may, not for any small part of the race, but for Man. I am driven to pursue Science and Art at all costs. Science shall be my wife, Art my mistress.'[35] Later in life, he would always carry his painting kit whenever he went for international conferences.

Did Bhabha also flirt with the communist philosophy? An acquaintance recalled that Bhabha told him in an unguarded moment that, in Cambridge, he had had a choice of becoming a communist, an artist or a scientist.[36] Soviet double agents like Guy Burgess, Kim Philby and others were at Cambridge with Bhabha. They were both in Trinity College, as was E. Herbert Norman, who was born and raised in Japan and went on to become a successful Canadian diplomat. Tragically, Norman committed suicide after he was accused of being a communist in the 1950s. What was a fact, however, was that Norman recruited students from the Commonwealth for the Communist Party under the influence of Rupert John Cornford (one of the great grandsons of Charles Darwin), who was an alumnus of both King's College and Trinity College. Bhabha and Norman, who were the same age, must have met in Cambridge at some stage.

There is no evidence to suggest that Bhabha was ever an active communist, but he did join the Socialist Club and was influenced by John Desmond Bernal, the noted crystallographer who propagated the doctrine of the social function of science. Bernal was an active communist and a member of the Communist Party of Great Britain. He held that science was rooted in society and moulded by it. He believed that modern science had the potential to not only solve the problems of poverty and unemployment but could also bring untold benefits to humanity if scientific development was consciously planned for the entire society. This influence was apparent in Bhabha's future career and the role he played in building science and technology in independent India.[37] Cambridge was the crucible that allowed Bhabha to explore ideas about not only science, industry and the arts but also freedom and democracy.

After he returned to India, Bhabha moved in intellectual circles that had socialist leanings in Bangalore and Bombay. When P.C. Joshi was the general secretary of the Communist Party of India there was a liberal phase in which a number of intellectuals, writers and artists were associated with the party. Bhabha was associated with the Friends of the Soviet Union and for a time was considered a communist sympathizer. He used to subscribe to the party paper *People's War*.[38]

THEORETICAL PHYSICS BECKONS

Soon after his arrival in Cambridge, Bhabha wrote to his father about his growing interest in physics. His disinterest in mechanical engineering and overindulgence in extracurricular activities took their toll. When the results were announced at the end of the first part of his degree, Bhabha found that he had got a third class. This distressed his father greatly, and he gave his first-born an ultimatum: he was to redeem himself by getting a first class in the second part or be recalled to India. Bhabha managed to negotiate even from a position of weakness—he would oblige with a first class if his father agreed to finance him for two more years to do the Mathematical Tripos. His father was well aware of his son's growing obsession with mathematics and physics, having received letters from him expressing his strong dissatisfaction with engineering. In one such impassioned letter written on 8 August 1928, Bhabha wrote:

> I seriously say to you that business or a job as an engineer is not the thing for me. It is totally foreign to my nature and radically opposed to my temperament and opinions. Physics is my line [...] I am burning with a desire to do physics [...] I have no desire to be a 'successful' man or the head of a big firm. There are other intelligent people who like that and let them do it [...] I, therefore, earnestly implore you to let me do physics.[39]

In September 1929, Bhabha went to Rugby, a small town in Warwickshire, for practical experience. He wrote to Seervai from there, expressing his dislike, even contempt, for what he was doing.

> I go as a workman to B.T.H. works [...] The labour is quite unintelligent and soon becomes boring. I am quite wasted on a job that any unintelligent man could perform, possibly better because he would attach more importance to it [...] I strongly object to this mechanization of man. I despise all businessmen, industrialists and engineers as such. What they do with great self-importance is nothing great. Any man of average intelligence could do it. With these views and my temperament, how can you expect me to be an engineer? [...] My father has been very reasonable and good, and has promised me one more year at Cambridge after I have finished my engineering Tripos for the pursuit of physics, perhaps two. Then if I do well, I may do research at Cambridge and stick to Physics.[40]

Bhabha wrote in some detail about how he started work at 7.30 a.m. and continued till 5.00 p.m.

> When one goes in the morning, one has to pass through some gates where one gets a round disc with one's number on it. The gates close punctually at 7.30 as a whistle blows and everyone has to be in by then [...] Then you have to go to the department where you work and put the disc into a machine within 2 minutes past 7.30. If you are later than that you lose half an hour.[41]

It is not difficult to imagine what effect such a routine would have had on the exceptionally intelligent and artistic Bhabha. It is obvious that he thought engineering was beneath him. He had strong views and was not afraid to air them. A common thread in all his letters is that engineering required average intelligence and that he would be wasted on something so monotonous, indeed, pedestrian. Just before his examination, Bhabha wrote to his father, and made a statement that remains applicable to this day. 'No exam was ever a test of original or creative ability in anybody. However, don't let that disturb you; you shall get the results you want.'[42] Bhabha kept his promise by securing a first class in his Mechanical Tripos in 1930.

His father kept his part of the deal and Bhabha was admitted to the Mathematical Tripos at Cambridge. His eagerness to pursue mathematics is understandable in the context of the great influence that the subject wielded on the development of mathematical physics in England at the time. In fact, physics was first taught at Cambridge as part of the Mathematical Tripos. The path to theoretical physics was through mathematics and the young Bhabha looked forward to sharing a platform with the great creative minds of the West. Physics was the medium for him to channel his true passions. Unlike mechanical engineering, with its earthly aspirations, physics allowed Bhabha to ruminate on what he might have seen as the greater mysteries of existence. It was through this final switch to physics that the young aristocrat found his professional niche. It was this Bhabha that returned to India in 1939: an artist–scientist with a distinctly occidental aesthetic.[43]

Chapter 5

THE QUANTUM REVOLUTION

> My father has been very reasonable and good, and has promised me one more year at Cambridge after I have finished my engineering Tripos for the pursuit of physics, perhaps two. Then if I do well, I may do research at Cambridge and stick to Physics.
>
> —Letter to Homi Seervai, 11 September 1929.[1]

One of Cambridge's great strengths over the years has been its tradition of mathematics. From the mid-eighteenth century, all Cambridge undergraduates had to take the Mathematical Tripos examination irrespective of their main subject of study. The word 'tripos' refers to the three-legged stool on which university examiners used to sit as they questioned the students. In the late eighteenth century, the role of the tripos examiner was downgraded, but the name stuck. Originally used to refer to the examinations mathematical candidates took to earn their degrees, later, the word was extended to examinations in other fields—a Classical Tripos, a Natural Sciences Tripos, and so on. The Mathematical Tripos was a feat of at least endurance, if not skill. Candidates would sit for four days, take a week's break and then return for four more days. The Tripos, which went back in its earliest form to 1730, had always been difficult. Over the years, as it grew more demanding, its importance increased, and it took on the lustre of tradition.

The written mathematical exam was a challenge in its own right with the ablest students competing for the accolade of 'wrangler', a term that originated with the fact that candidates had to argue with the tripos examiner.[2] The students who gain first-class degrees in mathematics are called Wranglers. At the top of the rankings is the Senior Wrangler, the top mathematics undergraduate at Cambridge, a position once described as 'the greatest intellectual achievement attainable in Britain', given that the Mathematical Tripos is famously difficult. Below the Wranglers are the Senior Optimes, or second class students, and the Junior Optimes, or third class students. The lowest scoring of the Junior Optimes used to be given the Wooden Spoon by his friends, which was solemnly lowered from the Senate House gallery.

The Senior Wrangler became a celebrity, at least around Cambridge. Many Senior Wranglers went on to become international figures in mathematics, physics and other fields—John Herschel, Lord Rayleigh, Sir Arthur Stanley Eddington, John Edensor Littlewood, Frank Ramsey and Jayant Narlikar. The first Indian to become Senior Wrangler was Sir R.P. Paranjpye (film director Sai Paranjpye's grandfather) who went on to become the principal of Fergusson College in Poona, and the vice chancellor of the University of Bombay. Between 1748 and 1909, the two colleges with the highest number of Senior Wranglers were Trinity and St John's with 56 and 54, respectively. Bhabha's college, Gonville and Caius, was a distant third with 16.

However, the precise order of merit meant far less than it seemed. Second Wranglers like James Clerk Maxwell, J.J. Thomson, Lord Kelvin and the famous economist Alfred Marshall seemed to distinguish themselves more than Senior Wranglers. While the examination for the Mathematical Tripos was tricky and challenging, it had little that would interest serious mathematicians. The problems posed in these examinations were dated, and there was a preoccupation with Euclid and Newton. Philosopher–mathematician, Bertrand Russell, who made numerous contributions to the field of mathematics in his career, was Seventh Wrangler in 1893. He later wrote that preparing for the examination 'led me to think of mathematics as consisting of artful dodges and ingenious devices and as altogether too much like a crossword puzzle'.[3] With the Tripos over, Russell swore he would never look at mathematics again, and, at one point, it is said that he even sold all his mathematics books.

Cambridge's tradition in mathematics was complemented by new knowledge in the sciences. The need for the practical training of scientists and engineers was emphasized by the success of the Great Exhibition of 1851 and the requirements of an industrial society. In 1871 the chancellor of the university, William Cavendish, the seventh duke of Devonshire, a landowner and industrialist whose ancestor, Henry Cavendish discovered hydrogen, agreed to fund a laboratory provided the university promised to establish a chair in 'experimental' physics. The laboratory was inaugurated in 1874 and was named the 'Cavendish Laboratory' in his honour. The new laboratory was to become a flagship of scientific research in the twentieth century.

THE CAVENDISH LABORATORY

The Cavendish Laboratory is arguably the most distinguished scientific institution in the world. Since its establishment in the late nineteenth century, it has produced some of the most consequential scientific advances of all time, including the discovery of the electron in 1897, the isotopes of the light elements and the splitting of the atom in 1919, the proton in 1920, the neutron in 1932, the structure of DNA in 1953 and the discovery of pulsars in 1967. More than 20 Cavendish or Cavendish-trained physicists have won the Nobel Prize for either physics or chemistry.

Its first head and the first Cavendish professor of experimental physics, James Clerk Maxwell, postulated the new theory of electromagnetism in 1864. A wizard in mathematics, Maxwell realized that the behaviour of electricity and magnetism could be summarized in precise mathematical terms. One law stated that a moving magnetic field could create an electric field. Another law stated the opposite, that a moving electric field could create a magnetic field. Then, Maxwell came up with an idea for the ages. What if a changing electric field created a magnetic one that then created another electric field that then created another magnetic field, and so on? The end product of this process would be a moving wave, where electric and magnetic fields turn into each other. When he calculated the speed of this moving wave, he found that it was close to the speed of light—a remarkable coincidence. He then posited that this was indeed light—light was an electromagnetic wave.[4]

Maxwell's arrival at the Cavendish Laboratory set the stage for a century of remarkable achievements. With developments as far-reaching as those of Newton some 200 years earlier, the work done at the Cavendish forever changed our view of the world, opening up the atom and then its nucleus for closer inspection. As the nineteenth century drew to a close, the world seemed to sense that the time had come for radical new ideas and innovations. Maxwell did not live to see his theories fully confirmed by experiment, but his practical legacy was designing and equipping the new laboratory. He died in 1879, aged only 48, and was succeeded by Lord Rayleigh, who was responsible for setting up a systematic course of instruction in experimental physics.

Thomson succeeded Lord Rayleigh in 1884 and began the revolution in physics that led to the discovery of quantum mechanics in the 1920s. In 1895, during Thomson's long tenure, the university allowed students from outside Cambridge to study for the new degree of Doctor of Philosophy. Among the first generation of physics graduate students were Rutherford and C.T.R. Wilson, who, along with Thomson, were to win Nobel prizes for their research. The discovery of the electron by Thomson, the invention of the cloud chamber by C.T.R. Wilson, and the discovery of artificial nuclear fission by Rutherford are examples of the extraordinary advances in experimental technique that ushered in what we know as modern physics.

THE BIG, BLUFF NEW ZEALANDER

In 1919, Thomson was succeeded by his former student, Rutherford. It was during his tenure that Francis Aston discovered the isotopes of the chemical elements, Patrick Blackett first photographed artificial nuclear interactions, Chadwick discovered the neutron, and Cockcroft and Ernest Walton carried out the experiment that produced the first controlled nuclear disintegrations induced by accelerated high energy particles. They also experimentally proved, for the first time, Einstein's famous equivalence, $E=mc^2$.

By the end of the nineteenth century, most scientists considered the atom to be the most fundamental form of matter. In fact, its name comes from the Greek word *atomos*, which means indivisible. In 1897, Thomson discovered the electron (he called them corpuscles)—the first-known subatomic particle and one which had a negative charge. This was a startling discovery, since the atom had all along been considered the smallest form of matter. The atom was not indivisible after all. With atoms no longer being considered the indivisible pinpoints of matter that their Greek nomenclature implied, attention shifted toward their inner structure. If the electrically neutral atom contained negatively charged electrons, it also had to contain a compensatory positive charge. The man who discovered this positive charge—the atomic nucleus—was Rutherford.

The identification of the electron meant that there was a subatomic world to explore. Rutherford was the person, who more than anyone else, shaped this world and helped gain an understanding of the structure of the atom. He dominated both atomic and radioactive research in the first three decades of the twentieth century and was probably the greatest experimental genius of all time, his closest competitor being Michael Faraday in the nineteenth century. Bhabha's years at Cambridge coincided with the golden years of the Cavendish where, in the words of Samuel Devons, a former professor of physics at the University of Manchester, 'Rutherford the individual and Cavendish the institution became fused in one, and together radiated a brilliance rarely matched.'[5] It is, therefore, hardly surprising, that Bhabha should have found himself strongly attracted to physics at the time.

Born in New Zealand in 1871, Rutherford was awarded the 1851 Exhibition Scholarship in 1895. He was Cambridge University's first research student recruited from outside the university. Rutherford's great contribution was the new insight into the atom, which he developed in a historic series of experiments, the greatest being his discovery that the cloud of negatively charged atomic electrons was electrically matched by a positive charge concentrated in a tiny nucleus deep at the heart of the atom. In 1898, he took a chair at McGill University in Montreal where between 1899 and 1900, he broke up radioactivity into alpha, beta and gamma rays. With his collaborator, the English radiochemist, Frederick Soddy, he discovered that certain elements transform radioactively into other elements. He was awarded the 1908 Nobel Prize for chemistry, which was ironic since he was in the habit of making rude remarks about chemistry being a mere sub-branch of physics. 'All science is either physics or stamp-collecting.'[6]

A supremely self-confident man, Rutherford ran his laboratories as if they were personal fiefs. But with his booming voice and boundless enthusiasm, he had an uncanny feel for experiments that could be important and how to do them. He passed this knowledge on to his assistants, whom he called his 'boys'. These 'boys' i.e., Bohr, Blackett, Chadwick, Pyotr Kapitza and Cockcroft, went on to become Nobel laureates.[7] In some ways, Rutherford himself embodied that heroic age. In 1909, he decided to investigate the internal structure of the atom and devised a series of experiments with the help of his German assistant, Hans Geiger (who gave his name to the counter which is used to measure radioactivity). Rutherford set up an experiment in which a beam of alpha particles from a radioactive source was fired at a thin sheet of gold foil. When the alpha particles passed through the gold foil they struck a fluorescent screen producing tiny flashes of light.

As the experiments continued, Rutherford found that some of the alpha particles went straight through, some were deflected slightly to one side and a few bounced back. The atom, obviously, consisted of a tiny but extremely concentrated centre, which carried a positive charge. Based on his results, Rutherford posited that the atom consisted of a very dense but extremely minuscule nucleus (a name he coined) that was surrounded by a number of negative electrons, which circled it in fixed orbits under the influence of its attraction.

During the First World War, Rutherford worked with the Admiralty on submarine detection but continued his research on the side. In April 1919, he published a paper, which, even if he had done nothing else, would have earned him a place in history. It was innocuously called 'An Anomalous Effect in Nitrogen', but its import was unmistakable. Rutherford found that nitrogen atoms bombarded with alpha particles were converted into a form of oxygen, with the ejection of a hydrogen nucleus (a proton, another term he coined). This process involved a change in the nucleus of the atom and it heralded the dawn of nuclear physics.[8]

In the conclusion to this paper, he somewhat hesitantly posited that 'the long-range atoms arising from the collision of [alpha] particles with nitrogen are not nitrogen atoms but probably atoms of hydrogen [...] If this be the case, we must conclude that the nitrogen atom is disintegrated.'[9] Rutherford had, in fact, achieved the first artificial transmutation, breaking off protons from the nucleus through this bombardment, becoming the first person to change one element into another. The discovery was also significant because it enabled the nucleus to be studied in a new way. Rutherford also predicted the neutron, which Chadwick later discovered in 1932. This period from 1919 to 1932 was a golden decade for physics. Rutherford himself was aware of the times they were living in. At a meeting of the British Association in 1923, he startled his colleagues by suddenly shouting, 'We are living in the heroic age of physics!'[10] Someone once asked Rutherford how it was that he always managed to keep on the crest of the wave. 'Well,' Rutherford responded, 'that isn't difficult. I made the wave, why shouldn't I be at the top of it?'[11]

THE QUANTUM ATOM

Rutherford's model of the atom was certainly aesthetically pleasing since it looked as if the smallest unit in the world worked in the same way as the solar system. The micro and macro worlds mirrored one another! His conception may have been an inspired one but, unfortunately, it didn't work. There was nothing to stop the electrons from falling into the nucleus. According to the laws of classical physics, the electrons would simply radiate away all their energy in a millisecond and collapse into the nucleus.

The man who attempted to rescue Rutherford's model was the Dane, Niels Bohr. To explain how he managed to solve this problem, it is necessary to understand the laws of physics. At the turn of the twentieth century, the laws of physics were based on the fundamental assumptions of classical physics that saw the universe as operating strictly on mechanical lines, in absolute time and space. There were two forms of energy—the one possessed by particles in motion and the one that travelled in waves—and the two were mutually exclusive. According to Maxwell's theory, light travelled in waves with different frequencies, producing different types of radiation. At the time, these laws were immutable: this was the way the world worked.[12] But things were to change soon.

Planck's Quanta

In December 1900, the German physicist, Max Planck, was trying to account for what is known as 'black body radiation'. One of the anomalies that had perplexed classical physicists was the so-called 'ultraviolet catastrophe'. Since a black body absorbs all frequencies of light, it should, when heated, radiate all frequencies of light. This, however, is not what happens. It emits frequencies at the lower range, which only gradually increase with the intensity of the heat. As we know, a heated body emits low frequency red light. Next, as it gets hotter, it emits a higher frequency orange and then yellow. But according to classical physics it should emit an equal amount of all frequencies. However, if this did happen, we would all suffer severe ultraviolet burns every time we sat in front of a fire. But as we know the 'ultraviolet catastrophe' doesn't happen.[13]

When physicists tried to calculate the radiation emitted by hot, vibrating atoms by applying the work of Newton and Maxwell, the results defied expectations. Planck found that to obtain results, which agreed with observation, he had to assume that the radiation was not emitted continuously as had been previously believed, but that it was emitted in discrete amounts, which he called 'quanta' (from a Latin word meaning 'how much?'). Planck worked out that the size of the quanta varied in proportion to the frequency of the radiation, expressing it in his famous formula '$E=hv$', where E is the energy value of the quantum and v is the frequency of the radiation. The h is a fundamental constant, now named Planck's constant after its discoverer. Planck seems to have realized that he had made a pathbreaking discovery. He told his son that he believed he had made a discovery of equal significance to those of Newton. It might seem like a grandiose claim to make but he was merely stating the sober truth.

Planck's idea was so revolutionary that even he had difficulty in believing it. However, five years later, in 1905, Planck's 'quantum theory', as it came to be known, was clarified by Einstein, who used it to explain the photoelectric effect, another anomaly in classical physics. When ultraviolet light strikes certain metals, it causes electrons to be emitted. However, these electrons don't behave according to the laws of classical physics. The rate of their emission depends upon the frequency of the bombarding light rather than its intensity. The higher the frequency, the more electrons are dislodged. Back then, this phenomenon could only be explained if light was regarded as quanta, (with higher frequency light consisting of larger quanta).

Light could be seen behaving either as a wave or as a stream of particles. Einstein's individual packages of light were christened photons by the American chemist Gilbert Lewis in 1926. Einstein's theory was rejected by many of his contemporaries, but it was verified by Robert Millikan's work on the photoelectric effect and by Compton's discovery of the Compton Effect—the scattering of photons by electrons. Millikan's aim in starting his experiments had, in fact, been to prove Einstein wrong; but he only ended up proving him right, and along the way, also derived a very accurate measurement of Planck's constant. Einstein, Millikan and Compton all went on to win Nobel prizes for their work. Thus, it was concluded that light is made of photons, which are quanta or particles. However, each photon creates fields surrounding it (electric and magnetic) that are shaped like waves and follow Maxwell's equations.

Bohr's Model

Bohr showed for the first time that the reason atoms are stable is due to quantum physics. Bohr's model melded classical theory (the idea of orbiting electrons) and quantum theory (the idea that energy is emitted or absorbed in discrete quanta). Born in Copenhagen on 7 October

1885, Bohr grew up in a privileged family. His father Christian Harald Bohr was a professor of physiology at the University of Copenhagen and was nominated twice (in 1907 and 1908) for the Nobel Prize for his work on the physiology of respiration. Bohr's Jewish mother belonged to a wealthy Danish banking family. Bohr received a PhD in physics in 1911 for his thesis titled *An Investigation into the Electron Theory of Metals*.

During his investigations for his thesis it was clear to him that at the subatomic level the old assumptions of classical physics were not valid. To understand the working of the inside of an atom would need an entirely different kind of physics. So, he headed for the source, setting off for Cambridge to study with J.J. Thomson. But his association with Thomson proved to be a big disappointment. This was, in part, because of his imperfect English and the fact that his research interests did not mesh with those of the Cavendish at that time. Thomson, who was in his mid-50s by then, no longer greeted new ideas with the enthusiasm of the past.[14] Bohr didn't help matters when he insisted that some of the great man's calculations concerning the electron were wrong. Bohr, during an informal discussion with Thomson, was trying to explain a point on the theory of metals. To his mortification, Thomson dismissed his ideas as rubbish and then proceeded to say the same thing himself using different words.[15]

Bohr knew that he had to have a better knowledge of English and, to this end, sat down with the works of Charles Dickens and an Engelsk-Dansk Ordbog[16], looking up every word he didn't understand.[17] It didn't help much considering Bohr's method of composition remained painfully slow; he was always searching for more precise meaning. Later, while writing his papers he gave dictation to his wife who corrected and improved his English; for many years, she effectively worked as his secretary. In fact, Bohr disliked writing and avoided it as much as he could. He could complete his doctoral thesis only because his mother took the dictation. He wrote slowly and in an almost indecipherable scrawl, finding it difficult to think and write at the same time.[18] When this technique was extended to the lecture hall, it proved to be a severe handicap to both speaker and listener alike. It was, however, counterbalanced by his obvious enthusiasm and the brilliance of his ideas. Nevertheless, understanding a lecture by Bohr became something of an intellectual feat.[19]

In October 1911, Bohr attended the annual Cavendish dinner and heard a speech by Rutherford. He was entranced by the bluff, middle-aged New Zealander. There seems to have been an immediate connection on both sides: Rutherford took to the young Dane, impressed with the fact that he played football (Rutherford had played rugby for his college in New Zealand).[20] A month later, Rutherford invited him to come and work with his research team at Manchester University. Bohr accepted enthusiastically and moved to Manchester in March 1912. There, he worked out the first quantum model of the atom based directly on Rutherford's model.

Bohr returned to Denmark in the summer of 1912, married his fiancée and took up a junior teaching appointment at Copenhagen University. He always had very happy memories of Manchester and soon began to regard Rutherford as a father figure and a role model. He could also not forget the atmosphere of the laboratory in Manchester—one of camaraderie and fruitful argument, where young and old, both listened to each other. Later, Bohr replicated this atmosphere in his own institute. It is perhaps no exaggeration to say that it was this kind of congenial and free interaction that revolutionized twentieth century science.[21]

In Copenhagen, Bohr wrote three papers on the structure of the atom. All of these papers were published before the end of 1913 and formed the basis of the work for which he would receive the Nobel Prize in 1922. He maintained close ties with Manchester, corresponding with Rutherford and his colleagues about his progress. His final paper on atomic structure underwent several revisions.[22] Bohr's trilogy was published in the *Philosophical Magazine* as 'On the Constitution of Atoms and Molecules'. The first appeared in July 1913 and the second and third in September and November, respectively; the trilogy presented ideas on the possible arrangement of electrons inside atoms. He pursued this subject throughout the next decade, trying to use the quantum atom to explain the periodic table and the chemical properties of elements.[23]

Einstein's explanation of ultraviolet light striking certain metals was in the same territory as Bohr's thesis on the electron theory of metals. Bohr, too, had concluded that subatomic particles didn't obey the laws of classical physics. Although, like most others, Bohr did not believe in light quanta, it was clear to him that the atom was in some way regulated by the quantum. However, in September 1912, he had no clue as to how it was linked to the problem of atomic structure.

Bohr began rebuilding Rutherford's model to ensure that electrons did not radiate energy as they orbited the nucleus in the model. Unlike classical physics, which did not restrict an electron's orbit, Bohr restricted them to special orbits. An electron could occupy only a few select orbits, the stationary states, out of all possible orbits allowed by classical physics. The bottom rung of this atomic energy ladder is 'n'=1, when the electron is in the first orbit at the lowest energy level which is called the ground state. If the electron occupies any other orbit, it is said to be in an excited state. Later, 'n' was designated as the principal quantum number that designates the series of stationary states an electron can occupy and the corresponding set of energy levels.

Bohr's research received a shot in the arm when Hans Hansen, his friend from his student days in Copenhagen, who had studied spectroscopy at Göttingen University, asked him if his work on atomic structure shed any light on spectral lines. If any element in its gaseous form is intensely heated, it glows. And when the light it emits is examined with a spectroscope, it breaks down into its component colours. These show up as bands of coloured lines in the overall spectrum of colours. Each element has its characteristic emission spectrum. The number, spacing and wavelengths of the spectral lines produced by the atoms of any element are unique and can be used to identify it. Each line in the emission spectrum can be assigned a precise number in accordance with its wavelength. These appeared to repeat, much like a harmonic sequence.[24]

The formula for this sequence was discovered by Johan Balmer, a Swiss mathematics teacher at a girl's school in Basel. Balmer had been told about the four spectral lines of hydrogen by a friend of his when he complained he had nothing interesting to do. Intrigued, he set about finding a mathematical relationship between the lines where none seemed to exist. In the 1850s, Anders Ångström, the Swedish physicist, had accurately measured the wavelengths of the four lines in the red, green, blue and violet regions of the visible spectrum of hydrogen. In June 1884, Balmer found a formula that reproduced the wavelengths of the four spectral lines. With the help of a constant, discovered experimentally, Balmer not only generated the four known spectral lines but also predicted the existence of a fifth, unaware that Ångström had already discovered and measured its wavelength. The two values, experimental and theoretical, were in near-perfect agreement.[25] When more lines in the hydrogen emission spectrum were later discovered, Balmer's formula led to accurate results.

When Bohr came across Balmer's formula in a school textbook, he had a eureka moment. He recognized that Balmer's formula could be written in another way using Planck's constant *h*. The great revelation was that the spectrum of an element is linked to quantum theory. Electrons jumping between different allowed orbits produce the spectral lines emitted by an atom. But how does this happen? Bohr studied the simplest element of all—hydrogen. The single electron in a hydrogen atom could, in theory, circle the nucleus in any number of different orbits. The greater the radius, the faster the electron orbits, consequently, the greater its energy. Smaller orbits have lesser energy. In case the electron moves to a smaller orbit, it has to emit energy in the form of light. The energy emitted by an atom forms its spectrum. Bohr was convinced that the precise energy bands that made up the spectrum were somehow linked to the different orbits of the electron.

He realized that to emit the energy required, the electron would have to move between certain fixed orbits. When the electron has a fixed orbit, the atom remains in a stationary state. Only when it moves between these fixed orbits does it emit the energy needed to produce a particular band in the spectrum. Bohr knew that the fixed orbits of these stationary states would have to be linked to Balmer's formula.[26] He also discovered that it is impossible to say where an electron is during a jump. The transition between orbits and energy levels has to occur instantaneously to prevent the electron from radiating energy continuously as it travels from one orbit to another. In Bohr's model, the electron disappeared in one orbit and instantly reappeared in another. It was a curious mixture of Newtonian and quantum ideas but it agreed precisely with Planck's original theory of light. It correctly predicted various properties of the hydrogen atom, such as its radius and explained the production of spectral lines.

When the first part of his trilogy was published in July 1913, Bohr had been appointed to a lectureship at Copenhagen University. It provided little job satisfaction since his major responsibility was to teach elementary physics to medical students. At the beginning of 1914, he set about trying to establish a new professorship in theoretical physics for himself. Rutherford wrote a strong letter of recommendation and the faculty of the university, too, welcomed the decision. But the authorities procrastinated and ultimately it was Rutherford who provided an escape route. He offered Bohr a position at Manchester and in September 1914 the Bohrs sailed for England. Bohr stayed in Manchester till May 1916, when the position in Copenhagen finally materialized.

Despite the successes of Bohr's studies, there were problems that the quantum atom could not address. The answers it gave for atoms with more than one electron were not corroborated by experimental data. It predicted spectral lines that did not exist. In 1892, it was discovered that the red alpha and blue gamma Balmer lines of the hydrogen spectrum, thought to be single lines, were split in two. Bohr's model could not explain this 'fine structure', as the splitting was called. This problem was eventually solved by Arnold Sommerfeld, a professor of physics at Munich. Over the years, Sommerfeld had turned Munich into a thriving centre for theoretical physics, mentoring some of the brightest minds in physics, including Werner Heisenberg, Hans Bethe, Gregor Wentzel, Rudolf Peierls and Wolfgang Pauli. Like Bohr, Sommerfeld loved skiing and invited students and colleagues to his house in the Bavarian Alps to ski and discuss physics.

Sommerfeld applied Einstein's relativity to the movement of electrons and realized that these orbits must be elliptical. Bohr had only allowed for circular orbits to simplify his model; Sommerfeld decided to lift this restriction. The quantum number 'n' in Bohr's model specified a stationary state, a permitted circular orbit and the corresponding energy level. Something more was required to encode the shape of an ellipse and Sommerfeld introduced 'k', the 'orbital' quantum number. This change accounted for the 'fine structure' but failed to account for the Zeeman effect due to which a magnetic field splits a single spectral line into several separate lines or components. Sommerfeld came to the rescue again. He suggested that the orientation of an orbit was the missing component and introduced what he called the 'magnetic quantum' 'm' to quantize the orientation of the orbit.

Bohr's model (also sometimes called the Bohr–Sommerfeld model) was making progress, but it still didn't have all the answers. Sommerfeld had come to Bohr's rescue with two new quantum numbers but his contribution still could not explain what was called the 'anomalous' Zeeman effect, a situation where the spectral lines split up into four or more instead of just two or three lines. Before we see how Pauli explained this through his famous exclusion principle in January 1925, let us examine some of the other important developments in quantum physics.

Louis de Broglie's Waves

The next big step in quantum physics came in 1924, when the French physicist Louis de Broglie posited in his doctoral thesis at Sorbonne Université the idea that just as electromagnetic waves could be described in terms of particles all material particles such as electrons could be described in terms of waves. His PhD guide, Paul Langevin, was nonplussed and referred the thesis to Einstein, who pronounced it to be sound work and not just mathematical sleight of hand. Louis de Broglie's theory was independently experimentally confirmed by Clinton Davisson and Lester Germer in the US, and by George Thomson (the son of J.J. Thomson) in Aberdeen. Davisson and George Thomson shared the Nobel Prize in 1937. J.J. Thomson received the award for proving that electrons are particles; George, his son, was awarded the prize for proving they are waves, which says something about how quantum theory flies in the face of common sense.[27] Bohr himself observed, 'For those who are not shocked when they first come across quantum theory cannot possibly have understood it.'[28]

Einstein and Bose

Einstein made another contribution in 1924 when he received in June of that year a paper, as noted before, from S.N. Bose. This paper derived Planck's blackbody radiation law by treating radiation as if it were a cloud of gas and then applied a statistical method to analyse it. Bose's new insight was in treating any two photons that had the same energy state as indistinguishable and, therefore, not to be treated separately in statistical calculations. Einstein helped in getting Bose's paper published and wrote three of his own. He applied S.N. Bose's counting method later called Bose–Einstein statistics to actual gas molecules, thus becoming the primary inventor of quantum statistical mechanics.[29] Einstein developed S.N. Bose's concept further and predicted what is known as the Bose–Einstein Condensation.

Pauli's Exclusion Principle

Bohr showed how quantum theory was essential to the understanding of subatomic phenomena and, in doing so, came up with the first comprehensive picture of a stable atomic structure. His model was far from perfect and raised as many questions as it answered, but it showed that the way forward must involve quantum physics. One of the best features of the model was that it provided the basis for how and why some elements react with one another to form compounds and others do not.

In 1924, Wolfgang Pauli, then a pudgy 24-year-old set out to explain the anomalous Zeeman effect. Considered a genius comparable to Einstein, Pauli was born in Vienna in 1900, the year that quantum mechanics was born with Planck's announcement of the energy quanta. Pauli's father was a physician and chemistry professor at the University of Vienna, and his godfather was the physicist and philosopher of science, Ernst Mach. When he found himself bored during class, Pauli would read Einstein's papers on relativity. By the age of 20, Pauli, then a student of Sommerfeld at Munich, had published papers on relativity and written an encyclopedia article on it, which greatly impressed everyone, including Einstein himself.

Having learned classical mechanics and relativity, Pauli was at first disconcerted by quantum mechanics. After receiving his doctorate in 1921, he spent some time in Göttingen and then Copenhagen, before taking up a position at the University of Hamburg in 1923. He delivered his first lecture there on the periodic table of elements, which he found unsatisfactory because the structure of the atomic shell had not yet been understood. Bohr had proposed that electrons could occupy only certain quantized orbitals but he had not been able to explain why all the electrons in an atom didn't simply crowd into the one with the lowest energy state. There was also no convincing explanation of the structure of the periodic table.

Pauli who was working on an explanation for the anomalous Zeeman effect, was convinced that the two problems were somehow related. In late 1924, inspired by the work of Edmund Stoner, who was doing his doctorate under Rutherford, Pauli boldly suggested that a fourth quantum number be added to the existing three that were being used to describe an electron's quantum state. The first three quantum numbers made sense physically, since they related to the electron's motion around the nucleus. But unlike the other numbers, Pauli's new number could only have two values. He called his new quantum property a 'two-valuedness not describable classically'.[30] It was this 'two-valuedness' that doubled the number of electron states and explained the splitting of spectral lines in the anomalous Zeeman effect. In January 1925, Pauli announced his famous exclusion principle, which states that no two electrons in an atom can have the same set of four quantum numbers.

Pauli's proposed fourth quantum number puzzled physicists at the time because no one could explain its physical significance. Pauli was also bothered by the fact that he couldn't logically explain the exclusion principle or derive it from other laws of quantum mechanics. Nonetheless, the principle worked—it became central for explaining the structure of the periodic table and other properties of matter.

What was known by then was that each shell of electrons was arranged such that the inner shell always contained just one orbit, whereas the next outer shell contained four. Pauli showed that no orbit could contain more than two electrons. Once it had two, the orbit was full and other electrons were forced or excluded to the next orbit. This meant that the inner shell (one orbit) could not contain more than two electrons and that the next shell (four orbits) could not contain more than eight. Pauli was, thus, able to explain why all the electrons in orbit around the nucleus didn't just crowd in on the inner shell.

Pauli was a ruthless critic and very hard to please. Feared for his sharp and incisive criticism of new and speculative ideas, Pauli's truculent style of scientific discourse gave rise to legends. He had once ticked off a colleague: 'I do not mind if you think slowly, but I do object when you publish more quickly than you think.' On another occasion, he ridiculed a paper by saying, 'It is not even wrong.'[31] He was notoriously rude to people he considered fools. Pauli, a perfectionist, was a kind of honest supreme judge, an inquisitor defending physics. He had no hesitation in calling someone an incompetent idiot in public if he spotted an error in his work. Even Bohr was not safe from his sarcasm. In a letter to Bohr's wife, he wrote: '[...] you wrote two weeks ago that Niels would answer my letter on Thursday, but you didn't say which Thursday. However, a reply written on any other day of the week would be equally welcome'.[32] His colleagues dubbed him the 'conscience of physics', and physicists were relieved if they passed the 'Pauli test'. Pauli's rudeness to his collaborators was legendary and it was something that was accepted as a part of his personality. The only person with whom his behaviour was deferential was his teacher, Sommerfeld.

Pauli also had such an amazing propensity to cause accidents that scientists began to believe that even to have him come close to one's laboratory meant doom for the experiment. A legend attached to him was the so-called Pauli effect—a kind of evil eye. It was said that when Pauli appeared near a laboratory, something dreadful happened. Something would break, the glass would explode or some other mishap would take place. One story involving German physicist, James Franck, working in Göttingen was as follows: Franck arrived one morning to find his laboratory in a complete mess: failed equipment and broken glass on the floor. His instant reaction was to send a telegram to Pauli asking him where he was the night before. Pauli responded, 'Travelling from Zurich to Berlin'. The train passed through Göttingen![33]

Spin

Late in the summer of 1925, two Dutch graduate students at Leiden University, Samuel Goudsmit and George Uhlenbeck, realized that the number Pauli had proposed as the fourth quantum was an intrinsic property of the electron that they called 'spin'. Spin is the intrinsic angular momentum of a body. An electron can move up and down, back and forth, and side by side—movements that physicists call 'degree of freedom'. Uhlenbeck believed that Pauli's new quantum number meant that the electron had an additional degree of freedom. The fourth quantum number implied that the electron was rotating.

In classical physics, spin can point in any direction but what Uhlenbeck was proposing was quantum spin—a two-valued spin, spin 'up' or spin 'down'. He visualized two possible spin states of an electron spinning either clockwise or anti-clockwise along a vertical axis as it orbits the nucleus. As it does this, the electron generates its magnetic field and acts like a tiny subatomic bar magnet. Uhlenbeck and Goudmit showed that electron spin can be either plus or minus half, values which satisfied Pauli's condition that the fourth quantum number be 'two-valued'.[34] But whatever it is, spin does not correspond in any way to the notion of an electron spinning on its axis—it is a purely relativistic quantum property with no equivalent in classical physics.

In a broader sense, spin is an essential property influencing the ordering of electrons and nuclei in atoms and molecules, giving it great physical significance in chemistry and solid-state physics. Likewise, spin is an essential consideration in all interactions among subatomic particles, whether in high energy particle beams, low-temperature fluids or solar winds. Indeed, many, if not most, physical processes, ranging from the smallest nuclear scales to the largest astrophysical distances, depend greatly on interactions of subatomic particles and their spins.

Matrix Mathematics

We have seen that Bohr showed how electrons can make a 'quantum jump' from one orbit to another, in the same way that climbing up or down a ladder requires us to move each foot high or low enough to reach the next rung. In the early 1920s, physicists learned how to account for these jumps, and the behaviour of simple atoms, like hydrogen, could be worked out mathematically. The development of actual quantum mechanics—the mathematical theory—happened between 1924 and 1927. Initially, there were two seemingly different approaches: matrix mechanics developed by Heisenberg in 1925 and wave mechanics developed by Erwin Schrödinger, a professor at the University of Zurich, in 1926. Both systems worked, but their interrelation was unclear. It was difficult to believe that two such different theories could be describing the same quantum reality. However, it soon became clear that these were distinct aspects of a single theory that came to be known as quantum mechanics.

Heisenberg was a student of Sommerfeld at Munich. When Sommerfeld travelled to America, he arranged for him to study at Göttingen under Born, who recognized Heisenberg's talents at once, calling him as gifted as Pauli. After completing his doctorate from Munich, Heisenberg went to Born in Göttingen. Born was convinced that the mishmash of quantum rules and classical physics that characterized the Bohr–Sommerfeld model had to give way to a consistent new theory that he called 'quantum mechanics'. Heisenberg visited Bohr in Copenhagen in March 1924 and was drawn to him both for his scientific abilities and personality.

Heisenberg's conversations with Bohr convinced the former that something fundamental had to give. He thought that he knew what that might be as he tried to solve the long-standing problem of the intensities of the spectral lines of hydrogen. The Bohr–Sommerfeld model accounted for the frequency of hydrogen's spectral lines but not how bright or dim they were. Heisenberg decided to separate what was observable from what was not. He abandoned the idea of electrons orbiting the nucleus of an atom since the orbit of an electron around the nucleus was not observable. In fact, Pauli had expressed doubts about the usefulness of electron orbits even before Heisenberg decided to adopt this new strategy. He decided to ignore anything that was not observable focussing his attention only on those quantities that could be measured in the laboratory: the frequencies and intensities of the spectral lines associated with the light emitted or absorbed when an electron jumps from one energy level to another.[35]

He devised a method of tracking all possible electron jumps that could occur between the different energy levels of hydrogen, recording them in the form of arrays. He constructed an abstract model consisting of an infinite series (called a Fourier series) of harmonic oscillators, each characterized by an amplitude and a frequency. He had identified each oscillator—each term in the Fourier series—with a quantum number 'n' to an orbit characterized by a quantum number 'm'. The result was an infinite table of terms in the Fourier series, organized into columns and rows, each term representing a transition from an initial state to a final state.[36]

Heisenberg now assumed that he could calculate the intensities of the spectral lines as the squares of the amplitudes of the terms that appeared in the table. More generally, he calculated the intensity of a spectral line resulting from any quantum jump as the sum of the products of the amplitudes for all possible intermediate jumps.[37] But there was a hitch. He discovered that when he multiplied two arrays together, the answer was dependent on the order in which they were multiplied. Ordinarily, 5x6 and 6x5 both give 30, a property of mathematics called commutation. Puzzled by this result, he proceeded to calculate the energies to check if the law of conservation of energy, an imperative for any new theory of the atom, was valid. His calculations showed that the energy principle held true for all the terms. Heisenberg hastily wrote a paper and sent it to Born, who finally solved the mystery of the multiplication rule: it was the rule for multiplying matrices, something Born had been taught as a student.

Born worked with his student Pascual Jordan to recast Heisenberg's calculations into the language of matrix multiplication. On being told that the mystery of the multiplication had been solved, Heisenberg said, 'I do not even know what a matrix is.'[38] Heisenberg soon learnt matrix mathematics and, with Born and Jordan, published a paper that presented the first logically consistent formulation of quantum mechanics. While the three had been busy developing the theory, Pauli successfully applied matrix mechanics to reproduce the line spectrum of the hydrogen atom, providing the first concrete vindication of the new theory, upstaging Heisenberg who had been unable to do so.

Working independently, 23-year-old Paul Dirac, a senior research student at Cambridge and a protégé of the respected Cambridge physicist, Ralph Fowler, sent a paper titled 'The Fundamental Equations of Quantum Mechanics' to the *Proceedings of the Royal Society*. It contained the nuts and bolts of quantum mechanics and was completed nine days before 'the three-man paper' was finished. A proof copy of Heisenberg's paper had found its way to Fowler. Unimpressed, he had passed it on to his young student who had immediately started work on it. Dirac sent a copy of his paper to Born who was in Boston at the time. Born was surprised to find that 'everything was perfect in its way'.[39] Dirac also sent a handwritten copy to Heisenberg and was disappointed to learn that his work had been anticipated by Born and Jordan. Heisenberg was, nevertheless, full of praise. We will return to Dirac and his contributions in the next chapter.

Pauli and Dirac moved quickly to use the new matrix mechanics to derive key features of the hydrogen emission spectrum, notably the Balmer formula. The message was clear that the way forward was by ridding oneself of the conceptual baggage of classical physics, and focussing on what could be measured and observed in the laboratory. But many older physicists were unfamiliar with the use of matrices that used the strange multiplication rule that lay at the heart of Heisenberg's new theory. However, the direction of future development seemed set. But the Young Turks of the quantum revolution were about to be upstaged by a 38-year-old Austrian physicist called Erwin Schrödinger.[40] Within a matter of a few months, he offered physicists an alternative that they eagerly embraced.

Schrödinger's Wave Mechanics

Schrödinger secured his doctorate in physics from Vienna University in May 1910. After obligatory military service, he returned to the university and qualified as a privatdozent[41] in 1914. After four years of military service during the First World War, he returned to Vienna in 1917, becoming an associate professor in 1920. He secured a professorship in theoretical physics at the University of Zurich in

1921 after brief visits to Jena, Stuttgart and Breslau. Praised for his breadth of knowledge and versatility as a scientist, he was yet to make a really noteworthy contribution to physics and watched with dismay as a younger generation of physicists overtook him. But his luck would soon change.

In October 1925, Schrödinger had read a paper by Einstein that referred to de Broglie's thesis on wave–particle duality. Intrigued by this stamp of approval, he acquired a copy of the thesis and read it. In late November, he gave a talk in Zurich on how de Broglie associated a wave with a particle and how he could obtain the quantization rules of Bohr and Sommerfeld by demanding that an integer number of waves should be fitted along a stationary orbit.[42] But there was no wave equation (de Broglie had never tried to derive one for his matter waves) and Schrödinger knew that he had to find one. The physics of a wave—any wave from sound to electromagnetic—has an equation that describes it.

His starting point was the well known equation of classical wave motion. Into this equation, he substituted the relationships between the particle mass (more specifically, particle momentum) and wave frequency that de Broglie had deduced. He also realized that for his equation to be meaningful, he had to make some assumptions about the form of the electron wavefunction—the mathematical description of the electron as a wave. He assumed that the functions were single-valued, finite and continuous.[43]

Not very long after he started his quest for such a wave equation, Schrödinger thought he had devised one. But when he initially applied it to the hydrogen atom, the equation gave wrong answers. He had not taken relativity and spin into account. Later, he found a relativistic wave equation but since it did not include spin, it failed to give results in conformity with experimental data. With Christmas approaching he set about finding an equation that did not include relativity. He knew it would fail for electrons travelling at speeds close to that of light but he felt that, for his purposes, such a wave equation would do.

He was also facing marital problems at the time. His marriage had been a relatively 'open', one with both he and his wife, Annemarie (affectionately called Anny) indulging in extramarital affairs. He wanted to escape for a few weeks and left Zurich for his favourite Alpine resort, Arosa, for a rendezvous with an ex-lover. We don't know the name of his girlfriend or what influence she had on him, but when he returned to Zurich on 8 January 1926, he discovered wave mechanics.[44] A serial womanizer, Schrödinger would, after fleeing Nazi Germany, scandalize Oxford and then Dublin when he set up home with his wife and another mistress under the same roof. As we shall see in a later chapter, he was also at one point contemplating becoming a professor of theoretical physics at Allahabad University.

In Zurich, he found that his wave equation not only successfully produced the series of energy levels of the Bohr–Sommerfeld hydrogen atom but, more importantly, also did away with the ad hoc additions of that model. In his derivation, the only ansatz concerned the constraints he had placed on the form of the electron wavefunction. The rest, including the quantum numbers, their interrelationships, the hydrogen energy levels and the Balmer formula, was the result of mathematical logic.[45] All the previous tinkering and tweaking that sat uneasily emerged naturally from Schrödinger's equation. Even the mysterious quantum jumps appeared to be eliminated by the smooth and continuous transitions from one permitted three-dimensional standing wave to another.[46]

In the six months from January to June 1926, Schrödinger produced six creative papers on the new wave mechanics, born from what his friend, the great German mathematician, Hermann Weyl, called 'a late erotic outburst in his life'.[47] Schrödinger would sometimes take refuge in the bohemian nightlife of Zurich in the company of Weyl and Dutch physicist Pieter Debye. It was a complex web of extramarital relationships. Anny fell in love with Weyl; and Weyl's wife Helene Joseph was having an affair with Swiss physicist, Paul Scherrer.[48]

Schrödinger's new wave mechanics offered physicists a familiar and reassuring alternative that explained the quantum world in terms that were closer to nineteenth-century physics than Heisenberg's abstract matrices. Schrödinger used differential equations, an essential part of every physicist's toolkit. The two theories were mathematically equivalent but the nature of physical reality that lay beyond the mathematics was very different. For Schrödinger, there were no quantum jumps, only smooth continuous transitions from one standing wave into another. As part of his interpretation of the wave function, he introduced the concept of a 'wave packet' to represent the electron, since he believed that the electron was only particle-like but was not actually a particle, despite the overwhelming evidence in favour of it being so. His interpretation ran into difficulties almost immediately since it failed to make physical sense.

Schrödinger believed that wave mechanics allowed the restoration of a classical picture of physical reality, one of continuity, causality and determinism. Born disagreed. 'Schrödinger's achievement reduces itself to something purely mathematical', he told Einstein, 'his physics is wretched'.[49] Schrödinger's interpretation left many questions unanswered. It could not account for the photoelectric and Compton effects. How could a wave packet possess an electric charge? Could wave mechanics incorporate spin? If his waves did not represent real waves then what were they?

It was Born who answered these questions. He was appalled by the fact that Schrödinger's wave mechanics had rejected particles and quantum jumps. While he acknowledged the richness of Schrödinger's formalism, he rejected the Austrian's interpretation. It was necessary, he said, to retain the formalism and fill it with new physical content. Convinced that particles could not be abolished, Born found a way to weave them together with waves using probability to explain his new interpretation of the wave function.[50] In his attempt to save particles, he posited an interpretation that challenged a fundamental tenet of physics—determinism.[51]

When an electron slams into an atom, it can be scattered in almost any direction. Born made a startling claim. He said that when it comes to atomic collisions, physicists could not answer the question 'what is the state after the collision?' The answer to this question lies in probability. The best that physics could do was calculate the probability that the electron would scatter at a certain angle. The wave function, then, has no physical reality, existing only in the realms of the possible.

Taking Born's lead, Bohr argued that until it is observed or measured, an object like the electron does not exist anywhere. Between one measurement and the next, it has no existence outside the abstract possibilities of the wave function. It is only when it is actually measured that the 'wave function collapses', as one of the 'possible' states of the electron becomes the 'actual' state and the probability of other possibilities becomes zero.[52] Thus, Born introduced a new kind of probability—quantum probability—to physics.

Even though Einstein had been the first to introduce probability into quantum physics in 1916 when he explained the spontaneous emission of light-quanta, he could never wrap his head around quantum mechanics, which defined reality in terms of indeterminacy, uncertainty and probability. He was very reluctant to give up complete causality. He wrote to his friend Born in December 1926: 'I, at any rate, am convinced that He is not playing dice.'[53] (It is pertinent to note that Einstein did not use the word 'God' but 'He' because he never believed in a personal God). This is perhaps one of the most famous quotations associated with Einstein. When he repeated this in his famous debates with Bohr on the subject, Bohr is supposed to have said, 'Einstein, stop telling God what to do.' Decades later, in an interview, Stephen Hawking stated: 'Not only does God play dice, but... He sometimes throws them where they cannot be seen.'[54]

Einstein believed that describing reality required firm predictions followed by direct observation. But individual quantum interactions cannot be observed directly, leaving quantum physicists with no choice but a probabilistic prediction of events. Bohr argued that the mere act of observing the atomic realm changes the outcome of quantum interactions. According to him, quantum predictions based on probability accurately described reality.

Einstein, on the other hand, felt that physics must provide a deterministic and observer free, description of reality. He could not imagine a universe where electrons roamed free, unfettered by causality. At the 1927 and 1930 Solvay Conferences on Physics in Brussels, Einstein engaged in profound discussions with Bohr and others about his conviction regarding classical determinism versus the statistical causality of quantum mechanics. He retained his belief in a deterministic philosophy till the end of his life. The quantum theory, which was discovered and refined in the first three decades of the twentieth century, is considered to be the most successful theory of physics ever devised. Its concepts underpin much of the twenty-first century technology that we take for granted. The list includes the laser, the transistor, (and thus the microchip), the electron microscope, magnetic resonance imaging (MRI), GPS and computers.

While Einstein accepted that quantum mechanics was an experimentally confirmed empirical theory, he thought its structural features made it inherently incomplete. Einstein changed the universe: his Theory of Relativity established him as the greatest scientific mind after Newton and his name has become an eponym for genius. He even made major contributions to quantum theory, but in the end he was unable to accept the implications of his discoveries. Isolated from the new generation's exciting work on quantum theory, he was ignored by most working scientists, his ideas opposed by even his closest friends. He wasted over a quarter of a century in the search for a chimaera—his quest for an overarching theory to explain the workings of the universe through his Unified Field Theory. However, he was also unwittingly the inspiration for one of the most stunning breakthroughs in quantum theory—the uncertainty principle.

Heisenberg's Uncertainty Principle

Heisenberg's most famous and disruptive contribution came in 1927 when he declared that it was impossible to know the precise position of a particle and its precise momentum at the same instant. The more precisely the position of a particle is measured, the less precisely it is possible to measure its momentum. The very act of observing something affects the observation. Not only that, Heisenberg went as far as to say that a particle does not have a definite position or path until we observe it. This is famous as the uncertainty principle, one of the most perplexing aspects of quantum physics.[55] When Heisenberg came to India in 1929, he visited Presidency College, Madras, in October to deliver a lecture as Sommerfeld had done the previous year. Chandrasekhar borrowed a car and took him sightseeing in Madras.

The two young men spent the whole day together, and Chandrasekhar wrote to his father that by merely talking to Heisenberg for one day he was able to learn a lot about physics.[56]

Chapter 6

THE ROMANCE WITH PHYSICS

I am doing engineering now, but it is not very interesting [...] However, I am determined to do physics and astrophysics eventually. Nothing will stop me [...] The mere circumstances that there is an opening in steel or hydro will not change my course [...]

—Homi Bhabha to Homi Seervai, 2 August 1928[1]

The only theorists who were internationally famous in the early 1930s at Cambridge were Ralph Howard Fowler and his former student, Paul Dirac. Fowler, a mathematician, who agreed to become Bhabha's PhD supervisor in 1932 (in the absence of anyone else) was the first to apply quantum mechanics to astrophysics. He was at the heart of a circle of brilliant scientists and was responsible for introducing Dirac, who was then his student, to quantum theory. Later, he introduced Dirac, Heisenberg and Bohr to each other. He was married to Eileen, the only child of Rutherford who was his close friend and colleague.

Fowler was a 'big man both physically and in the world of science'[2], and was in the words of English astrophysicist, Arthur Milne, 'a man of winning charm who inspired friendships'.[3] Born into a privileged background, just like Bhabha, he was a keen sportsman who played golf, rugby and cricket with distinction. He was also an accomplished rock climber and bridge player. An affable, easygoing man with a famously hearty laugh and open smile, he was much liked by all who knew him. He was also the thesis guide for Chandrasekhar who, unlike Bhabha, was 'self-effacing to the point of invisibility'.[4] After his first meeting with him in September 1930, Chandrasekhar wrote to his father in October 1930: 'Mr. Fowler—strong (big) healthy, middle-aged man, quite happy, full of the joy of life. He says "Splendid" very often.'[5] At that first meeting, Chandrasekhar was so nervous that he fell down a whole flight of stairs.

Fowler, it appears, took full advantage of this meekness. He was frustratingly difficult to meet, always absent from Cambridge, or busy in his study at Trinity College or in the library. Chandrasekhar would spot him going into the library, and wait patiently for him to emerge, hoping he would invite him to his office for a chat, but Fowler never did.[6] He would see other people like Dirac go in to talk to Fowler, but Fowler would just pass him by, smile courteously and walk away. When this happened on numerous occasions, he stopped trying. Chandrasekhar would never have thought of approaching Fowler directly. He was, at the time, too much of an introvert and too in awe of people like him, Rutherford and Arthur Eddington. Chandrasekhar, unlike Bhabha, was the quintessential introvert.[7]

When Fowler left Cambridge on a sabbatical at the beginning of the second term, Dirac became Chandrasekhar's official advisor. The latter wrote home expressing his disappointment with Dirac's introvert personality. 'Dirac is a lean meek shy young 'Fellow' (FRS) who goes slyly along the streets. He walks quite close to the walls (like a thief!), and is not at all healthy. A contrast to Mr Fowler...Dirac is pale, thin and looks terribly overworked.'[8] But Chandrasekhar loved Dirac's lecture that he heard for the first time in the autumn of 1930 when he arrived in Cambridge. Chandrasekhar remembered seeing Dirac's book *The Principles of Quantum Mechanics* displayed at a bookstore near the hostel where he stayed. The first thing he did on the morning after his arrival was to buy Dirac's book.[9]

Even after he had had studied the book thoroughly, Chandrasekhar, attended the entire course four times because he enjoyed Dirac's virtuosity in presenting his ideas. Dirac presented his ideas with the utmost clarity and economy. In a radio interview many years later, Chandrasekhar said it was like listening to great music. 'It is not unusual for one to listen more than once to a Beethoven symphony and discover something new every time'.[10]

Dirac was perceptive enough to realize that Chandrasekhar was a kindred spirit. He invited Chandrasekhar for tea in his rooms in St John's about once a month and on some Sundays used to drive him out to fields outside Cambridge where they would go for long walks on the Roman Road.[11] Chandrasekhar recalled amusing behaviour by Dirac on these walks. There were periodic dips on the Roman Road. When they came to each of them, Dirac would try the experiment of going down with as little exertion as possible and use that energy to find out how far he could climb on the other side. He always used to say, 'converting potential energy into kinetic energy' as he went down, and 'kinetic energy into potential energy' as he went up.[12]

DIRAC AND HIS EQUATION

The theoretical advances of Bohr and others in the mid-1920s came to be known as the Copenhagen interpretation of quantum mechanics. It posited that there is no single underlying reality that is independent of our observations. As noted, the inability to know the underlying reality meant that there was no strict determinism in the classical sense. Schrödinger's famous equation precisely described the motion of particle waves. His equation, powerful as it was, had the limitation of being nonrelativistic—it said nothing about the speed of light, special relativity and how electrons interact with light via Maxwell's equations. It treated space and time separately and did not include spin. It also lacked elegance and was difficult to handle mathematically with the calculations being tedious and time-consuming.

It was left to Dirac to write a wave equation that was compact and elegant, and obeyed Einstein's special relativity by merging space and time. Dirac became Lucasian Professor (a chair made famous in more recent times by Stephen Hawking) at Cambridge in 1932. Born in 1902, the son of a Swiss immigrant to Britain, he became a legend for his intellect as well as for his unnerving reserve. Dirac, who was tall and thin with a pale complexion, got his French name from his Swiss–French émigré father. He was famous for his pathological reticence, something that started early in life thanks to a tyrannical father who insisted on speaking only French at meals. Since his father thought Dirac should learn French, he was allowed to eat with him. He couldn't speak much French either and thought it prudent to stay silent, a reticence which remained with him all his life. His colleagues at Cambridge jokingly defined a unit called a 'dirac', which was one word per hour. Later, Dirac was fluent in both English and French, but his lack of normal interaction at home altered his personality forever.

All theoretical physicists are mathematically gifted, but Dirac, who brought rare mathematical gifts to physics, was the only physicist who pure mathematicians thought could also excel in their rarefied field. He made many outstanding contributions, but the most famous one came in 1928 when he discovered the equation that bears his name. The new quantum theory was one of the twin pillars of the new twentieth century physics. The other was Einstein's Special Theory of Relativity, which pointed out that the dramatic implications of the velocity of light emitted by a source at rest being the same as that emitted by a moving source.

It was clear that Heisenberg and Schrödinger had also made impressive advances. Yet, at first sight, the way they had presented their new ideas appeared so different that it was not immediately clear whether they were the same theory expressed differently or whether they were competing theories. As we have seen, important clarificatory work followed, to which Max Born at Göttingen and Dirac at Cambridge contributed significantly. It soon became apparent that there was a single theory whose mathematical articulation took a variety of equivalent forms.

At Cambridge, Dirac, worried by the apparently conflicting quantum pictures of Schrödinger and Heisenberg, sat down and wrote *The Principles of Quantum Mechanics*, which explained the new quantum mechanics for the first time, expressing it in a self-consistent mathematical form. Although it is about physics, it has no diagrams, refers to no explicit experiments, has no references and includes no suggestions for further reading. After Cambridge University Press had dismissed the book as unworthy of publication, it was published by Oxford University Press in 1930.

In his greatest achievement, Dirac married quantum mechanics and Einstein's Theory of Relativity in the form of an exquisite equation to describe the electron. In 1927, he proposed his quantum theory of emission and absorption of radiation. This was followed, in 1928, by the famous Dirac equation that accounted for relativity and incorporated the phenomenon of electron spin and magnetic moment. Dirac's fully relativistic electron equation explained everything in a neat mathematical package with no need for extras. More importantly, the equation also predicted something new. What was puzzling was that it seemed to have two solutions, one positive and one negative, in the same way, that a number such as nine has two square roots—in this case, 3 and -3. The equation was predicting the existence of negative energy states for electrons. Dirac realized that his equation was predicting the existence of an unknown particle with the same mass as an electron but with positive charge.

To explain this, Dirac proposed in 1930 that all the negative energy states were filled with one electron each, in accordance with Pauli's principle. An unoccupied negative energy state, i.e., 'a hole', would behave as a charged particle with an electric charge opposite to that of an electron. This 'hole' theory of Dirac predicted the existence of 'antimatter' in physics. It predicted that the electron had to have a counter particle, which Dirac called 'antielectron', but which was subsequently renamed the positron. We now know that for every kind of particle, there is a kind of antiparticle—an antiproton, an antineutron, and so on. That Dirac's equation led to the discovery of antimatter, a topic that played no part at all in the original motivation for the equation, is an example of the long-term value of a really fundamental scientific idea.

The positron, first observed experimentally by Carl D. Anderson in 1932, was found to have the same mass as an electron.[13] The success of Dirac's prediction is considered one of the most outstanding triumphs of theoretical physics. It won him the 1933 Nobel Prize for physics, which he shared with Schrödinger. Dirac wanted to refuse the prize, as it would distract him from his work. He accepted it when Rutherford advised him that if he refused, the attendant publicity would be a greater distraction.

Dirac's book has been the classic introduction to quantum mechanics for generations of students and is still used today. It remains a continual source of inspiration for researchers. Dirac's greatest influence on students at Cambridge was through his course of lectures on quantum theory, the script for which was provided by his book. For many years, it was the first course in quantum theory that Cambridge students could attend. The lectures followed the pattern of his book closely. In his lectures, Dirac copied material from his book onto the blackboard.

Dirac's utter lack of emphasis on his considerable personal contribution was striking. Sir John Polkinghorne, who was a professor of mathematical physics at Cambridge from 1968 to 1979, had the privilege of attending his lectures at Cambridge. Dirac, Polkinghorne said, was 'a kind of scientific saint, in the purity of his mind and the singleness of his purpose. The lectures enthralled one by their clarity and the majestic unfolding of their argument'.[14] There was more to the lectures than the printed page can convey. It was this book that Bhabha read and learnt his quantum physics from. The book later became Einstein's companion. He took it with him on vacations for leisure reading, and when he came across a difficult quantum problem, he would mutter to himself, 'Where's my Dirac?'[15]

Dirac rarely accepted individual students. He had almost no supervision skills and believed that his only obligation was to point his students towards an interesting theoretical concept and then to oversee any work they produced as a consequence; it was up to the student to take almost all the initiative. Only the brightest could flourish under such a regime. Dirac realized this and therefore showed no interest in recruiting apprentices.

One of those he supervised was the astronomer, Fred Hoyle, who rose to become the Plumian Professor of Astronomy at Cambridge. When Hoyle was told he needed to find a supervisor, Dirac took him on, partly because he was amused by the prospect of a relationship between a supervisor who did not want a student and a student who did not want a supervisor.[16] Several of the finest young minds sought his guidance, including the Indian mathematician Harish-Chandra and the Pakistani theoretical physicist Abdus Salam. It also appeared that the great majority of Dirac's successful students were foreigners.

Dirac and his work significantly influenced Bhabha. Dirac was Bhabha's idol and much of his early work was inspired by Dirac's contributions. When Bhabha registered as a student in mathematics, he decided to change his name from Hormusji to Homi, the name he would keep for the rest of his life. However, university authorities did not accept this decision at the time. It was only in 1959, when he was awarded the Doctor of Science *Honoris Causa*, that they acknowledged that he preferred to be known and was commonly referred to as Homi Bhabha.

Before him, C.V. Raman had also shortened and simplified his name from Chandrasekhara Venkataraman to Chandrasekhara Venkata Raman, thus introducing the shorter form by which he is now famous. When Bhabha joined Cambridge, it was famous for experimental physics but was not particularly renowned for theoretical physics, which was dominated by Bohr in Copenhagen, Heisenberg in Germany and Planck in Switzerland. There were very few who were world-class in both: Enrico Fermi who went on to build the first atomic pile in Chicago in 1942 was possibly the only exception. Thus, in a sense, Bhabha was immediately made marginal by his choice of field.[17]

EXPANDING HORIZONS

Fowler agreed to be Bhabha's guide but the relations between the two appear to be a little strained. Given that both were extroverts, and Bhabha too laughed with a big guffaw and had his favourite expletives, this tension is difficult to understand. Bhabha received several scholarships to continue his studies. The first was a Salomons Studentship in engineering, which was followed by a Rouse Ball Travelling Studentship (1932–1934) after he completed his Mathematical Tripos in 1932, with distinction. Bhabha used the latter to travel to Europe in 1932 and meet some of the big names in the continent: Bohr in Copenhagen, Pauli in Zurich and Fermi in Rome. Bhabha visited Bohr's Institute for Theoretical Physics in Copenhagen, which was the hub of theoretical physics research at the time. Bohr was the presiding deity in Copenhagen just the way Rutherford was at the Cavendish.

In 1921, a year before Bohr became the first Dane to win the Nobel Prize, the government built for him the Institute for Theoretical Physics to which the cream of the world's physicists flocked. Later, the government gave him the Carlsberg Palace as his residence. Set in the grounds of the local Carlsberg brewery and surrounded by acres of immaculate gardens, this was a grace-and-favour residence, a gift of the government, which offered it whenever it became vacant to the person considered the most distinguished living Dane.

Unlike Einstein, who was a loner, Bohr taught endlessly, compulsively and perhaps most of all in conversation. He was arguably the greatest teacher and mentor in twentieth century physics; some of the brightest and best young physicists flocked to him in Copenhagen, many of who became Nobel laureates, including Pauli, Dirac, Heisenberg, the Dutch physicists Hendrik Anthony 'Hans' Kramers and Hendrik Casimir, Lev Landau and George Gamow from Russia, and the chemists Linus Pauling and Harold Urey. Chandrasekhar who visited Bohr's institute about the same time as Bhabha did called it an 'intellectual aristocracy' of breadth and magnitude hardly ever written about before.

The athletic Bohr was a very good soccer player (he was a goalkeeper) and his younger brother, Harald, was even better, representing Denmark at the 1908 London Olympics and winning a silver medal. Later, Harald became an eminent mathematician founding the field of almost periodic functions. Bohr, who climbed staircases two stairs at a time and who was also an expert table tennis player, took everyone walking, hiking, skiing, motorcycle riding, talking physics all the time. There is a photograph of Bhabha in skis, though it is not known how well he could ski. (The chairman of Tatas, J.R.D. Tata, with whom Bhabha would share a close relationship later in life was an expert skier who skied regularly in Europe).

During the 1920s, the centre of gravity of quantum physics shifted to Copenhagen largely due to Bohr. A large man who spoke painfully slowly, he was generous and avuncular, having none of the instincts for a professional rivalry that could sour relations. What struck everyone who knew him was his great modesty; despite his great intellect and renown, he remained rather shy and diffident.[18] By the time Bhabha met him, Bohr was a heavyset man in his late 40s with unruly black hair that used to be smoothly combed back, a high domed forehead and heavy features. One of his large hands carried his trademark pipe. Never known to be articulate, he spoke softly in heavily accented English; listeners had to strain to hear him.

It was said that in Munich (where Sommerfeld held court) and in Göttingen (under Born's influence), one learned how to calculate. In Copenhagen, you learned how to think. New ideas were welcomed no matter how bizarre. In fact, the crazier the better. Everyone interacted with everyone. Students, visitors and distinguished scientists all added their contributions to the intellectual crucible that was Bohr's institute. Ideas were thrown up and debated with no holds barred. There was no place for annoying British understatement.

Bohr's famous response to the wildest ideas was, 'That's a crazy idea. The trouble with it is that it's not crazy enough!'[19] Cambridge, with its stiff upper lip, was extremely dull in comparison. Bohr liked interacting with young scientists and was always on the lookout for bright students who were unafraid to argue. He was impressed by the confident young Bhabha, who must have soaked in the atmosphere. Many years later, Bohr visited Bhabha's TIFR in Bombay. Bhabha also visited the Dutch physicist, Kramers, a protégé of Bohr at Utrecht University.

After Copenhagen, Bhabha visited Zurich to meet Pauli and it was there, at the Eidgenössische Technische Hochschule (ETH), that he completed his very first research paper on the role of electron showers, titled 'On the Absorption of the Cosmic Rays'. It was completed in July 1933 and was published in the *Zeitschrift für Physik* in October of that year. Cosmic rays are very high energy particles arriving from the depths of space and incident on the earth's atmosphere at all places and at all times. Cosmic ray research led to the discovery of many fundamental particles of nature and ushered in the era of elementary particle physics in man-made accelerators. Even a hundred years after their discovery, the sources of these particles and the mechanism of their acceleration largely remain a mystery.

The Dutch physicist, Hendrik Casimir, who was known for his

work on superconductors and the Casimir effect (named after him) had worked as an assistant to Pauli in Zurich after completing his PhD. Casimir and Bhabha were exact contemporaries (both were born in 1909), and the former recounts that when Bhabha came to work under Pauli in November 1932, he brought the usual letter of recommendation from Fowler. It was more a letter of complaint though because Fowler used it to express his great annoyance with his protégé. Asking Pauli to be 'as brutal as you like', he said that Bhabha was 'opinionated and unruly' and needed a 'strong hand'.[20]

Pauli was amused at Fowler's comments on Bhabha and this made him immediately well-disposed to the young man. Casimir recounts a bemused Pauli showing him the letter and repeating over and over again, 'I can be as brutal as I like.'[21] Notwithstanding Pauli's echolalia, he and Bhabha ended up becoming good friends. Casimir and Bhabha also became good friends and would often have their meals together and visit a pub afterwards. Casimir observed, 'Also he had the knack of becoming befriended by interesting people (and with very pretty girls). My wife liked to refer to him as the fairy-tale prince.'[22] Bhabha gifted Casimir a beautifully carved picture frame for his marriage with the advice that it should 'hold the picture of Archimedes Casimir', the less than euphonious sounding name Bhabha thought Casimir should give his first-born son. Casimir said that his son was very grateful that he did not take Bhabha's advice![23]

Casimir's version of events is, however, not the only one. His account does not take into consideration the role played by the German physicist, Gregor Wentzel, who was appointed Schrödinger's successor as professor of physics at the University of Zurich in 1928, the same year that Pauli took over the chair of theoretical physics at the ETH in Zurich. These two former students of Sommerfeld became the joint leaders of physics in Zurich, which was at the time a very vibrant scientific community.

According to Wentzel, Pauli, after reading Fowler's letter of introduction, refused to have anything to do with Bhabha. However, Bhabha still managed to produce his paper on cosmic rays, despite the hostile atmosphere. When he tried to show it to Pauli, he was told, 'If you did this, I am not interested.' At his wit's end, Bhabha went to Wentzel, the other senior theoretical physicist in town. Familiar with Pauli's eccentric behaviour, Wentzel took the paper and read it. He immediately realized its worth and assured Bhabha that he would convince Pauli of its merit. At first, Wentzel's attempt to explain Bhabha's work to Pauli was met with the expected 'If Mr. Bhabha did it, I am not interested.' It was a response Wentzel was prepared for. 'For just a moment,' he suggested, 'imagine that I have done it.' Pauli relented, and things improved after that. Bhabha and Pauli became friends and corresponded with each other regularly after Bhabha departed from Zurich. Bhabha remained forever in Wentzel's debt.[24]

Fowler's attitude, and that of many others, can be explained, in part, by the fact that the wealthy, cultured and extremely talented Bhabha, who could engage with Europeans on their own terms, was an ambivalent figure who did not fit any stereotype. As a former protégé recalled many years later, 'Bhabha displayed none of the diffidence that many Indian scientists felt in front of White men. This was inspiring to many of us.'[25] A highly westernized Parsi, Bhabha's tastes and style reflected a modernity that was both contemporary and European in its essence. It is, therefore, hardly surprising that his contemporaries abroad didn't know what to make of him; in any case, he certainly didn't fit the category of 'colonial' scientist. No one had expected to find a colonial scientist in a western laboratory who was indistinguishable from a western scientist. As Itty Abraham has observed, Bhabha was wealthy, cultured, a man of the world and unlike most scientists.

> A Renaissance figure, he drew the attention and admiration of all those he met, a condition that only added to the anxiety of how to understand this figure—was he a colonial scientist or wasn't he?—and hence how to understand themselves [...] *Ultimately, of course, race is the issue, but this was difficult for the European scientists to acknowledge explicitly* (emphasis added).[26]

Casimir, in an oral history interview with Charles Weiner, for the History of Quantum Physics project, recounted that Bhabha had learnt almost all his quantum physics from Dirac's *The Principles of Quantum Mechanics*. Having read little else, he was unaware that some of the propositions were controversial and had been contested. Casimir compared him to the savage in Aldous Huxley's *Brave New World*, who had learnt English from reading a set of Shakespeare's collected plays and, as a result, only spoke an archaic form of English. As Itty Abraham observes, 'This was how Bhabha was represented in the West—almost speaking physics but not quite, physically present but betraying his unscientific origins, a child of the world who had yet to come to full maturity.'[27]

However, confidence was one thing that Bhabha never lacked. The German physicist, Otto Robert Frisch, who mistook him for an Italian, mentioned that Bhabha saw him reading a book by Edgar Wallace on the train from Berlin and even before they had introduced themselves asked him if he was a physicist. Frisch, in a double-edged comment, narrates how 'a really good scientist knows how to draw correct conclusions from the wrong assumptions.'[28] When Bhabha and Frisch got to know each other better, Bhabha introduced the latter to Beethoven's late quartets. Frisch, as if to put Bhabha in his place, would complain that the gramophone ran fast and played in the wrong key—the implication being that Bhabha did not realize that the quartets, which he claimed to know so much about, were being reproduced incorrectly. It is not clear if Frisch really meant what he said or was doing it only to annoy Bhabha, since he said, 'I irritated him by complaining that his gramophone ran fast and played them in the wrong key.'[29] Perhaps, he was unable to tolerate the fact that an Indian could teach him how to appreciate Beethoven.

Frisch seemed to enjoy snubbing Bhabha. In another episode, he narrated his response to Bhabha's desire to learn how to use a Geiger counter.

> But I was amused when one day he casually asked for instruction on how to use a Geiger counter: he was to travel to India by boat next week and wanted to measure the variation of cosmic rays with latitude. I told him the story of the boy who wanted to be a baker and was told he would have to serve a three years' apprenticeship: one year to run errands of the baker's wife; one to clean out the oven, and last to learn how to bake the bread. He smiled and got my point.[30]

Physicist Victor Weisskopf who had also met Bhabha at Bohr's institute was more generous. 'Once he pointed out some special points in the late Beethoven quartets that I, who had been raised with this music, had not appreciated until he called attention to them.'[31] He said that he and Bhabha were part of the 15 or so scientists who made up what they called the 'inner circle' at Bohr's institute. The others included George Placzek, George Gamow, Felix Bloch, Evan James Williams and Chandrasekhar.

AN UNDERCURRENT OF RACISM

Itty Abraham observes that even when many of the comments about Bhabha were positive, a hostile modifier was always added later as if to expose the real Bhabha. The German-born American physicist, Walter M. Elsasser, was greatly impressed by Bhabha's paintings but thought them extraordinarily sombre, given his liberal use of dark colours. However, Elsasser felt Bhabha's efforts at theoretical physics were not as impressive, hinting that perhaps he would be better served by becoming a professional painter.[32] Such remarks cannot be separated from their undercurrent of racism. There was also a healthy dose of colonialism involved. It was subtly expressed but it existed all the same. It was an essential element in the British attitude towards Indians. As Rudyard Kipling, the poet of imperialism, wrote, 'East is East and West is West and ne'er the twain shall meet.'

Chandrasekhar, who was in Cambridge with Bhabha, also faced this problem. He recalled a story that Raman had told him about his visit to Cambridge in the 1920s. Rutherford was his host and as they walked around, Raman saw young men playing tennis and remarked, 'Do they ever do any work?' Rutherford laughed and said, 'My dear Professor Raman, we don't want bookworms. We want governors for our empire.'[33]

The mathematician, J.E. Littlewood, described an unpleasant episode involving the mathematician Srinivasa Ramanujan's election as a fellow at Trinity in 1919. Ramanujan had already been elected as an FRS, so election to Trinity should have been a mere formality. Yet, there was a great deal of opposition that stemmed from the colour of his skin. One voter bluntly said that he would not want a 'black man' as a fellow. Ramanujan did become a fellow at Trinity, but that was purely because he was an FRS, and you simply couldn't reject an FRS. Such crude and overt racism vanished 14 years later, and Chandrasekhar became a fellow at Trinity despite Fowler's pessimism. However, Chandrasekhar was neither offered a university position, although such openings were available, nor did any of his colleagues think of recommending him for an FRS.[34] It was left to Raman to propose both Bhabha and Chandrasekhar, who became fellows in 1941 and 1944, respectively.

What added to the alterity was the fact that the flamboyant Bhabha was extremely attractive to European women, not something that could be expected of a theoretical physicist from one of the colonies. Bhabha was described as handsome, cultured, wealthy and proud, and pretty women were infatuated with him. Polish physicist Leopold Infield, who was a Rockefeller fellow at Cambridge (1933–1934), observed, 'He was [...] so handsome and with such a fine figure that women were mad about him. I myself saw how a beautiful Canadian model lost her mind at his appearance.'[35] Bhabha continued to charm women throughout his life. Many years later, one of J.R.D. Tata's favourite stories was about how Bhabha had stood up two attractive Air India hostesses in Paris. He had invited both of them for dinner but failed to show up because he had been so involved with his work. The Tata chairman teased him by saying that it was a sign that he was finally ageing.[36]

A VISIT TO FERMI

After Zurich, Bhabha visited Fermi in Rome in early 1934. By then, he had been awarded the Isaac Newton Studentship. The brilliant Nobel-Prize-winning Italian physicist, Enrico Fermi, who was an exact contemporary of Pauli, Heisenberg and Dirac, was also a prodigy. He made the first atomic pile in Chicago in December 1942 (something which is discussed later in the book). He was rarely wrong, and his colleagues in Rome used to joke that Fermi was infallible, like the Pope. He acquired the nickname 'the Pope of Physics' early on in his career an appellation that persisted, deservedly, throughout his lifetime.

The Viminal Hill, one of ancient Rome's proverbial seven hills, rises gently from the flat expanse linking the Forum to the Colosseum. The villa at 89A, Via Panisperna (one of the many streets that wind up the hill) housed the physics department of the University of Rome when Fermi became a professor of theoretical physics in 1927. The building had three floors and the basement had a serene atmosphere due to the palm trees and bamboo thickets in the garden surrounding it. Apart from Fermi, Emilio Segrè (who went on to win a Nobel prize himself), Edoardo Amaldi and Franco Rasetti worked there.[37] These four formed the nucleus of what came to be known as the Via Panisperna boys. Later, they were joined by the brilliant Ettore Majorana, considered to be as much of a genius as Fermi.

Fermi's self-control was legendary. He was always calm, clear-headed and self-possessed. When others were rattled, Fermi was at his coolest. A famous story refers to the construction of the first atomic pile described above. Some of the scientists in Chicago feared a runaway chain reaction or some other nasty surprises in case Fermi had gotten his math wrong. Fermi, having started the reaction, made some quick mental calculations, reinserted the rods used to stop the reaction and calmly invited everyone to have lunch. Work resumed in due course.[38]

Another characteristic of Fermi was his amazing lucidity. Segrè narrated how Fermi wrote a physics textbook, *Introduction to Atomic Physics*, without consulting any reference books and with a pencil that had no eraser. Fermi is known for neutron research, for the Fermi–Dirac statistics, discovered independently in 1926, and for discovering the weak force in nature. His name is memorialized in the Fermilab, fermions (a class of particles), the Fermi (a unit of length) fermium (a radioactive element) and the Fermi Prize, among others.[39] We will encounter Fermi again later in the book.

◆

Bhabha was in Cambridge when he first made Nehru's acquaintance. The two must have known each other by 1938, since we find Nehru writing to the future defence minister of India, V.K. Krishna Menon, about Bhabha. When and where they first met is not known, but Indira Gandhi recalled that she first met Bhabha when she was travelling with her father to Marseilles by ship in 1938. It is also possible that the young Bhabha was introduced to Nehru at the house of his uncle, Sir Dorabji Tata, where many Congress leaders used to gather. Bhabha was one of two persons outside the family who called Nehru 'bhai'. The other was Jayaprakash Narayan.[40]

Bhabha and Krishna Menon had met in England but their relationship had never been particularly warm. Krishna Menon had a large fan following among students from wealthy families in India, who went to study in England in the mid-1930s. They included Mohan Kumaramangalam, Rajni Patel, Jyoti Basu and Nikhil Chakravarty. But Bhabha was never in awe of him. Nehru wrote to Krishna Menon in October 1938 about Bhabha.

> If you go to Cambridge, get in touch with Homi Bhabha who is a Fellow of Caius College. He is a brilliant physicist and has an intelligent appreciation of political events. I met him today for a few minutes as he was going to England on his way back from India. He has some information about some really good scientists who are at a loose end because of the anti-Jew movement...[41]

Krishna Menon and Bhabha never really got on well, even though both would become part of Nehru's inner circle. Both had king-sized egos, but apart from this, they had completely different personalities. A large part of the lack of closeness could also perhaps be attributed to ideological differences. Krishna Menon was too leftist for Bhabha's liking and Krishna Menon, in turn, felt that Bhabha was too right of centre. Menon was also against the development of nuclear weapons. Their paths often crossed later, but they never managed to ever like each other.

Chapter 7

COSMIC RAY RESEARCH

> It would be therefore more logical and also shorter to call the new particle a meson instead of mesotron.
>
> —Article in *Nature*, 18 February 1939[1]

Cosmic rays are very high energy particles that arrive from the depths of space and are incident on the earth's atmosphere. They are not rays of light but consist of high energy particles, such as protons, electrons and atomic nuclei. The outer atmosphere of the earth is continuously bombarded with photons and other subatomic particles. Cosmic radiation is quite different from the alpha and beta radiation emitted by radioactive nuclei—cosmic ray particles have much higher energies than alpha and beta rays and are more thinly spread. These two factors in particular have made the nature of cosmic rays much more difficult to discover.

When these cosmic ray particles reach the earth's atmosphere they collide with the molecules of air, triggering a domino effect wherein each collision produces more particles, each with lesser energy. The set of ensuing collisions produce a shower of secondary particles, most of which are lost in subsequent collisions but, among which, tiny particles (now called muons) survive and hit the ground. As we shall see, Bhabha pioneered the study of these showers and worked out the details of such interactions and the structure of an 'air shower' with Walter Heinrich Heitler of the H.H. Wills Physical Laboratory at Bristol.

IN THE BEGINNING

Cosmic rays were discovered in 1912 by the Austrian scientist, Victor F. Hess, who during the course of his balloon flights into the earth's atmosphere observed that there was a direct relationship between the intensity of the radiation and an increase in altitude. This led him to conclude that there was a powerful radiation originating in outer space that entered the earth's atmosphere and diminished as it reached the ground. Hess's experiments required personal attendance, which endangered the safety of the scientists.

During the First World War, Erich Regener, in Berlin, designed a series of self-recording electroscopes which could be sent up in unmanned balloon flights. His work made Berlin the centre of cosmic ray research in the world. The young Debendra Mohan Bose, the great J.C. Bose's nephew and student, came to Regener's laboratory in Berlin in 1914 on a travelling fellowship.[2] D.M. Bose, had a postgraduate degree from Christ College, Cambridge, where he had been taught by J.J. Thomson and C.T.R. Wilson. It was from the latter that he learnt how to build a cloud chamber, the device that enabled scientists to see the tracks of energetic charged particles for the first time. D.M. Bose returned to Calcutta as a lecturer in the City College and soon attracted the attention of Sir Asutosh Mookerjee, who offered him a Sir Rashbehari Ghose Professorship in physics. The Ghose Professorship came with a two-year travelling fellowship that brought him to Regener in Berlin and subsequently he was awarded a PhD for his work on cloud chambers. D.M. Bose was stranded there due to the First World War, but the Germans let him continue working in the laboratory, since Regener and Planck were guarantors for the young Indian, who was technically an enemy alien. He returned to India in 1919 but discontinued his work on cosmic rays. It would take him another two decades to return to this field of study.[3]

In the mid-1920s, Robert Millikan, the winner of the 1923 Nobel

Prize, who was famous for verifying Einstein's photoelectric effect, entered the field. With the help of his student, Ira Sprague Bowen, he invented a lightweight electrometer and an ion chamber that could be sent up in unmanned flights, communicating data to the ground using wireless technology developed during the First World War. Millikan was a forceful personality, a skilful popularizer and an impressive lecturer. He had a knack for memorable phrases: it was Millikan who coined the term 'cosmic rays'.

Dmitry V. Skobeltsyn, another young physicist working in the USSR, first discovered the tracks in a cloud chamber associated with cosmic rays in 1927–1928. A magnetic field exerts a force on an electrically charged particle so that its path becomes curved, the curvature being more for slower or lighter particles and less for fast or heavy ones. In some of the photographs Skobeltsyn took, he noticed a few tracks that were almost straight. These indicated values of momentum and energy that were much greater than those observed for an electron from any source known at the time. The real revolution in experimental cosmic rays came with the invention of the Geiger–Müller counter in 1928. The small and rugged instrument could be easily sent up in balloons obviating the need for risky manual flights.

Skobeltsyn did not follow up on his discovery, but in 1930, Carl D. Anderson, a student of Millikan's, started building a more powerful cloud chamber than the one used by the Russian. Millikan, who was also president of the California Institute of Technology, insisted that Anderson remain there to build a cloud chamber for studying this corpuscular radiation from space. Anderson discovered that cosmic rays contained both positively and negatively charged particles in about equal numbers. Millikan, who believed that the cosmic ray particles were electrons knocked from atoms by primary gamma rays, insisted that the positive particles must be protons knocked from atoms by high energy gamma rays. Anderson, on the other hand, thought that the tracks could be due to electrons moving upwards through the chamber rather than downward moving positive particles.

To settle the debate, Anderson placed a lead plate across the middle of the chamber. It turned out that both Anderson and Millikan were wrong. Anderson found a positive particle that was much lighter than a proton, whose ionization and curvature suggested a mass similar to that of an electron. He observed tracks of a particle whose $^{e}/_{m}$ ratio[4] was equal in magnitude but opposite in sign to that of an electron. This could not be the proton since its $^{e}/_{m}$ ratio is about 1,836 times smaller than an electron's. Anderson had, in fact, as noted before, discovered the positron—the antielectron that Dirac had predicted but whose existence had been received with incredulity by other scientists like Pauli and Heisenberg. This was the first observation of an elementary particle that does not reside inside the atom.[5]

The name 'positron' was suggested by an editor of *The Physical Review*, where Anderson published his paper. Within a year, Pauli and his student, Weisskopf, had written the first quantum field theory by recasting the tentative theory of the electron field introduced in 1928 by Pascual Jordan and Eugene Wigner. This marked the beginning of the new quantum electrodynamics (QED).[6]

BHABHA'S SEMINAL CONTRIBUTIONS

Bhabha Scattering

Dirac, as noted before, was Bhabha's teacher and hero, and the young physicist started thinking deeply about the new QED. At that time, however, a number of singularities[7] plagued the theory leading many to flatly reject it. Bhabha investigated positron interactions in a number of his papers using Dirac's 'hole' theory. He completed his third research paper, which he had been working on in Cambridge in collaboration with H.R. Hulme in Rome while working in Fermi's physics department. This paper dealt with the 'Annihilation of Fast Positrons by Electrons in the K-Shell' and was published by the Royal Society on 15 October 1934.

Cosmic rays have two components—soft and hard. The soft component is relatively less penetrating and consists of electrons and photons; the hard component is more penetrating. Bhabha's early work dealt with the soft component i.e., positron theory. Bhabha developed a theory of this process. He then turned to the reverse process, where electromagnetic energy materializes into an electron and a positron. The faith in the correctness of Bhabha's scattering formulas is such that they are now routinely used to calibrate beams in large accelerators using positron or other antiparticle beams.[8] His paper titled 'The Scattering of Positrons by Electrons with Exchange on Dirac's Theory of Positron' was published by the Royal Society on 2 March 1936. It was to become the crowning achievement of this group of papers on positron interactions and the process is now known as 'Bhabha Scattering'. This work got Bhabha instant fame and new respect in the eyes of his contemporaries.

From 1937, Bhabha held the Senior Studentship of the Exhibition of 1851 in theoretical physics. It was the first time that an Indian student had been awarded this studentship in an open competition. He could now continue his research work in Cambridge. In 1935, Bhabha had been awarded a doctoral degree in mathematics for his thesis, 'On Cosmic Radiation and the Creation and Annihilation of Positrons and Electrons'.

Bhabha-Heitler Cascade Theory

Bhabha's work on positron scattering would have ensured his lasting reputation, but he was far from finished. In what was his most fertile period at Cambridge, he made another important contribution in collaboration with Heitler, who was famous for propounding (with Fritz London), the quantum theory of the covalent bond. The two met when Bhabha visited Bristol in 1936. Heitler was an assistant to Max Born at the Institute for Theoretical Physics at the Göttingen University. When Adolf Hitler came to power in 1933, Born arranged for Heitler, who was a Jew, to get a position as a research fellow at the University of Bristol with the British physicist, Nevill Francis Mott.

Both Bhabha and Heitler had a common interest in high energy particles. It may be recalled that Bhabha's first publication, written from Zurich where he was visiting Pauli, was 'On the Absorption of the Cosmic Rays' in which he discussed the role of electron showers. Since that time, he had devoted himself to positron processes in QED. Heitler had also been working on similar subjects and had, in collaboration with Bethe, calculated the mean free path an electron could take in the presence of a mix of oxygen and nitrogen nuclei (i.e., air) to be about 2 km using QED methods. This, however, could not explain the large number of electrons at sea level. Many contemporary scientists felt that Bethe and Heitler's calculation was wrong because they had used QED, which was considered flawed at the time.

Bhabha and Heitler's cascade theory solved this problem. The incident high energy electron is scattered by a nucleus, which slows it down, effectively putting a brake on an electron while diverting its path. The decrease in the kinetic energy of an electron appears as *bremsstrahlung* (braking radiation). This high energy photon later produces an electron–positron pair. An electron thus produced creates another braking radiation photon. When a positron annihilates an electron, it gives rise to a pair of energetic photons and this process continues till the energy falls below a given threshold. This cascading effect produces a shower of electrons, positrons and photons, which widens as it moves downwards.

In a cloud chamber, the whole shower is compressed because lead has a very high density, but in nature where the primary electron is incident on the upper atmosphere and the air has very low density compared to lead, the shower can reach the ground and extend over a large area. The Bhabha–Heitler model, to a large extent, successfully explained experimental observations. These air showers form the bedrock on which all cosmic ray studies are based. Bhabha went on to refine this theory upon returning to India in 1939. In fact, J. Robert Oppenheimer and J.F. Carlson had also been working on the same problem on the other side of the Atlantic, and had arrived at the same conclusions with very similar methodology. However, they lost priority since their work appeared a year later.

MESOTRON OR MESON?

Anderson was unsure about the origin of positrons in cosmic rays. The answer was provided by Patrick Blackett and Giuseppe 'Beppo' Occhialini working at the Cavendish. Positrons were created by the cosmic radiation itself. Their many photographs provided dramatic evidence that positrons are produced by the collision of cosmic rays. After this initial success, the duo also observed positive and negative electrons, which originated in the material immediately above the chamber. They were seeing pair production for the first time. The tracks in the cloud chamber showed particles with positive or negative charge but which were far more penetrating than electrons or positrons.

Anderson also noticed particles that curved differently from electrons and other known particles when they were passed through a magnetic field. They were negatively charged and curved less sharply than electrons but more sharply than protons for particles of the same velocity. It was assumed that the magnitude of their negative electric charge was equal to that of the electron. So, to account for the difference in curvature, it was assumed that their mass was greater than an electron but smaller than a proton. At the California Institute of Technology, in 1936, Anderson and his colleague, Seth Neddermeyer, were convinced that the penetrating particles were something new. In November of that year, they publicly presented the evidence that these particles had a mass between that of an electron and proton. They called them 'mesotrons', based on the Greek word for 'middle'. This is the particle that we now know as the muon. J.C. Street and E.C. Stevenson's cloud chamber experiment confirmed the existence of muons in 1937.

In November 1938, Anderson and Neddermeyer wrote to *Nature* about their views on the name of the new particle, suggesting the word mesotron (intermediate particle).[9] Anderson calculated that the mass of the muon would be more than 200 times that of an electron, and as with a positron, there already existed a theory that could account for the mesotron, or so it seemed. One year before the discovery of the mesotron, a Japanese theoretical physicist, Hideki Yukawa, working in Kyoto, had developed a theory of the strong force that binds the atomic nucleus.

Yukawa hypothesized that the nucleus, which contained protons and neutrons, was held together by a different kind of force (the strong force), with properties different from electromagnetic and gravitational forces, the only two types of force known at the time. In modern physics, all forces are thought to be 'mediated' by a particle. The more massive the mediating particle, the smaller is its sphere of influence. Yukawa estimated that the particle responsible for the strong force would be about 250 times the mass of the electron and about $^1/_7$ as much as a proton. When the mass of Anderson's mesotron was found to be only 20 per cent away from his prediction, Yukawa claimed the particle as his own. Oppenheimer and Robert Serber at Berkeley also seemed to think that Yukawa's particle had been found and promoted his theory in the West.

For a few years, it seemed that the mesotron was the particle Yukawa had predicted. However, this only lasted until three young Italians—Marcello Conversi, Ettore Pancini and Oreste Piccioni—made a surprising discovery in a remarkable experiment that they had begun secretly in Rome during the Second World War.[10] They set up their apparatus in a makeshift laboratory from a basement near Vatican City where they were hiding from the occupying Germans. Their experiment clearly showed that Yukawa's particle was not Anderson's mesotron but something different.

The complete picture was only revealed in 1947 when a particle perfectly fitting Yukawa's description, known as the pion, was finally discovered by Cecil Powell and his group at the University of Bristol using photographic emulsions. In the late 1940s, the development of special photographic emulsions that could be carried in balloons made the first images of the interactions of high-altitude cosmic rays possible. These emulsions were especially sensitive to high energy particles whose presence could be detected by a line of dark specks that they formed on the developed emulsion.

Powell, who had perfected the use of emulsions, had been a student of C.T.R. Wilson at the Cavendish Laboratory and moved to the University of Bristol in 1928. He found that the photographic technique was simpler and gave more accurate results than a cloud chamber. At the end of the war, he extended the technique to study cosmic rays at high altitudes. By that time, Occhialini had also come to Bristol on Blackett's recommendation. In May 1946, an emulsion using eight times the silver bromide normally used was developed and this was used to good effect by Powell and Occhialini. The latter took some plates to the Pic du Midi Observatory in France, and the results stunned physicists by revealing a whole new world. Trapped in the emulsions examined in 1947 was evidence for a new particle. Like the mesotron, it had a mass between that of an electron and a proton, but it was slightly heavier than Anderson's particle. The tracks revealed that the new particle could decay into a particle like the mesotron. Powell and his team had at last discovered Yukawa's particle—the particle we now know as the pion.[11]

Soon afterwards, the mesotron was renamed the muon. The muon is a result of pion decay. However, for 10 years, it had remained hidden and was mistakenly believed to be Yukawa's particle. Atomic nuclei from outer space hit the upper atmosphere and produce pions. These soon decay to produce showers of negative and positive muons. However, muons are not stable particles. Negative muons decay into electrons, neutrinos and antineutrinos. Positive muons decay into positrons, neutrinos and antineutrinos. All this happens when muons are at rest. However, when muons are in motion, like all particles, they experience the subtle slowing of time inherent in Einstein's Theory of Relativity. As we shall see later, it was to Bhabha's great credit that he recognized this effect, something that tends to get forgotten. For the two particles with intermediate mass, the more general term meson was adopted to refer to any particle in such a mass range. Yukawa received the Nobel Prize in 1949; Powell was honoured the following year.

So, where does Bhabha figure in all this? As we have seen, Bhabha's paper on cascade showers with Heitler dealt with the soft component of cosmic rays that were composed of electrons and positrons and could be explained by the Dirac model. In 1937, Bhabha, as noted before, was awarded the 1851 Exhibition Scholarship for continuing his work in theoretical physics. In the same year, he examined the 'penetrating component' of cosmic rays. Bhabha found that cosmic radiation consisted of ionizing particles that penetrated more than a metre of lead. Since it was believed that no electron could make its effects felt through more than about 15 cm of lead, it became clear to Bhabha that the phenomenon he was witnessing could be attributed to two factors. Either the theoretical formulas for the energy loss of fast electrons broke down above a certain critical energy or the penetrating component did not consist of electrons.

At the end of his paper, Bhabha argued that 'the majority of the penetrating particles must have masses nearer to a hundred times the electron mass rather than ten times the same'.[12] In this bold conjecture, Bhabha was predicting the existence of a meson (Anderson's mesotron) purely by analysing cosmic rays. He was working independently of both Anderson and Yukawa and knew nothing about the developments in this field. When he predicted the meson in October 1937, he was not aware of Yukawa's particle, which was brought to his attention by Heitler when he sent his first note on meson theory for publication. It was natural for Bhabha to also identify Yukawa's particle with the one indicated in his own research.

Bhabha suggested that the particle be called meson and not mesotron. In a footnote to an article he sent to *Nature*, which was published in February 1939, he gave his reasons for the recommendation.

> The name 'mesotron' has been suggested by Anderson and Neddermeyer [...] for the new particle found in cosmic radiation with a mass intermediate between that of electron and proton. It is felt that 'tr' in this word is redundant, since it does not belong to the Greek root 'meso' for middle, the 'tr' in neutron and electron belong, of course, to the roots, 'neutr' and 'electra'. In these circumstances, it seems better to follow the suggestion of Bohr and to use electron to denote particles of electronic mass independent of their charge. It would be therefore more logical and also shorter to call the new particle a meson instead of mesotron.[13]

Bohr also seems to have had a role in approving the name of the new particle. Initially, it seems that he had called the particle a 'yukon', a name which did not appeal to the Americans. Millikan wrote to Bohr giving his views on the subject. Bohr, in his reply, which was published in the *Physical Review*, said that at a meeting in Copenhagen where Pierre Auger, Bruno Rossi, Fermi and Heisenberg were present, everyone had agreed that it should be called the mesotron.[14] It seems that, originally, Anderson and Neddermeyer had suggested mesoton as the name of this new particle, but Millikan had insisted that it be called mesotron.

Bhabha had written to Bohr in mid-December 1938 about his footnote in the article in *Nature*, saying that Dirac and other physicists at Cambridge all preferred meson, but in case Bohr did not approve of it, he would change it in the proof. It is not known what Bohr's reply was (if there was one), but, somehow, the name meson gained sufficient currency to stick. The Swiss–British physicist Nicholas Kemmer recalled that, initially, there was some confusion. Papers on the particle discovered by Anderson—now called the mu-meson—referred to the mesotron, which had been Anderson's original term, while papers on Yukawa's theory referred to the meson. The original name given by Yukawa—'heavy quantum'—was felt to be provisional and too general for permanent use. Informally, the particle was known as the yukon, but Kemmer did not recollect this term ever being used in print. Kemmer said that it was at the home of Swiss-born chemist and nuclear physicist, Egon Bretscher, at Cambridge that Bhabha, Maurice Pryce and himself decided to use the word meson, which finally prevailed against the mesotron because it was also accepted by Powell.[15]

Time Dilation

Albert Einstein in his Special Theory of Relativity showed that as a body approaches the speed of light, time appears to lengthen. For example, to an object moving at 98 per cent of the speed of light, one second becomes five seconds long. This phenomenon is called time dilation. The effects of speed on time are insignificant at speeds encountered in daily life, but they assume significance as the speed of light is approached.

Bhabha's association with the meson would not be complete without mentioning his contribution to describing the effect of time dilation associated with it. At that time, as noted, it was not known that there were two particles—the pion and the muon. Bhabha thought of these particles in an unusual way—not as elementary particles but as tiny rapidly moving clocks. Each tick of this clock carried it closer to the moment of its decay. But, according to Einstein's Special Theory of Relativity, clocks in motion ticked more slowly than those at rest. This is known as relativistic time dilation.

Special relativity suggests that, for an observer in an inertial frame of reference, a clock moving relative to him will be measured to tick slower than a clock that is at rest in his frame of reference. The faster the relative velocity, the greater the time dilation among the two, with the rate of time reaching zero as one approaches the speed of light. This causes massless particles that travel at the speed of light to be unaffected by the passage of time. Thus, Bhabha was the first to point out that cosmic rays, moving at close to the speed of light, are the best place to verify Einstein's Special Theory of Relativity. Mesons, which would normally decay within a distance of half a kilometre, can travel over 10 km down to the earth's surface because relativistic time dilation increases their decay lifetime. It was Bhabha's great insight to posit that cosmic ray muons constituted just such high speed 'clocks' and that, if relativity was correct, their rate of ticking would also be slowed in such a manner. '[...] the U-particle may be described as a "clock", and hence it follows merely from considerations of relativity that the time of disintegration is longer when the particle is in motion.'[16]

When Einstein formulated his theory, there had not been the slightest possibility of testing it experimentally, since it would have meant accelerating a clock to impractical velocities. The notion of the slowing down of time around the speed of light, while widely accepted, had remained unproven until then. Bhabha's insight provided the first experimental test of Einstein's prediction by measuring the variation of the lifetimes of these mesons in the laboratory frame. It is now a textbook example of how one applies the Special Theory of Relativity but has become so commonplace that its originator has been forgotten. Bhabha, together with a select band of British scientists, including Kemmer, Herbert Fröhlich and Heitler, was among the pioneers on meson theory outside Yukawa's group in Japan.[17]

Bhabha's confidence in the existence of a penetrating particle with a much heavier mass than that of an electron won him an admirer and friend in the British physicist, Patrick Blackett. At a weekend conference in Manchester, Blackett refused to believe that there could be a particle heavier than an electron in the hard component of cosmic rays. But Bhabha stuck to his guns and, later, Blackett was forced to concede that he had been wrong. Blackett worked under Rutherford at the Cavendish for 10 years and the latter asked him to use a cloud chamber to find visible tracks of this disintegration. By 1925, Blackett took thousands of photographs showing tracks of ionized particles. Eight of these were forked, and this showed that the nitrogen atom-alpha particle combination had formed an atom of fluorine, which had then disintegrated into an isotope of oxygen and a proton.

The London-born Blackett served in the navy and won the Nobel Prize in physics in 1948. He left the Cavendish to join Birkbeck, a workingmen's college, after disagreements with Rutherford over the teaching load at the Cavendish. He was also actively connected with India, becoming Nehru's advisor on the research and development needs of the Indian armed forces. From the late 1940s, he was a frequent visitor and advisor on military and civil science for 20 years. Blackett met Bhabha, 12 years his junior, in Cambridge when Bhabha was studying under Dirac. The two knew each other well and became fast friends later in life.

THE KAPITZA CLUB

Bhabha had made his mark and was being noticed. The appreciation of his peers was reflected in the fact that he was asked to join the Kapitza Club, which was a group of physicists who met informally in the 1920s and 1930s in Cambridge. The group was founded by the Russian physicist Pyotr Leonidovich Kapitza, a favourite of Rutherford, in October 1922. Kapitza, was a colourful character who did not speak any language well, including it is said, his own. But this did not prevent him from amusing his colleagues with card tricks and narrating amusing stories in 'Kapitzarene', a language that roughly consisted of Russian, French and English in equal parts.[18]

Kapitza noticed what he considered to be the excessive deference of British physics students to their seniors. Appalled at the formal, excessively hierarchical structure of Cambridge, he started a club where rank did not matter. It was a weekly seminar on a topical subject in physics. It was an invite-only club, and its membership was both limited and coveted. The rules of the club were that a student could become a member only by giving a talk and that his membership would be withdrawn if he missed a few meetings. The talks usually took place in Kapitza's room in Trinity College on Tuesday evenings after dinner. The speakers, normally volunteers from the club's members, spoke with the aid of a piece of chalk and a blackboard mounted on an easel.[19]

Kapitza frequently opened discussions with deliberate howlers, so that even the youngest member could speak up to correct him, thereby loosening the grip of tradition. Bhabha spoke at this club on 1 February 1938 on 'The Heavy Electron'. Cockroft has made a special mention of this lecture in his memoirs. The membership of this elite club was important in ways that went beyond scientific work. Bhabha's acquaintance with the best scientific minds in the world proved to be invaluable to him as an institution builder and science organizer later in his life. Kapitza left Cambridge in 1934 never to return on a permanent basis. He was not allowed to return by the Soviet authorities, and his original equipment, which he used in his high magnetic field experiments at the Mond Laboratory, Cambridge, was sold to the Soviets.

It was rumoured that many revolutionary discoveries were first announced informally in these meetings. It is said that James Chadwick (when he discovered the neutron in 1932) announced his results initially in a meeting of the Kapitza Club. After a very brief address (he had worked continuously for 10 days), he said: 'Now I want to be chloroformed and put to bed for a fortnight.'[20] The club, ceased to exist in the spring of 1958, superseded by seminars. It was reconvened on 10 May 1966, when Kapitza visited Cambridge, after a hiatus of over two decades. This reunion allowed some of the club's surviving members, including Dirac and Cockcroft, to not only meet one last time but also to have Kapitza formally close it. The venue was a common room in Gonville and Caius, where the participants sipped fine wines, in contrast to the meetings 40 years before, when there had been only dishwater coffee on offer.[21]

THE AUSTIN WING

The Cavendish Laboratory had a tradition of hosting an annual dinner that was organized by its students every year. The last such dinner before the Second World War started was held on 19 December 1938. Herbert Austin, the first Baron Austin, who had announced a major benefaction of £250,000 to build what came to be known as the Austin Wing of the Cavendish Laboratory in 1936, was the chief guest. Sir Maurice Vincent Wilkes, the British computer scientist, recalls Bhabha's role at this dinner. 'This occasion provided the young Bhabha with unrivalled opportunities for his high spirits. His most striking contribution was a full-sized portrait of Lord Austin which he had painted himself and which he ceremoniously presented to Lord Austin at the dinner. I was told that it was painted on brown paper.'[22]

Wilkes recalls 'another frolic' the same evening which he felt had every sign of originating from Bhabha. It had to do with the fact that the word 'wing' has two meanings in English; it is both an architectural term and also describes a part of a car, particularly an Austin car. As Lord Austin had presented the building known as the Austin wing to the laboratory, it was thought fitting to present him with the other sort of 'wing' in return.[23] Accordingly, a wing from an old Austin Seven car, duly painted, was presented to him at the dinner. What Lord Austin thought of this gift remains unknown.

Sir William Lawrence Bragg, who had by then succeeded the late Rutherford as head of the Cavendish, has his own story to tell of this dinner.

> At the Cavendish dinner in December 1938, Lord Austin was our chief guest. He stayed with us and I remember that the boys were very worried because our car was a Morris [...] At the dinner, Bhabha unveiled an oil painting which he had done (without a model) of me, in the grand eighteenth-century style, with velvet curtains partly concealing a vista.[24]

This painting, which is slightly damaged, is currently located in the museum storeroom of the laboratory.

Another Indian, Rappal Sangameswaran Krishnan, a protégé of Raman, of whom we shall hear more in a succeeding chapter, was in Cambridge at this time and it is possible that he met Bhabha for the first time at this dinner. The construction of the new building was completed in 1940 but it was used for war work. It was handed back to the university in 1945. The structure of DNA was discovered in the Austin Wing by Francis H.C. Crick and James D. Watson, for which they and Maurice H.F. Wilkins were jointly awarded the 1962 Nobel Prize for Medicine.

◆

The esteem that Bhabha was held in by his seniors is highlighted by an interesting episode. Chadwick wanted to hire a theoretical physicist for the post of reader at Liverpool where he was setting up a new physics department. Just before his departure for India, and on the eve of the Second World War, Bhabha applied for the post. He was turned down because Chadwick considered him too good for the job.

> Bhabha was too good. The post required some teaching to be done, and the quality of students was not the quality of students in Cambridge, and I thought that much of it would be drudgery to a man like Bhabha, who was a most exceptional man. He was a painter and a poet and had extremely wide interests...not merely interests but far more than that—and I did not feel that however much I like him it was fair.[25]

However, this is only part of the story. Chadwick also rejected Heitler for the same post since, as he put it, 'Liverpudlians would never understand why they were being taught physics by a German during the Second World War.'[26] Bhabha's appointment presented a slightly different predicament. If Bhabha was good enough to teach Liverpudlians physics then, presumably, the Indians were also competent enough to run their own country.[27] The position ultimately went to Maurice Pryce.

◆

Blackett offered Bhabha a position at Manchester, but the latter left England in the summer of 1939 to holiday in India. Before leaving, he met K.S. Krishnan in Cambridge in June that year. They had been introduced by Chandrasekhar and the three Indian physicists had lunch together on 22 June 1939.[28] Krishnan and Bhabha later served on the AEC together and became good friends later in life. Bhabha visited

Krishnan at his home in New Delhi, enjoying a typical South Indian dinner with good Carnatic music playing in the background on the latter's tape recorder.

Bhabha found himself stranded in India with the outbreak of the Second World War in September 1939, forcing him to replant his career in unfamiliar soil. That it flourished in the thinnest of soils is part of our story. Bhabha's scientific research at Cambridge was of the highest class. Had he done just this and little else, he would have been remembered as a physicist of some importance—certainly not as revolutionary as Einstein, and perhaps not associated for perpetuity with a single shining discovery but as someone who substantially contributed to his field by the accretive power of daily work.

His career overlapped tidily with the bloom of Indian independence and with the effort of the new state to harness the power of science and technology in the service of the nation—efforts that Bhabha masterminded, and which met with substantial success. Thus, Bhabha is remembered more as a science administrator par excellence, even though his scientific contributions in the early part of his career were of the highest class. He was, as we shall see later, nominated for the Nobel Prize in physics on five occasions but never won the coveted prize. He did, however, have somewhat better luck as a nominator of the prize.

Chapter 8

RETURN OF THE NATIVE

> If you complain about being isolated in America you can imagine what it's like in India. There is not a single person with whom I can talk any fundamental problem.
>
> —Letter to Wolfgang Pauli, 4 June 1943[1]

When Bhabha set out for his vacation to Bangalore in the summer of 1939, he would never have guessed that the Second World War would make a return to Cambridge impossible. C.V. Raman took Bhabha under his wing at the IISc and he was appointed as special reader in its physics department in November 1939. He started teaching in January 1940 and five months later, the Sir Dorabji Tata Trust gave him a grant of ₹5,000 to support his research on cosmic rays.

It is clear from Bhabha's correspondence during this time that he did not think that his stay in India would be a permanent one—it was a pause necessitated by the war, a temporary interference in his plans, all of which were centred on being abroad. At that time science institutions in India were of little interest to him: they were below par and with hardly the kind of environment that would make a young man passionate about science.

The return to India was crippling for Bhabha. The IISc was no

Cambridge or Princeton. Indian science was both underdeveloped and underfunded. Bhabha, so used to being at the vanguard of progress, found himself at the rear. While researching at the Cavendish, the pinnacle of British science, Bhabha had been granted access to this apotheosis of western progress. We can only imagine how gratifying those Cambridge days must have been for him: he could remove himself from the shackles of his brown skin and Indian birth, and join the relatively egalitarian universalism of the western scientific community. As Keshav Dimri has observed, 'Here was the engine of empire and power, and there was Bhabha, close enough to feel its heat.'[2] Bhabha's return marked a turning point in his life and made way for new thinking about scientific institution building, citizenship and art.

Judging by his own description, Bhabha's decision to stay in India was more pragmatic than ideological. In fact, had a position in a famous university been forthcoming, perhaps he may not have stayed at all. Bhabha's strong desire to be in Europe or America is hardly surprising. He was, at least at that stage, an unabashed Anglophile. That his love of the West should extend beyond just its art was only to be expected. By the early 1940s, the Indian independence movement had acquired great momentum. So had the Second World War, and it was not the ideal time for discussions on science or for conducting scientific experiments. While he missed the lively discussions at the Cavendish and longed to be part of the international scientific community, he managed to work on relativistic equations of higher spin, which came to be known as the Bhabha–Corben equations.

Bhabha had already begun some work on this problem with H.C. Corben before leaving Cambridge and he continued working on it in Bangalore. We saw earlier that particles of matter have a property known as spin. Quantum theory dictates that particles are constrained to only spin at certain rates specific to each kind of particle.[3] Electrons and protons have spin ½ and photons have spin 1. Bhabha wanted to generalize Dirac's work to the case of spinning point particles.[4]

A preliminary note on this work was sent to *Nature* on 17 March 1940, his first from Bangalore. The definitive paper on the Bhabha–Corben equations was titled 'General Classical Theory of Spinning Particles in a Maxwell Field'.[5] As the mathematical physicist William Penney wrote, 'The theory was constructed with great generality and mathematical exactness within the framework of the Special Theory of Relativity. While this work at Bangalore was less recognized internationally, it appeared to give him greater intellectual pleasure than the earlier and more celebrated work done whilst in Europe.'[6]

One of Bhabha's correspondents at this time was Pauli. In April 1940, Pauli wrote to Bhabha voicing concern over Bohr and his institute in Copenhagen. 'Let us hope they are not menaced personally—but in any case, the Germans will eat every food [*sic*] of this country (paying it with some printed banknotes),' he wrote. He also said that Kemmer was in England but was being treated well despite being a German citizen and 'Aryan'. Pauli wanted to know if there was some position for Kemmer in India. He ended with, 'If you are going to America, I think Berkeley and Pasadena would be the best places.'[7]

Pauli wrote again in May 1940, saying that they still had no news of Bohr. He told Bhabha that he had received an invitation for a visiting professorship at Princeton and would try to go there as soon as possible but was not too sure he would be able to succeed.[8] In another letter written in mid-October, Pauli asked Bhabha about his plans since there was confusion in people's minds about them. He said that he had heard contradictory news about Bhabha coming to America. Bloch had written to him to tell him that Bhabha had accepted an invitation from Millikan to come to Pasadena, whereas Oppenheimer had told him that it had been called off. He said that it made no sense for him to invite Bhabha to Princeton without knowing his plans and that he would be very glad to be in the same place as Bhabha and to work with him.[9]

In November 1940, Bhabha wrote to Pauli about his calculations on the scattering of mesons by the spin of heavy particles.[10] Pauli was in Princeton at the time and, in the same letter, Bhabha expressed his desire to work with Pauli there or, for that matter, in Berkeley. There is a hint of desperation in his plea. He seems stifled in Bangalore with no one with whom he could discuss physics.

Bhabha said that nothing would please him more than being able to work with Pauli at the IAS in Princeton. If that didn't work out then he would be happy to go to the California Institute of Technology or to Berkeley if a definite offer was made to him. He asked Pauli to be so good as to intimate this to J. Robert Oppenheimer whom he unfortunately did not know personally.[11]

Bhabha did get to know Oppenheimer well but that was only after the Second World War had ended. Oppenheimer, who was born five years before Bhabha, is best remembered today as the father of the atom bomb. Bhabha also got on famously with Kitty, Oppenheimer's wife. Oppenheimer went to Harvard, where he completed a four-year chemistry programme in three years, graduating summa cum laude in 1925. He went on to study subatomic physics at the Cavendish but didn't last long there. Unsuited for performing experiments, he was clumsy in the laboratory and experienced a mental breakdown.

He had a happier time at Göttingen University in Germany, where theoretical physics dominated. This was great for 'Oppie' (a nickname he picked up there) who was awarded his PhD in 1927 with distinction. His next goal was to bring what was then called 'new physics' back to the US. Back home, he taught at two schools: the University of California at Berkeley and the California Institute of Technology. He attracted many of the smartest students, who all seemed to adore him: they went so far as to follow him when he took his summer retreat. Bernard Peters, who came to TIFR in August 1950, was Oppenheimer's student.

Oppenheimer made an original contribution to early quantum mechanics and published one of the first theoretical models of a black hole, both contributions that tend to be overlooked. Oppenheimer, a chain smoker, was a charismatic teacher who inspired a generation of American physicists. He was the director of the IAS, Princeton, for almost 20 years, when giants like Albert Einstein, John von Neumann and Kurt Gödel were in residence. The tall, nervous Oppenheimer who moved with an odd gait, his head always a little to one side, one shoulder higher than the other, was a highly sophisticated, cultured man. (Bhabha had a left shoulder tic and was also known for his general cultivation and sophistication). 'Oppie' studied Sanskrit and read the Bhagavad Gita as extra-curricular activities. But he was also cold, elitist and arrogant. And it was ultimately this arrogance that proved to be his undoing.

Pauli's reply took five months to reach Bhabha. This was not surprising given the distance involved and the dislocation caused by the war. But Bhabha was very keen to continue the epistolary connection with Pauli. He liked to be at the centre of things, and his scientific exile in Bangalore (as he initially viewed it) resulted in his being out of touch with the vibrant scientific community to which he had once belonged. He hoped Pauli would write to him regularly, as he valued their correspondence very much.

They continued to correspond and, in one of his letters, Pauli thanked Bhabha for the drawing the latter sent him. Pauli then confided in Bhabha that he wrote poems and short prose pieces as a hobby. 'If you have your painting and drawing so I have hobby to write poems or short stories in prosa [*sic*]. However, I don't want to go public with it.'[12] In the same letter, he hoped that the war would not disrupt Bhabha's activities too much. At Princeton, he said, there were only a small number of aliens who continued to do scientific work because 'physicists get other things to do'.[13] A few months later, in August 1942, he reiterated his views. 'Physics is going to be more and more reduced in this country in favour of military work. It is not a pleasant time and nobody can predict how long the war will last.'[14]

In April 1940, Saha invited Bhabha to deliver a lecture at Calcutta University. He said that he and his students were very keen to meet him and have a full discussion on all matters concerning cosmic rays. Saha, it appears, wanted to familiarize the young Bhabha with the physics scene in Calcutta. Bhabha accepted the invitation and delivered the lectures in December 1940. Saha published a short account of Bhabha's lecture series in *Science and Culture*, an article that was vetted by Bhabha's father, Jehangir. The handwritten document with his notes is still intact and available at the TIFR archives.

There is as yet no evidence of Bhabha's nationalist obligation as official accounts often claim. At this time, all he was looking for was a quick return abroad, where all the action was. He confided in Millikan that Mysore University was thinking about creating a special chair for him. 'Its main function will be to promote research in the University and there will be considerable funds available for experimental research should I decide to undertake any.' 'Moreover',

Bhabha added, 'there will be no routine teaching attached to this chair, which for India is a very welcome exception.'[15]

In an undated letter to Millikan (perhaps written in June 1940), Bhabha expressed a wish to work with him at the California Institute of Technology. Millikan visited Bangalore in 1940 and met Bhabha, who wrote: 'There is no place I should like to visit with more eagerness than the United States, and I sincerely hope that some definite opportunity may arise in the not very distant future to make it possible. In particular, I should find working at your Institute very stimulating and trust that an occasion may occur not long hence which will enable me to do so.'[16]

In March 1941, he wrote to Millikan about the paucity of foreign exchange due to restrictions imposed by the Second World War. He said that the restrictions would not allow him to take sufficient money out of the country to meet his expenses in the US. However, he was confident about getting a passport even though the war had made it a more difficult proposition.[17] Bhabha yearned for the end of the war so that he could freely resume his scientific activity. Rumours about him leaving for the US had reached Cambridge and the cosmic ray physicist, Hugh Carmichael, wrote to him from the Kew Observatory asking him to come back to England and 'apply your brains to applied mathematical war science in this country? Revolutionize weather forecasting for example!'[18]

GETTING STARTED

As noted earlier, by mid-1940, Bhabha had been assured of a grant from the Sir Dorabji Tata Trust to start a unit for cosmic ray research. He wrote to Pauli about how he had the funds for high-altitude flights. He told him that he wanted to repeat some of Millikan's experiments. This was a significant divergence from the work for which Bhabha was known. At Cambridge, he had been the odd-man out—a theoretical physicist among a plethora of very famous experimentalists. As Itty Abraham observes, 'Now Bhabha was engaging directly in experimental work, expanding his repertoire of experience and expertise, and working towards providing the experimental proofs of his theoretical work: closing a scientific circle, so to speak.'[19]

Part of the reason for this switch was the fact that the experimental data that had been available to him in Cambridge was no longer at his disposal. This deficiency had initially made his work in Bangalore more mathematical, until, in 1943, he turned to experiments himself. At least some of the inspiration for this must have come from Millikan's visit to Bangalore from December 1939 to March 1940. Millikan conducted cosmic ray experiments using the balloon technique with his collaborators, Victor Neher and William Henry Pickering from the California Institute of Technology. He was nearly 72 at the time, and had won the Physics Nobel in 1923 for his determination of the charge of an electron. He was chairman of the executive council of the California Institute of Technology and director of the Norman Bridge Laboratory.

Millikan and his colleagues conducted a series of rubber balloon flights in Bangalore, Agra and Peshawar to study the latitude variation of primary cosmic rays near the geomagnetic equator. Closer to the equator, there are fewer background low energy particles to confuse the photographic plates carried by the balloons. The Indian Meteorological Department (IMD) was responsible for the hydrogen supplies and laboratory facilities for the Millikan experiments. Flights were launched in early 1940 from the heart of Bangalore. Millikan's visit was a wonderful opportunity for Bhabha and his colleagues to familiarize themselves with the balloon technique and conduct their own experiments.

In Bangalore, Bhabha initiated activities on cascade theory in collaboration with the Department of Mathematics of Central College, particularly with K.S.K. Iyengar, who had a Mathematical Tripos from Cambridge, and B.S. Madhava Rao. Bhabha took advantage of the availability of a B-29 aircraft in Bangalore, which had been brought there during the Second World War.[20] The first subject he decided to study experimentally was the latitude effect on mesons. The soft component of cosmic rays, consisting of electrons, positrons and gamma rays was described accurately by cosmic ray shower theory. The hard component of cosmic rays, consisting mainly of mesons, was much less understood and needed experimental observations. He studied the variation of the penetrating component as a function of altitude in collaboration with S.V. Chandrasekhar Aiya of the Department of Communication Engineering at the IISc. He also constructed a cloud chamber of 12-inch diameter to study the scattering properties of mu-mesons in collaboration with M.S. Sinha.[21]

In order to devise an experimental setup to study the hard component of cosmic rays, it was necessary to devise procedures to discriminate between the soft and the hard components experimentally. Bhabha studied this problem in 1943. Based on his detailed work on cosmic ray showers with S.C. Chakrabarty, he concluded that the usual method of separating the two components of cosmic rays by interposing absorbers such as lead of different thicknesses and measuring the cosmic ray absorption was not sufficiently reliable. Bhabha devised a new method (Bhabha method) for this purpose.[22]

With Chakrabarty, Bhabha further explored cascade theory, improving on the theoretical work he had done with Heitler to bring it closer to the observational data that had become available. He was closely following the work done by American physicists, Hartland Snyder and Robert Serber, and his work in collaboration with Chakrabarty can be considered an analytical continuation of their work. The result was a paper titled 'Calculations on the Cascade Theory with Collision Loss', which was published in the *Proceedings of the Indian Academy of Sciences* in 1942. It was republished in a slightly different form under the title 'The Cascade Theory with Collision Loss' in 1943, in the *Proceedings of the Royal Society*.

SARABHAI AND RAMANNA

Assisting Millikan during his time in Bangalore was a young man from Ahmedabad who would go on to achieve greater things: Vikram Sarabhai. Ten years younger than Bhabha, Sarabhai also completed his PhD from Cambridge. He was finishing his Mathematical Tripos at Cambridge when the Second World War broke out. He returned to the IISc in 1940 and Raman offered him a position at the IISc. Sarabhai was interested in cosmic rays but it is pertinent to note that he did not attach himself to Bhabha, who was already an international figure, preferring instead to work alone. He authored six papers on the latitude effect[23] during the six years of the war. When the war ended, he returned to Cambridge to collect his degree and submit his PhD thesis.

Bhabha and Sarabhai bonded well despite their age difference, sharing as they did patrician backgrounds. Both were good-looking—Bhabha with his swarthy complexion and brooding mien, and Sarabhai with his delicate features and almost translucent complexion. The scientist, S. Ramaseshan, who was Raman's nephew and biographer, said that Sarabhai had the skin of a baby—it actually glowed.[24]

Bhabha and Sarabhai were multidimensional men with interests outside science and a liking for the good things in life. After work, they would visit the posh West End Hotel to meet friends, mainly local, self-styled intellectuals. This group included an attractive Sri Lankan woman, Anil Marcia de Silva, who was a journalist, biographer, art historian and founding member of the Indian People's Theatre Association (IPTA), which was closely allied with the Communist Party of India. The daughter of George E. de Silva, a Sri Lankan politician, her free and liberal lifestyle invited censure in Sri Lanka, which was perhaps, too conservative to countenance such behaviour. She set up the People's Theatre in Bangalore in 1941 and assisted in the formation of IPTA in 1942 in Bombay. Bhabha suggested the name 'People's Theatre', inspired by Romain Rolland's book on the same theme.

These apparently playboy-type outings scandalized the staid students, many of whom were conservative Tamil Brahmins, and also made them envious.[25] Bhabha stayed at the West End whenever he visited Bangalore in later years. He and Sarabhai attended classical music concerts on Sunday morning at Basavangudi. It was in Bangalore that Sarabhai met Mrinalini Swaminathan, an accomplished Indian classical dancer whose commitment to her art rivalled his commitment to science. Sarabhai and Bhabha decided to host a dance performance to raise funds for charity. Sarabhai approached the acclaimed Bharatnatyam dancer, Ram Gopal, for help and the latter suggested Mrinalini. However, this was not their first meeting. Sarabhai had met

Mrinalini when she had been a teenager, but he had been put off by her apparent frivolity at the time.[26]

Bhabha also knew Mrinalini well—they were tennis partners. In fact, according to Vikram and Mrinalini's daughter, Mallika Sarabhai, Bhabha introduced her father to her mother.[27] Many years later, Mallika narrated, 'On their first evening out, Papa clarified to Amma, "I am committed to science and never want to marry." To his surprise, Amma looked relieved and said, "Thank God. I am so tired of being proposed to. I too want only to be a dancer. Now we can be friends."'[28] Destiny, though, had other plans: they were married six months later. The close ties between Bhabha and Sarabhai continued long after they had both left Bangalore. Almost two decades later, Sarabhai would create the space programme under the umbrella of the DAE. When Bhabha died in a plane crash in the Swiss Alps in January 1966, Sarabhai inherited his mantle as head of the atomic energy programme.

Bhabha also embarked on his own journey of cultural exploration, visiting the temples at Halebid and Belur, marvelling at their architecture. During his five years in Bangalore, he gradually established a deep feeling of identity with his country. As he became conscious of the great cultural heritage and tradition of India, he also began to realize the possibility of economic prosperity and social change based on science and technology as demonstrated in the West.[29] Later, when he moved to Bombay, he continued to attend concerts of Indian classical music. A society in Bombay used to arrange South Indian classical music programmes by top musicians like M.S. Subbulakshmi, G.N. Balasubramaniam and others, in the 1950s. Bhabha also attended concerts and recitals at the Shanmukhananda Hall. He is said to have told nuclear scientist, M.R. Srinivasan, 'We can have hundred scientists like you and me but there will be only one M.S. Subbulakshmi.'[30]

It was also through music that, in 1944, Bhabha met a young man for the first time, one who would go on to play an important role in Indian atomic energy: Raja Ramanna. Ramanna was an accomplished piano player and had played with the Mysore palace orchestra in the past. Its old German conductor, it was claimed, had been interned during the war for nursing pro-Nazi feelings. Alfred Mistowski, an examiner from the Trinity College who had been asked to stay on due to the war, oversaw rehearsals.[31] Ramanna's favourite composition was Beethoven's Concerto No. 4 in G Major. He had, in fact, received a scholarship to go to the Trinity College of Music in London in 1940 and had even been sanctioned a grant of ₹7,000 by Maharaja Krishnaraja Wadiyar IV of Mysore with whose orchestra he had played in the past. He never received the money though. After the Maharaja died, his aides claimed to know nothing of the promise.[32]

It was Mistowski who awarded Ramanna the overseas scholarship and also taught him harmony, counterpoint and orchestration. Ramanna used to visit Mysore from Tambaram, where he was doing his BSc (Hons), during his holidays. While he was in Mysore, Mistowski introduced Ramanna to Bhabha and his mother, who had been spending their vacation in the state guest house there. Bhabha and his mother came down from their rooms every evening, formally dressed, to listen to gramophone records. Ramanna recalled in his memoirs that his meeting with Bhabha would determine the course of the next several years of his life.[33]

The Bhabha family's connection with the Mysore court continued. Bhabha corresponded with Maharaja Jaya Chamarajendra Wadiyar, with whom he shared a love for music. The Maharaja invented ragas based on western tonal vocabulary and wrote Carnatic compositions for western instruments.[34] He also agreed to give a subvention of £10,000 to the Philharmonia Concert Society in London to put the Philarmonia Orchestra on a firm basis.

In one such letter, the Maharaja thanked him for postcards of the sets designed by Bhabha for Mozart's *Idomeneo,* calling them charming and saying that they brought out pictorially the spirit of the opera.[35] They also shared a mutual admiration for the composer, Nikolai Medtner, who is now recognized as one of the most significant Russian composers for the piano. Walter Legge, a classical music record producer and the founder of the Philharmonia Concert Society, travelled to Mysore in 1947–1948, and met the royal who promised to pay for the recordings of the Medtner piano concertos and some of his own compositions.[36]

Music was to remain Bhabha's solace all through his life. In his later years, when he was immersed in developing TIFR and India's atomic energy programme, he would relieve his tension and find fresh stimulus by listening to music at night. The role music played in his life is best described by his brother Jamshed. 'What religion, with its forms of ritual and prayer, is to so many, music was to him. In time of stress, it gave him solace and strength, and all the time, music was a source of immense stimulus and inspiration to him.'[37]

HE'S A JOLLY GOOD 'FELLOW'

In the seventeenth century, even as England was being racked by civil war, plague and fire, it was witnessing a revolution in science. It was the end of a world ruled by superstition and magic, and the beginning of a new method of thinking based on scientific experiments and proof. At the heart of this Renaissance were the founding fathers of modern western science: the Royal Society. It grew out of a series of meetings of 'natural philosophers' in Oxford and London that included Francis Bacon, William Harvey, Sir Christopher Wren, Robert Hooke and Sir Isaac Newton. The society was granted a royal charter by King Charles II as 'The Royal Society', a name by which it has become famous. This initial royal favour has continued and since then every monarch has been its patron. The Royal Society (formally The Royal Society of London for Improving Natural Knowledge) is the UK's national academy of sciences.

It is the oldest national scientific institution in the world: election as a Fellow was, and still is, considered a great honour. Bhabha was elected a Fellow in March 1941, almost a hundred years after the first Indian, and fellow Parsi, Ardaseer Cursetjee Wadia, gained admittance in May 1841. Ardaseer was an Indian shipbuilder and engineer belonging to the famous Wadia ship-building family in Bombay, being the great-grandson of Lovji Nusserwanjee, the master shipbuilder who relocated from Surat and was the founder of the famous Wadia family. He introduced several novel technologies to the city of Bombay, including gas lighting, the sewing machine, steam pump-driven irrigation and electroplating.

In July 1940, Raman proposed Bhabha's name for election to the Royal Society. Only Fellows could nominate scientists for this honour.

Raman had been elected in 1924 and was keen to put Indians on the scientific map of the world. He had been instrumental in getting K.S. Krishnan elected in 1940. After Bhabha, he proposed his nephew, S. Chandrasekhar, who was elected in 1944. He also nominated Suri Bhagvantam and R.S. Krishnan, but their candidature did not find favour. Till 1954, Raman and Bhabha remained on cordial terms, but subsequently, they grew apart. Bhabha hardly met Raman after that, something which hurt the elder scientist very much. Bhabha excelled in pencil portraits and there is an excellent one of Raman by him. Raman described it as a picture of one scientist by another. 'No,' replied Bhabha, 'it is a picture of one artist by another.' Air India brought out a calendar of Bhabha's portrait sketches, which included pictures of the British physicist P.M.S. Blackett, artist M.F. Husain, and a mountain villager.[38]

Raman asked Dirac to second Bhabha's nomination. Writing to Dirac, in 1940 he praised Bhabha, calling him a physicist of exceptional ability whose work was of the highest quality. But Dirac knew that already and duly obliged. Others who supported Bhabha's nomination from personal knowledge included Sir A.S. Eddington, William Lawrence Bragg, Sir Harold Jeffreys, Mark Oliphant, D.R. Hartree and Max Born. His supporters from general knowledge included K.S. Krishnan and Birbal Sahni. Given Raman's strained relations with Saha, it is entirely understandable that Saha, who otherwise would have been a logical choice for supporting Bhabha's nomination, was not asked.

Bhabha was elected as a Fellow on 20 March 1941, becoming the eighth Indian after Wadia, Ramanujan, J.C. Bose, Raman, Saha, Sahni and Krishnan to gain this distinction. He could now append those three magical letters 'FRS' to his name; they would now appear on his scientific papers and letters addressed to him. 'It was the ultimate mark of scientific distinction. Younger scientists lusted after it, older scientists lamented their lack of it'.[39] Bhabha was only 31 and achieved the distinction on his very first nomination. It meant a lot to him. Upon his election, the Government of India conferred on him the status of professor as a 'personal distinction'. Raman had already elected him a Fellow of the Indian Academy of Sciences.

The statutes of the Royal Society normally required a newly elected

Fellow to appear for his admission on or before the fourth ordinary meeting of the society after the day of the election. The new Fellow, after making all the payments, presented himself to the ordinary meeting to subscribe his obligation by signing in the charter book of the society. He was then introduced to the president who would formally admit him. When Krishnan had been elected in 1940, he had been unable to gain formal admittance due to the war and had written to the society, asking for an extension of the period of his formal admission. Bhabha, too, faced the same problem after his election.

In August 1941, Bhabha was offered the chair of physics at Allahabad University, but he declined. He hated the drudgery of daily teaching and tedious administrative duties. One of the reasons could also have been the fact that there had been an indication that he would be given a full professorship in Bangalore on the strength of his recent election to the Royal Society.[40] The position in Allahabad had also been offered to Krishnan and S. Chandrasekhar, both of whom had refused. Even Schrödinger had been considered a strong candidate for the post but he took up an appointment as a professor of theoretical physics in Dublin. In fact, so hopeful were the Allahabad authorities of getting him that Amarnath Jha, the vice chancellor, had announced somewhat prematurely that the university had been able to secure the services of an eminent scientist whose greatness was universally recognized.

Chapter 9

BANGALORE

[...] it is a great comfort to us to feel that we have such loving parents and such a nice home to which we can always return [...]

—Letter to his parents, 27 November 1938[1]

In March 1941, while Raman was away in the US for five months, Bhabha held charge of the physics department at the IISc. In November, the institute committed ₹40,000 to cosmic ray research, even as the Sir Dorabji Tata Trust increased its grant for the same area of research to ₹60,000 over a period of five years. On 20 January 1942, Bhabha officially accepted professorship and stewardship of the IISc's cosmic ray unit. As professor, he was allowed the luxury of a personal research group: he was allowed the services of an experimental physicist and had four postgraduate students working under him. This was something that even a professorship in England would not necessarily have brought.

THE PHYSICIST WHO BECAME A MATHEMATICIAN

Bhabha's cosmic ray unit was set up in one of the verandahs adjoining the library of the institute. He started work there with a small group of scientists and students. One of them was the mathematician

Harish-Chandra who started out as a theoretical physicist but found pure mathematics more to his liking. Inspired to take up theoretical physics after reading Dirac's classic book as a young undergraduate in 1940, he was encouraged by Krishnan who was the professor of physics at Allahabad University at the time. On Krishnan's recommendation, Harish-Chandra went to Bhabha in Bangalore in 1943 as a J.H. Bhabha Memorial student, a scholarship named in memory of Bhabha's father, Jehangir.

Raman, who had been Harish-Chandra's examiner during his MSc, took an active interest in him from the start. In the mid-1940s, Bhabha was often absent from the institute, since he was busy with establishing TIFR in Bombay. So, Harish-Chandra saw more of Raman than he did of Bhabha. Since there was a shortage of accommodation in the hostel, Harish-Chandra lived with the Kale family for the first six months, an arrangement suggested by Bhabha. G.T. Kale, a botanist by training, had got a position as a librarian at the IISc following his return from the International Institute of Agriculture in Rome after the outbreak of the Second World War.

Kale's Polish wife, Hadasa, who had been Harish-Chandra's foreign language teacher at Allahabad University, was appointed as lecturer for foreign languages at the IISc, teaching French and German for scientific reading.[2] The couple had a daughter Lalitha (Lily), who went on to become Harish-Chandra's wife. To 10-year-old Lily, Harish-Chandra's serious absorption in science was a source of merriment, and she took great pleasure in trying to distract him from his work. When Harish-Chandra returned from the US in 1953, Lily, who had grown into a strikingly beautiful young woman, became his wife. It was a whirlwind courtship, and his proposal was terse and oblique. He asked her if she had a passport, 'because you'll need one if you come back with me'. It worked! It was often remarked that the two made a strikingly handsome couple.[3]

During his time in Bangalore, Bhabha was mainly interested in two subjects: the classical theory of point particles and the theory of particles of higher spin, both of which had their origins in papers by Dirac. Bhabha and Harish-Chandra worked jointly on these areas, and Bhabha sent their papers for publication to the Royal Society.

Since the mail during the war was very slow, Bhabha requested Dirac to correct the proofs of the two papers. This was how Dirac came to know about Harish-Chandra. Subsequently, Dirac accepted him as his student in Cambridge on Bhabha's recommendation. This not only brought Harish-Chandra to Dirac's notice but also gained for the latter the hyphen in his name, which was first placed there by a copyeditor, which he decided to retain.

The result of Bhabha's collaboration with Harish-Chandra was published in 1944 in the *Proceedings of the Royal Society* as 'On the Theory of Point Particles'.[4] The famous mathematician Robert P. Langlands, Harish-Chandra's long-time associate and friend observed, 'The mild-mannered, gentle Krishnan inspired in Harish-Chandra not only respect but also an affection that never abated. For the boisterous, egoistical Raman and his achievements he had also, in spite of the difference in their temperaments, a high regard, but his own ascetic nature did not allow him to perceive the virtues accompanying the high-living Bhabha's extravagance.'[5] However, Bhabha did recognize the young man's promise and sent him off to study with Dirac as soon as the wartime conditions permitted, for Harish-Chandra was on board the ship bound for England before the war in the Pacific had ended. In fact, he was on the ship when Hiroshima was bombed.

The taciturn Dirac, as was to be expected, left Harish-Chandra to his own devices. The two met only once a term, but Dirac remained the only person for whom Harish-Chandra expressed unreserved admiration. He attended the lectures of mathematicians J.E. Littlewood and Philip Hall and was increasingly drawn to pure mathematics. While at Cambridge, he attended a lecture by Pauli and pointed out a mistake in Pauli's work. Subsequently, the two became lifelong friends. Whenever Harish-Chandra showed Dirac his work, the latter returned it to him without any comment. This greatly disheartened the young man, and he went to Geneva to work with Pauli.

Later, when Krishnan met Dirac, the latter wanted to know where Harish-Chandra had learnt his mathematics, since he would like to send some students there to study. Upon hearing this, Krishnan said, 'Why did you drive him to Pauli with your studied reticence on his work?'[6] Dirac replied that he did not want his critical remarks to disturb or distort the flow of ideas of a creative mind. Young Harish-Chandra, he thought, would have taken them too seriously to think independently. Among Dirac's students at that time was another brilliant student who would go on to make a big name for himself—the Pakistani physicist, Abdus Salam.

Harish-Chandra received his degree in 1947, and in the same year, he visited the US. Dirac went to Princeton for one year and Harish-Chandra worked as his assistant during this time. In Princeton, convinced that he did not have that 'mysterious sixth sense' that one needed to succeed in physics, he decided to abandon the subject for pure mathematics.[7] He was greatly influenced by the mathematicians Hermann Weyl and Claude Chevalley. The period from 1951 to 1963, which he spent at Columbia University, was his most productive. He was also in close contact with André Weil during this period. Like Bhabha, Harish-Chandra also sketched and painted but gave up painting later in his life. An excellent copyist, he was especially fond of the Impressionists. In his last year, often too ill to work, he spent many hours with reproductions of the paintings of Paul Cézanne and Vincent van Gogh.

Harish-Chandra made a pioneering contribution to harmonic analysis on semi simple lie groups. This subject, which hardly existed before the Second World War, became one of the major mathematical edifices of the second half of the twentieth century.[8] When he died in 1983, he was the IBM John von Neumann Professor at Princeton, widely considered India's second greatest mathematician of the twentieth century, after Ramanujan. Today the Harish-Chandra Research Institute, Allahabad, an autonomous institute funded by the DAE, is a research institution focussing on mathematics and theoretical physics.

A PART OF THE 'ADAMS' FAMILY

Being elected an FRS and winning the Adams Prize for the year 1941–1942 were the two distinctions that gave Bhabha the most joy. The Adams Prize is one of the Cambridge University's oldest and most prestigious prizes. Named after the mathematician, John Couch Adams, it was endowed by members of St John's College. It commemorates Adams's role in the discovery of the planet Neptune through the calculation of the discrepancies in the orbit of Uranus. Previous prize-winners before Bhabha included James Maxwell, J.J. Thomson and Ralph Fowler. In more recent times, it has been won by Roger Penrose and Stephen Hawking. In 1948, Bhabha was also awarded the Hopkins Prize of the Cambridge Philosophical Society.

The prize valued at about £300 when Bhabha received it was awarded every even year for the best essay on a subject pertaining to pure mathematics, astronomy or any other branch of natural philosophy. Any person who had at any time been admitted to a degree at Cambridge could compete for this coveted award. Research scholars considered it to be one of the highest distinctions of distinguished scholarship and exceptional ability at exposition. If the high standards of the prize were not met, which was often the case, it was withheld that year. Between 1850 and 1913, it was awarded only 15 times!

Bhabha was using the theory of orthogonal groups to explain the interactions of elementary particles. In July 1942, Dirac wrote to Bhabha about applying for the prize. The letter, which is available in the TIFR archives, arrived by airmail and had to be redirected to Bangalore, since it bore Bhabha's Bombay address. In it, Dirac brought him up to speed with developments. He said that the Council of the Senate, which had met a few days ago, had agreed to accept Bhabha's essay for the Adams Prize, provided it was dispatched no later than 31 December. He wanted Bhabha to register it so that there would be evidence of the date of dispatch. The ordinary postcards, Dirac said, might be illegible. He wanted Bhabha to write to the registry (the address being the University Registry, Cambridge) as soon as soon as he got the letter, saying definitely whether he intended to compete.[9]

Bhabha won the prize for his essay, 'The Theory of the Elementary Physical Particles and their Interactions'. Bhabha was also writing a book on cosmic radiation, a promise he had made to Fowler in 1938, before he left for India. It was to be part of a series of international monographs on physics published by the Oxford University Press. Bhabha kept procrastinating till 1941, but he spent the first nine months of 1942 either writing or researching it. He managed to finish half of it and sent it to Cambridge on 29 December 1942 as he wrote in a letter to the paleobotanist Birbal Sahni.[10]

Sahni had wanted to publish some of Bhabha's lectures, but Bhabha declined, since he had already committed to writing a book on the same subject.[11] In February 1943, Fowler wrote to Bhabha expressing his confidence that it would be a most important book and would play an important part in the development of the next step in theoretical physics. Bhabha later extended the scope of the book to include a separate monograph on spinor algebra. The book proved to be stillborn—one of the few things that Bhabha set out to do but did not complete.

MEHRANGIR

After Bhabha returned to India he first stayed in Bangalore and then in his family home, Mehrangir, at 12, Little Gibbs Road—behind the Hanging Gardens in Malabar Hill, Bombay. 'Mehrangir', a portmanteau word that combined Bhabha's parents' names—Meherbai and Jehangir—was a three-storeyed building, in whose design Bhabha played an important role. Mehrangir was filled with Regency-style furniture, exquisite silverware and a profusion of dinner and tea sets of the finest English porcelain. The house was strewn with rare Persian carpets, fine Lalique chandeliers, a vast collection of records and numerous paintings.

One of these paintings was Bhabha's portrait of his mother wearing an intricately embroidered Chinese Gara sari[12]. He had reproduced all the details to perfection in the portrait, which hung on the wall at the head of the dining room of the house. An evening at the Bhabha household was always an event to remember. Conversations there were always stimulating and the subjects often challenging, but you would come away with a feeling of an evening well spent.

Mehrangir boasted of visits by some of the most prominent personalities of those times, including Nehru. In 1960, the family even entertained Queen Elizabeth II. Jamshed Bhabha spent all his adult life at Mehrangir till his death in 2007. His prodigious estate was bequeathed to the National Centre for the Performing Arts (NCPA), his favourite offspring, indeed his life's work and his mission.

One of the dignitaries who visited Mehrangir and enjoyed Bhabha's hospitality was Gustav Born, a British army medical doctor and son of Max Born, who stayed with them in December 1943. 'My parents and his family were very close,' he recalled. He remembered Meherbai fondly. 'That is another, indelible set of memories: the very fine mother; a delicious Parsee girl with whom I could have fallen in love. I still have her portrait photo.'[13] Gustav had also stayed with Bhabha at his 'big and bright' Cunningham Road bungalow in Bangalore on two occasions between early 1944 and mid-1945. In his words: 'You approached on the drive and there were wide portals—a wide view into his wide, high living room study. And Mussolini-like (it's the only likeness!), Homi sat behind the desk many paces away from the entrance.'[14]

Gustav also recalled how he had played the violin at a musical party that the Bhabhas had thrown.

> I usurped one of the two violin parts in the Bach D Minor Double Concerto, the other part being played by the delightful leader of the Bombay Symphony Orchestra by the name of Mehli Mehta. He had brought his charming wife and their almost new born son. And so I was introduced to Zubin Mehta, at the very outset of his admirable career, at the Bhabha house.[15]

Gustav seems to have got his facts and Zubin's age wrong; in 1943, Zubin was seven years old. Even Zarin, his younger brother who was two years younger, would have been five, by no means a babe in arms. When a famine devastated India in 1951, violinist Yehudi Menuhin contacted Nehru, volunteering to perform in India in early 1952 to raise money for the famine victims. In Bombay, Mehli Mehta was given the task of arranging Menuhin's tour, a job in which he was assisted by Bhabha.[16]

Mehrangir, it appears, did not get its name immediately. On 3 September 1941, Bhabha wrote to his father on the subject: 'I still think that Mehrangir or Meherjahan are the best names for our home of which I prefer the first. Let us decide on that. Jamshed likes that too. After all, the name should correspond to the fact that you have both played an equal share in building of this happy home.'[17] In December

1941, Meherbai wrote to say that 'we have accepted the name of our house that you and Jamshed have selected' and expressed the hope that 'we shall live in it for many, many years with Health, Wealth and Happiness'. This was followed by expressing every mother's hope. 'We are longing to see our darling boys get married to nice Parsi girls and our grandchildren playing about in Mehrangir. I am praying for it every day [...] and I hope my prayers will be heard and we shall always live happily together for many years. May God fulfil all our wishes. Amen!'[18] Bhabha remained a bachelor; Jamshed did marry, but had no issue.

As early as November 1938, Bhabha had written to his parents: '[...] it is a great comfort to us to feel that we have such loving parents and such a nice home to which we can always return [...] and it is a great comfort...' In the same letter he hoped that it 'shall be as elegant as we can afford.'[19] In March 1945, Meherbai wrote that it had been exactly six years since they had moved into their home, 'How glad we were to have a house of our own and what trouble father took over it...'[20] The two brothers also corresponded with each other on the interiors of the house as is apparent from a letter Jamshed wrote to Bhabha in 1943, 'For my bedroom windows, I would have had white Venetian blinds made by Sunderdas. They look so handsome; I am ordering two more for my sitting room...'[21]

But the happy family of four was broken up by the demise of Jehangir, Bhabha's father, in July 1942. Even though he had been a semi-invalid after a heart attack in 1941, it still came as a major shock to the family, and one from which they took some time to recover. It threw Bhabha's work schedule out of kilter for many months, and he confided in Sahni in January 1943 that he had been burning the midnight oil to make up for lost time. He wrote that his father's untimely death had upset his work for several months, and he had had to make up for lost time by working late into the night till 29 December. He added that he had come to Bombay for a four-week-long holiday. Bhabha also said that although it had been six months since his father died, the passage of time had in no way lessened the acute sense of loss they all felt at his death. Their family of four had always been so close that whether in work or in pleasure, they constantly missed the presence of the one who had hitherto shared everything with them. Bhabha called his father both an exceptional husband and father, attributing whatever ability he had to the enormous pains his father had taken to develop his talent.[22]

In May 1942, Bhabha wrote to Pauli telling him about how the war had put paid to his hopes of any new experimental work. 'The threat of invasion in South India is quite considerable, and I am afraid that scientific work is being much hindered by the war situation. It is almost impossible to start on any new experimental work now, and I am taking the opportunity to concentrate on my book [on Cosmic Rays].'[23]

In early June 1943, Bhabha wrote to Pauli from Ooty, his first letter in nearly a year and also the first letter to him since his father had expired. Bhabha confessed that his father's demise had been 'a great blow to my family who had always been in the closest touch with each other'.[24] He complained about being alone and isolated. 'If you complain about being isolated in America you can imagine what it's like in India. There is not a *single* person with whom I can talk about any fundamental theoretical problem.'[25] Pauli wrote back in September with his condolences and voiced a similar sentiment of isolation. 'Please write to me as often as possible, the mail connection with [*sic*] here is very good. I am rather isolated and life is becoming more dull. Dirac and Heitler write me sometimes.'[26]

RESEARCH IN BANGALORE

In August 1943, British physicist Maurice Pryce wrote to Bhabha asking him to apply for the Wykeham Professorship of Physics at Oxford. An alumnus of Trinity College, Cambridge, Pryce went to Princeton, where he worked with Pauli and John von Neumann, supported by a Commonwealth Fund Fellowship. He obtained his PhD working under Born and Fowler and returned to Trinity as a Fellow in 1937. In 1939, he was appointed reader in theoretical physics at Liverpool under James Chadwick.

In the letter, Pryce, who was in Canada at the time assisting John Cockcroft in the war effort, had asked Bhabha to apply for the post, saying it would be very good for English theoretical physics if he returned. He said that he had also applied but did not think his competition would prove to be serious.[27] Bhabha never applied, and it was Pryce who was invited to the professorship after the war ended. Bhabha, in a note he wrote for Nehru on TIFR in January 1954, did not fail to mention (and attach as an appendix) the letter from Pryce, asking him to apply for the professorship at Oxford. Bhabha wrote in his note to Nehru, 'Prof. Pryce pointed out the shortage of outstanding theoretical physicists in England and wanted me to return to England by applying for the Wykeham Professorship at Oxford. I replied to him that I was already committed.'[28]

Bhabha was trying his best to demonstrate his conscious rejection of opportunities both in the country and outside it to buttress his arguments and commitment to his institute for fundamental research. Was he trying to show what a great patriot he was? Perhaps not, but this provided enough fodder for hagiographers to sing paeans of praise about Bhabha's great sacrifice for his country. As Indira Choudhury observes, 'Little wonder then that Bhabha's avowed nationalism became a part of institutional lore, obscuring the social and political context of the foundation of the institute.'[29] As we have seen, Bhabha did receive offers in India but these appointments did not fit in with his grand vision of an outstanding physics institute, something he was not slow in communicating to Sir Sorab Saklatvala, chairman of the Sir Dorabji Tata Trust, when he wrote to him for funding the TIFR, a story that is told in a subsequent chapter.

In early 1943, Bhabha was the president of the physics section of the ISC at Calcutta. Immediately after the congress, he was invited by Saha to conduct a course on collision processes in gaseous media, so, he stayed on in Calcutta. According to Robert S. Anderson, 'Saha had not failed to see the potential in Bhabha's knowledge and connections and sought to bring him closer into his orbit.'[30] By 1944, Bhabha was in a commanding position. He had received both international and national recognition for his work. Independence was near, and the political leadership had already envisioned an important role for science and technology in economic development and nation building. Bhabha knew that the time was auspicious and would allow him opportunities for institution building. He saw himself in the role of a leader, an agent of change, someone who could have a transformative impact on society.

Badanaval Venkatasubba Sreekantan who joined Bhabha at the TIFR in 1949 recalled that when Bhabha was in Bangalore, he used to come to Central College and give lectures on cosmic rays and other subjects. According to B.V. Sreekantan, it was during this period that Bhabha realized that the type of research being done in the country was not of a very fundamental nature. He also realized that the conditions were not suitable in the universities, and even for that matter at the IISc, for doing front-ranking research in the new areas of cosmic rays and nuclear physics. The type of work that was being done at the IISc was mostly on spectroscopy, a legacy of Raman. Bhabha realized that there was need for a place where fundamental research in these newly emerging areas could be pursued.[31]

In Bangalore, Bhabha found a collaborator in the mathematician, B.S. Madhava Rao, who had worked with Born at the IISc. The two published a paper on the scattering of charged mesons, which was published in the *Proceedings of the Indian Academy of Sciences* in 1941. His other collaborators and students at that time included R.C. Saxena, M.S. Sinha, D. Basu and S.V. Chandrasekhar Aiya. Bhabha authored or co-authored 24 papers while he was in Bangalore. The research output of his unit was commendable given the restrictions of the war and the highly explosive political situation in the country, which made it impossible to carry out experiments with any consistency. Later, Bhabha would look back on his days in Bangalore, calling them six very happy and fruitful years of his life.

At the Cavendish, Bhabha had seen how science was done at the highest level. When he came back to India, he did not always try to compete directly with his European counterparts, but instead he looked for projects where India was at an advantage, for example, conducting experiments that could only be performed near the equator. There were also deep mines in which experiments could be conducted. Bhabha was not always explicit about these, but through his choices, he made them a part of his scientists' way of thinking.[32]

When Bhabha started cosmic ray experiments, the latitude effect of mesons was one of the first problems he explored. Bhabha's experimental work started in July 1943, but rubber balloons were unaffordable then because of the war, which forced him to improvise. G.V. Vasudevachar, who joined as Bhabha's laboratory assistant, recalled that they used to send telescopes mounted with lead pieces and Geiger counters in a Dakota plane for high altitude studies from the airport in Bangalore. Later, Bhabha took the indispensable Vasudevachar with him to TIFR. Bhabha also approached the US Air Force and had Geiger-Müller telescopes flown at 30,000 ft. On 26 and 28 December 1944, two flights were made at altitudes ranging from 5,000 to 30,000 ft, which measured the high-altitude intensity of penetrating particles at the equatorial latitude for the first time.[33]

In 1945, experiments to measure the penetrating power of cosmic rays at 40,000 feet were conducted with special equipment flown on a US aircraft, the C-87 Transport Liberator. The intensity of cosmic rays after they had passed through 30 cm of lead was measured at that height for the first time and the experiments showed a sixfold increase at that altitude. In September of that year, Bhabha wrote to the Ministry of Labour and National Service in England, asking for the release of a corporal K.B. McCabe, a radar mechanic with the the Royal Air Force (RAF), to join TIFR as a technical assistant.

McCabe's knowledge of radio engineering and electrical circuits would enable him to build apparatus for the experiments Bhabha had planned. The request was acceded to, and McCabe came to Bombay. He was responsible for, among others things, a complete overhaul of the Wilson Chamber at TIFR. Working with A.B. Sahiar, he drastically modified the chamber so that it could withstand the hot and humid climate of Bombay while incorporating all the latest features in the process. Two years later, when Blackett visited the institute, he advised the scientists on the techniques of the Wilson Chamber.

By this time, although Bhabha was well into experimental physics, he continued his theoretical work as well. The papers he published after returning to India were more mathematical and deeply introspective. As M.G.K. Menon observed, 'There was a powerful mathematical streak in Bhabha's mental makeup, and he was deeply conscious of the aesthetic beauty inherent in exact mathematical solutions.'[34]

A.V. HILL'S VISIT TO INDIA

Although Bhabha had been elected an FRS in 1941, he was not formally admitted until 3 January 1944. The admission ceremony, which was held at St Stephens College, New Delhi, was a historic one since it was the first meeting of the Royal Society to be held outside the UK. A specially prepared sheet of parchment bearing the historic 'Obligation' was brought to India by Sir Archibald Vivian Hill, the biological secretary of the Royal Society. The purpose of the Delhi ceremony was to admit four Indian Fellows who had been admitted to the Royal Society: Birbal Sahni (in 1936), K.S. Krishnan (in 1940) Bhabha (in 1941) and the S.S. Bhatnagar (in 1943). None of them had been able to attend a meeting of the society to confirm their admission.

Sir Henry Dale, the president of the Royal Society, suggested that the most suitable occasion would be the ISC in January 1944. He wrote to Birbal Sahni, the president of the National Academy of Sciences at Allahabad, informing him that Hill was being deputed as the Royal Society's emissary to strengthen the bonds between Indian scientists and their British counterparts. Hill's mandate was to advise the Government of India on organizing scientific efforts in the country. His visit was testimony to the pressing demands of war, and a realization of the importance of scientific activity for the Empire. It was also proof that the government was finally ready to listen to the nationalist elite, a group that it had earlier ignored. The impact of the Quit India movement and rising nationalism in 1942 required a solution.

Influential circles in London believed that evaluating Indian scientific capability by an eminent scientist would placate at least some part of the Indian political and intellectual elite.[35] Hill was an English physiologist who was one of the founders of the discipline of biophysics. He shared the 1922 Nobel Prize in medicine for measuring heat production during muscle contraction and recovery. During the First World War, he was asked by the Ministry of Munitions to advise them on how to train anti-aircraft gunners. Hill provided accurate range tables for anti-aircraft guns with the help of a team of men that came to be called 'Hill's Brigands'. They included Bhabha's PhD guide, Ralph Fowler (who was wounded at the time), Douglas Hartree, a mathematician and physicist, and the future astrophysicist, Arthur Milne.

In March 1940 Hill was deputed to the British Embassy in Washington on what has been called 'Hill's Mission' to promote war research in the still neutral US. His visit was partially motivated by the chemist Henry Tizard's desire to 'bring American scientists into the war before their government'.[36] Hill was also a member of the war cabinet's scientific advisory committee. As an independent Member of Parliament for Cambridge University, he had spoken for more effective relations with Indian scientists in February 1942. (Hill was also the brother-in-law of a very famous Cambridge personality: the economist John Maynard Keynes).

Hill had never been to India and was thus seen to be starting without the existing baggage of previous relationships. Given the charged political atmosphere of the times, it became necessary in official circles to garner support from unofficial and apolitical groups in India to mobilize popular opinion in favour of the government and counter the criticism of nationalist leaders.[37] Hill was certainly the right man for the job. He possessed a comprehensive overview of the war effort and knew how to organize science. Wartime diplomacy and intelligence gathering was not uncommon among scientists during the war. Hill, it was expected, would inform, and if needed, placate his audiences, as well as glean information that would be useful to the government.

When Hill arrived in New Delhi on 9 November 1943 after a nine-day journey, he discovered that Indian science was a curious mixture of ego clashes and complicated individuals and relationships. The Indian scientific community was hopelessly divided between the National Academy of Sciences, Allahabad (established by Saha in 1930 and led by the Indian Science Congress Association) and the Indian Academy of Sciences, Bangalore (established by Raman and led by the Association of South Indian Scientists). The Royal Society's correspondence had been only with Saha's academy and Raman was understandably quite upset. As a senior statesman and the person who had nominated three of the four Fellows who were to be admitted, he justifiably felt that he had a right to be more than just a passive spectator. He asked Bhabha, Sahni and Krishnan to register their protest. The four of them (including Raman) wrote a letter to Hill on 2 December 1943, telling him that a function of the kind being proposed, which affected only the Fellows and was being held for the purpose of obtaining their signatures on the parchment, should be unconnected with the meeting of any other society or organization.[38]

Hill was upset at this development since he had intended to use the ISC as a platform to cajole Indian scientists to contribute to the war effort. The importance of his visit can be gauged from the fact that Prime Minister Winston Churchill had written a letter to each of the Fellows for the occasion. In the end, only Bhabha and Bhatnagar signed the parchment, and were formally admitted to the fellowship in the presence of Lord Wavell, the viceroy. Krishnan and Sahni chose to stay away out of loyalty to Raman. Bhabha, in a letter to Sahni before the investiture, attempted to justify his and Bahtnagar's attendance. He thought they should attend because 'Raman's not attending will not be such an unfriendly gesture as our not attending.'[39] It is not clear if Bhabha discussed this with Bhatnagar, since there is no reference to him in the letter.

Lord Wavell while inaugurating the ISC had said that if India was to play 'the part in the world to which her size, her population, her history and her position entitle her, she too must make every possible use of scientific advancement'.[40] Even before Hill arrived, Wavell had been obliged to pay more attention to science and industry. The government decided in late 1942 to enhance the funding and powers of the CSIR. At Bhatnagar's insistence, four scientists were appointed and, more importantly, it was agreed that the CSIR should not be a government department but should be registered under the Societies Act and resources should be made available from a separate fund of the government. This autonomy was meant to protect it from excessive government interference. However, with a budget of only ₹10 lakh there was virtually nothing left for research after the fixed costs had been taken care of. This funding was insufficient to create the new kind of industries in which people would invest.

During Hill's five-month-long trip, he toured all the major cities, laboratories and factories and met with all the leading scientists and industrialists in India. The response of the scientific community in India to Hill was 'cautiously optimistic if not uniformly enthusiastic'[41] even if it was seen as 'industrial espionage' in the local press, and an attempt to sabotage rather than encourage industrial production in India. Extolling the virtues of Indian science, Hill wanted all the available resources to be used in the war effort and called attention to the indifference and arrogance of the imperial regime in London. He agreed with Indian scientists that science-based industrialization was the need of the hour and the answer to India's poverty.[42]

Hill wanted all research to be under government control. He favoured the establishment of a centralized government department for coordinating all scientific and industrial research and concentrating all executive power within itself. Jahnavi Phalkey has argued that while the British presence in India was under threat, continuing British interests in the region was not. 'A favourably inclined Indian political leadership in control of scientific research might later prove an ally in safeguarding British interests.'[43] Surprisingly, both Bhatnagar and Saha supported Hill's recommendations despite their centralizing import.

Bhatnagar realized that if the Department of Scientific and Industrial Research was to function as a government department under a civil servant, then his wings would be clipped. He resolved the predicament very neatly: he demanded that he be made the secretary in charge of the department and threatened to resign if his demand was not met. He wrote to Hill, 'I hope you continue to be of the same opinion as before that if they don't give me the position of at least an additional secretary I should resign.'[44] The government agreed, and this set a precedent for a similar arrangement for Bhabha when he became secretary of the DAE.

SCIENTIST AS INSTITUTION-BUILDER

While at Cambridge, Bhabha gave no indication that he would ever return to India on a permanent basis. However, by the end of the Second World War, he had a change of heart. In hagiographic accounts he became Bhabha the patriot who sacrificed his brilliant career in the West to build indigenous science in India. His decision to finally remain in India went hand-in-hand with the desire to create his own institute for fundamental research.

When Bhabha returned, India was, technologically speaking, still a backwater. True, Raman had won a Nobel and a few others had become members of the Royal Society, but these scarcely added up to a national environment conducive to the practice of science. Indians regarded their country to be inferior in terms of modern development and looked up to the West for inspiration. Like Bhabha, there were many others who went abroad. Endowed professors at major universities required the holders to have a foreign qualification. The new recruits to the Indian Civil Service (ICS) were trained in England. This feeling of inferiority remained even after Independence, and India's first scientific adviser to the prime minister was the Englishman, Patrick Blackett. To this day, there is greater prestige in publishing in western and not Indian academic journals.

Bhabha's decision to stay back in India has been a subject of debate. Did he stay back because he had no job waiting for him in England, or anywhere else for that matter? We know that by 1942 Bhabha had written to Blackett about building a new school of physics, for no other reason than reducing his extreme isolation from 'western' science. In the same letter, he said that he very much looked forward to being able to return to England. It would be reasonable to assume that even though he had no job offer in hand, Bhabha would not have had much difficulty in securing a position in keeping with his ability and credentials.

Back in India, the colonial government had made it clear to him that his assistance in the war effort would not be accepted. Bhabha wrote to Blackett about the rebuff.

> One wishes one could take more share in helping on the cause in which one believes. I would certainly do whatever I could if I were in England. But in India this is unfortunately not possible as the attitude of the Government is as die hard as ever. The mis-rule would astonish you.*

Bhabha probably felt that if he had been in England, the British government would have allowed him to aid the war effort, since so many of his Cambridge colleagues could vouch for him. However, it could not have escaped Bhabha that it was the colour of his skin rather than his scientific abilities that had caused the rejection. Despite his wealth and western outlook, when it came to the crunch, his racial origins barred the way. As Itty Abraham has observed, Bhabha faced the contradiction between the myth of the universality of science in theory and its parochiality in practice. 'Even as he offered himself to science again and again, his national, colonial origins emerged unerringly and denied him the access that he felt he had in every other way—intellectually, culturally, and experientially.'†

In March 1944, Bhabha wrote what has now become a famous letter to Sir Sorab Saklatvala, chairman of the Sir Dorabji Tata Trust. In his letter, he said that he had concluded that when appreciation and financial support were forthcoming 'it is one's duty to stay in one's country and build up schools comparable to those that other countries are fortunate in possessing.'‡ However, his actions do not

---

*Abraham, Itty, *The Making of the Indian Atomic Bomb: Science, Secrecy and the Post-Colonial State*, Orient BlackSwan, Hyderabad, 1999, p. 65.

†Ibid. 43.

‡Chowdhury, Indira and Ananya Dasgupta, *A Masterful Spirit: Homi J. Bhabha, 1909-1966*, Penguin Books, New Delhi, p. 111.

appear to be those of a person acting from a grim sense of duty; on the contrary, they appear to be the acts of someone who has found a sense of purpose.* His five years in Bangalore had allowed him to re-identify with his country and had made him aware of the role that he could play in the development of India, which was on the threshold of Independence.

In hagiographic accounts, Bhabha's decision to settle down in India has often been portrayed as 'lofty idealism' and 'incipient nationalism'. One can argue that the mainspring of this decision was a mixture of both patriotism and ambition. Secure in family wealth and an international reputation, it is possible that Bhabha found an aptitude for institution building on a grand scale—institutions that he knew would be free to flower in the relative scientific vacuum of India. The coming of Independence, along with the international recognition that Bhabha had received made it possible for him to imagine a more constructive role for himself in India. Upon returning to Bangalore, his involvement with the narrative of a developing India deepened. What he had in mind was a symbiotic relationship between fundamental science and its technological applications.

Given Bhabha's connections with wealthy and powerful people, he was also shrewd enough to realize that these contacts could make his idea a reality. The future showed that his ability to raise money was exceptional: roadblocks that would have deterred a lesser person were surmounted with ease thanks to his connections and background. One does not know if the idea to establish the TIFR was born out of one glorious epiphany or a gradual realization over several years, however, given the cosmopolitan circumstances of his life, it must have become clear to him that he was the right man in the right place at the right time.

---

*Greenstein, George, 'A Gentleman of the Old School: Homi Bhabha and the Development of Science in India', *The American Scholar*, Vol. 61, No. 3, 1992, JSTOR, https://tinyurl.com/428vjnnr. Accessed on 11 August 2022.

India was a blank canvas waiting for a skilled artist, and Bhabha may have decided that he was going to be that artist. His acute sense of isolation and the feeling of being cut off from the enriching intellectual environment of Europe only made him more determined to create such an atmosphere in India. Bhabha was clear that an atmosphere conducive to science that encouraged scientific thinking had not been fostered in India.

The world of science fundamentally changed due to the Second World War. Before the war, physicists from all over Europe had moved freely between countries—physics had transcended the narrow boundaries of territoriality. However, after the war, the nationality of a scientist started to matter and information could no longer flow as freely between national borders as it once had. After the incorporation of scientists in the war effort it became obvious that nationalism and politics could not be separated from science and its practitioners. It is difficult to say if Bhabha realized this fully, but, in the end, given his decision to remain in India, this was of limited relevance.

Chapter 10

AN IDEA AND A WISH

> The scheme I am now submitting to you is but an embryo from which I hope to build up in the course of time a school of physics comparable with the best anywhere. If Tatas would decide to sponsor an institute such as I propose through their Trusts, I am sure that they would be taking the initiative in a move which will be supported soon from many directions and be of lasting benefit to India.
>
> —Letter to Sir Sorab Saklatvala, 12 March 1944[1]

TESTING THE WATERS

Bhabha had first broached the topic about his institute with Tata chairman, J.R.D. Tata, with whom he shared a warm relationship. Tata, only five years his senior, was known to be receptive to new ideas. On 19 August 1943, Bhabha wrote to him saying that lack of proper conditions and intelligent financial support had hampered the development of science in India. He believed that those who chose a scientific career were not being given the necessary equipment, facilities and environment for conducting research. The administrative burden and teaching load did not leave them with enough time and energy for research. To make matters worse, the poor working conditions and financial prospects made the brightest opt for administrative or commercial careers.[2]

In his letter, Bhabha pointed out that if Indian science was to progress, far greater financial support was needed for 'pure' or fundamental research, something that was unlikely to yield any immediate economic return. He gave the example of the USSR, which emphasized using science and technology for solving economic and social problems while strongly supporting fundamental research.[3] He also suggested what P.C. Mahalanobis's biographer, Ashok Rudra, called 'brain irrigation'—the large-scale use of eminent foreign professors, even for short visits, which would improve scientific standards in the country and attract eminent scientists from all over India to work at the institute.[4]

In his reply on 2 September 1943, J.R.D. admitted that there was scope for rendering valuable service to the cause of scientific research in India. He thought that if Bhabha put up a concrete proposal backed by a sound case, there was a very good chance that the Sir Dorabji Tata Trust, and, perhaps, the Sir Ratan Tata Trust, would respond. 'After all, the advancement of science is one of the fundamental objects with which most of the Tata Trusts were founded, and they have already rendered much valuable help in a new way, I am quite sure that they will give it their most serious consideration.'[5]

Bhabha was also friends with Rustum D. Choksi, a trustee of the Sir Dorabji Tata Trust, and his brother, Jehangir D. Choksi, the legal advisor to the Tatas, who were both 'immersed in the ethos of liberal nationalism'. Their sister, Phiroza A. Wadia, a married lady who was popularly known as 'Pipsy', went on to become Bhabha's companion and partner, and, as was commonly believed, a little more.

Rustum Choksi was a professor of Latin and English at Wilson College in Bombay when he succeeded Dr Clifford Manshardt as a trustee of the Sir Dorabji Tata Trust in 1941.[6] He became the managing trustee in 1958 and served in that position till 1980. J.R.D. asked him to continue teaching so that he could remain in touch with the younger generation. An erudite, self-effacing man with considerable personal charm, he not only helped Bhabha shape his proposal but also gave him the lowdown on what the trustees were thinking. His patient persuasion played no small role in giving Bhabha what he wanted. As J.R.D. Tata said, it was he who set the tone and gave the leadership to the trust.

Rustum Choksi 'epitomized an interesting, if paradoxical, hybridity in manner, speech and thinking that characterized a section of the nationalist elite of that period'.[7] His student, the eminent economist, V.K.R.V. Rao, who later became education minister of India, recalled that Choksi was always dressed in a white khadi suit with a khadi tie that had an unusually large knot, and his clothes had an appearance of studied casualness that both thrilled and amused them. Rao said that he combined Byronic handsomeness without the Byronic gloom and said that he was greatly influenced by Choksi's friendly disposition that put him at ease and succeeded in bringing out the best in his nature.[8]

Choksi was also the chairman of the council of the IISc and 'created an atmosphere of freedom in which viewpoints could be freely expressed. On the rare occasions when discussions tended to become acrimonious, the Professor would drop in a gentle word or a soothing phrase, and harmony would return'.[9] He was opposed to the idea of the trust being reduced to a lending institution. Indeed, the trustees had always been concerned about the need to make a distinction between its grant-making activities committed to social development ('constructive philanthropy') and run-of-the-mill donations to alleviate distress. From the beginning, it was clear that the trust would not fritter away its energy and resources on many projects but focus on supporting the founding of a few institutions that individuals with foresight were unable to establish due to a paucity of resources.

Bhabha placed a formal proposal before Saklatvala after the matter had matured in careful discussion with Choksi who guided Bhabha on how to present his scheme to the trustees, offering advice on the style and tone of the proposal. On 17 February 1944, Choksi wrote to Bhabha asking him to submit a scheme to the trustees along with a personal letter to Saklatvala. He advised him to send the scheme as a draft, which Bhabha could finalize once he had received Saklatvala's opinion. He also asked Bhabha to restrict the time frame of the grant to five years.

I would urge, keep the total of recurring expenditure at ₹42,000 per year. It will grow. You will, of course, enlarge as you think. Fit from the need for such research, point out Bombay's backwardness, deep water experiments, and all else that's cogent. Suggest you don't name it, or call it an Institute for the present. Inst. sounds big to some...[10]

Choksi informed Bhabha that he had sounded out the trustees and graded their response, like a school assignment.

Have seen bosses D, T, S and M in that order, 3rd and 4th [are] somewhat non-committal. Following algebraic symbols denote reaction in the same order—b+, a, c+, c. All the details you shall have later—perhaps in person. May suggest in your letter that you'd come to Bombay to unfold, explain to Sir S and co-trustees, if they wish.[11]

D stood for Sir Ardeshir Dalal, the tall, well-groomed, disciplined, methodical, neat and precise ICS officer who, in 1928, became the first Indian to be appointed municipal commissioner of Bombay. A stern and sometimes forbidding presence, he joined the Tatas in 1931 and was made the director-in-charge of Tata Steel. He struck terror whenever he went to Jamshedpur. More feared than liked, he adopted a perpetually serious mien and 'no humour ever escaped his lips and seldom did his cigar yield place to a smile'.[12] Dalal was a highly competent man who was known for his earnestness, application and industry. Bhabha recalled having a quiet dinner with Dalal at the Great Eastern Hotel in Calcutta, telling Russi Mody that Dalal 'discussed nuclear physics with me as if he was my equal.'[13] During the Second World War, Lord Wavell had invited Dalal to join as member for post-war reconstruction and head the new planning and development department as a part of the Viceroy's Executive Council.

T referred to J.R.D. Tata and M was Dr John Matthai, who was an economist by training and served as the minister for railways and transport in 1946 in the first cabinet. Later, Matthai became the finance minister, but he resigned in 1950 after differences with Nehru over the increasing power of the Planning Commission. Known for his unimpeachable integrity, 'Honest John' was a stickler for protocol and precedent. After resigning from government, he rejoined the Tatas as the director-in-charge of Tata Steel for the next seven years. He was also the first non-Parsi chairman of the Sir Dorabji Tata Trust. His nephew, Dr Verghese Kurien, is generally recognized as the architect of India's White Revolution and was the founder of the Institute of Rural Management Anand. Matthai's son, Ravi, was an educationist and the first full-time director of the Indian Institute of Management, Ahmedabad.

Bhabha seems to have been upset at Choksi's restriction of the grant to only five years in his letter. On 26 February 1944, Choksi wrote to reassure him that no damage had been done:

> Am sorry if this five-year business has proven disturbing. Please set it right out. I haven't said a word about it in my four talks here [...] But do send only to SDS for the present. It is just possible that if J.T. receives a copy, or mentions it, your stock may sink a trifle with S. On the other hand I'm fairly certain that J.T. will respond splendidly without any preliminaries. So cut out the time factor altogether. Take it for granted that things just go on, and emphasize that you'll stay in India, and plan to build up something big. Say what you've said about these short-term uncertainties [...] I am sorry if this 5 year biz has damped and perplexed.[14]

There seems to have been an undercurrent of tension between J.R.D. Tata and Saklatvala, given Choksi's advice to send it only to Saklatvala first and that, if J.R.D. advocated the case, his stock would 'sink a trifle with S'. This is not surprising considering the fact that Saklatvala was an overlooked contender for the post of chairman, losing out to his younger rival. Sorab Saklatvala was a nephew of the founder Jamsetji Tata, but he had been overlooked for the chairman's job at Tata, which went to J.R.D., possibly because he was considered 'old and not too bright'.[15] Choksi, who wanted very much to help Bhabha, was trying his best to see that he navigated the waters of the internal politics (if it could be called that) of the trust with care.

Choksi clarified and amplified the aims and objectives of the trust. In two covering notes on trust policy enclosed with Bhabha's proposal, Choksi admirably enunciated the kind of work he thought a trust should be doing. The first note was entitled 'Note on Trust Policy' and expressed the kernel of the philosophy that he had shaped for the trust. As Indira Chowdhury has observed, 'Perhaps no single figure better articulated the conceptual shift in philanthropy in twentieth-century India than Rustum D. Choksi.'[16] The second note dealt with the specifics of Bhabha's proposal. In the first note, Choksi focussed on the character of a trust's giving and said that it was important for it to maintain its pioneering character.[17] In the second note, while advocating Bhabha's institute, he reiterated the pioneering aspect of the trust's work and emphasized the need to maintain that reputation.[18]

On 12 March 1944, Bhabha wrote his now famous letter to Saklatvala from Bangalore:

> The scheme I am submitting now is not one which has been hastily conceived. It has been germinating in my mind for nearly two years, and I recently discussed it at length with Prof. A.V. Hill, Senior Secretary of the Royal Society apart from being an eminent scientist himself, is one who has a great and intimate knowledge of the organization of science and scientific institutions in England, and the many valuable suggestions he made have been incorporated in the scheme as it stands.
>
> There is at the moment in India no big school of research in the fundamental problems of physics, both theoretical and experimental. There are however scattered all over India competent workers who are not doing as good work as they would do if brought together in one place under proper direction. It is absolutely in the interest of India to have a vigorous school of research in fundamental physics, for such a school forms the spearhead of research not only in less advanced branches of physics but also in problems of immediate practical application in industry. Moreover, when nuclear energy has been successfully applied for power production in say a couple of decades from

now, India will not have to look abroad for its experts but will find them ready at hand. I do not think that anyone acquainted with scientific development in other countries would deny the need in India for such a school as I propose.

Prof. Hill when he was in Bombay, repeatedly stressed the fact that all research has in the beginning to be built around a suitable man, and at the present moment there is no one else in India able to do the type of research proposed.

It might at first be supposed that the absence of a good school of physical research in Bombay at the moment would make it an unsuitable place for the object I have in mind. This is not so. The best and the most promising students desirous of studying theoretical physics or cosmic rays, who for the last three years have been sent to me in Bangalore from all parts of India, would come to Bombay instead. I am convinced that within five years we could make Bombay the centre of fundamental physical research in India.

It was while I was on holiday in 1939 that the war broke out and stopped my return to my job in Cambridge. For some time after that I had the idea that after the war I would accept a job in a good university in Europe or America, because universities like Cambridge and Princeton provide an atmosphere which no place in India provides at the moment. But in the last two years I have come more and more to the view that provided proper appreciation and financial support are forthcoming it is one's duty to stay in one's own country and build up schools comparable with those that other countries are fortunate in possessing.

The scheme I am now submitting to you is but an embryo from which I hope to build up in the course of time a school of physics comparable with the best anywhere. If Tatas would decide to sponsor an institute such as I propose through their Trust I am sure that they would be taking the initiative in a move which will be supported soon from many directions and be of lasting benefit to India.[19]

Bombay was one of the most progressive cities of India and philanthropic organizations other than the Tatas could also be counted upon to provide funds. Bhabha also wanted to move back to Bombay for family reasons. This suited the trust because the Tatas had already been criticized by some narrow-minded Parsis for establishing the IISc in Bangalore and not Bombay, where the majority of the community resided.

There were also some institutions of note in the city. The Royal Institute of Science which Bhabha had attended had been set up in 1920 for research and advanced teaching in the sciences. For engineering, there was the Victoria Jubilee Memorial Technical Institute. Wilson College and Elphinstone College were two other institutions of repute at that time. These institutions would provide Bhabha with personnel and help with infrastructural needs in the early stages of establishing his institute.

Bhabha's proposed institute was based on a model that focussed on the individual—in this case, him. Citing the example of the Kaiser Wilhelm Society, which believed in first picking an outstanding man and then building an institution for him, Bhabha suggested that the trust fund an institute that would be based around him. It had never been done before in India and Bhabha reassured Saklatvala that he had discussed the model with A.V. Hill in New Delhi, who had expressed the same views. He added that outstanding schools of research had been built in the UK in the same way, and he saw no reason why the same could not be done in India.

On 24 March 1944, Bhabha wrote to J.R.D. Tata updating him about the progress. He told J.R.D. that he had sent a letter to the chairman of the Sir Dorab Tata Trust with a scheme for the foundation of an institute of advanced physical research in Bombay that would, no doubt, come to him in due course. He explained that he had written the letter with a good many blushes, but in order to convincingly put forward a scheme that he believed to be in the interest of science he had had to throw modesty to the winds. The labour, he said, had got him thinking philosophically, and he had been amused to note how different the codes of behaviour were in different professions. If a scientist said that he had produced the theory of this or that, or indeed, the only correct and accepted theory, people immediately began to think that he must put on his hat with a shoe horn. And yet, no one in the least was startled if a businessman said that he produced the best beer in the world or the best soap. Indeed, he could proclaim it on a huge board and ruin the beauty of a mountain landscape, but few people would think it was immodest. Such was the relative nature of moral codes.[20]

Then, in a not-so-subtle dig at Raman's huge ego, Bhabha wrote, 'Perhaps it is well that scientists labour under this code of impersonality and modesty, for otherwise the view of visitors to this [...] hotel might conceivably be distracted by an enormous board fifty feet by a hundred on the dam opposite with the words "Raman Effect—the greatest scientific discovery in the world".'[21]

In scientific circles, Raman's ego was a famous one and, perhaps, it was something that everyone took a dig at if they got the chance. J.R.D. Tata, in his reply on 5 April 1944, told Bhabha that his letter had been circulated to the trustees and would be discussed in the next council meeting. About Bhabha's light-hearted observations, he responded with characteristic humour. 'You don't need to blush too much about what you wrote in your letter. As hardened business men we are not easily shocked by such mild forms of trumpet blowing!' Then, as if unsure about having made his point strongly enough, he added with his own hand: 'In fact it was almost inaudible.'[22]

Regarding the differences in ethics between various professions he observed:

> The answer to your philosophical wanderings into the difference of ethics between the various professions seems to me to be a simple one. Industry is not a profession and there is nothing personal in the advertising of his product by an industrialist. It is the former and not the latter that is boosted, while the professional man could only advertise himself. The ethical code of a manufacturer should prevent him from exaggerating or lying about his goods in his advertising, not from advertising at all.[23]

J.R.D. Tata was against unethical self-enrichment by businessmen and, more importantly, also practised what he preached, commenting that 'an ethical life is part of an economic life'.[24] In what was a mild dig at Bhabha, he said: 'As a matter of fact, the scientist does advertise in a subtle way when he puts a lot of letters after his name, such as "F.R.S.", when he publishes his discoveries!' J.R.D. joined Bhabha in the fun at Raman's expense, 'I think you are crediting Raman with undue modesty. His advertisement would be on a much bigger hoarding than 50 ft. by 100 ft. and would proclaim that the Raman Effect was the greatest scientific discovery of the Universe—not merely the world!'[25]

Even before Bhabha had written his letter in March, J.R.D. Tata had, in a note circulated that month, communicated his approval of the scheme. It had the approval of Saklatvala and Sir Homi Mody. Choksi, while convinced that the proposal was worthy of support, outlined what he thought were the two main issues before the trust. Firstly, how appropriate was it to support something that was, when it began, only a one-man affair? Secondly, the duration and trajectory of the trust's financial support were something the trust had to take a call on.

R.M. Lala, in his study of the Sir Dorabji Tata Trust, has said that J.R.D. tried to interest a senior industrialist friend in the scheme, spelling out what he thought the institute would do for the country. He said that while advanced physics and mathematics were abstract subjects that were unlikely to produce material results soon, nuclear physics, which was still in the realm of pure science, would within a relatively short period of time make a new and inexhaustible source of motive power available to man.

J.R.D. then went on to say, somewhat inaccurately, that Bhabha and other Indian scientists had already contributed substantially to the subject. He said that the rest of the world was fully invested in pure research, and it was vital that India did not miss out.[26] When I asked R.M. Lala about the identity of the industrialist, he merely smiled and said it was confidential. When I prodded to find out if it was the Godrej family,[27] he refused to comment. 'You are free to draw your own conclusions' was all he said.[28] While we can only speculate about who Lala had in mind, it is certain that J.R.D. had requested a fellow Parsi industrialist, Ness Wadia, for a grant for the endowment of a chair in astrophysics at the TIFR. Wadia replied that his family trust could not be used for anything except housing for poor Parsis.[29]

Choksi said that given Bhabha's aims for a school of physics, it was proper for the trust to provide moderate support for the project in the initial stages. He carefully avoided any reference to Bhabha or his merits, preferring to make his case by highlighting his aims. Bhabha was already receiving a grant of ₹20,000 for his cosmic ray unit and his request for ₹45,000 per annum for the new institute was a 'moderate request'. As for the 'one-man show', Choksi argued that earlier, too, the trustees had supported Manshardt to build the Tata Institute of Social Sciences (TISS). The new institute would, in all probability, grow much faster than TISS. Given the place science and technology had in nation-building, funds would be much easier to come by. The balance sheet of the trust showed a healthy surplus, and Choksi wanted the trustees to take full responsibility for the project over the next five years.

THE APPROVAL

On 14 April 1944, the trustees discussed the proposal while Bhabha waited outside in the anteroom—on the tiptoe of expectation, breathless between transports of joy and dread. However, it ended well. The trustees were persuaded by J.R.D. Tata's argument that fundamental research was necessary even though there were many other pressing economic problems before the country. What the trustees needed to discuss, J.R.D. said, was not the necessity for such an institute but whether the trust would be able to take on the commitment of launching another major institution and support it over a period of several years.

The trust minuted as follows:

> After discussion of the financial implications of Dr Bhabha's proposal as embodied in his letter, the Trustees decided to undertake the responsibility. They were of the opinion that this responsibility should be shared from the outset with the Bombay University and the Bombay Government—both in respect of finance and administration. It was remitted to Professor Choksi under the general direction of the Chairman to explore the possibilities of such cooperation with the University and the Government. Dr Bhabha was present at this stage of the meeting and entered into

the discussion. He readily accepted the Trustees' views regarding the sharing of the responsibility.[30]

Echoing the vision of the industrial house it was born out of, this decision of the trust was in keeping with the Tatas' vision (based on the founder Jamsetji's ideas) of defining and providing what India needed. With Independence on the horizon, the trust now also saw its role as contributing to nation-building. Rather than supporting conventional institutions, it would encourage and nurture new institutions that would advance the development of the nation. Choksi had stressed the pioneering nature of the institute, and it was this emphasis on supporting the unconventional that became the nub of the argument in support of Bhabha's institute. The trust's support of Bhabha is usually couched in terms of support for science and encouragement of its worthy practitioner, but as Indira Chowdhury points out, it was also a result of the 'reconceptualization of philanthropy' in India.[31]

It is also crucial to remember the motivations of J.R.D. Tata, the man who gave the institute his wholehearted support from the start. He had missed out on a university education, something that rankled him all his life. Like Bhabha, he was to enrol in Gonville and Caius for mechanical engineering and was even sent to a crammer in Suffolk to brush up his English before he went to Cambridge. He confided in his father, Ratanji Dadabhoy Tata, that he was incapable of holding a serious conversation in English, his French background being completely useless for his future.

Unfortunately, the French passed a law of conscription that made it compulsory for all French boys to do two years of military service on turning 20. As a French citizen, J.R.D. Tata was obliged to enlist, and Cambridge had to be postponed. As it turned out, he never managed to get there, and throughout his life, was a little self-conscious about this lack of academic achievement. This background coupled with his belief in Bhabha's mission and the benefits that he thought would flow from the new institute made him a pillar of support for Bhabha from the beginning. Choksi, after all, did say that 'JT' would respond splendidly. He was not wrong in his assessment.

In many ways, J.R.D. Tata's unflinching confidence in Bhabha's proposal set the stage for building a pillar of science in India. It was not just the funds that J.R.D. helped provide, but the time and energy he gave to the institute in its early years were equally important. Old timers remember him coming to the institute with Bhabha on Sunday mornings to discuss its new building.[32] In his foreword to the TIFR's Silver Jubilee Commemoration Volume in 1970, J.R.D. wrote that he was convinced that only brilliant and dedicated men could take the country forward and to do this science and technology was indispensable. He said that the appearance of Bhabha, with his prophetic vision and immensely talented mind, seemed truly predestined.[33]

J.R.D.'s belief in the new institute reflected his faith in the ability of science and technology to develop India. In January 1944, Dalal, Matthai and J.R.D. were signatories to what became known as the Bombay Plan. Between 1943 and 1944, J.R.D. hosted a group of eminent industrialists and technocrats in Bombay. The guest list was a distinguished one: G.D. Birla, Kasturbhai Lalbhai, Sir Purshotamdas Thakurdas, Sir Shri Ram, Dalal, A.D. Shroff and Matthai. It was no ordinary gathering, nor was their one-point agenda—a plan for the economic development of India. Their brainchild was a ₹10,000-crore, 90-page economic plan that became famous as the Bombay Plan (more formally 'A Plan for the Economic Development of India').

Drafted by Matthai, the plan had one predominant objective: how to double the per capita income (₹65 in 1931) in 15 years, providing for an increase in population of 5 million per annum. It was a 15-year blueprint for India's post-war economic development, which sparked off debates and embarrassed the British government, proposing, as it did, rigorous measures of state control, stressing that economic development of the kind proposed could be possible only if it was based on a strong central directing authority.

An immediate consequence of the publication of the plan was that Lord Wavell requested J.R.D. for the services of Dalal to start the Department of Planning. The Bombay Plan can also be seen as part of the industrialists' belief in the importance they thought would be due to them, and the role they expected to play in nation-building. By participating in national planning, they saw themselves as being partners in this process. Playing a part in this optimism and expressing the zeitgeist of the times was the role that science and technology would play in the process of building up industrial infrastructure and its contribution to the development of the newly independent nation. This feeling was reflected in the mission of the philanthropic organizations of the Tatas in general and the Sir Dorabji Tata Trust in particular. Bhabha's scheme struck a resonant chord with the trustees, many of whom had been signatories to the Bombay Plan. They saw it as an effective way of creating conditions for nurturing a scientific culture and, ultimately, using the power of science to revolutionize Indian industry.

The trust also prudently suggested that the responsibility should be shared from the outset with the Bombay University and the Bombay government. Bhabha was to give a final form to this suggestion when he formulated a tripartite arrangement between the trust, the Government of Bombay and the Government of India (operating through the DAE). This 'Bhabha formula' has been employed with notable success in other cases as well. The institute received financial support from the Government of India (from its second year) through the CSIR and the Ministry of Natural Research and Scientific Research. Today, the institute is funded by the Government of India through the DAE.

The background that led to such an arrangement can be found in an offer of professorship that Bhabha received from the Royal Institute of Science in Bombay shortly after his scheme was approved by the trust. The director of public instruction of Bombay, S.N. Moos, had approached Bhabha, aiming to create a chair for him and activating the physics department of the institute. While he supported Bhabha's idea of starting a separate institute, he felt that his larger project would have to wait until the end of the year. In the meantime, they could begin by creating a special chair at the institute in which the Tata Trusts could also be invited to assist. Bhabha declined the offer, since the trust had already accepted his proposal for setting up a separate institute.

Subsequently, Bhabha sought to capitalize on this offer in a different way. He suggested that the funds that the Bombay government wanted to commit to the chair (₹25,000 a year) could be more effectively used if the government cooperated with the trust in a joint venture. In a letter to the governor of Bombay, Sir John Colville, he wrote that he had been unable to accept the government's earlier offer but believed that the two proposals—the government's and his—could be worked together as one project with more lasting benefit than would otherwise be achieved. Bhabha's suggestion was accepted by the government and negotiations between the trust and the Bombay government were taken up and proceeded favourably from the beginning. As Indira Chowdhury has observed, Bhabha's proposal presented the opportunity of partnership between private enterprise, the government and the university system through philanthropy.[34]

Bhabha made it clear to the trustees that his institute would not play a direct role in Indian industry. Any benefits that would accrue would be indirect. This emphasis on fundamental research shaped the agenda of the new institute—it would concentrate on long-term visions and not instant returns. While Bhabha, guided by A.V. Hill's thinking, believed that the government should provide financial support to scientific institutions, he did not think that it was wise for them to be placed under government control. Hill, in a lecture to the ISC at New Delhi, had said: 'Many of these independent scientific institutions in Great Britain nowadays are receiving substantial state support: but nearly always when this is done a buffer of some kind is interposed to prevent Government support from becoming Government control.'[35]

Thus, the TIFR was run by a governing council on which the Sir Dorabji Tata Trust, the Government of Bombay and, later, the AEC were represented. When the trust sanctioned the proposal, pending the passing of final orders by the government, a provisional council was created to take all the necessary steps for the establishment and maintenance of the institute. Saklatvala became the first chairman of TIFR with a provisional committee consisting of Matthai, Bhabha and Moos, who represented the Bombay government.

The newly instituted council met for the first time on 18 May 1945 in Saklatvala's room in Bombay House. The members of the provisional committee were joined by the man who had made everything happen—Rustum Choksi. He had not become a member of the council yet, but he sat in on the meeting that day and remained actively involved with the day-to-day functioning of the fledgling institute. Tentative proposals for the budget of ₹80,000 for the year 1945–1946 were passed. The available funds were sourced from the initial Tata grant of ₹45,000 per year along with ₹25,000 from the Government of Bombay, which was further augmented by a grant of ₹10,000 from the CSIR.

This interest of the Union government was largely thanks to Bhatnagar. Bombay University chose not to join the alliance, but the interest of the central government boded well for the new institute. The governing body of the CSIR under Dalal agreed to provide a block grant of ₹75,000 for the year 1946–1947, subject to the condition that adequate representation on the council of the institute was given to the CSIR. Bhatnagar took his seat on the council and this block grant increased gradually with the normal increase in salaries.

The director and one professor were appointed, and sanction was accorded for one post of reader in experimental physics, one research assistant and two studentships; a registrar, librarian and the necessary clerical staff were also sanctioned. The institute commenced its work on 1 June 1945 at the IISc in Bangalore. A beginning had finally been made, but this was just the first tile in the TIFR mosaic—the saga continues in the following chapters.

Bhabha wrote to Pauli about his new venture. Pauli wrote back on 12 July 1945, voicing his interest in the institute. He said he was very interested to hear that Bhabha was shifting from Bangalore to Bombay. He wanted to know more about the new institute for fundamental research. Pauli added that the prospect of seeing the institute once for himself was stimulating. He also told Bhabha that the IAS at Princeton had offered him a professorship on the occasion of Einstein's retirement, but that he had decided nothing until then.[36]

Chapter 11

CATHEDRAL OF SCIENCE

> An institute, be it a laboratory or an academy, has to be grown with great care like a tree. Its growth in terms of quality and achievement can only be accelerated to a very limited extent. This is a field in which a large number of mediocre or second-rate workers cannot make up for a few outstanding ones. And a few outstanding ones always take at least 10 to 15 years to grow.
>
> —Presidential Address, National Institute of Sciences of India, 1963[1]

The TIFR started functioning in the cosmic ray unit of the IISc, but this was always going to be a temporary arrangement, since Bhabha planned for the institute to be based in Bombay. This was easier said than done because he needed suitable premises to house his institute. Once again, Bhabha's family connections came to the rescue. His aunt, Cooverbai Panday, agreed to rent half of her bungalow, Kenilworth, to Bhabha for his institute. Half of the old house, about 6,000 sq. ft, was rented for ₹500 per month.[2]

It was a homecoming of a different kind: Meherbai revealed that Bhabha's office on the first floor of the bungalow was also the room where he was born! Bhabha proudly noted the advantage of having begun in medias res: 'Work has inevitably been built up first and the permanent buildings have come up afterwards.'[3] The bungalow is now long gone, having made way for an apartment complex, also named Kenilworth, that accommodates the staff of the DAE. Contractors completed the construction of Kenilworth in 1962–1963, and the DAE allotted flats on 1 April 1964, some two years before Bhabha's demise.

There is a story that after the TIFR shifted from Kenilworth, Bhabha's aunt sold the building, and the new owner converted it into a hotel with the same name. When it was put up for sale, Bhabha acquired the building for the AEC and flats were made for the staff. The shape of the building, the odd layout of the flats, covered ducting and air conditioner openings all lend credence to the veracity of the story.[4]

Bhabha needed his Bangalore equipment to start work in the new premises. In July 1945, the provisional council of the TIFR, during its second meeting, found a way to shift Bhabha's equipment from Bangalore to Bombay. Bhabha had discussed the matter with J.C. Ghosh, the director of IISc, and the trust offered ₹50,000 as a grant to the IISc in return for the old cosmic ray equipment and the personnel to man it, along with some new instruments. The laboratory assistant, G.V. Vasudevachar, and the glassblower, H.L.N. Murthy, also moved to Bombay. Vasudevachar recalled bringing the equipment from Bangalore to Bombay by train and then hiring a *tonga* from the Victoria Terminus (now Chhatrapati Shivaji Maharaj Terminus) to Pedder Road.

When the move to Bombay was planned, Bhabha called Vasudevachar and asked him to relocate with him. Vasudevachar said that he would have to ask his father. His father said, 'When Bhabha is asking—why can't you say "yes", why ask me?'[5] That clinched it. However, he told Bhabha that he had no place to stay in Bombay. Bhabha promised to arrange suitable accommodation. Vasudevachar stayed at Kenilworth, where Bhabha's aunt looked after him very well. After his marriage, he occupied the outhouse of the bungalow.[6] Bhabha's mother, Meherbai, was no less caring. Vasudevachar recalled an incident when they were at the Central Observatory in Bangalore to launch balloons. All the hotels were closed that day and Meherbai had asked him if he had eaten breakfast. When she was told that the hotels were closed, she drew Bhabha's attention to the fact. 'Homi, come here. See people here have not had food. Do something.'[7]

Murthy was an obscure glassblower from Bangalore who worked in Raman's laboratory. Later, he established his shop called the Mysore Scientific Glassblowers. Bhabha immediately recognized his genius and asked him to join the TIFR in January 1946. Murthy was not merely a glassblower—he was a genius at fabricating precision equipment. One of the first things he did after joining TIFR was make a Geiger–Müller counter. Bhabha arranged to send him to Bristol University for a laboratory arts course. Bhabha was told that the university needed no payment to keep Murthy because he had proved to be so useful. Bhabha then sent him to Metropolitan Vickers to learn precision mechanics. After he returned, Murthy was instrumental in training a large number of glassblowers and mechanics from the AEET as well as other institutions, and in strengthening the laboratory workshop base of some of the finest scientific institutes in India.[8]

N.R. Puthran, a young accountant, joined the institute's staff of six around that time. He went on to become the registrar of TIFR. Since the nearest tea shop was some distance away, Bhabha's aunt catered to the refreshment needs of the staff from her kitchen. On Parsi New Year's Day, the staff feasted on traditional Parsi delicacies. Puthran, who worked for Bhabha for two decades, recalled that in all those years, he could not recall a single occasion when Bhabha lost his temper or raised his voice.[9]

THE BEGINNING

The institute was inaugurated on 19 December 1945 by the governor, Sir John Colville, at a simple ceremony. The guests included J.R.D. and Naval Tata, Jehangir D. Choksi, Syed Ahmed Brelvi, Sir Rustom Masani, Dr Nazir Ahmed, Sir Henry Knight and Dharamsey Khaitan, among others. Saklatvala welcomed the governor to the 'modest temporary premises' and noted the shadow of war that still hung over the world. He said that although the institute did not have a permanent building, work in experimental physics and theoretical physics had already started. He then reiterated the trope about an institute being built around a brilliant individual: 'An Institute of this character requires first, men of calibre and then, suitable buildings and equipment.'[10] It was clear that the trust had great faith in Bhabha's abilities. At that time, a great scientific institution was defined not so much by its work but by its reputation in the eyes of the public, which was based on the intellectual achievements of its personnel.

Bhabha used his speech at the inaugural ceremony to not only express his sentiments but also for the edification of his audience. His speech, while free of technical jargon, was abstract and complex and not meant for the layman. He began by describing the progress of physics in the world and the lines along which fundamental research was proceeding. He also thanked the trust for helping in the more mundane matters of administration, which freed up the scientists' time to do research from the very beginning.[11]

Bhabha briefly summarized the six elementary particles known back then: the electron, proton, neutron, meson, neutrino and photon. He said that there had been considerable progress in mathematically describing the behaviour of the known elementary particles. James Maxwell described the photon through a set of equations in the middle of the nineteenth century, and Dirac's equation did the same for the electron in 1928. What was important, Bhabha said, was the fact that, in both these instances, theory successfully predicted new phenomena and pointed the way to experimentation.

He also told the audience about his recent research. His argument also had social relevance—the promise of science revolutionizing everyday life must have appealed to the luminaries gathered there. What Bhabha was trying to convey was that nobody could predict what benefits would accrue from fundamental research, since studies aimed at understanding nature had led to unanticipated applications of enormous significance, including unexpected benefits that improved aspects of daily life.

Bhabha revealed that the cooperation of the commanding officers of the 84 Air Depot of the US Army Air Force at Bangalore had made it possible to measure the intensity of cosmic rays penetrating 30 cm of lead at an altitude of 40,000 ft. Later, the guests were shown the apparatus that had been carried on the Liberator plane that had made the experimental flight from Bangalore.[12]

In his speech, Colville welcomed the establishment of the institute for the 'enlivening and original influence' that it would exert on applied scientific work, which he hoped would increase as the industry expanded. The Government of Bombay formally approved the establishment of the institute in January 1947. It was resolved that the institute would be called the Tata Institute of Fundamental Research. It was to be controlled by a council consisting of two representatives of the trust, one representative of the Government of Bombay, and the director of the institute. So long as the government or the CSIR made annual grants, the central government would be entitled to appoint a representative to the council. The resolution provided for the payment of annual block grants of ₹45,000 and ₹25,000 for three years in the first instance by the Sir Dorabji Tata Trust and the Government of Bombay, respectively. Regarding the capital cost for the site and the new buildings, it was agreed that the trust and the Government of Bombay would commit up to ₹4 lakh and ₹2 lakh, respectively.

In 1946, the CSIR appointed an Atomic Research Committee for promoting research on atomic energy. On the recommendation of this committee, the CSIR gave a grant of ₹26,000 to the TIFR in 1947–1948 since most of the theoretical and experimental work in nuclear physics was being done there. This grant was meant for training a team of scientists in general techniques of nuclear physics, with special reference to a high energy accelerator. In 1948, the block grants from the trust and the Bombay government were raised, and, in 1951, the grant from the trust was increased to ₹1 lakh. In 1948, the Ministry of Natural Resources and Scientific Research entered the picture and sanctioned a block grant of ₹1 lakh for 1948–1949, with increases in the following years.

BHABHA'S INVITATION TO S. CHANDRASEKHAR

Bhabha had written to Chandrasekhar in April 1944 to sound him out about joining TIFR, even before it had been established. A.V. Hill had been goading Bhabha to get Chandrasekhar to India. 'There are not so many of his quality in India that you can afford to give him away to the Americans,'[13] he said. In his letter to Chandrasekhar, Bhabha congratulated him on his election as an FRS and proceeded to tell him about the plans for an institute. He also tried to ascertain

Chandrasekhar's interest in accepting a suitable position at TIFR. 'Your own line of research would certainly come within the scope of the scheme and if it were possible to create a Chair for you would you be prepared to consider it?'[14]

Bhabha went on to tell Chandrasekhar about the emoluments, duties and service conditions he could expect. The offer also included long vacations that would enable him to maintain his foreign contacts. Bhabha completed his pitch by telling Chandrasekhar that he was unlikely to get more favourable conditions for conducting his research in India. Bhabha told Chandrasekhar that, until recently, he had felt that after the war, he would accept a good position in some university in Europe or America because a place like Cambridge had an atmosphere that no place in India had at that moment. He said that he had recently come to be of the view that if proper appreciation and financial support were forthcoming, it was the duty of people like them to stay in the country and build up outstanding schools of research such as those some other countries were fortunate enough to possess.[15]

In conclusion, Bhabha clarified that this was only an informal approach on his own initiative, but if he received a positive response, he would try to get the promised chair created for Chandrasekhar. A month later, Chandrasekhar responded from Chicago, expressing his interest in the proposal. He said that a professorship in the institute under the circumstances Bhabha had described came so very close to the ideal position he could have wished for himself that in the event of such a chair being offered to him, he would have no hesitation in accepting it. He added that he fully shared Bhabha's view that it was their duty to stay in the country and build up outstanding schools of research such as the ones other countries possessed.[16]

In this letter, Chandrasekhar seems almost to have accepted Bhabha's offer. In the concluding part, he made a reasonable request. 'There is only one condition I would require to be met before I accept such a position and that is that the travelling expenses of my wife and myself entailed in the transfer from Yerkes to Bombay be paid by the proper authorities.'[17]

It was not as if Chandrasekhar had not received offers to return to India in the past. In August 1935, his uncle, C.V. Raman had offered him an assistant professorship at the IISc, with a promise that '[...] at the end of your 5-year tenure you can look forward to a full Professorship being created for you [...]' Raman had sent a copy of his letter to Chandrasekhar's father, who, in turn cabled Chandrasekhar, saying, 'My advice keep off his [Raman's] orbit.'[18] In his next communication, his father explained why he wanted Chandrasekhar to reject Raman's offer. He said that it was his ardent desire for at least one 'Madrasi' to be able to do something in the world of physics on his own efforts and not be indebted to Raman who thought he was the 'kingmaker' of professors and doctors of science. He did not want the world to think that Chandrasekhar had gained from his relationship with his uncle. 'If you will honor my sense of self-respect at least if not your own, you may not accept the offer.'[19]

Chandrasekhar did not need his father to tell him this. While he respected his uncle's brilliance in physics, Raman was not a role model for him. Raman often spoke in contradictory terms, revelling in controversies and sensationalism. For the quiet, introverted Chandrasekhar, he was an irritation. Just a few months before offering his nephew the job, Raman had belittled his pursuit of astrophysics as a waste of time in the backwaters of science. He had said that he would not have such a person anywhere near Bangalore. He believed that if Chandrasekhar changed his field and worked in nuclear physics instead, he might make better progress. Such reports had infuriated Chandrasekhar and therefore the rejection of Raman's offer was a foregone conclusion in his mind.[20] Nephew and uncle shared a strange relationship. Despite Raman's poor opinion of astrophysics, he nominated Chandrasekhar for both the FRS (in 1944) and the Nobel Prize in physics (in 1957).

In 1944, around the time that Bhabha offered him a position at TIFR, Chandrasekhar was sounded out about the directorship of the Kodaikanal Observatory. He had no interest in administrative matters nor was he interested in observational work, his métier being pure theory. He wrote back saying that if it was free of administrative duties and observational work, he might consider the offer. Nothing came of this offer and a similar one made after the war.[21]

While Chandrasekhar's letter of May 1944 showed his interest in Bhabha's offer, his letter written in August 1945 showed that his initial resolve had weakened. He expressed reservations about breaking all links with the US and returning to India permanently. Instead, he offered to come to India for a year, on a leave of absence from Chicago, to test the waters, as it were. If he liked what he saw, he would accept a permanent position. Bhabha, in his reply in October, accepted his terms and gave him a detailed description of TIFR and its activities. In May 1946, Bhabha sent a telegram inviting him to accept the chair in astrophysics at a salary of ₹1,250 per month for five years and full travelling expenses by sea for him and his wife. He was expected to join before 26 July 1946.

It had not been easy for Bhabha to get to this stage. Chandrasekhar's appointment meant an additional expense of ₹15,000 per year, and the institute's budget of ₹80,000 could hardly be expected to cover this sum. A grant of ₹75,000 from the CSIR for the year 1946–1947 saved the day. However, Chandrasekhar was not destined to come to TIFR. On 2 April 1946, he wrote to Bhabha about the emigration problems he had been facing. He apologized for the delay in replying to Bhabha's letter written the previous year. He said that since he had last written to Bhabha, some difficulties regarding emigration had arisen and he had been waiting and hoping that he would be able to give a definite reply. However, he said, the situation was still very confused.[22] The problem was with the issue of re-entry permits to permanent residents in the non-quota category and the stringency of the issuance of new visas. He said that without clarification from the State Department about the confusing situation around re-entry permits, it would not be possible for him to plan to leave the US.

In addition, Chandrasekhar said that the Chicago University had expressed its reluctance to grant such a lengthy leave of absence and wanted him to wait for a while. It also appeared that the various things he had set out to accomplish that year would not be completed. 'However, it is not impossible that things might clear up by the time you visit this country when I hope we can discuss the matter. I am sorry to leave the matter in this unsettled state, but I hope you understand my position.'[23]

The matter seems to have rested there till Chandrasekhar visited

India in 1951, and Bhabha reiterated the offer. Chandrasekhar recalls that left to himself, he may, perhaps, have been tempted by the offer, but he was advised against it by many people, including Raman, who asked him not to play 'second fiddle' to Bhabha. The guileless Chandrasekhar didn't appreciate that remark; ultimately a paucity of time prevented him from making a decision. The matter came up again when Bhabha visited Fermi in Chicago in 1953. He renewed his invitation and left it to Chandrasekhar to decide. Once again, he could not be persuaded to take it seriously.[24] In the same year, Chandrasekhar became a US citizen. The return to India now seemed even less likely.

Chandrasekhar may not have accepted Bhabha's offer, but they continued to be friends. He made it a point to visit TIFR whenever he came to India. Physicist G. Venkataraman recalls attending a colloquium by Chandrasekhar and the wonderful way Bhabha introduced his Cambridge colleague. In November 1968, when the TIFR auditorium named in memory of Bhabha was being inaugurated by Indira Gandhi, the subject of Chandrasekhar's return came up once again. Ali Yavar Jung, the governor of Maharashtra, narrated how as vice chancellor of Osmania University he had also been trying to attract Chandrasekhar while Bhabha had been trying to do the same. 'But neither of us could catch him; he eluded us both,' he said.[25]

THE POLYMATH OF THE DECCAN QUEEN

The mathematician and polymath Damodar Dharmanand Kosambi was the first to join the TIFR faculty. Kosambi had delivered a series of nine lectures on path geometry at Bhabha's cosmic ray unit in Bangalore. Bhabha thought that Kosambi was an ideal candidate for TIFR, since he was both a capable teacher and researcher. Although TIFR at that time, was mostly about physics, Bhabha thought that Kosambi would be a useful addition to the institute because his mathematical research was of great interest to physicists and was used by them regularly.

Bhabha had been drawn to Kosambi in the first place due to his flair for mathematics. He wrote to Saklatvala in May 1945, advocating Kosambi's case, calling him one of the best mathematicians in India.

He referred to Kosambi's work on tensor analysis and path geometry—the latter was a generalization of tensor analysis that Kosambi had largely developed himself.[26] Before writing to Saklatvala, Bhabha had written to Rustum Choksi, in March 1945, praising Kosambi's lectures in Bangalore and advocating a position for him at TIFR.[27] Dr M. Raziuddin Siddiqi had offered Kosambi a research institute that was being planned at Hyderabad, under the aegis of the Osmania University. Bhabha knew this, and ultimately, succeeded in getting Kosambi to TIFR on 16 June 1945, just two weeks after he himself was appointed director.

Kosambi wore many hats: mathematics professor, historian, archaeologist, epigraphist, polyglot, numismatist, Sanskrit scholar, Indologist and Marxist. Widely regarded as one of the founding figures of contemporary Indian historiography, Kosambi quantified numismatics and used statistical inference to study Indian history. He foregrounded the place that technology occupied in the evolution of Indian society by connecting material remains with textual evidence to create an entirely new understanding of India's history. He made fundamental contributions to Indology and the study of prehistory, and his translations of the poetry of Bhartrihari are considered definitive.[28] He was also considered to be a bit of a maverick, a reputation, which as we shall see, was not entirely undeserved. His intellectual journey touched upon a variety of subjects: Kosambi's ability to stalk confidently over a large part of the map of human scholarship gave him his renown as a polymath.

His father, Dharmanand, was a Buddhist scholar. Kosambi inherited his passion for learning, his sharp versatile intellect, and above all, his humanism from his father. On his second visit to Harvard University, as visiting faculty he was entrusted with the task of editing an important Pali text and Dharmanand took the young Kosambi along with him. Thus, in 1918, Kosambi was catapulted from a local Poona school to the Harvard Grammar School and later, the Cambridge High and Latin School in Cambridge, Massachusetts. When his father returned to India in 1922, Kosambi stayed back in the US, completing his school in 1924. He returned to India, but the difference in the educational system meant that he could not secure admission to a college. Consequently, Kosambi went back to the US, where he enrolled at Harvard College in 1926. Norbert Weiner, who went on to become a renowned mathematician at the Massachusetts Institute of Technology (MIT), was one of his close friends. Both shared a passion for mathematics.

Kosambi graduated summa cum laude from Harvard College but was unable to secure a scholarship for postgraduate studies. This was partly because of the Great Depression and partly because his mathematics teacher, the famous G.D. Birkhoff, was unsure about Kosambi's commitment to the subject. Kosambi valued the Renaissance type of versatility, which entailed wide-ranging knowledge without sacrificing depth. Though Kosambi excelled in his studies, his refusal to specialize went against him: Birkhoff did not offer Kosambi a fellowship for further research in mathematics at Harvard.[29]

On returning to India, he taught mathematics at the BHU. In December 1931, Kosambi joined the AMU, at the invitation of the mathematician André Weil, who headed the mathematics department there. Weil also recruited the mathematician, T. Vijayaraghavan, who had just returned from Oxford. From Aligarh, Kosambi sent articles to various journals in India and abroad, including one about the 'Bourbaki theorem', which was a mathematical prank. His paper, 'On a Generalization of the Second Theorem of Bourbaki', was written at the provocation of Weil—a parody passed off as a serious contribution to a provincial journal.

Kosambi's special area was path geometry and tensor analysis, and he published eight papers in English, Italian and German while at AMU. However, his stay in Aligarh proved to be short-lived. University politics led to the departure of both Weil and Vijayaraghavan. Kosambi left in 1932 and returned to Poona, where he taught mathematics at Fergusson College for the next 12 years. Raman had visited Aligarh in 1931, and although there is no record of their meeting each other, the Nobel laureate seems to have formed a very high opinion of Kosambi's abilities. Certainly, Raman's subsequent actions seem to confirm this. In 1934, when he founded the Indian Academy of Sciences in Bangalore, he elected the mathematicians Vijayaraghavan and Sarvadaman Chowla. The following year, he appointed Kosambi, along with P.C. Mahalanobis and V.V. Narlikar (father of the famous astrophysicist Jayant Narlikar).

In Poona, Kosambi flowered, setting the scene for his stupendous interdisciplinary knowledge. He made valuable contributions to genetics, statistics and other branches of knowledge. In genetics, his work on chromosome mapping was believed to advance the prevalent chromosome theory and was used by professional geneticists.[30] In 1940, he undertook a study of punch-marked coins, weighing some 12,000 coins and laying the foundations of scientific numismatics in India, distinct from epigraphy and archaeology. This led him to the study of Sanskrit, which he mastered by taking up Bhartrihari's three *shatakas* (centuries) of epigrams. He undertook text criticism by studying 400 manuscripts. He also acquired unquestionable proficiency in Pali and Prakrit. From Sanskrit, it was a natural progression to the social setting of its literature—ancient Indian history. Kosambi wrought an intellectual revolution through his redefinition of the nature and scope of history.[31] He was also an active Marxist.

In Fergusson College, exam-oriented students, who were unable to understand the mathematics he taught, complained about his pedagogy. Eventually, the college was forced to ask Kosambi to leave because students found his lectures incomprehensible. Kosambi was famous for his daily commute on the Deccan Queen train from Poona to Bombay. He switched from commuting every day to only over the weekends, and then reverted to a daily commute. The two-way train journey was more than seven hours. In addition, he walked a few kilometres from his house to the Poona station with a haversack full of books to read on the journey. He would read as many as three mystery novels in just one leg of the journey and donated hundreds of science fiction and mystery novels to the common room in TIFR.

It is said that his book *Introduction to the Study of Indian History* was written on the Deccan Queen during these lengthy commutes. It has been noted with some amusement that the preface to the book provides the Deccan Queen as his address. Since he spent the train journey reading and writing, he preferred to travel first class, prompting his Sanskrit collaborator, Daniel H.H. Ingalls, to comment that the Marxist Kosambi travelled in first-class comfort while an American capitalist, like himself, travelled second class.[32]

Eminent historian, Romila Thapar, recalls meeting Kosambi on the train when she was a young research student. She says it was a memorable occasion for her, both because of the opportunity to talk with him and because the journey itself was a journey into the past. 'For Kosambi every tree, every stone, every hamlet along the route was the source material of history.'[33] Since Kosambi also had to travel a few kilometres once he reached Bombay, first to Kenilworth on Pedder Road and then to the Old Yacht Club (OYC) at the Gateway of India, and finally to Navy Nagar, beyond the Afghan Church, he joked that TIFR was just a waiting room between his train journeys.[34]

In the early years, Bhabha and Kosambi shared a relationship of respect and easy camaraderie. Bhabha called him by his nickname 'Baba'. Both he and Kosambi were recognized as university teachers for the course of studies leading to a PhD degree from Bombay University. Kosambi shared updates about his work with Bhabha, and, at least in the early years, was very involved with the development of the institute. He functioned as the acting director from June to October 1946, when Bhabha was away in England and the US, something that is discussed in the next chapter.

EARLY RESEARCH AT THE TIFR: BHABHA EQUATIONS

Bhabha's first research paper from TIFR dealt with 'Relativistic Wave Equations for the Elementary Particles'[35] and appeared in an issue of *Reviews of Modern Physics* to commemorate Bohr's 60th birthday. As noted, Dirac announced his relativistic wave equation in 1928, which described the behaviour of electrons with spin ½ and also successfully predicted the existence of positrons.

Dirac's success led physicists to devise a formalism for elementary particles with an arbitrary spin. Richard Duffin, Nicholas Kemmer and G. Petiau had given similar first-order wave equations, describing particles with spin 0 and spin 1. Later, Dirac, Markus Fierz and Pauli proposed relativistic wave equations for particles having any integral or half-odd integral spin. An unsatisfactory feature of the equations proposed by Dirac, Fierz and Pauli for spin greater than 1 was the presence of subsidiary conditions. These subsidiary conditions created difficulties when one considered these particles in interaction with electromagnetic fields. Bhabha, therefore, investigated general first-order relativistic wave equations without any subsidiary conditions. Such equations have come to be known as Bhabha equations.[36]

While Bhabha's equations are equivalent to Dirac–Fierz–Pauli equations for spins corresponding to 0, ½, and 1, for every case of spin greater than 1, they are different. The primary difference is that elementary particles with higher spins are capable of several states of spin and mass instead of having a unique value as the Dirac–Fierz–Pauli equations do.[37] This work required Bhabha to delve into the sophisticated mathematics of group theory. According to Virender Singh (a former director of TIFR), Bhabha was among the few physicists of his time who was at home with group theory.[38] His essay, 'The Theory of the Elementary Physical Particles and their Interactions', for which he was awarded the Adams Prize in 1942, contained a good deal of the theory of orthogonal groups.

◆

When India became independent, the day was celebrated quietly with a simple flag-hoisting at Kenilworth. Vasudevachar recalled that Bhabha had given him a flag the night before to hoist on the terrace of the institute, and he had done so. When Bhabha arrived and asked about the flag, he was told that it had already been unfurled. Bhabha remarked, 'No! On the first day of independence, you don't unfurl the flag—you raise the flag.'[39] Subsequently, they raised the flag and sang the national anthem.

Independence changed the relationship between scientists and the State. Bhabha expected his scientists to eschew narrow nationalism and be a part of the universe of science, which offered ways of belonging to a world uncircumscribed by national boundaries. In early January 1948, the mathematician F.W. Levi joined Kosambi at the TIFR. A Jewish refugee from the Nazi regime, Levi was an expert on algebra, especially the theory of groups. He taught at Leipzig University from 1920 to 1935, until the Nazi government dismissed him because he was a Jew. In 1935, he accepted an offer as head of the mathematics department at the University of Calcutta.

According to the mathematician, Raghavan Narasimhan, who did his PhD at TIFR, Levi had an important influence on the development of twentieth century mathematics in India, especially by introducing modern algebra at the Calcutta University. TIFR, too, was happy to have him, since it could admit students for research on modern algebra and the theory of functions. In August 1947, Levi delivered lectures at TIFR and his arrival in 1948 gave a fillip to Kosambi's and Bhabha's plans for mathematics at the institute. Bhabha also invited Dirac's student, the Jaffna-born mathematician and physicist, Christie Jayaratnam Eliezer, to TIFR. But Eliezer, who was at the time teaching in Cambridge, declined, citing his commitments in Ceylon.[40] However, significant progress in mathematics was only made when the mathematician K. Chandrasekharan joined in 1949, something that is discussed in a later chapter.

Chapter 12

THE ATOMIC IMPERATIVE

> The Commission noted that it had recognised the Tata Institute of Fundamental Research as the only laboratory of the Commission for fundamental research in atomic science. In view of this decision the Commission would not set up another laboratory of its own for fundamental research in atomic physics.
>
> —Historical note on TIFR on the occasion of the foundation stone-laying ceremony of the institute's new building on 1 January 1954[1]

LAYING THE FOUNDATION

The story of the nuclear programme starts with the formation of the Atomic Energy Committee that started functioning in 1946 with Bhabha, Bhatnagar, and Saha as members. Membership does not appear to have been exclusive, with geologist Darashaw Nosherwan Wadia, D.M. Bose (a nuclear chemist), Nazir Ahmed (a nuclear physicist) and K.S. Krishnan also participating occasionally. Bhabha had been advocating the formation of such a committee and it is likely that his pressure on Sir Ardeshir Dalal, who was the president of the CSIR's governing body, may have been the reason it was set up under the umbrella of the CSIR. The name of the committee was changed to the Atomic Energy Research Committee (AERC), since nuclear physics was a new subject and India had little experience with it. A Board of Research on Atomic Energy, with just Bhabha, Bhatnagar and Saha as members, was also set up. It met with the committee in Bombay and disbursed the funds sourced from the CSIR on behalf of the committee.

The AERC first met on 15 May 1946, with Bhabha as chairman in Bombay House, the Tata headquarters. The committee was exploring uncharted territory, since there really was no plan on how to organize nuclear research as a State-funded activity. The idea was to formulate a general policy to develop atomic research in India in the shortest possible time. The committee agreed that since India did not have the luxury of large-scale funding for nuclear research, like the US and the UK, it was necessary for all important research in atomic physics to be concentrated at one centre in the country. This centre, it was agreed, would be TIFR. With Bhabha at the helm of the committee, there could have been no other choice.

The committee decided that the Palit Laboratory at the University College of Science at Calcutta would receive grants for a cyclotron. The Bose Institute, also in Calcutta, was to receive grants for research on transuranium elements. The TIFR would get funds for a 200-million-volt betatron[2] with a 10-member team to operate it. Bhabha had been in touch with both Cockcroft and the Australian physicist Mark Oliphant on the subject of nuclear physics and meson research. At the Cavendish, particle accelerators were at the centre of discussions—the end of the war had resulted in the excitement of men returning to their laboratories after five years.

This was also the beginning of the marginalization of Saha in matters concerning atomic energy and research in India. Slowly and steadily, Bhabha started taking centre stage. This was certainly not to Saha's liking, and as we shall see later, he became a trenchant critic of both Bhabha and his style of functioning. Bhabha's plans for making TIFR the only centre also extended to refusing permission to Raman to introduce nuclear physics at IISc, something which is discussed later in this chapter.

The anticipation of conducting further research on mesons, and the acceptance of particle accelerators to study elementary particles, had convinced Bhabha of the need to explore options for TIFR. In December 1945, he wrote to his friend from Cambridge, Oliphant asking if these machines would work in a tropical climate without air conditioning. More importantly, he inquired about the impact of Bombay's humidity and salinity on a machine like the cyclotron. If Bombay's climate was a problem, he was willing to locate it in Poona, which was drier than Bombay.[3]

Oliphant replied saying that the equipment would work just as well in tropical conditions with air conditioning being a convenience rather than a necessity. He asked Bhabha to check with the Navy and the RAF, who had operated radar equipment during the war. We don't know if Bhabha did as advised but it is very probable, since war surplus material played a very important role in the equipment made by the early experimental groups at TIFR. Oliphant volunteered to send the drawings for any of the machines if Bhabha so wanted. If Bhabha was to get anywhere with his research on the meson, a betatron was a must.

THE EMPIRE SCIENTIFIC CONFERENCE

A.V. Hill's visit to India led to two 'goodwill missions' of Indian scientists and industrialists to the UK, the US and Canada, which enabled them to see the scientific and industrial research laboratories of the Allies. The Indian Scientific Mission in 1945 included seven scientists and administrators from important scientific institutions across India. The group included, Ahmed, Bhatnagar, J.C. Ghosh and Saha. A notable exclusion was Raman, whose absence attracted criticism.

The first mission spent two months visiting various facilities in the UK before moving on to the next stop on their itinerary, which was the US. The members visited several laboratories in the US, including Ernest Lawrence's Radiation Laboratory, under strict vigilance. Work on the atomic bomb was well under way at the time and secrecy was paramount. Even so, Saha had his ear to the ground. In a report of this visit, which appeared in *Science and Culture*, he said that they had vaguely heard of the atomic cities and passed close to some of them. The visit of the Indians invited suspicion and Saha was even questioned by the Federal Bureau of Investigation (FBI) on his knowledge about nuclear fission and its use in the war effort. Many in the group, including Saha, then visited the USSR in June 1945, on the occasion of the 220th anniversary of the Russian Academy of Sciences. The mission generated much hope for post-war reconstruction in India, contributing significantly to the debate on State-led science organization.

The second goodwill mission took the form of participation in the Empire Scientific Conference. Soon after the first meeting of the AERC, in May 1946, Bhabha travelled to the UK to attend the conference. The Second World War had brought together scientists of the British Empire, especially those from the dominions, and Britain wanted to continue this fruitful collaboration in a world order that was getting increasingly nationalized. This conference would be held in close association with the conference of the Royal Society. The meeting lasted for five weeks with the first three weeks being led by the Royal Society, and the other two by the British Commonwealth Official Scientific Conference. The attending delegations came from across the globe.

The Royal Society Conference was declared open by George VI on 17 June 1946. It took place for a week each in London, Oxford and Cambridge, with laboratory visits in the afternoons. The idea was to encourage a free and frank discourse without the fetters of bureaucracy and nationality. The Indian delegation was led by Bhatnagar. Raman chose not to attend, refusing to recognize the National Institute of Sciences as representative of Indian science. The other members of the Indian delegation, apart from Bhabha, included J.C. Ghosh, Krishnan, Mahalanobis, Saha, Sahni, and D.N. Wadia. Bhabha led the discussions on cosmic ray research and Saha contributed to discussions on the dissemination of scientific information within the Empire.

Another celebration of science took place nearly a month later in the form of the Newton Tercentenary celebrations. Saha thought that as far as science was concerned, it was a rather 'flat affair'.[4] He was disappointed that there was hardly any discussion on nuclear physics. Bohr, Christian Møller, Léon Rosenfeld, Pauli, Schrödinger,

Born, Kemmer, Heitler and Bhabha, all presented papers. Saha thought the result was very disappointing because everybody seemed to have a different theory to explain one phenomenon, and they left everything else unexplained. It appeared that they did not even understand one another. Saha wrote to his former student D.S. Kothari calling the atmosphere trifling. He said that the Indian scientists working on meson theories shone in the reflected light of their western counterparts and were arrogant, as a rule. Saha added that he was happy he had not wasted his time on such barren enterprises.[5]

Saha also informed Kothari that Bhabha was planning to go to Zurich to work with Pauli, Wentzel and Fierz. It does not need much imagination to deduce that Saha's 'arrogant' worker was Bhabha.[6] The souring of their relationship had begun with the disagreements on the Board of Research on Atomic Energy's decision to make TIFR the primary centre for atomic research, effectively ending Saha's hopes of establishing his own institute as a nodal centre. They soured further when Bhabha refused to support Saha's application to the UNESCO for funds for *Science and Culture*. Bhabha felt that such an application had to be forwarded through formal channels—an application from a private individual would not be supported. Saha said that he had learnt nothing new in the meeting. The Commonwealth of Science was presented as an advantageous alliance, but nuclear research was not on the agenda. Raman, too, expressed his disappointment with the Royal Society, which he thought was tending to become more a political than a purely scientific body.

Bhabha met Powell, Cockcroft and Oliphant while he was in England and then, travelled to the US. It is likely that he tried to purchase a betatron from the General Engineering and Consulting Laboratory (GEL), New York, on this trip. Even if he had, nothing would have come of it due to the restrictive McMahon Act, which prohibited the export of dual-use technologies and whose aim was to keep secret any knowledge relating to nuclear research, even from the countries that had been allies during the Second World War.

Kosambi corresponded with Bhabha regularly during this period, keeping him abreast of routine matters as well as details about the setting up of the workshop and using war surplus material (procured from the RAF) for cosmic ray experiments. Kosambi also told Bhabha about the tardy progress of his own research.[7]

Kosambi's commitment to building a School of Mathematics is apparent from the enthusiasm in his letter to Bhabha written in early July 1946. He stressed the need to train people in branches of mathematics, such as topology, which had no specialists in India. Such training would have to come from abroad, possibly the US. Kosambi knew that Harish-Chandra had worked with Bhabha in Bangalore and wanted to know if it would be worthwhile to have Harish-Chandra spend time with Pauli in Princeton (Bhabha was in Princeton at the time). References to 'our institute' show the sense of ownership he felt and, in July, he wrote in a jocular vein that '[...] the Directorship weighs very lightly on my unworthy shoulders'. Kosambi had a biting wit, which is on display in the last part of his letter. 'I hope [...] that your health has not suffered in any way under the strain of travel, of British rationing and cooking, and the company of your fellow delegates.'[8]

From Princeton, Bhabha visited London and met Julian Huxley, the first director general of the UNESCO. Bhabha had been appointed chairperson of the 'Sub-commission on Natural Sciences' at the first General Conference of UNESCO in Paris in 1946. Huxley shared plans for a mathematical institute in Asia with Bhabha. There were plans—as Bhabha discovered from Huxley's secretary, Joseph Needham—to set up two institutes, one in India and the other in China, under the aegis of UNESCO. These institutes would be given a grant of ₹100,000 each. The recurring grant from UNESCO would be small and Needham had wanted to know if the institute should be located in Calcutta or Bombay. Bhabha asked Kosambi for his views on how the new institute would fit in with TIFR and whether he was personally interested in the scheme.[9] Calcutta and Bombay must have been obvious choices to Needham because of the presence of Mahalanobis and Bhabha, respectively, in the two places. In February 1953 Huxley offered Bhabha the directorship of UNESCO. Huxley wrote: '[you] are eminently indicated for the post—an eminent scientist, with experience of administration, an artist, with a great knowledge and appreciation of the arts and at home both in the West and the East.'[10] For reasons that are obvious, Bhabha turned down the offer.

Bhabha was excited about the UNESCO support because he thought that it would provide the institute with modern apparatus, like a differential analyser for calculating wave functions and for solutions of differential equations in astrophysics. He thought that such a machine would be of great help if Chandrasekhar joined the TIFR. Kosambi was also enthusiastic, realizing that such a machine would be useful not just for Chandrasekhar but also for him and Bhabha. A grant would also enable TIFR to purchase Vannevar Bush's new electronic calculating machine as well as an IBM machine.

Although Kosambi thought that they would face stiff competition from Mahalanobis and the ISI, he felt that a differential analyser would not be of much use to them. There seems to have been competition between Bhabha and Mahalanobis on who would get the first computer. As it turned out, Mahalanobis procured the first computer, but Bhabha built the first one in India. Kosambi had already been running a project on the Kosmograph and was interested in continuous calculators and integrating machines. He wrote to Bhabha about having the Fourier analyser redesigned. He also felt that the TIFR would greatly benefit from the computers and analysers that the mathematics institute would make possible. He, however, cautioned Bhabha about the lack of trained personnel.

There seems to have been no master plan for a mathematics department and many of the exchanges between Kosambi and Bhabha show that both were thinking aloud, trying to figure out what would work. Kosambi saw the UNESCO institute as an adjunct to TIFR, making expensive equipment, which would otherwise have been beyond the institute's budget, available. As acting director, he even discussed this plan with the council of the TIFR in August 1946. The council wanted to wait for a formal communication from UNESCO. It never came and there was no further talk about such an institute.

Bombay witnessed communal riots on the eve of Partition and the political situation continued to be a cause for concern even after. Kosambi wrote to Bhabha in September about how work was suffering because many of the tools required had to be procured from the affected areas.[11] In November 1946, he wrote to Bhabha again, bringing him up to date with the developments. Kosambi said that Bombay was yet to recover from three months of looting and rioting. He had seen a considerable amount of both, having visited the affected areas at least twice a week.

Writing about the proposed new site in Colaba for the institute, he said that Morarji Desai, who was the revenue minister, was a 'hard nut to crack', but a man of his word and very straight.

> This is all very nice as far as it goes but you see that we cannot undertake building operation this season without a proper and secure title. The principal difficulty with the Government of Bombay is its leaden foot, which is further shackled by strong chains of red tape. To dictate a letter which you or I could get out of the way within seven minutes it takes two or three interviews at the Secretariat and even then, all we get is that 'the matter will be attended to in due course'.[12]

In the same letter, he asked Bhabha if Chandrasekhar was going to join TIFR since '[...] we have to make up our cadres reasonably soon'. Kosambi went on to speak about his own mathematical work, which he called 'experimental mathematics' unfit for publication at the time. He did not forget to remind Bhabha that he was doing Bhabha's work for him. 'This is of course in addition to signing all kinds of notices, letters and applications which should normally have fallen to *your* lot.'[13]

In 1946, Kosambi was elected a Fellow of the Indian National Science Academy and received the Bhabha Prize (named after Bhabha's father, Jehangir Hormusji Bhabha) the following year. Bhabha also had him elected president of the mathematics section of the 34th ISC that was held in Delhi in December 1947, where he spoke on 'Some Applications of the Functional Calculus'.

But, even with two mathematicians on the rolls, research in mathematics made little progress. There were hardly any students who enrolled for mathematical research. The real growth in the mathematics department came only after the arrival of Chandrasekharan from IAS, Princeton, in 1949. His genius for organizing mathematical research and his contacts led to international recognition for the School of Mathematics at TIFR. In a letter written in January 1948, Kosambi identified Chandrasekharan as the person suitable for developing a strong mathematics department at the institute. Eleven years younger than Bhabha, Chandrasekharan had studied at Presidency College, Madras, and in a combination of talent not usually found, showed equal promise in both mathematics and English literature. He was advised to take up mathematics, since it was thought that there may not be much of a future in English literature after Independence.

Chandrasekharan took up mathematics and worked for his thesis under K. Ananda Rau, a student of G.H. Hardy and Ramanujan's contemporary at Cambridge. Apart from Rau, R. Vaidyanathaswamy (at the newly created department of mathematics at Madras University) and Father Racine at Loyola College, who got his PhD under Élie Cartan, were also in Madras at the time. Besides Chandrasekharan, the 'Madras School' produced other distinguished mathematicians, including S.S. Pillai, S. Minakshisundaram, P. Kesava Menon and Kollagunta Gopalaiyer Ramanathan. Chandrasekharan's mathematical interests were centred on analysis and analytic number theory. When the American mathematician Marshall Stone, who was professor of mathematics at Chicago University, visited Madras in 1944, he was keen to meet the best Indian mathematicians. After making the acquaintance of Chandrasekharan and Minakshisundaram, he was impressed enough to invite them to IAS, Princeton. Minakshisundaram, who was seven years older than Chandrasekharan, had also done his doctorate under Rau.

Kosambi had some reservations about Chandrasekharan's mathematical work, calling it 'respectable' but not 'great'. However, Kosambi thought that he was still the best candidate for the job. It was not the most peaceful of times in Bombay though, and Kosambi was worried that Chandrasekharan would find riot-torn Bombay an uncongenial place to return to.

ELIMINATING COMPETITION

We have already noted that Bhabha wanted to establish the primacy of TIFR, if not ensure that it was the only institute for nuclear research in India. The episode with Raman and his protégé R.S. Krishnan is a case in point. In 1938, Raman, who was heading the physics department at IISc, sent R.S. Krishnan to the Cavendish to train in nuclear physics, hoping that he would return to India and establish the field in Bangalore. R.S. Krishnan was admitted as a research student at Trinity College, Cambridge. Bhabha was also in Cambridge at the time and it is likely that the two met for the first time, as noted before, at the Christmas dinner in December 1938. R.S. Krishnan was awarded a PhD in nuclear physics in 1941 for his work on artificial radioactivity.

He returned to Bangalore in late 1941 and was appointed a lecturer at IISc in August 1942. His intention was to establish a particle accelerator laboratory to study artificial radioactivity and fission. R.S. Krishnan struggled for five years to establish nuclear physics in Bangalore. He failed, or more accurately, was not allowed to succeed, and, by 1947, had left the field altogether. Bhatnagar diplomatically said that he was not against any other laboratory and was open to a free and frank discussion on the subject. He, however, was complicit in scuttling R.S. Krishnan's plans. On 19 July 1947, he wrote to Bhabha saying that IISc should not be allowed to establish a chair in nuclear physics because it would be in conflict with the development of TIFR as the centre of nuclear research. He was worried about IISc, CSIR and TIFR competing among themselves when the funds came from the same source.

Bhatnagar sent Bhabha a letter for Rustum Choksi, who represented the Tatas on the council of IISc. He wanted Choksi to uphold the decision of the CSIR to develop TIFR as the centre for nuclear research. 'I am not against the creation of new Chairs in Physics but I am particularly interested in Bangalore developing other branches because nuclear physics has been taken up at Bombay and unnecessary rivalry in Tatas own organization is not good.'[14] In his reply, Bhabha said that it was the considered view of the Atomic Research Committee, that all large-scale programmes in the future were to be concentrated at one centre. Since Bangalore was not an existing centre of nuclear research, he could not support the creation of a centre there, since it would only be a waste of money and manpower.[15]

Meanwhile, the council of IISc decided to set up a committee to consider R.S. Krishnan's scheme for starting a nuclear physics laboratory at the institute. A committee was formed to deliberate upon the question of a nuclear physics laboratory at IISc. It met on 31 July 1947 at the Royal Institute of Science in Bombay. Those present at the meeting included Bhabha, J.C. Ghosh, C.V. Raman and H.J. Taylor, who was a nuclear physicist and missionary who had an MSc from the University of Sheffield. In 1927, he was appointed by the United Free Church of Scotland as an educational missionary to Wilson College, Bombay. He held the post of professor of physics there for 27 years, building up a great tradition of active research in his department, and producing some top physicists in the 1930s and 1940s. He was a fellow of the Bombay University and a member of the Senate.

From 1933 to 1935, Taylor was at the Cavendish as a research student under Rutherford. He was fortunate to be there during its heyday, a period that coincided with one of the most creative and fruitful periods in the history of physics.[16] There, he worked on the photographic emulsion technique for the detection of nuclear particles, gaining his doctorate for this work. He continued work in this field after returning to Bombay, taking part in high altitude balloon research on cosmic rays. He was the second person to join TIFR, after Kosambi. Bhabha had requested him to work part-time at the institute as an experimental physicist. In 1937, Raman had elected him to the Royal Academy.

Taylor also helped in the war effort and was on a committee to handle scientific questions arising from the armed forces, especially the navy in Bombay.[17] In December 1948, Bhabha prevailed upon Taylor to accept a part-time appointment as professor of experimental physics at TIFR. Taylor retained his responsibilities in Wilson College, since he realized that this was never going to be a permanent arrangement. However, he held on for three years and was there when the physicist Bernard Peters came to Bombay. Taylor was responsible for selecting young candidates for TIFR and visited Madras and other places to interview them. One of the candidates he recruited was Ranjan Roy Daniel. Taylor got along very well with Bhabha and was invited to Mehrangir. Like Bhabha, he was a man of many gifts. A born researcher, a powerful speaker and a gifted writer of prose and poetry, he was an example of great learning, lightly worn. He also had a good sense of humour. Bhabha took him swimming, presumably to the Willingdon Sports Club.

The meeting of the committee ended in an impasse. Both Bhabha and Taylor opposed the plans for a nuclear laboratory at the IISc, and J.C. Ghosh asked them to submit a written report elaborating their position. While the decision to reject the Bangalore proposal had not yet been taken, the fact that Bhabha and Bhatnagar had discussed inviting R.S. Krishnan to join TIFR leaves no doubt about what the end result was going to be. The Bhabha–Taylor report, and a written response from R.S. Krishnan, were published in the council minutes of 8 September 1947. The authors adverted to the huge capital expenditure and argued that the very nature of the field and its fundamental character put it in a different category and not one that could be taken up on a small-scale.[18]

Jahnavi Phalkey has argued that while it is easy to explain the rejection of the Bangalore proposal in terms of competition, it would do well to remember that there was a similar debate in the UK about the role of Harwell, the atomic energy establishment of that country. Cockcroft wanted Harwell to be the one good facility for nuclear research. James Chadwick thought that injury to the universities apart, a focus on basic research at Harwell would deviate from its primary responsibility, which was to work on questions of nuclear energy alone. British universities were comparatively smaller than the ones in the US and certainly much less equipped to carry out large-scale nuclear research. Bhabha's doubts about a nuclear research facility in Bangalore could well have been informed by similar concerns.[19]

After the IISc council meeting ended, Bhabha invited R.S. Krishnan to join TIFR. The latter refused politely, citing his continuing research in solid-state physics and his poor health. He knew that Raman was to retire the next year and he was the frontrunner to succeed Raman as the head of the physics department. This meant at least some freedom in organizing research, unlike at TIFR, where he would be led by Bhabha. On Cockcroft's recommendation, Krishnan was appointed as Raman's successor and he shaped the department into an important centre for solid-state research in India.

INVITING MAX BORN

Within a few years of setting up TIFR, Bhabha was heading the AEC, which was established in 1948. He was also in charge of the AEET and the pressure of building two institutions simultaneously took its toll on his research output. After 1954, Bhabha published no research papers; his book on elementary particles was never completed, and hence never published. While he was fully conversant with the scientific trends of the day, he did not have the time to pursue new lines of research himself. This lack of research output was incompatible with his status as the head of a premier research institution, and Bhabha seems to have been aware of this. Theoretical physics, Bhabha's main field, had a slow start in the early years at TIFR. From 1950 to 1959, only seven students completed their PhDs under Bhabha, including R.P. Thatte and B.V. Sreekantan, the latter being director of TIFR from 1975 to 1987.

Recognizing his limited research output, Bhabha made every effort to create an atmosphere congenial to research. The most important part of this quest was to attract scientific leaders to the institute, eminent academics, who would not only contribute to the atmosphere of research but, in fact, lead it. One of the outstanding scientists Bhabha extended an invitation to in October 1947 was Max Born. We have already noted Bhabha's friendship with Gustav, Born's son, when he was posted in India during the war. Bhabha had heard of Born's impending retirement from Edinburgh (he was the Tait Professor of Natural Philosophy) and wanted to get him to Bombay. In a letter to Born written on 21 October 1947, Bhabha assured him of full freedom in his research, one assistant, and as many research students as he desired. Bhabha also promised to keep his teaching burden light and limited to a few lectures a week. The other perks included an air-conditioned office and the holidays he would be entitled to.

In his reply to Bhabha, Born said that the news of his retirement was untrue: he planned to continue in Edinburgh till he superannuated. By the time he retired, he wrote, he would be over 70 and far too old to be of any great use. He told Bhabha that he had been offered the chair at Dublin, which would allow him to continue till he was 75. Schrödinger, Born said, wanted him to accept the offer, but he had decided against it because there was no provision of a pension in case he was forced to retire earlier due to ill health.[20]

Born, however, did not turn Bhabha down completely. He offered to come as a visiting professor a year or so later.

> I think it would be possible for me in a year or so, to get leave of absence, say for one or two terms and to come to your Institute as a visiting Professor [...] I could not, of course, ask at once for another such extended leave, but perhaps in a year or two this would be possible, if you really think you would like to have me in Bombay.[21]

This particular attempt to find a suitable prestigious leader for theoretical physics at TIFR failed, but Bhabha had not given up on the idea of persuading a big name to visit and spend some time at his institute.

MEETING JOHN VON NEUMANN

In January 1948, Bhabha visited Princeton for a day and had a long conversation with the mathematician John von Neumann about calculating machines. Born in 1903, von Neumann was a pure mathematician of extraordinary ability. When he was just 22 years old, he helped lay the mathematical foundations of quantum mechanics. He moved to America in 1930, and realizing early on that the Second World War was imminent, studied the mathematics of ballistics and explosions. He lent his expertise to the American armed forces and the Manhattan Project. At Los Alamos, where the atomic bomb was developed, von Neumann determined the arrangement of explosives that would be required to detonate the more powerful 'Fat Man' device by compressing its plutonium core.[22] He also was one of the founders of game theory—a field of mathematics devoted to understanding conflict and cooperation. After the bombing of Japan, he turned his attention to computers, and his musings on the parallels between the working of the brain and computers helped trigger the birth of artificial intelligence and the development of neuroscience. His seminal paper from 1946 described the basic logical scheme for computers, now called von Neumann architecture. As von Neumann's biographer Ananyo Bhattacharya observes, 'Look around you and you will see Johnny's fingerprints everywhere.'[23]

Like Bhabha, von Neumann was born into wealth and privilege. He too stood at the intersection of the humanities and sciences. His father was an amateur poet and his mother believed that music, art and related aesthetic pleasures played an important part in human lives. A child prodigy, it is said that, among other mental feats, he could recall verbatim pages from books or articles he had read in any of five languages.[24] John von Neumann, who liked flashy Cadillacs (but drove atrociously), enjoyed annoying neighbours, such as Einstein and Kurt Gödel, at IAS in Princeton, by playing German marching tunes loudly on his office gramophone.[25] He was a bon vivant who hosted sparkling parties at which he loved telling jokes and reciting risqué limericks in a number of languages. He ate so heartily that his wife once joked that he could count everything except calories.[26]

Both Bhabha and Kosambi were interested in acquiring a calculating machine for the institute but the scarcity of foreign exchange was an important concern. Bhabha wrote to von Neumann expressing the view that it would be preferable to put such a machine in an institute where it would be used only part of the time so that other institutions might also have a chance of using it.[27] In his reply, von Neumann agreed with Bhabha.

> They will at first necessarily be rather rare and, therefore, the ones which exist should be used on as broad and as varied a basis as possible. From every possible point of view, it would be a great mistake to tie them down to any one particular problem and to any kind of routine production [...] I think it would be very desirable to have the machine run under a system of organization

which makes it accessible to many scientific groups and guarantees a wide and varied use.[28]

Meanwhile, Bhabha had been making his own inquiries about Chandrasekharan with von Neumann while he was at Princeton. The latter told him that Chandrasekharan had been working under Salomom Bochner[29] and the latter had praised his recent work calling it '[…] as good as had been done in this line by anybody during the last few years'. Writing to Kosambi from New York, in January 1948, Bhabha was certain that Chandrasekharan would be offered a suitable appointment at an American university but thought that if a good enough offer was made to him in India, he would prefer to return.[30]

Bhabha also discussed the issue with Weil who thought that Chandrasekharan would be an asset to the institute. Bhabha believed that Chandrasekharan would accept a reader's post in the pay scale of ₹500–800, if it were offered to him. However, Bhabha did not offer the post to Chandrasekharan immediately, perhaps somewhat put off by Kosambi's assessment of the kind of mathematical work he would do. But he seems to have almost made up his mind and broached the subject with Chandrasekharan when he was at Princeton. In February 1948, he raised the issue of his appointment at a council meeting, and Chandrasekharan was offered the post of reader the same year. He joined only the next July, though, because his term at IAS was extended by a year. His return would prove to be an inflexion point both in his career and for mathematics at TIFR.

There is an unauthenticated story about the advice von Neumann gave Chandrasekharan in Princeton when he heard that he was joining Bhabha in Bombay. Chandrasekharan and von Neumann were taking a walk together when they spied Bhabha walking with Einstein at a distance. The latter asked Chandrasekharan if it was true that he was planning to join TIFR. When Chandrasekharan responded in the affirmative, von Neumann gave him some advice: 'That man is as good a physicist as any, but do not let him intimidate you—stand up to him.'[31] Bhabha and Chandrasekharan were known to have differences of opinion and the latter left TIFR in 1965, shortly before Bhabha's untimely demise in January 1966. Perhaps Chandrasekharan did take von Neumann's advice seriously. We will return to Chandrasekharan and his contribution to mathematics at TIFR in a later chapter.

In December 1948, Bhabha arranged a UNESCO deputation to the US and UK for Kosambi. In a letter to the Ministry of Education in April 1948, Bhabha recommended Kosambi's deputation to study the latest types of machines and to work with mathematicians like von Neumann, who was designing such a machine. When the visit materialized, Bhabha even organized a party for Kosambi at Mehrangir before he left. Kosambi's brief included investigating the possibility of getting a computing machine for TIFR and attracting new talent, in particular, the mathematicians Chandrasekharan and Minakshisundaram. During the winter session, Kosambi was visiting professor of geometry at Chicago University and delivered a course of 36 lectures on tensor analysis. He also visited the IAS, Princeton, where he had discussions with Einstein. His extended visit helped considerably in networking with leading figures, like Stone and J. Robert Oppenheimer. However, there is no record of his involvement in designing calculating machines when he returned to India.

If Kosambi's correspondence with Bhabha is any indication then, as early as 1948, Kosambi had started getting a feeling that he was no longer fitting in with the main purpose of the institute. He wrote to Bhabha, who was in the US at the time, telling him that the institute needed people whose work was connected with modern physics, people like Harish-Chandra and others. He went on to add: 'Then I and my strong department of mathematics will not be a real help to the main purpose of the Institute.'[32] But their relationship at that time was still a very cordial one and Bhabha appreciated Kosambi's knowledge of Sanskrit. In his reply, Bhabha, while discussing Kosambi's forthcoming visit to the US, said: 'Oppenheimer would be doubly glad to have you at Princeton, not only because of your mathematics but also as he pretends to read Sanskrit and is therefore tickled to death that you have edited the poems of Bhartrihari.'[33]

A large part of Kosambi's attention at this point was taken up with applied statistical analysis. Bhabha, however, was very clear about the primary role of mathematics at the institute. While Kosambi was still publishing papers in mathematics, this was also the period when he was beginning to turn increasingly towards history and archaeology.

Kosambi was a versatile scholar, who would have fitted in very well in an academic institution that promoted the study of both the liberal arts and the sciences. However, he was cast more in the mould of a solitary genius working in isolation and does not appear to have had the leadership qualities required to helm a group. The gradual shift in his research direction, the rapid professionalization of the school of mathematics at TIFR from 1949 onwards, and his Marxist tendencies did little to endear him to the establishment. Gradually, his relationship with Bhabha deteriorated to the point where the two avoided meeting each other. Kosambi continued to lecture and work on mathematical problems, but as we will see later, it was all to end on an unhappy note.

Chapter 13

EARLY BIRDS

> The fine location of the Royal Institute of Science, its handsome building and the enthusiasm of some of the members of the staff made it a real pleasure to work there [...] I remember that it was in the main lecture hall that I first heard of cosmic rays, the subject which was later to become my own special field of study.
>
> —Note on the Royal Institute, 12 September 1945[1]

By the time Bhabha started a major experimental programme in cosmic rays at TIFR, cosmic ray research had entered its second phase the world over. Cecil Powell had discovered the pion as the parent of the muon in 1947 by exposing the newly developed high sensitivity nuclear emulsions at Jungfraujoch in Switzerland. In December of the same year, George Rochester and Clifford Butler discovered the so-called V-particles in an experiment using a cloud chamber. At around the same time, H.L. Bradt and Bernard Peters discovered that cosmic rays contain heavy nuclei. Also, more importantly, the act of meson production was caught both in nuclear emulsions and multi-plate cloud chambers. With these developments, the new directions of cosmic ray research became clear.

It was obvious that for a careful study of the particles produced in high energy nuclear interactions, attention would have to be focussed on nuclear emulsions and cloud chambers. These studies, which needed to be carried out in a variety of settings, needed specialized detector systems. These systems required the development of a variety of technologies: plastic balloon fabrication technology, fabrication of Geiger-Müller counters, plastic scintillators, multi-plate cloud chambers, pulsed electronic circuits and, later, even a digital computer. Thanks to Bhabha's organizational genius, this was done in record time at TIFR itself. The Indian electronics industry was non-existent at the time, and importing material was not an option due to the shortage of foreign exchange.

Bhabha, showing great foresight, initiated activities using both nuclear emulsions and cloud chambers. R.R. Daniel had set up a first-rate nuclear emulsion laboratory at Kenilworth, which was later moved to the basement of the OYC. The cloud chamber, which A.B. Sahiar had brought from Bangalore and set up at Kenilworth, was also moved to the new premises. Several other cloud chambers were built over the next decade and, in December 1954, two multi-plate cloud chambers (named Rani and Maharani) were set up one above the other in the laboratory at Ooty in the premises of the Raj Bhavan by Sahiar and others.[2]

There are two parts to pursuing science: doing great research and building an institution that does great research. When scientists become heads of institutions, their managerial duties include hiring the right people, mentoring them, suggesting new experiments and offering insights based on their experience. Bhabha did all these and more. When Steve Jobs' biographer Walter Isaacson asked him what Jobs' best product was, he expected Jobs to say Macintosh or iPhone. However, Jobs said that more important than creating a good product was the creation of a team that could continually make such products.[3] Many years before Jobs, Bhabha's actions too were informed by a similar philosophy.

In April 1947, the physicist and mathematician Prahlad Chunilal Vaidya became one of the earliest members to join TIFR as a vacation student.

He was no greenhorn, however, having studied under Professor V.V. Narlikar, who taught mathematics at BHU. Vaidya had been inspired by Narlikar's lecture on general relativity in Bombay in 1937 and wrote to him expressing interest in working with him. He became Narlikar's private student, riding a bicycle twice a week to BHU to meet him at his residence. In 1942, Vaidya's work on general relativity led him to develop what is now known as the 'Vaidya metric'. Consider a massive, spherical star radiating large amounts of energy in outer space. How does the gravitational field around it behave? Vaidya worked on this problem and his solution, the Vaidya metric, was an important contribution in this field.

On his return from Banaras, he worked as a lecturer in a college in Surat for three years before receiving Bhabha's offer to join TIFR. He accepted with alacrity. On the day he joined, Bhabha asked him to speak at a colloquium later in the month. Vaidya recalled attending lectures by Bhabha, the mathematician F.W. Levi, and R.C. Majumdar of Delhi University. He also learnt tensor analysis from Kosambi. Unfortunately, Vaidya had to leave Bombay in 1948 because of accommodation problems. Bhabha had planned to escalate the matter to the level of the prime minister, but his reassurance that he could arrange something did little to convince Vaidya, who left for Anand and continued his career in Gujarat for the rest of his life, becoming vice chancellor of the University of Gujarat in 1978.

Bhabha's pedagogy did not involve spoon-feeding. Students were provided with a list of books and publications that they were expected to read and discuss among themselves. Peer learning was a significant aspect of practising science at the institute and Vaidya noted the role of the Saturday Club, founded in September 1947, in this regard. It was a discussion group, where the latest developments in theoretical physics were discussed. It also provided a forum for familiarization with the work of foreign visiting professors before they began their series of lectures at TIFR, an example being the work of Mexican physicist, Manuel Sandoval Vallarta in 1947.

Lecture notes constituted an important pedagogic tool as well. Students were asked to take notes during the lectures, and once approved by the lecturer they were circulated in a cyclostyled form, becoming an important learning resource. As early as 1949, Bhabha stopped delivering full courses of lectures. Students were taught by a long stream of eminent visitors—a galaxy that included top names like Paul Dirac, Wolfgang Pauli, Gregor Wentzel, Robert Serber, Léon Rosenfeld, George Gamow and Robert Marshak. When top foreign scientists visited, students were nominated to write their lecture notes. This served to greatly reduce the 'distance' between the foreign visitor and the students and also built up their self-confidence. Thus, B.M. Udgaonkar wrote the lecture notes for Pauli in 1951, E.C. George (later E.C. George Sudarshan wrote the notes for Marshak in 1953 and K.K. Gupta for Wentzel in 1956. These notes, when printed, became a scientific resource for students all across the country).

Udgaonkar, who joined TIFR in 1949 as a research student under Bhabha, recalled that TIFR was different not only in its pedagogy but also in the way candidates were interviewed. When a visiting Sri Lankan physicist asked him a question on a subject that he was not expected to have studied in his MSc, Bhabha immediately rescued him by turning the line of questioning to topics that he was familiar with. Udgaonkar says that this represented an object lesson in how to conduct an interview: test a candidate on subjects he knows, instead of trying to trip him on subjects he is not conversant with. The idea was to test knowledge not expose ignorance. When Udgaonkar joined, Bhabha asked him to maintain a logbook of what he was reading. It was clear to him that while the institute would provide the necessary resources, the effort was to be his.

Udgaonkar wrote a few research papers on the theory of elementary particles and was sent to Saclay, France, in 1953 for training in reactor physics. He recalled that Bhabha did not put his name on the two papers they did together. Udgaonkar observed, 'This was a lesson for me, especially as the practice is very common in our country.'[4] He returned from France two years later, and, by 1959, his focus was entirely on high energy physics. With Bhabha's encouragement, he went to Berkeley in 1960 to work with Geoffrey Chew, who became famous for his bootstrap theory of strong interactions. He collaborated with Murray Gell-Mann and Virendra Singh, then a student at Berkeley. He published several papers on the bootstrap approach to hadron dynamics. Upon returning to India in 1963, he took charge of the theoretical physics group, which soon acquired an international reputation.

According to Udgaonkar, realizing the deficiencies in the syllabi of universities, Bhabha had tried to modernize the physics syllabus at Bombay University. Despite his eminence, Bhabha could not persuade the authorities to introduce changes. He had even assured the university that students who secured high marks after the syllabus was modernized would be assured of jobs in the atomic energy programme. When there was no response, Bhabha started the Atomic Energy Training School.[5]

THE FARMER'S SON FROM PUNJAB

Like Udgaonkar, another early bird at TIFR, was Piara Singh Gill, who joined in June 1947. His was undoubtedly a unique story. The son of a farmer from a village in the Hoshiarpur district of Punjab, Gill sailed to the US in 1928 in search of his fortune. He struggled for many years, doing odd jobs for a living but managed to complete his bachelor's and his master's degree from the University of Southern California[6] before joining Chicago University for a PhD under Arthur Compton.

In the 1930s, Compton discovered the 'latitude effect' of cosmic rays and despatched Gill to Tasmania with an ionization chamber to gather data. Between July 1937 and September 1938, Gill made 15 trips between Vancouver and Hobart. Compton may have discovered the effect but it was Gill's data that went a long way in proving it beyond doubt. Gill, along with Marcel Schein, also studied the size frequency distribution of cosmic ray bursts, a phenomenon associated with extensive air showers, a subject that was also Bhabha's area of interest. Gill also showed that the pion has spin 0.

In March 1940, Gill was awarded his PhD and Compton offered him a research fellowship at Chicago, but he was keen to return to India. Compton generously gifted Gill the equipment he had built in Chicago and, in April 1940, he sailed for India. He had some difficulty in getting his equipment cleared by customs, and Saha came to his rescue. They, subsequently became lifelong friends.

In August 1940, Gill started teaching at Forman Christian College, Lahore. In his autobiography, Gill wrote about how he first met Bhabha in 1940, when the latter was a reader at IISc in Bangalore.

> Dr Bhabha asked me to give a talk on bursts, the well-recognised work I had done with Dr Schein. After my lecture, Bhabha suggested that I go back to the States as there would hardly be any future for me in India. I took this as friendly advice; however, the events which followed made me suspect that Bhabha was keen to keep me out of India for other reasons.[7]

In the summer of 1945, Gill organized a scientific expedition to the Baralacha La Pass, which connects Lahaul in Himachal Pradesh to Ladakh, to conduct experiments on meson production by non-ionizing cosmic rays at 17,000 ft. The results showed no production of mesons. When these experiments were later repeated at Lahore using the Mosquito bombers of the RAF, meson production was detected at 20,000 ft.[8] After the war, entire squadrons of British warplanes were sitting idle at Indian airbases. With Saha's help, Gill was able to persuade some of the pilots to take him on high altitude flights where he could expose photographic plates to cosmic rays.

Bhabha was also performing similar experiments, and if Gill is to be believed, wrote to the Air Force authorities asking if a plane could be provided for high altitude studies since his was the *only* institute doing worthwhile cosmic ray work. On that occasion, at least, his request was declined and he was asked to coordinate with Gill. It was very clear to Gill that Bhabha wanted TIFR to be recognized as the sole institute for cosmic ray research.

Gill also accused Bhabha of using the unpublished data from his expedition that he shared with Bhabha without due acknowledgement.

> I shared my results with Dr Bhabha [...] In acknowledging receipt of my results he wrote to me on May 3, 1946, "...We recently detected the same hump at 500 millibars in the curve of mesons penetrating 30 centimetres of lead. You will see this in the paper which will appear this month in the proceedings of the National Institute of Science." In that paper, however, no acknowledgement was made of my results.[9]

In 1946, the government selected six scientists to visit laboratories in America and Europe for a period of six months and Gill was one of them. In the US, he was invited by Compton, who was then Chancellor of the University of Washington in Saint Louis to deliver a lecture on his work on cosmic rays in India. He spoke on his work in the Himalayas and often referred to Bhabha's research on cosmic rays. However, Gill, when he was dining with Compton that evening, discovered that Bhabha had not been as forthcoming. Bhabha had been in Saint Louis 10 days before him and had met Compton but had studiously avoided any reference to Gill's work on the same subject. Gill observed, 'Bhabha also tried to persuade Compton to advise me to join him in Bangalore. After listening to talks by Bhabha and me on the same subject, Compton was of the opinion that India was a large country, and I should stay as far away from Bhabha as possible.'[10]

However, fate decreed otherwise. The Partition forced Gill to relocate from Lahore and, in June 1947, he became the professor of experimental physics at TIFR (with Nehru's intervention, if his version is to be believed). Gill had already conducted several experiments on the azimuthal variation of cosmic rays on the basis of Manuel Vallarta's theory. Soon after joining TIFR, he launched experiments to measure the azimuthal variation of cosmic rays in Bombay. In December that year, Gill started a series of balloon flights from the Juhu aerodrome. It grabbed the attention of the press and a full-page article and an interview with Gill appeared in the pages of the *Sunday News of India* on 25 January 1948.[11]

This set the cat among the pigeons. While there can be no doubt that the story of a farmer's son who had become a scientist made for interesting reading, TIFR authorities decided that they would not countenance such self-promotion. Bhabha was away in Washington DC at the time, and Rustum Choksi was in charge. Choksi not only conveyed his sense of outrage to Bhabha in a letter, but also communicated his distress to Gill, reminding him that it had been made amply clear to him in the past that 'no publicity of this kind could be given except through the authorities of the Institute'.[12]

Gill, in his defence, said that he had made it very clear before joining the institute that he was to have the freedom to publish 'when and where I please without referring it to the Institute authorities'.[13] He also appears to be a little miffed that Choksi who was only the acting director had thought it necessary to place the matter before the council despite his clarification. He wanted Bhabha's views on the subject before responding to the points raised by Choksi.

Gill pointed out that he had nowhere made public the results of his experiments and asserted his right to talk about his work done at Lahore and in the US. Choksi viewed the article on Gill as cheap publicity, something which was to be discouraged in an institute that was trying to build an academic culture. While scientific articles and results were peer reviewed, there was no such check on newspaper articles. Thus, while publication was central to scientific work, publicity in the media was perceived as undignified and unbecoming of scientific work.

The TIFR council met in late February 1948 and the matter was discussed at some length. Predictably, it concluded that such publicity was not in the best scientific tradition and was likely to harm the image of the institute. The views of the council were communicated to Gill by Saklatvala; no punitive action had ever been contemplated, the rap on the knuckles in the form of a reprimand was considered sufficient to get Gill to toe the line. In fact, at the same meeting a decision was taken to send Gill to the Carnegie Institute in Washington to work with Vallarta for six months from 1 April to 30 September 1948. However, Gill cut short his visit and returned to India in July and resigned from his post at TIFR. In his autobiography, Gill said that he found himself out of place with Bhabha's autocratic style of functioning.[14]

The premature return and resignation led to further unpleasantness. There seems to have been a problem in the acceptance of his resignation. Gill, who knew Nehru well, wrote to him about it. Nehru did not reply, since the letter—by accident or design—ended up with Bhatnagar, who sent Bhabha a telegram asking him to send Gill to Delhi, since the Prime Minister wanted to meet him. Bhabha did not inform Gill about this development. Gill finally decided to go to Delhi on his own and met both Bhatnagar and Nehru in the latter's office. Nehru asked Bhatnagar to accompany Gill to Bombay and ensure that his resignation was accepted at the council meeting, which he wanted convened the very next day. Gill says that on the plane, Bhatnagar made many derogatory remarks about Bhabha and asked him to remain firm.[15]

At the council meeting held in Bombay House, Saklatvala told Gill that they had decided to accept his resignation but that he should apologize to Bhabha for returning prematurely from the US. Gill refused to apologize, saying that he had successfully completed his work and deserved to be congratulated. He then walked out of the room. Bhatnagar, who was a little perturbed at the turn of events, cautioned Gill. 'They are very powerful people. They will ruin you unless you write a letter of apology as directed by the Chairman.'[16] That same evening, they took the Frontier Mail to Delhi. Before boarding, they were met by TIFR's registrar, who wanted Gill to sign a letter of apology drafted by him. Gill declined to sign the letter but drafted one of his own. In it, he said that he was sorry if his return from the US earlier than planned had caused any inconvenience to the director. The next day he apprised Nehru of the developments.[17]

Later, Gill was appointed Officer on Special Duty (OSD) in the AEC by Nehru with a promise of better remuneration. The appointment letter, however, indicated no such increase, and K.S. Krishnan told Gill that Bhabha had opposed the pay hike. Zakir Hussain, who was the vice chancellor of AMU at the time, invited Gill to head the department of physics there. Consequently, Gill left to join AMU and the 15 years he spent there were the most fruitful years of his life. In 1951, he became the director of the High-Altitude Laboratory at Gulmarg, which was a collaborative venture with AMU and the University of Jammu and Kashmir.

When AMU requested for funds from the AEC, the request was refused on the grounds that it was unsuitable for such research and that the Prime Minister had refused financial assistance on the advice of experts. Gill was director of the High-Altitude Laboratory from 1951 to 1971. In 1963, ironically, Bhabha started his own high-altitude research laboratory in Gulmarg. It was not such an unsuitable site after all! However, there was no cooperation between his and Gill's laboratory. Obviously, Bhabha resented any kind of competition and was not above trying to subdue it.

In 1959, the Central Scientific Instruments Organisation was established to produce, design, service and repair scientific instruments and train technicians. Gill was its director from 1963 to 1971. After retiring, he started a business of manufacturing magnetic heads for tape recorders. Eventually, he sold the unit and migrated to the US, where he lived with his daughter until he passed away at the ripe old age of 90.

THE LAWYER'S SON WHO BECAME A SCIENTIST

One of Bhabha's first students at TIFR was Alladi Ramakrishnan. When Ramakrishnan was a BSc student at the Presidency College in Madras, he had attended a lecture by Bhabha in 1941 on meson theory and was greatly inspired by it. He wanted to pursue a career in science, but he needed to convince his father first. As the son of Sir Alladi Krishnaswami Ayyar, a prominent Indian lawyer who helped draft the Indian Constitution, a lucrative career in law awaited him. However, Ramakrishnan's mother convinced her husband to allow their son to pursue a career more suited to his natural inclination.

It was on one of his visits to Delhi with his father in 1947, that Ramakrishnan got to meet Bhabha. Sir Alladi had been invited by Bhatnagar to a dinner to meet Bhabha. Since he was unable to attend, he suggested that Ramakrishnan go in his place. At the dinner, Ramakrishnan told Bhabha how his lecture in 1941 had inspired him to change his career. Bhabha invited Ramakrishnan to meet him in Bangalore and start working under him.[18]

Eventually, Ramakrishnan moved to Bombay to work with Bhabha at Kenilworth as one of the first members of TIFR. Finding accommodation for students was a serious problem at this time. Ramakrishnan took up temporary residence at the Taj Mahal Hotel but such luxury was out of reach of other students. Aware of the influence Ramakrishnan's father wielded in New Delhi, Bhabha asked Sir Alladi to help resolve the accommodation issue. He wrote about the RAF and army hutments in Bombay, which were vacant: 'It seems monstrous to me that red-tape in our government should be such as to let hutments remain vacant when housing is so short. If you could really help us in getting some of these released it would go a long way in helping our students.'[19]

Bhabha introduced Ramakrishnan to cascade theory and the problem of cosmic radiation, the study of which required the probabilistic analysis of the distribution of a discrete number of particles in continuous energy space. Ramakrishnan approached this problem in a different manner, introducing a new technique, that he called the method of product densities. However, Bhabha preferred to pursue the problem in his own way. Bhabha was a master of limit processes and he approached this problem from that angle, using techniques from measure theory.

Ramakrishnan's approach was the mathematical equivalent of Bhabha's prolonged limiting process. Although Ramakrishnan had provided a much simpler solution, Bhabha preferred his own approach. As Ramakrishnan recalled, 'At that time Bhabha was under the influence of Kosambi who could not appreciate the simplicity of my formulation.' Ramakrishnan left TIFR in 1949 and joined Manchester University, where he completed his PhD under the statistician M.S. Bartlett. The solution he had arrived at while working with Bhabha was published in a major journal.[20]

Although Ramakrishnan left TIFR due to lack of recognition of his work, he never held a grudge against Bhabha. He never criticized him publicly or in writing. On the contrary, he was grateful that Bhabha had given him a good problem to work on and acknowledged this on many occasions. However, he was disappointed that Bhabha had not recognized his ideas. He admired Bhabha and even after leaving TIFR, he tried his best to maintain good relations with his mentor. In a keynote address that he delivered at the inauguration of a conference in Madras in 1998, Ramakrishnan spoke about the episode: 'Bhabha summoned me as his research student and gave me the greatest gift a Guru can give a Sishya: the unsolved problem of cosmic radiation.'[21]

Ramakrishnan attended a conference in Edinburgh in 1949, where he met Bohr, Heisenberg and other famous physicists. This led to him being invited to several top European universities. After his PhD, Ramakrishnan returned to India and joined the Madras

University, first as a reader, then as professor of physics. In 1956, at a high energy physics conference at Rochester, Ramakrishnan met J. Robert Oppenheimer, then director of the IAS, Princeton. When Oppenheimer inquired about Ramakrishnan's plans for the future, the latter told him of his desire to visit the institute. Within a few months, Ramakrishnan received an invitation to visit Princeton for the academic year 1957–1958.[22]

This visit was a turning point in his career. At Princeton, Ramakrishnan got the opportunity to listen to the lectures of leading physicists of his generation. A year later he returned to India, filled with a desire to induct talented students into theoretical physics and expose them to the latest advances in the field. He wanted to create the same electrifying experience for his students at Madras University that he had experienced at Princeton. Since officials remained unresponsive, he started an advanced lecture series, titled the 'Theoretical Physics Seminar' at his home, Ekamra Nivas. Ramakrishnan invited eminent scientists passing through India to talk to his group of budding theoretical physicists. He had access to this calendar, due to his ties with Bhabha and a few eminent European scientists. His wife graciously hosted the visitors and the students.[23] Ramakrishnan was perhaps the first to introduce stochastic methods in particle physics and can be considered the father of the Monte Carlo simulation[24] work being done today. Unfortunately, his contribution is never acknowledged.

B.V. SREEKANTAN AND THE KOLAR GOLD FIELDS

B.V. Sreekantan, who did his PhD under Bhabha, knew him from the time he had delivered lectures at the Central College in Bangalore. Sreekantan specialized in electronics for his MSc and joined the communication engineering department of IISc in August 1947. However, his heart was in physics. As a student, he had been interested in fundamental research and wanted to join Bhabha to do theoretical research. Bhabha called him for an interview in 1947 at the West End hotel in Bangalore. Bhabha, who was rarely punctual, turned up two hours late, and went straight for breakfast. He never returned to the lobby, leaving Sreekantan hurt and frustrated.

Later, Bhabha sent him a letter asking him to apply for a PhD. He was interviewed in Bombay in 1948 and soon became Bhabha's student. In fact, he was interviewed thrice on the same day. The first interview with Bhabha was on experimental physics. When he entered the room for the second one which was on theoretical physics, Bhabha told the others, 'I have examined him in physics. He is quite good. Now you examine him in mathematics.'[25] During the third interview, Bhabha asked him what he wanted to do. Sreekantan was inclined towards theoretical physics but left the decision to Bhabha, who wanted him to be an experimental physicist. It was the start of a long and illustrious career for Sreekantan, who went on to become director of the TIFR.

Sreekantan observed that Bhabha appreciated merit and would not tolerate underperformance. 'But if somebody has to perform, he had to be given time and he would do that, but if he did not perform even then, of course Bhabha would not tolerate. Mediocrity does not help, particularly in scientific research and advancement of a country.'[26] Sreekantan noted that Bhabha abundantly utilized his international contacts and the TIFR had '[...] a stream of foreign visitors and experts in various fields giving lectures, participating in discussions and criticising our programmes when they were not right. All this helped in evaluating our activities and laying a pathway for the future.'[27]

Bhabha wanted Sreekantan to develop a cloud chamber. The latter knew nothing about cloud chambers, but, after two months of reading and discussions with Bhabha and others, he completed designing a cloud chamber. When Bhabha was working on the programme details of the first international conference on elementary particles in Bombay in December 1950, he called Sreekantan and said, 'Looking at some of the papers that I have received, it is desirable for you to start work in Kolar Gold Fields [KGF] and work on underground cosmic rays.'[28]

The Kolar mines had been in use since the second century BC. By 1905, there were a dozen different mines, of which the deepest was the Champion Reef, whose shafts reached a depth of 3,200 m. John Taylor, an Englishman, had set up a mining company in the village in 1880 and it soon became a bustling township 90 km to the east of Bangalore. The idea of using the Kolar mines for physics experiments was not new. As early as 1914, Herbert E. Watson, a chemist working in the newly established IISc had sent his assistant down one of the mines to collect rock samples and look for signs of radioactivity. Since the results at the time had not been encouraging, further attempts had been abandoned.

By 1922, H.E. Watson had left for England, but Bhabha was aware of his efforts when he returned to Bangalore from Cambridge. The germ of an idea may have been planted then, but it was more than a decade later that he sent B.V. Sreekantan down the Champion Reef mine with a sensitive Geiger-Müller counter to check for radioactivity. By 1948, the miners were digging at far greater depths than in 1914.[29]

In the winter of 1951, Sreekantan, along with Raja Ramanna and Bibha Chowdhuri, the first woman in high energy physics, went to Kolar and asked for permission to set up scientific experiments. Arthur Taylor, the son of John Taylor and managing director of the mining company, gave them permission to set up their laboratory in the mines. Not only that, he provided them with a fully furnished bungalow and domestic staff for free. Sreekantan was also allowed to use the club. He observed, 'I don't know why, but we were treated like officers.'[30]

On Bhabha's suggestion Sreekantan had already conducted an experiment on the lifetime of muons stopping in various absorbers. Subsequently, he asked Sreekantan to check whether there were other penetrating charged particles with masses different from muons. Sreekantan started working on muons at the KGF and performed experiments at depths of 8,000 ft. His experiments also aimed to study the properties of muons and their heavy electron nature at extremely high energies underground.

Sreekantan's first experiment intended to find the lifetime of muons and he decided to use Geiger-Müller counters to detect them. This also required a variety of electronic devices to be built, including triggered cathode ray tubes, coincidence circuits and pulse electronics with better than microsecond capability. These were made out of war surplus components: valves, resistors, condensers and oscilloscope tubes, all of which were sourced from Bombay's flea market on Muhammed Ali Road, called Chor Bazaar! Simultaneously, the production of Geiger-Müller counters also began. In fact, the earliest laboratory at the TIFR was set up with war surplus equipment.

After many failures, R.V.S. Sitaram successfully developed two different portable Geiger-Müller counters with three different sensitivities. Sitaram, who joined TIFR in 1947, after completing his MSc from BHU became the founding director of the Society for Microwave Engineering and Research, which was set up under the Department of Electronics in 1978. H.L.N. Murthy, the glassblower, produced a large supply of self-quenching Geiger-Müller tubes. The TIFR had a large workshop from the beginning. India had no industry that could provide and maintain such equipment and, therefore, training scientific staff to build and service such equipment was inescapable. Bhabha observed in his report on TIFR for 1952–1953, 'Here in India, where there is a less developed industry to fall back on, a well-equipped workshop is essential.'[31]

The TIFR workshop, which was a significant part of the institute, originated in July 1946 in the large dining room at Kenilworth. At that time, the workshop had only a lathe and a drilling machine with two mechanics and a supervisor. It was expanded upon moving to the OYC, with the addition of various types of lathes and other machines. Welding and carpentry units were also added at this time. However, there were many failures and progress was slow. This was understandable considering India was a technological backwater. Suresh Sawant, who joined the workshop in 1964, commented on the versatility of the workshop, which produced everything from antennae to chairs for the auditorium.

> Whenever Dr Bhabha came to the workshop—he would show us how we should do things. He would make drawings and show us. The Bhabha trolley used in TIFR was designed by him—he gave the drawings and designed the trolley himself. He loved the patterns of the grains on the wood. If the plywood didn't have the grain pattern he would ask us to send it back.[32]

The workshop also undertook precision jobs for cloud chamber experiments and when Patrick Blackett visited TIFR in 1947, he advised the scientists on the techniques of the Wilson Chamber. Bhabha's international contacts also helped. In 1961, a large number of precision machines were added, and the workshop played an important role in carrying out precision jobs for cloud chamber experiments. It also made special types of cameras. The control desk and panels of the first nuclear reactor, Apsara, were also made there. It also offered training programmes and took on apprentices, who were absorbed into the institute and AEET.

Sreekantan, along with S. Naranan (and later P.V. Ramanamurthy), measured the intensity of particles at various depths and different angles using Geiger-Müller counters. It was considered one of the more sophisticated experiments in cosmic rays at the time, and the value that Sreekantan got for the lifetime of positive muons was very close to the modern value.[33] Sreekantan recalled that when Bhabha came to Bangalore, he wanted to visit the mines and see Sreekantan's experiments. It happened to be Christmas Eve and the mines were closed, but the superintendent made special arrangements so that Bhabha could see Sreekantan's experiments.

Between 1951 and 1954, Sreekantan, S. Naranan and P.V. Ramanamurthy, set up a series of underground experiments to measure the flux of muons as a function of depth. These experiments were primarily intended in studying the properties of muons to corroborate the 'heavy electron' nature postulated at that time. Underground experiments at KGF, however, came to an abrupt halt in 1954 as the gold stocks in the mines seemed to be exhausted, and the news of the imminent closure of the mines spread.

By 1953, Sreekantan analysed his data, and it comprised his PhD thesis, which was about the intensity and angular distribution of muons at different depths. He submitted it to Bombay University. In 1954, a few months after he had submitted his PhD thesis, Bhabha asked him to summarize it in a talk. Bhabha, who had many irons in the fire, did not have the time to read his thesis. Sreekantan duly obliged. Two months later, while Sreekantan was at MIT, Bruno Rossi, who was the external examiner for his PhD thesis, made a similar request. Sreekantan repeated the talk he had given Bhabha, based on which Rossi sent his assessment to Bhabha.

Sreekantan received his PhD a few months later, with the examiners calling his experiments well-conceived, carefully executed and intelligently analysed. The experiments at KGF continued for three more years. They also took a multi-plate cloud chamber down to the mines but were asked to leave by the mining authorities, who were contemplating shutting the mines. Subsequently, they shifted the cloud chamber experiments to an old abandoned railway tunnel in Khandala in Maharashtra. When it became known that the Kolar mines were not closing they returned to conduct experiments at greater depths. The experiments continued after M.G.K. Menon joined, and TIFR started collaborating with foreign scientists, details of which are discussed in a later chapter.

◆

Another stalwart who joined TIFR was S.S. Dharmatti who was a pioneer of nuclear magnetic resonance (NMR) research in India. The technique of NMR was independently discovered in 1946 by groups led by Felix Bloch at Stanford and E.M. Purcell at Harvard. Dharmatti, who had worked with Felix, returned to India in 1953 and started an interdisciplinary research group at TIFR where physicists and chemists studied NMR. The NMR spectrometers at TIFR were housed in the military barracks at Navy Nagar, Colaba. When the new premises of TIFR were formally inaugurated by Nehru in January 1962, Bhabha made it a point to show him the NMR machine, explaining its concept and applications. The use of NMR techniques led to the invention of MRI, a non-invasive diagnostic technique that is routinely used in medical investigations today.

STORMING A MALE BASTION

Bibha Chowdhuri, who obtained her MSc in physics from Calcutta University in 1936, was one of the very few women scientists of her time. She was one of the pioneers of cosmic ray research in India who made a significant contribution to the discovery of mesons using photographic plates. The first Indian woman in the field of high energy physics, her ambitious trajectory was a relative rarity at a time when women were still discouraged from studying physics.

After obtaining her MSc degree, she joined Calcutta University to do research in physics under the guidance of D.M. Bose, who was then the Palit Professor of Physics. When D.M. Bose joined the Bose Institute as its director in late 1937, all his research students, including Chowdhuri, moved there with him. She stayed there for four years, from 1938 to 1942.

The annual meeting of the ISC was held in Calcutta in 1938 and was attended by luminaries from around the world. They included Sir Geoffrey Taylor, the leading expert on cosmic rays in Cambridge, and Walter Bothe, an expert on particle tracks who went on to win a Nobel Prize in 1954. Bothe spoke of the possibility of using photographic plates as detectors to record the tracks of charged particles, an idea pioneered by Marietta Blau and her student Hertha Wambacher at the Institute for Radium Research, Vienna. D.M. Bose and Chowdhuri, who attended the session, took to the idea and went high up in the Himalayas from Darjeeling to expose their plates to cosmic rays. During the war, they could only access half-tone plates, since full-tone plates were reserved for war reconnaissance work. They did see meson tracks on their plates but it is highly unlikely that they saw pions, as half-tone plates did not have the experimental resolution to distinguish pions, which leave much shorter tracks. However, they proved beyond doubt that photographic plates could be used for cosmic ray studies.[34]

Chowdhuri worked in the field of cosmic rays and conducted pioneering experiments at Darjeeling between 1938 and 1940, establishing a method of mass measuring of mesons in photographic plates by the multiple coulomb scattering and range method. She published three papers consecutively in *Nature* on the subject. However, she could not continue her investigations due to the non-availability of emulsion plates during the Second World War.

Cecil Powell had discovered the pion using D.M. Bose and Chowdhuri's technique, something which he acknowledged in his book. In fact, the measured mass of the particle by Bose and Chowdhuri was very close to the accepted value measured by Powell, who had used improved full-tone plates.[35] After the war, Powell not only had access to high-quality customized photographic plates coated with a far higher density of emulsion but also the facility of using balloons and aircrafts. Blackett played an important role in getting plate manufacturers to produce high-density plates for the Cold War, and

Powell, who had helped in the war effort, had easy access to them. Brazilian physicist César Lattes's suggestion of adding more boron to the emulsion worked wonders. However, the 1950 Nobel Prize went to Powell alone. D.M. Bose and Blau were forgotten—racism and gender bias perhaps playing a role in this omission. In fact, the citation noted Powell's development of the photographic method for studying nuclear processes and did not specifically mention his discovery of the pion.[36]

With no access to the kind of resources that Powell had, D.M. Bose and Chowdhuri discontinued their work when the war started. The latter left the Bose Institute and, in 1945, joined Blackett's cosmic ray research laboratory at Manchester University, where she completed her PhD. The thesis was examined by Lajos Jánossy and John G. Wilson. After finishing her PhD, she decided to spend some time in Paris with Louis Leprince-Ringuet in his laboratory, which was famous for the photographic emulsion technique. She needed funds to do so and approached the Government of India. The application was accompanied by a testimonial from J.G. Wilson and was referred to Bhabha for his opinion. He, in turn, asked J.G. Wilson in December 1948 for his confidential opinion about Chowdhuri's ability, and if any special advantage would be gained by a stay of six months in Paris. Bhabha argued that Chowdhuri had worked on air showers, whereas the project in Paris involved work on emulsion techniques. He wanted 'people who are really on top of a certain branch of a subject, however narrow, rather than people who have a smattering of a large number of subjects but are incapable of doing any first-class work in any of them on their own.'[37] He seems to have forgotten that Chowdhuri had done world-class work on emulsion plates.

After Bhabha's letter, Wilson's assessment of Chowdhuri changed and reflected the gender stereotyping prevalent at the time. Although Chowdhuri had done pioneering work, which was subsequently acknowledged by Powell, Wilson did not view her as a first-rate physicist. His assessment was either unfavourable or a case of damning with faint praise. He thought that she could make good progress under close guidance and worked satisfactorily although her physical background was not strong. He agreed that under no circumstances should she be 'encouraged to go with a new field of work.'[38] Bhabha seems to have laid great store by the opinions of foreign scientists. After her brief stay in Paris, Chowdhuri joined the experimental physics group at TIFR in November 1949 as a research fellow working on cloud chambers. It is interesting to note that Blackett recommended Chowdhuri strongly in his letter to Bhabha in March 1949. 'I do feel you have here a very able worker with a balanced judgement and enough experience to do really good work in the future.'[39]

Given the prevalent gender inequality, it is no surprise that Chowdhuri received limited encouragement to assume a leadership role. The TIFR, like other institutions at the time, was very much a male bastion. She left TIFR in December 1953 to join the Bengal Engineering College in Kolkata. In 1956, she continued her research at the Physical Research Laboratory in Ahmedabad, where Vikram Sarabhai was the director. She published several research papers in international journals during her long career. However, despite her lasting contributions to research, she was not elected to any major fellowships nor did she win any prizes. Her name also hardly features in the writing on cosmic particle research done at the KGF.

Chowdhuri was a lifelong researcher and continued to publish until her death in 1991. Forgotten for many years, there has been some interest in her in recent times, with a German publishing house publishing a book about her in 2018.[40] In 2019, the International Astronomical Union (IAU) announced that a yellow-white star in the Sextans constellation, HD 86081, would be named 'Bibha', after her. The IAU chose the name at the end of a global contest. Another honour came her way in March 2020, when the Ministry of Women and Child Development established a chair in physics to honour her memory to be set up either at TIFR or at the Physical Research Laboratory (PRL), Ahmedabad. However, these posthumous crumbs of recognition are meagre compensation for a lifetime of being ignored.

THE OLD YACHT CLUB

One of TIFR's immediate problems was the shortage of space. That Kenilworth was never going to suffice was known to all. In only three years' time, the staff had multiplied threefold. The institute had started with three academic staff, one technical assistant, one stores clerk, one stenographer and two peons. In 1948, the staff grew substantially, with 13 academic staff, along with eight and five administrative and maintenance staff, respectively. The institute also had five students. There simply wasn't enough space to accommodate all these people and the need to find bigger premises assumed urgency.

Since there was no permanent site or premises, a temporary solution was found in the OYC on Apollo Pier Road, overlooking the Gateway of India. The OYC was handed over by the Bombay government in November 1948 and took nearly a year to be readied for its new residents. The ballroom became the library; and the office and the butlers' quarters became the accommodation for staff and students. The OYC housed the institute till 1962 and two major international conferences organized by TIFR were hosted there. According to John Cockcroft, it was 'a most picturesque site' and was also remarkable because it displayed the paintings Bhabha had collected on the staircases and the offices. He said that it was Bhabha's 'persistence of purpose' that enabled him to prevent his lovely site from being ruined by the construction of a naval dockyard in front of it.[41]

Ramanna, who joined the AEET in 1949, stayed in the servants' quarters. He recalled that it had a lot of bugs and he couldn't sleep. Bhabha, who knew that Ramanna played the piano, gave him two rooms, and he would listen to Ramanna play as he went to his office. Ramanna, who went on to become chairman of the AEC and India's *bombmeister* in 1974 during the Pokhran peaceful nuclear explosion (PNE) became minister of state for defence in 1990 in V.P. Singh's government.

Former AEC chairman, Anil Kakodkar, recalls that in the 1970s, the staff at the OYC noticed that the floor appeared to be shaking and sinking. This led to a scare that the building was unsafe and the staff was shifted into one of the new 'matchbox architecture' buildings that were in complete contrast to the architecture of the OYC. The abandoned building remained unused and started decaying, and there was even some talk of demolishing it. It was saved from this fate by heritage activists, but it was only in Kakodkar's time that it was decided to restore the building to its past glory.

It was during this process of restoration that the mystery of the shaking floor was solved. The floor was originally built as a dance floor and had springs underneath. Over the years the numerous cupboards and cabinets had made it lose (quite literally) its 'spring'. As the heavy cupboards were gradually removed, the springs expanded again and the floor regained its old nature. Ultimately, the building was not only restored completely, but the portion that had been dismantled earlier was also reconstructed.[42]

Chapter 14

THE SEARCH FOR A NEST

> TIFR should be located in Bombay and in close proximity of the University. The climatic conditions in different parts of Bombay vary very little and the most suitable place in Bombay for locating the Institute would be the Colaba of the Fort area, which is near the colleges mentioned above.
>
> —Letter to H.M. Patel, 11 April 1950[1]

THE QUEST FOR A SUITABLE SITE

Bhabha had always been aware that finding a site for TIFR was a pressing problem that would take time to resolve, since it would involve government permissions and an unconcerned bureaucracy. Soon after the positive response from the Sir Dorabji Tata Trust, Bhabha had written to Sir John Colville of the great stimulus that such an institute could provide to both the Royal Institute of Science and Bombay University if it was favourably situated in a central position. The Fort area of Bombay fitted the bill perfectly, given its proximity to both the Royal Institute and Bombay University. The search for a piece of land started in 1946, and simultaneously G.M. Bhuta and Otto Koenigsberger started drawing up preliminary plans for a permanent building.

Koenigsberger was Max Born's nephew and the chief architect and planner of Mysore state. He was recommended for this post by Born when he was teaching at IISc in 1935. Koenigsberger was invited to come to Mysore by Krishnaraja Wodeyar IV on the advice of his diwan, Sir Mirza Ismail. As a Jew, Koenigsberger was exiled from Germany in 1933, and he journeyed to India through Egypt, arriving in Mysore in 1939.

Bhabha and Koenigsberger struck up a friendship, and the latter stayed in Bhabha's home on Cunningham Road between 1944 and 1948. For Koenigsberger, who drew up the initial plans for the building at TIFR, Bhabha opened the door to India's industrial, cultural and political elite. Koenigsberger was a busy man—first, as the chief architect of Mysore State until 1948, and then, as the director of housing for the Government of India until 1951. He designed many buildings in Bangalore, and when it was time for TIFR's final plans, he was busy with the plans for Faridabad and Gandhidham.

The Tatas were Koenigsberger's private clients: he prepared the Jamshedpur Development Plan and produced the Bhubaneswar Master Plan. Koenigsberger also produced a development plan for Bangalore. He wrote articles on town planning and was the co-founder of the design and architecture periodical, *Marg*. By 1948, however, TIFR had grown considerably, and it was necessary to revise the floor area and Koenigsberger's plans. By the time the land became available in December 1953 Koenigsberger had been gone from India for more than two years.

Several sites were surveyed for TIFR, and there were a few promising ones: the Colaba Observatory site, another one next to the Victoria Swimming Baths, a piece of land in the Port Trust area and even along the Marine Drive. Given its proximity to the Royal Institute of Science, the Port Trust land was the most appropriate. The institute tried its best to procure that land but failed. Bhabha was away in Paris at the time and Kosambi was in charge. In November 1946, he wrote to Bhabha that the bureaucracy, with its edifice of rules and precedents, was the main problem. He added that, Desai, the revenue minister, was a tough nut to crack. The problem seemed to be with the title to the land, and it was obvious that the institute could not have commenced construction without a clear title.

Despite full support from Nehru and Bhatnagar, the power of the bureaucracy and its attendant red tape ensured that even after five years, there was no progress. In February 1950, Bhabha requested H.M. Patel, the defence secretary, for the plot of land within the Colaba reclamation which the defence ministry owned. He explained that the heavy equipment could not be moved frequently, nor could the expansion be carried out without a permanent site. Bhabha invited Patel to see for himself the work they were doing and said that some of the work they were doing on the electronics side would be of interest to his ministry.[2]

In April 1950, Bhabha sought to buttress his earlier arguments by invoking the recommendations of the report of the committee set up by the Syndicate of the Bombay University in this regard. He hoped that the recommendation of the university would strengthen Patel's hands. He wanted TIFR to be located close to the university to enable the latter to take advantage of TIFR's facilities. Patel either remained unconvinced or didn't consider the matter important enough because it disappeared into the innards of the bureaucracy. When Nehru visited Bombay in January 1951, Bhabha pulled out all stops to get a permanent site for the institute. He wrote to J.R.D. Tata, who was away skiing in the French Alps, about Nehru's visit, telling him that they had driven through the entire area so that Nehru could get an idea of the space available.

Perhaps not wanting to tangle with the Congress veteran, Morarji Desai, Nehru told Bhabha that it would be better to find a site that did not require them to approach the Government of Bombay again. He thought that it would not be too difficult to spare sufficient land for the institute from Block VIII, which was the property of the Government of India. Notwithstanding Nehru's support, the matter dragged on till the end of 1953, when finally, Bhabha had his land. On 9 December 1953, Nehru wrote to Bhabha, giving him the good news. 'I gather that the Defence Ministry have at last decided to release the Colaba site for you. They propose to do so by the 15 December.'[3] Bhabha, who had already suffered bureaucratic delay and obfuscation, was gratitude itself. Two days later, he wrote: 'But for your intervention this

Baby Homi at Bad Kissengen, 1913

Young Homi

Jehangir and Meherbai Bhabha

The Bhabha family, circa 1910s

The Bhabha family, circa 1920s

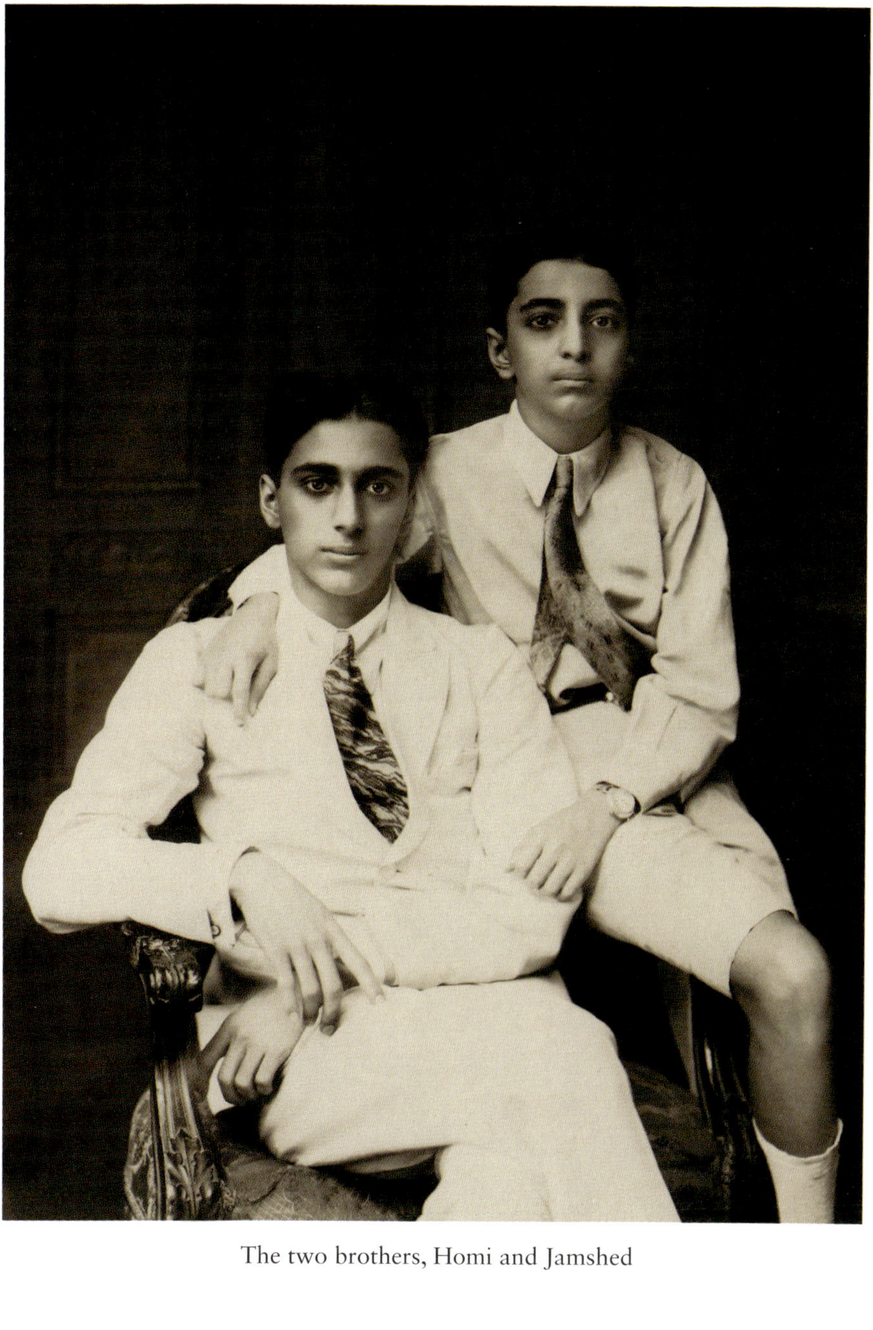

The two brothers, Homi and Jamshed

Bhabha and Meherbai

Jamshed Bhabha

Mehrangir

Self-portrait at the age of 17

Bhabha at work on his painting inspired by the Countess' aria, *Dovo sono i belli momenti* from Mozart's opera *The Marriage of Figaro*.

Painting inspired by the Countess' aria from Mozart's opera *The Marriage of Figaro*. It hung behind Bhabha's desk in his study at home.

Stage set design by Bhabha for Mozart's opera, *Idomeneo*, Cambridge, 1939.

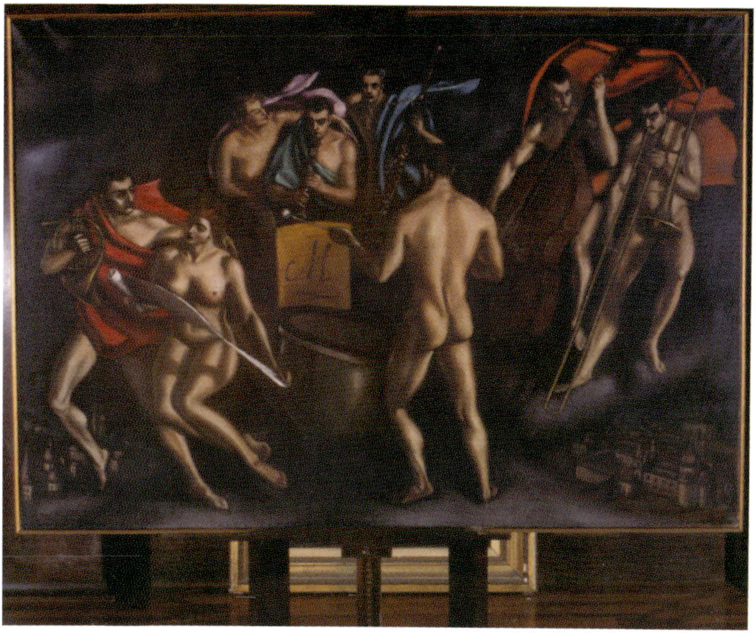

Angel's Concert: Homage to Mozart

Landscape with a bridge, Cambridge

Portrait of a lady

Sketch of C.V. Raman

Sketch of Niels Bohr

Sketch of Sher Gill

Sketch of Pipsy Wadia

Bhabha and C.V. Raman at the Indian Institute of Science, Bangalore.

From left to right, the Governor of Bombay, Sir John Colville, Sir Sorab Saklatvala and Bhabha at the inauguration of TIFR on 19 December 1945.

Bhabha (far left) at the Geneva Conference in 1955.

matter may easily have dragged on for yet another year and I cannot adequately express my appreciation of your interest in this Institute.'[4]

LAYING THE FOUNDATION STONE

Bhabha invited Nehru to lay the foundation stone of the institute on 1 January 1954. Besides Bhabha and Nehru, J.R.D. Tata, Desai, the then chief minister, and Bhatnagar made speeches to mark the occasion. After welcoming the guests, Tata thanked Nehru for supporting the institute and science in India in general.

> I feel that all directly and indirectly concerned with science in India owe a deep sense of gratitude to the Prime Minister and through him to the Government of India for the magnificent support and leadership in fact that they have given to science and scientific research in this country.[5]

He also thanked Desai, with whom he shared an uneasy relationship, saying that while they may have disagreed occasionally on some matters, Desai always had the respect and affection of the Tatas. Tata was all praise for Bhatnagar and his role in the whole affair. '[...] I would like, if I may, to single out for special thanks our good friend, the pilot, irrepressible, irreplaceable and fortunately indestructible, Dr Bhatnagar. Apart from the Prime Minister, I do not think that there can be a single man in the country who has done so much to foster and build up scientific research in this country.'[6] Sadly, Tata was way off the mark with the 'fortunately indestructible' bit. Bhatnagar died in his sleep exactly a year later on 1 January 1955.

Desai spoke briefly about the genesis of the institute and the role of the Tatas in it. He praised the Tatas' and Nehru's support of scientific research in the last few years. Desai's kind words for J.R.D. Tata were ironic in the light of his shabby treatment of Tata upon becoming prime minister more than two decades later in 1977. 'Mr J.R.D. Tata is a great friend of mine particularly because he has the capacity to take decisions in a friendly way and we are confident of the great work the House of Tatas has done for the country and for the State of Bombay particularly.'[7] In March 1977, J.R.D. Tata did not find a place in the reconstituted AEC; and in February 1978, his name was dropped from the reconstituted boards of both Air India and Indian Airlines, in effect a dismissal from the chairmanship of Air India.

Bhatnagar, in his address, said that the problem facing them had been whether to have a separate institute for nuclear studies or expand TIFR to include all nuclear physics and develop it into a '[...] really good laboratory for such investigations'. A lack of funds made him propose that TIFR should be expanded to include all scientific research of a fundamental character and named as a national institute. He also narrated an incident in which 'a distinguished young politician' asked him why he wanted to transfer such important practical activities to Bhabha, whose fame rested on his abilities as a theoretical physicist, a subject which, in the opinion of the politician had little to do with the material development of the country. Bhatnagar also said that the politician had tried to dissuade him from being party to an ambitious project that would not be completed in his lifetime.

Bhatnagar likened himself to an old gardener, planting a mango tree. '[...] I would not have mangoes if other men had not done what I was doing now.'[8] All his life, Bhatnagar said he had eaten mangoes from a tree that he had not planted. He said that he told the politician that he would give Bhabha all the help possible in laying the foundation for the future prosperity of the country. 'In furthering this project Dr Bhabha has worked with courage and devotion and the Government of India, the Bombay Government and the Tata Organization could not have chosen to give him a better new year present than the proposed Institute.'[9]

Bhatnagar expressed the hope that the project would be completed within the next few years.

> I am an impatient man and often cause annoyance to those who cannot keep pace with my speed which rightly or wrongly Bhabha calls 'jet propelled' [...] I can say with sincerity that one could not expect a more conscientious and considerate colleague than Dr Bhabha. Some people erroneously think that things can happen by magic wand when one the money is available. As one who has some experience of collecting funds, building, equipping and

manning of research laboratories, I can say there is no work which gives more headache than the task Dr Bhabha has now before him.'[10]

Bhabha, in his address, briefly traced the history of the institute and thanked Nehru and Bhatnagar. While appreciating the contribution of the latter, he said that his experience with TIFR had shown him the 'Herculean efforts that he must have put through to establish no less than eleven laboratories.'[11] In conclusion, he said, 'Anyway I now appreciate more than ever the tireless energy which characterises him.'[12] Bhabha also narrated the process of appointing the architects, designing the building and the close interaction between the scientists and the architect, which ensured that the end product was quite unlike any other laboratory building.

> There was constant talk between the architect and the scientists in the designing and at the last stage the architect actually asked the scientific workers to mark on the plans every piece of major equipment and even the furniture that was going into each place so that the areas and the rooms and the buildings should be very intimately correlated.[13]

While apologizing for the gap of eight years between the founding of the institute and the laying of the foundation, he said that in the end what really mattered was the work done at the institute. In conclusion, Bhabha said that fundamental research required a free atmosphere. The scientists must be free to think what they like and exchange their ideas with scientists in other parts of the world. He asserted that there would be no secrecy at TIFR and anyone who was genuinely interested could come and work there.

Nehru, in his speech, emphasized the latent benefits of fundamental research. Drawing a parallel with the electrical revolution, Nehru claimed that nuclear energy would bring untold benefits. The building of scientific and technological capabilities, he said, would also strengthen military capacity building. The Prime Minister reminisced about his role in the transfer of land. His speech also elicited a few laughs from those present.

Nehru said that a reference had been made about him in connection with his help in getting the land for the site. He said that he had helped a little but it had not been an easy matter. He added that the defence minister had to repeatedly confer with the prime minister on the subject. Ultimately, they had convinced each other and the decision had been unanimous.[14] The cause for the merriment was the fact that when the site was allotted to TIFR, Nehru also held the defence portfolio.

A CHANGE OF NAME?

At the time of the foundation stone-laying ceremony, Nehru faltered over the name of the institute. This was not mere happenstance. Bhabha had been under some pressure to rename his institute the previous year and at least five names were being considered, all of which included the word 'national'. Bhatnagar's laboratories were all called 'national laboratories' and there was pressure to either modify or include an additional label to the institute's name. The TIFR was a prime example of the complex interplay between public investment and private patronage in setting up a scientific institution. It was obvious that government support came with some expectations.

The Tata name had been associated with the institute from the beginning; the institute was now nine years old and Bhabha was never going to agree to a change in its name. In what ultimately proved to be a persuasive argument, he argued that the institute already had an international reputation in both physics and mathematics. He cited the praise of the Mexican physicist, Manuel Vallarta, who had called the experimental work using nuclear emulsion technique as the sensation of the Bagnères conference in 1953. The mathematics school, too, enjoyed an international reputation and was being referred to as the 'Princeton of the East'.

However, there was a time when Bhabha appears to have discussed the change of name with J.R.D. Tata, though the context had been different. Bhabha had wanted intellectual freedom for his institute, and such freedom extended to association with the Peace Movement or delegations to Moscow, which were a source of embarrassment for the Tatas, who had American business connections. Writing to

Rustum Choksi from London regarding Kosambi attending a peace conference, he said: '[…] there seems to be no alternative but to change the name of the Institute, a point we have often discussed and which Jeh strongly favours. Probably from all points of view the position would be more satisfactory if the Institute were to be called the Indian Institute of Fundamental Research with Tata Foundation written below it.'[15]

In his note on the history of the institute, written in 1954, Bhabha argued that the institute had reached a stage where scientists, students and academics from all parts of the world, including England, Israel, Japan and the US, applied for permission to work there. He wrote: 'All these achievements are associated in the minds of foreign workers with the present name of the Institute, and it would cause needless confusion to alter the name. Unless there are very strong reasons for a change, it is in the interest of the country and the Government that the present name should be retained.'[16]

Bhabha reasoned that just because the institute had taken the name of a business house did not make it less 'national' or render it ineligible for government funds. The Jamsetjee Jeejebhoy Hospital and the Goculdas Tejpal Hospital in Bombay had not been renamed, even though they were state institutions. The Cavendish at Cambridge and the Clarendon Laboratory at Oxford were publicly funded, but retained their original names. 'In all such cases, especially of famous institutions, the name of the laboratory acquires a significance of its own independent of any association with name of the founder. I am of the view that this is already the position with regard to the TIFR.'[17] Bhabha buttressed his arguments by saying that even the staff wanted the name to remain unchanged. In daily usage, at least in the early days, it was known as the Tata Institute.

At some point, Bhabha seems to have considered adding a second title to the institute to adequately reflect its primacy and association with the government. The phrases 'national research centre' or 'national research institute' were discussed in the historical note to Nehru on the occasion of the foundation stone-laying ceremony.[18] Bhabha was very keen that it be distinguished from national laboratories and university departments and wanted the separate institutional identity to reflect its numero uno status in scientific research in India. If at all any comparisons were to be made then those were to be with international institutions, TIFR being far superior to any other institute in India.

The matter, which was almost settled, seems to have cropped up once again in a discussion Bhabha had with Mark Oliphant, when he visited Australia in August 1954. Oliphant, too, was in favour of retaining a name that was distinctive rather than a nondescript one, and Bhabha wrote to Nehru citing Oliphant's views. He said that the fact that the government had provided a large part of the funds was no reason to change the name of the institute, particularly now that it had an international reputation. 'I look upon the name as one entity which exists in its own right, and no longer as something which achieves a meaning by association with the name of the firm.'[19]

On 11 January 1954, Bhabha wrote to Nehru thanking him for having agreed to lay the foundation stone. He said that his presence had been a source of great encouragement to the staff, who, he hoped, would apply themselves with even greater vigour. 'I cannot sufficiently express to you [...] our appreciation and gratitude for your interest and support, which have made it possible for us to do what we could to add to the progress of science in India.'[20]

A few weeks after the foundation stone-laying ceremony, Bhabha found himself in the honours list released on the occasion of Republic Day. The Padma awards had been instituted that year and scientists figured prominently in the honours list, with Raman being one of the three recipients of the Bharat Ratna and S.N. Bose one of the six recipients of the Padma Vibhushan. Bhabha was awarded the Padma Bhushan, along with Bhatnagar, J.C. Ghosh and K.S. Krishnan.

Chapter 15

THE FOREIGN HAND

> By his own work, and by his leadership of his group, Professor Peters has proved himself a most valuable asset to the Institute. As far as is known, he is completely wrapped up in his work and takes no active interest in politics.
>
> —Note to J.R.D. Tata, 18 May 1954[1]

In 1947, Bhabha suggested to the TIFR council that he would like to consider the possibility of inviting highly trained specialists from foreign countries to the institute. This reflected his desire to make his institute equivalent to Princeton or Cambridge. Many of the international scientists that Bhabha invited, including S.S. Chern, Frank Oppenheimer and Bernard Peters, were victims of persecution due to political upheavals in their own countries.

The mathematician Chern had been at the Mathematical Institute of the Academia Sinica in Nanking until the civil war in China intensified in 1948. In December 1948, Bhabha, along with Kosambi, wrote to Chern, inviting him to spend a year as a visiting professor at TIFR. Bhabha also offered to accommodate any close collaborators who chose to come with Chern.[2] However, by the time the offer reached him, Chern had already accepted J. Robert Oppenheimer's offer for a position at Princeton. He wrote back expressing his deep gratitude for the concern of his foreign friends, who had never failed him.

Another scientist to whom Bhabha made an offer a little less than a year later was Frank Oppenheimer, J. Robert Oppenheimer's younger brother, who he invited to Bombay for cosmic ray research. The younger Oppenheimer was being investigated for his communist links and had been called to testify before the House Un-American Activities Committee. Bhabha, after consulting his elder brother, made him an offer at the end of October 1949, which included a comfortable salary and a laboratory with two assistants. The initial tenure was to be one year. He would have full freedom to choose the experiments he wanted to conduct. However, Bhabha was not the only one to make an offer to Frank. Hans Bethe and Robert Wilson of Cornell University and Bruno Rossi from MIT also approached him.

Frank was worried about how long it would take to establish a good experimental laboratory and wanted a regular academic appointment. While he voiced these misgivings about a temporary offer to Rossi, he decided to accept Bhabha's offer anyway. Bhabha had promised him a tax-free salary and travel expenses if he stayed for at least a year. Frank's area of interest was high altitude cosmic ray investigations using cloud chambers. In place of the scarce helium, it was decided to use hydrogen. All things considered, Frank seemed all set to come to Bombay, but he had not reckoned with the US authorities. They denied him a passport and he wrote to Bhabha apologizing for the inconvenience, hoping that the offer would be renewed once his passport problems ended.[3]

THE OFFER TO BERNARD PETERS

The only offer in which Bhabha was successful was the one he made to Bernard Peters of the University of Rochester. A year younger than Bhabha, Peters was born in Poland. His Polish Jewish family had moved to Germany in 1912. A man with a colourful past, he had escaped Dachau and arrived in the US with his wife, Hannah, in 1934. They settled in New York where Hannah trained as a doctor and Peters worked in an import business. In 1937, when Hannah finished her medical degree, they went west. Hannah became a research fellow at Stanford and Peters worked as a longshoreman.

Bernard's meeting with Oppenheimer changed his life. The latter encouraged him to study at the University of California, Berkeley. Beginning his scientific career as Oppenheimer's student at Berkeley, Peters started work on cosmic rays. After completing his PhD in 1942, he spent three years working on the Manhattan Project at the Radiation Laboratory in Berkeley. Later, he joined the Rochester University, where he was assistant professor of physics. Peters then collaborated with the Minnesota cosmic ray group, which led to breakthroughs in the measurement of heavy primaries.

Along with H.L. Bradt of the Rochester University, Peters made a very significant discovery in 1948. They had flown sensitive nuclear photographic emulsions on balloons to high altitudes where one could hope to catch the primary particles of cosmic rays and not just the secondary products of air showers. Their results showed that the primaries contained heavy nuclei of the usual elements found on earth, proving for the first time that cosmic rays and the earth were made up of the same elements.[4]

Bhabha met Peters for the first time in 1948 and discussed the possibility of undertaking a joint project with TIFR that would take high altitude measurements of cosmic radiation at the magnetic equator. Peters recalled that his first contact with India occurred soon after it was established that the primary cosmic radiation bombarding the earth consists of a variety of nuclei. At equatorial latitudes, the geomagnetic field produces an asymmetry in the cosmic rays. It prevents positively charged particles at certain energy intervals from reaching the earth from the easterly direction, and simultaneously prevents negatively charged particles from reaching it from the westerly direction. This East–West effect could be investigated by measuring the arrival direction of complex cosmic ray nuclei in the equatorial stratosphere.

By then, Peters had already concluded that Bangalore was the best place to study cosmic rays because the experiments had to be performed near the geomagnetic equator. He wrote to Bhabha on 13 December 1948:

> If your plans for 1949 include a visit to this country, I hope I will have the opportunity of seeing you again. I was interested in rather fragmentary accounts of your far-reaching scientific planning in India [...] As a country and civilization India has always had a special appeal for me [...] Rather than visit your country as a tourist, it would interest me to spend a couple of years there working and participating in some useful activity.[5]

Peters met Bhabha again in New York in October 1949. He knew that Bhabha had already initiated a research programme on cosmic rays, including balloon flights into the stratosphere. Given his own interest in cosmic rays, he thought that cooperation between TIFR and Rochester University was a good idea. Bhabha, too, had already considered inviting Peters to TIFR. The minutes of the council meeting held on 9 March 1949 show that Bhabha had been considering the question of offering a visiting professorship to Peters who had signified his willingness to accept the appointment for a year or two provided his wife who was a trained and practicing doctor with research experience in gynaecology and cancer was able to obtain a suitable appointment in Bombay.[6]

Bhabha, in his reply, assured Peters of a position for him and his wife. Bhabha also told him that if he stayed for at least two years, he would also get travelling expenses to Bombay from the US and back. Before accepting, Peters also had some conditions, which included a research budget, travelling and transportation expenses and a compensation that was not to be less than ₹21,000 per year, of which, 30 per cent was to be paid in US currency. This annual salary was comparable to salaries in England. He also wanted the contract to provide for an option to attend one scientific conference abroad every year, the travelling expenses for which would be borne by the institute.[7]

Peters had a grant to the tune of $12,000 from the US Government for undertaking this research. Thus, apart from expenditure on travelling, the institute did not anticipate any substantial expenditure, since Peters was expected to pay a proportionate share of costs for the photographic plates. Overall, it was a good deal for Peters, and he knew it. He decided to accept Bhabha's offer, but it was hardly all smooth sailing.

Peters was denied a passport mainly because he had been denounced as a communist by J. Robert Oppenheimer in his deposition before the House Un-American Activities Committee. He had said that Peters was still being investigated for espionage at Berkeley during the Second World War. Oppenheimer had also named Peters's wife, Hannah, as a communist in a conversation with a senior FBI agent, who had been monitoring the Manhattan Project at Los Alamos.

Peters was aghast and wrote to Oppenheimer, asking the latter why he had named him the way he had. He requested Oppenheimer to intercede on his behalf to clear his name since he had done nothing wrong. Peters's passport was withdrawn as soon as he applied for the trip to India in December 1949. Oppenheimer, at the time, was chairman of the Advisory Committee to the Atomic Energy Commission. He had direct access to President Harry S. Truman and could have helped but chose to do nothing.

Bhabha and Oppenheimer were friends and often dined together in New York in the company of Oppenheimer's wife, Kitty. Oppenheimer, who was five years older than Bhabha, was teaching at Berkeley when Bhabha received his PhD at Cambridge. When Oppenheimer lost his security clearance in 1954, it was presumably on Bhabha's intervention that he was invited by Nehru on more than one occasion to visit and stay in India. Nehru invited Oppenheimer to visit India and even immigrate, if he so wished. Oppenheimer declined because he felt that it would not be proper for him to leave the US until he had been cleared of all charges. He feared that permission would not only be refused but also that it would only increase suspicion about him.[8]

It is not known if Bhabha realized the extent of Peters's predicament in Rochester before he came to Bombay, or if he had ever discussed the issue with Oppenheimer.[9] The reason for Peters' troubles had as much to do with his reputation as it had to do with the country he was to visit—India. It was the beginning of the Cold War and the US authorities thought that India was becoming too 'left of centre'. Peters was, by then, quite desperate. He wrote to Bhabha saying that he was willing to immigrate to India, but, without a passport, that was impossible. He was ready to go to India if the Indian government was willing to receive him.

Peters's woes only ended when his project was transferred to the full control of the US Atomic Energy Commission (US AEC) on 1 July 1950, which, soon afterwards, asked the State Department to approve Peters's passport. The US AEC advocated his case by saying that Peters was the only man who had the expertise to handle the project, and India was the only place in the world where this experiment could be conducted. In the end, a very relieved Peters arrived in Bombay on 31 August 1950, all set to finally start his research.

If his letters home are any indication, Peters was impressed with what he saw, and more than happy with his accommodation at the Taj Mahal Hotel, noting his inability to pay for such 'royal quarters'. On the day of his arrival, Peters wrote: 'My first impression about facilities, competence of people, and their character is extremely favourable, but so far based on only four hours of contact [...] Dr Taylor is the chief of the cosmic ray emulsion group, the division using photographic plates...'[10]

Peters' first visit lasted four months. Within a month of his arrival, he started work in earnest. On 29 September 1950, he wrote home. 'Taylor and I are flying to Madras on the 8 October, then drive to Bangalore, stopping at several places to pick out observation stations for theodolites. We will meet some of our crew in Bangalore, tranship equipment from there to Madras, and start flying on the 15.'[11] Peters wrote home regularly, narrating his exploits and their successes and failures.

AN INTERNATIONAL CONFERENCE

In September 1950, John Cockcroft organized an International Nuclear Physics Conference at the British Atomic Energy Establishment at Harwell. Bhabha attended and used the opportunity to see some of the laboratories there. Cockcroft wanted to showcase Britain's progress in nuclear research after the war. More importantly, it was done to gain credibility in the eyes of the international scientific community. Bhabha felt that a similar international conference would also mark a coming of age for Indian science and TIFR. It would be seen as one of the world's leading institutions. Optics were important even then!

Bhabha organized an International Conference on Elementary Particles from 15 to 22 December 1950 at the OYC. It was the first time that a conference of such a high scientific level was held in India. It was made possible by grants from various official and non-official organizations in India and abroad, and by a substantial subvention from the UNESCO, which enabled a number of distinguished scientists from abroad to attend the conference. Organizing such a conference was important from the point of view of credibility for the Indian scientific community.

In October 1950, Bhabha, after consultations with the International Union of Pure and Applied Physics (IUPAP)—of which he became the chairman in 1960—announced his own conference in December 1950. The IUPAP was to be the patron of the conference and TIFR the host. Earlier, such conferences had been organized in Krakow (1947) and Como (1949). The timing of the conference was such that invitees could also attend the ISC session in Bangalore, which Bhabha was to chair that year, immediately after the meeting. Bhabha acquired some fine photographs of Ernest Rutherford, J.J. Thomson and Paul Dirac from his old contact, Lettice Ramsey, in Cambridge for the conference.

Studies on both the experimental and theoretical aspects of elementary particles were discussed at the conference. Those present included Patrick Blackett, Gregor Wentzel, Bernard Peters, Léon Rosenfeld, Louis Leprince-Ringuet, Edoardo Amaldi, Rudolf Peierls, Walter Heitler and C.J. Eliezer. From the Indian side, apart from Bhabha, Saha, Sarabhai, K.S. Krishnan, D.S. Kothari, R.C. Mazumdar and Bibha Chowdhuri attended the conference. Bohr had also been invited but could not make it nor could Ralph Fowler or Cecil Powell. Discussions on the theory of elementary particles were led by Bhabha, who presented the relativistic theory in the latest form developed by him. Blackett presented some of his group's recent work on V-mesons, and showed some photographs of their tracks in a cloud chamber.[12] Though some studies had been made with powerful cyclotrons, cosmic rays remained the main source of high energy particles for most of the experimenters.

Some levity was provided at the end of the serious discussions during a social function, where Rosenfeld and Peierls read out their poem referring to many of the scientists present. Probably the only non-scientist to have made it there was the formidable Piroja 'Pilloo' Vesugar. 'Things would not work if Doctor Bhabha/ Was not tough as steel but resilient as rubber, / Here "chini" is the word for "sugar"/ The word for frankness is "Vesugar".'[13]

Vesugar joined the J.N. Tata Endowment for the Higher Education of Indians under Bhabha's father, Jehangir, in 1940 and guided its fortunes for 25 years. She was a formidable figure and seasoned scholars reportedly quaked in her presence. She was the secretary of the endowment and it was largely due to her stewardship that the Tata scholars were associated with intelligence and integrity in the eyes of the public. She was briefly a lecturer at Wilson College and an alumna of Lady Margaret Hall at Oxford.

Many Tata scholars came from humble rural backgrounds. She cared for Tata scholars as though they were her own children, generously giving her time and money to equip and train them. 'The "orientation" lunches at which hapless scholars fumbled with unfamiliar cutlery while she looked on censoriously, the escorted trips to Dean's, where male scholars were fitted out in suits under her hard-to-please eye, and sundry similar anecdotes conferred a near-legendary status on Ms Vesugar.'[14] However, invariably, there were also references to her kindness and generosity. Her numerous acts of kindness were common knowledge in Bombay House. Bhabha enlisted her services (part-time) on the administrative side of TIFR to avoid burdening himself with too many non-scientific details. H.J. Taylor remembered her as an able, charming, but slightly daunting lady. He said that Bhabha had called them all into his office one day to tell them about Vesugar. 'I hope that you will give her your cooperation. But it doesn't really matter, because she will elicit it.'[15] And indeed, she did.

At the concluding session, Blackett appreciated the way the conference had been organized. The conference was a resounding success. The invitees stayed at the Taj Mahal Hotel and experienced some of its famous hospitality. A fine report was produced at the Tata Press in Bombay and sent to laboratories all over the world. Bhabha and TIFR had arrived on the world stage, which had in fact been the aim of the conference, since Bhabha wanted TIFR to be seen as an institution of international repute.

PETERS RETURNS TO BOMBAY

Peters returned to the US after staying for four months in India. His stay convinced him of the suitability of TIFR for further research. He realized that the geographical position of India, combined with the facilities at TIFR, presented a unique combination for cosmic ray research. The geomagnetic field at low latitudes prevents the bulk of low energy cosmic rays from reaching the earth's atmosphere so that the very rare high energy processes can be studied without being swamped by background—a great advantage over the situation in the US and Europe.

In 1951, Bhabha invited Peters to join TIFR as a professor of experimental physics, and he arrived with his family at the end of that year. Peters had received a last-minute offer of tenure (an associate professorship) at Rochester University, but declined. He told Robert S. Anderson that the atmosphere in the US at the time with Senator Joseph McCarthy in full flow 'compared unfavourably' with the freer atmosphere at TIFR and this had led him to accept Bhabha's already attractive offer.[16] Peters' arrival at this time was a godsend for Bhabha because, in 1954, Taylor was transferred to Calcutta as principal of the Scottish Church College, a constituent college of Calcutta University.

Peters was not only getting a good deal at TIFR, where the financial and working conditions were good, but more importantly, there was no fear of political harassment. Much of the discomfiture that Peters faced resulted from administrators conflating cosmic ray research with efforts to develop a hydrogen bomb. Bhabha found employment for Peters's wife, Hannah, in the Health Physics Group at the AEET from 1952 to 1959. She recalled that Bhabha asked her to set up a medical division and organize a radiation protection service. She had no experience in this field, but Bhabha had a talent for using people's expertise and challenging them into finding solutions to problems. The Peters family lived at Kenilworth and Bhabha would often drive down at 7.00 a.m. and take Peters to Juhu, where the two had stimulating discussions on physics while taking a walk.

Hannah also noted Bhabha's love for plants, trees and landscaping of gardens.

> He brought plants from the countries he visited and tried to grow them in India. He told me that he brought the white bougainvillea from South America and planted it for the first time in India and it grew! And now it can be seen all over India. He used his botanical knowledge well when the new TIFR was being built in Colaba and the Atomic Energy Establishment was planned in Trombay. His dream was 'to create a Versailles' in Trombay, he told me.[17]

Bhabha introduced Peters to the painter V.S. Gaitonde, and Peters not only collected his paintings enthusiastically but also struck up a personal connection with the artist.

'Today, Gaitonde picked me up and we went to see his studio, a small place with a large terrace,' wrote Peters in a letter to his wife on one of his visits to Bombay after he had left India for good. 'He is a very sympathetic man, bachelor with no family; reads, plays music and paints. He has only three canvases at home, sells all he makes.'[18]

THE SEARCH FOR INTERNATIONAL TALENT

We have already noted how Bhabha failed to lure Max Born to Bombay. However, he did not give up on the idea of bagging a big international star for his institute—one who would be a beacon in leading research and be an embellishment as well. Bhabha was always conscious of the fact that a big name would add stature to TIFR, and as a direct corollary, to him as well. His next attempt to attract substantial international talent came in 1952 when he tried to get Richard Feynman, who had been a part of the Manhattan Project in Los Alamos, to come to Bombay.

Feynman, who was nine years younger than Bhabha, was known for his sense of humour and pranks. To the security staff at Los Alamos, he was a vexatious nuisance. He picked locks and cracked safes. Physicists tend to set the combinations of their safes using easily recalled mathematical constants. Feynman realized this and would open his colleagues' safes and leave them notes.[19] His exploits would be the subject of endless anecdotes at Oppenheimer's parties, fuelled by his lethal vodka martinis. Feynman's personality was unencumbered by dignity or decorum and he spent a great deal of time generating anecdotes about himself in which he usually came out looking smarter than the rest. In these stories, he was a rake, a gadfly, a clown and a naïf. English-born American physicist, Freeman Dyson, called him 'half genius half buffoon'. Part of Feynman's hostility to authority stemmed from anger at a personal tragedy: his young wife, Arline, had died from tuberculosis when he was at Los Alamos, since this was the era before antibiotics. Oppenheimer understood this and arranged for a place in a hospital close to Los Alamos so he could visit her frequently. In his memoirs, Feynman mocked every administrator he worked for with the exception of Oppenheimer.[20]

Bhabha wrote to Feynman that Peters was already in Bombay to convey that international scientists were already at TIFR and considered its atmosphere congenial to research. It was not the first time that he had broached the subject with Feynman. In a letter written in July 1952, Bhabha recalled their discussion from a few years ago about the possibility of his spending some time in India. Referring to Feynman's single status, Bhabha called him a sensible man, like himself, since his domestic problems were less complicated than those of others and asked if he would be interested in spending a few years at TIFR as a professor of theoretical physics.[21] In the same letter, he revealed details about some of the work done by Peters and his group. In the letter, he seems keen to tell Feynman that the research in TIFR at the time was very cutting edge and at par with that being done in other parts of the world.

Bhabha did not fail to mention his own research and publications, seeking, perhaps, to convey that he was still very much in the game. But, as we shall see, this was to be the case for only a few more years. The pressures of helming two important institutions left him with little time and energy for frontline research. Bhabha wrote his last paper on the meson in 1954, which was published in the *Proceedings of International Conference on Theoretical Physics, Tokyo*. By that time the arrival of high energy particle accelerators had also shifted the focus in the field.

Feynman took three months to respond. He thanked Bhabha for the offer but said that since he had now remarried, he was no longer a 'sensible man', like Bhabha, and, hence, could not join him as requested. While appreciating Bhabha's work, he said that he had no news of his own since 'I am becalmed in confusion'. He also wanted to know when they would meet and 'have the fun of arguing with each other again'.[22] Though there is no archival record, it is almost certain that the two met at the Tokyo Conference in 1953, as the proceedings included papers by both of them. Feynman's paper was on the 'Atomic Theory of Liquid Helium', and Bhabha had four papers on cosmic rays and the meson, including a presentation of the work of the experimental group at TIFR, which had been a major triumph for the institute's group at the conference at Bagnères-de-Bigorre in July that year.

Chapter 16

INTERNATIONAL RECOGNITION AND DISTINGUISHED VISITORS

> Professor Peters is distinguished for his work on cosmic rays, and he had made valuable contributions on the subject. The communication made by him at the International Congress on Cosmic radiation at Bagnères-de-Bigorre in July 1953 was the outstanding event of the conference and definitely established the group working under his leadership at the Institute as one of the strongest Groups in the world on the subject.
>
> —Note to J.R.D. Tata, 18 May 1954[1]

PETERS' BOYS

Yash Pal, a final year MSc student from East Panjab University's physics school, located on the campus of Delhi University, was one of the students who applied to TIFR based on an advertisement. He had planned to work on nuclear emulsion technique for his MSc dissertation, but there were no facilities available in Delhi for that kind of work. His interview at TIFR turned out to be a very friendly and cordial affair. After learning about his plans to work on nuclear emulsions, the interview committee suggested that he conduct his research at the institute. This was quite unexpected, since he had not completed his MSc, a necessary qualification for admission as a research student. However, much to his surprise and delight, he was told that he had been admitted as a research student and that he could continue his dissertation work at the institute. The permission from Delhi University was slow in coming, but he eventually made it to TIFR.[2]

Yash Pal recalls that he did not have any certificate for his matriculation, BSc or MSc degrees to show that he had passed these exams. He was very surprised when the TIFR authorities did not even bother to ask him for them. He settled in quickly but after the initial excitement wore off, the work seemed a bit routine and he wondered if he had made the right choice. The arrival of Bernard Peters put the spark back in him. He fondly recalled, 'Peters was a fantastic input not only in terms of science which he had done (and he was very famous already) but also a personality with which I resonated.'[3]

Peters was one of the early pioneers in the use of the nuclear emulsion technique and was responsible for giving the cosmic ray research at TIFR a direction and orientation. Yash Pal reminisced, 'I think Peters was responsible for introducing a certain way of selecting issues and problems in science for working on. You don't have to work on everything but work on something which has a meaning [...] It is not realised how much of a difference Peters made to this particular culture of science—experimental science particularly—at that stage in TIFR.'[4]

The TIFR group did balloon flights under Peters's guidance using the apparatus he had brought with him from Rochester. Balloons were flown from Madras and Yash Pal was stationed in Vellore as a theodolite observer to see which way the balloon went.[5] Some of the earlier experiments were performed from the campus of the Madras Christian College and one of the student volunteers was the physicist G. Venkataraman, who later wrote a brief biography of Bhabha.

Devendra Lal was a physicist from TIFR who Yash Pal fondly remembered and who became his lifelong friend. Lal joined TIFR after completing his MSc in physics from BHU in 1949, becoming part of the institute's very first group of PhD students, initially working under the tutelage of H.J. Taylor. When Peters relocated from the US, Lal became his student. Lal and Yash Pal worked together and also managed to have a lot of fun in the process. They realized that innovation was key and there are some interesting examples of their innovative thinking. Since the large Mylar balloons used for later cosmic ray research were not available in those days, the TIFR team had to make do with smaller meteorological balloons made of rubber. Thus, there would be 70, 80 or even 100 small balloons tied together that would carry the payload.

Unfortunately, these balloons were not of the same quality and some of them would burst as they ascended. The height finally achieved by the payload depended on how many balloons remained to take the payload up. The TIFR team did manage to send balloons to heights of 80,000–90,000 ft but they encountered another problem which was a direct consequence of balloons bursting. Not only did the burst balloon result in a loss of lift but the heavy neck of the burst balloon also became a dead load. If that dead load could somehow be removed then the payload could reach a higher altitude. They had to find a way to automatically detach a burst balloon from the payload. After a few hits and misses, Yash Pal and Lal found a way.[6]

The TIFR group was among the first groups in the world to develop emulsion detectors. One of Peters's innovations was using emulsions without glass backing, called stripped emulsions. The idea proved to be very successful and the team was able to trace the elementary particles they found back to the nuclear interaction from which they had been produced. Occasionally, while scanning through the emulsion plates, evidence of a strange particle was found—a charged hyperon (a class of unstable elementary particles heavier than the proton and the neutron) decaying or K-mesons decaying and so on. Yash Pal, Lal and Peters published several papers on the work. Yash Pal reminisced that they felt that TIFR was as good as any in the world in this area. They did not feel inferior, and found that they were recognized the world over very early in their life. Their work was praised at the conference at Bagnères-de-Bigorre in July 1953, which is discussed in the next section.

A WATERSHED CONFERENCE

The work of Peters and his group, which included Yash Pal, Lal, Sukumar Biswas and E.C. George Sudarshan, was praised at the conference at Bagnères-de-Bigorre, (a small town in the French Pyrenees) in July 1953. Organized by Patrick Blackett and Louis Leprince-Ringuet, this conference proved to be a watershed. It marked the beginning of subatomic physics and, more importantly, a shift from cosmic ray research to research at the new high energy accelerators. Many physicists who were using cosmic rays for their research now moved to accelerators. In terms of importance, this conference was in the same category as two other famous conferences: the Solvay Conference of 1927 and the Shelter Island Conference of 1948.[7]

Blackett and Leprince-Ringuet who were part of the IUPAP were present at the TIFR conference in December 1950. Blackett had been chosen president of the cosmic ray commission of the IUPAP and Leprince-Ringuet, the secretary. Besides Bhabha, the members of the 1952 conference included Carl Anderson, Gilberto Bernardini and Manuel Vallarta. It had been decided at the Bombay conference that the next one would be held in 1953. They chose Bagnères-de-Bigorre, a town close to the high-altitude French Observatory at Pic du Midi, located in the Pyrenees. The Pic du Midi Observatory was managed by the University of Toulouse and it co-sponsored the conference with IUPAP.

During the first three days, the experimental data on K-mesons and particles more massive than nucleons (hyperons) were presented by the experimenters. The experimental techniques, along with their limitations were also discussed. Peters's group presented three papers, creating a stir with their work on the K-meson (now known as kaon), a new kind of meson discovered in 1946 that decays into pi-mesons. Their work using the stripped emulsion technique revealed the characteristics of the K-mesons and drew praise from Leprince-Ringuet.[8] In 1953, Peters and his students, Yash Pal and Lal reported the discovery of a new particle that appeared to be the negative counterpart of the K-meson. They found evidence for it during a balloon flight from Bombay, having exposed their emulsion block for more than six hours at a height of 25 km. They also estimated the mass of this particle and showed that another meson, the tau-meson, and the K-meson had roughly the same mass.[9] The success of the group using equipment that had been fabricated indigenously, showed that Bhabha's vision for creating an atmosphere that encouraged original thinking had borne very fruitful results.

The conference was the zenith of the role of cosmic rays in particle physics. 'It was a conference characterised by great food and wines and greater physics, a truly memorable occasion [...] It was a conference at which much order was achieved out of a rather chaotic situation through nomenclature alone. For example, it was decided that all particles with a mass around 1000 m_e were to be called K mesons.'[10] The word hyperon was coined at this conference to describe particles with a mass greater than the neutron and less than the deuteron. Bagnères-de-Bigorre has an avenue called de l'Hyperon, perhaps the only place in the world where an avenue has been named after a fundamental particle.[11]

Peters's group had performed commendably, but the shadow cast by his past had not been erased. In 1954, an article in *The Times of India* was headlined: 'Tata Institute Expert: A Red'. It focussed on Oppenheimer's testimony against Peters.[12] This report attracted the attention of J.R.D. Tata, who asked Rustum Choksi for an explanation because the former did not want the institute to be involved in politics. In a handwritten note, J.R.D. asked 'if Peters was on a contract with the TIFR and if so, when it expired?'[13]

Bhabha responded on 18 May 1954, defending his foreign colleague. He clarified that Peters had been actively anti-Nazi in Germany and inevitably had come in contact with the Communist Party but had never been a member of that party. When he had deposed before the committee, he had not refused to testify on the grounds that he might incriminate himself, thereby showing himself to be a man of principle and character.[14] His note to J.R.D. Tata ended with praise for his professional contribution. He called his contribution at Bagnères-de-Bigorre the outstanding event of the conference. By his own work and by his leadership qualities, Peters, Bhabha said, had proved a most valuable asset to TIFR.[15]

A SHIFT IN FOCUS

By the time Yash Pal left TIFR in 1954 to pursue a PhD at MIT, particle accelerators had become available for research. With accelerators becoming increasingly capable of producing particles of higher and higher energy, Peters realized that TIFR's advantage in cosmic ray research was fast eroding, and he decided to focus on the radioisotopes produced by cosmic ray interactions in the atmosphere. In 1946, Willard Libby had discovered Carbon-14 produced in the atmosphere by cosmic rays and used it as a dating tool.

Peters wanted to investigate whether there were other such radioisotopes in nature that might be useful as tracers or dating tools. He asked for Bhabha's permission to start a geophysics group and it was readily given. Bhabha gave him a free hand and asked him not to worry about the cost. Lal decided to follow Peters into this new area of research, even though it did not have the same prestige as cosmic ray research. The TIFR group discovered the radioactive nucleus of beryllium (10), the number 10 denoting the number of nucleons in its nucleus. The normal nucleus has only eight particles. Since beryllium has a longer decay time of over a million years, it seemed promising for dating events that had occurred over a million years ago. Some amount of radioactive beryllium (10) produced in the atmosphere (due to the constant bombardment of cosmic rays and their collision with atmospheric nuclei) would come down on earth when it rained. If the amount of radioactive beryllium at a geological site was measured, one could estimate the age by simply calculating how much beryllium remained after a certain period of decay. The TIFR group finally detected and measured radioactive beryllium during the monsoon of 1955.[16] Later in his career, Lal became director of the PRL in Ahmedabad and was a professor at the Scripps Institution of Oceanography, University of California, San Diego. He was elected an FRS in 1979.[17]

By the late 1940s, Bhabha was all but lost to the world of pure science, with all his time and energy being consumed by the atomic energy programme. As Sreerup Raychaudhuri observes, 'In a way this was a pity, for the cosmic ray group punched way below its weight during the halcyon days of the 1950s, when cosmic rays were the chief laboratory for the discovery of new particles—one or two which were being found every year.'[18] But by that time, as noted above, a lot of the research effort was devoted to searching for exotic nuclides in cosmic rays. Peters left TIFR in 1959 to join Niels Bohr's institute at Copenhagen. After his departure, TIFR refocussed on cosmic ray efforts in the KGF, and this proved to be more successful, something which is discussed in a later chapter.

Perhaps Peters' greatest contribution was the growth of self-esteem among Indian scientists after his arrival. Indian scientists were now confident that they could do fundamental research comparable favourably with the best in the world. Speaking at the silver jubilee celebrations of the institute in April 1971, Peters commented on the success of innovation at TIFR. He said that from the start, the institute was not satisfied to purchase instruments for experiments or blindly copy foreign designs. 'It is precisely in those fields where new or improved techniques were developed, that the most interesting scientific results were obtained [...] In each case there was a genuine contribution of a technical nature which is responsible for the originality and the value of the results obtained.'[19]

Ranjan Roy Daniel, who worked with Peters on cosmic ray research, said that his legacy was a scientific temper and an enabling atmosphere in which free and objective human thoughts could thrive.[20] Peters was awarded the Padma Bhushan in 1985. He had been nominated for the Nobel in physics in 1952 by the French astrophysicist Alexander Dauvillier.

'BRAIN-IRRIGATION' AT THE INSTITUTE: PAULI, MARSHAK AND DIRAC

TIFR frequently invited visiting faculty. Bhabha's stature and his networks drew many of his western colleagues to India. One of the first big names to visit TIFR was Wolfgang Pauli, who visited in 1952–1953. He wrote to Bhabha on the last day of 1951, telling him that he proposed to stay in India from 1 November 1952 till the end of February 1953. Pauli wanted to know if it would suit Bhabha to have him there at that time and if Bhabha would be in Bombay during that time. He said in jest that he was afraid that Bhabha would be off representing India at the South Pole or preventing some atomic-energy committee (in Paris or elsewhere) from falling asleep, which he called an unwise action. He wanted an assurance and promise that he would not be left alone in Bombay during that time.[21]

Bhabha agreed, and Pauli wrote to him about sailing from London on 9 October 1952. He and his wife were planning to arrive in Bombay by the end of the month, and Pauli wanted to know if that would be convenient. He also wanted Bhabha to intervene with the Indian embassy to speed up their visas.[22] A month later he wrote back saying that he would be taking the ship sailing from London on 6 November and asked Bhabha not to be upset about the delay.

He proposed lectures for two hours every week but said that he could even make it three.[23] Pauli also agreed to conduct a seminar for a smaller group to discuss papers on theoretical problems. Bhabha had informed him that K.K. Gupta, Abdus Salam and, perhaps, C.J. Eliezer would also be there during his visit. Pauli said that he had done some new work and was looking forward to discussing it with Salam and Gupta. Pauli had a high opinion of Gupta's mathematical skill and thought that he would be the right person to assist Bhabha. 'The best for your Institute I believe would be to appoint him first temporarily and then to see later whether you wish to have him permanently.'[24]

In his next communication, Pauli voiced his concern and irritation about the delay in his visa and the lack of communication from the Indian authorities in Berne. He was unhappy that Bhabha had not given him a personal introduction to a higher official at the embassy and wanted his help and advice.[25] Pauli did get his visa and managed to come to Bombay. He delivered lectures and seminars and returned to Zurich via Rome.[26]

There is an anecdote that S. Chandrasekhar narrated to his biographer Kameshwar C. Wali in which Bhabha was guilty of indulging in what Chandrasekhar called 'exaggerated forms of personal prestige'.[27] Chandrasekhar recalls Victor Weisskopf telling him about an incident when Pauli and several foreign visitors were in India and

Bhabha made them commute in a bus while he followed them in a limousine all by himself. Pauli, reportedly, was so annoyed at this that he left India the next morning.[28] Since Pauli wrote to Bhabha immediately on returning to Zurich and asked him to thank the secretary, Mr Godbole, for sending his trunk and its key, it is highly unlikely that he left Bombay under the circumstances that the above anecdote would have us believe.[29] This story was later shown to be false. Raja Ramanna categorically contradicted it in public and personally confronted Chandrasekhar, whose unconvincing reply was: 'We are all old men; memory sometimes plays tricks.'[30] He didn't reveal why he chose to countenance hearsay. Did he also deep down consider Bhabha to be arrogant? We will never know.

Marshak's Visit to TIFR

After Pauli, the American physicist, Robert Eugene Marshak visited TIFR. He had completed his PhD on white dwarf stars from Cornell University under Hans Bethe when he was only 22. Marshak was one of many scientists recruited for the Manhattan Project and served as the chief deputy for theoretical physics under Bethe, interacting closely with J. Robert Oppenheimer.

Marshak believed that the free exchange of ideas could advance science and further the cause of world peace. One of the crowning achievements of his career was his organization of international conferences in the field of high energy particle physics, starting in 1950. Since the first conference were held in Rochester, the events continue to be known as the Rochester Conferences.

During his lecture tours, Marshak recruited outstanding graduate students many of whom went on to become leaders in the physics community. One such student was Sudarshan, whom Marshak recruited from TIFR. As we shall see later, Marshak, along with Sudarshan, developed the V-A (vector minus axial vector) weak interaction theory, an explanation for the fourth force of nature. Marshak first visited India in 1953 at the invitation of Bhabha and met Nehru. When Bhabha heard that Marshak was going to the Tokyo Conference, he asked him to stop by Bombay for a month.

Since Bhabha knew that the French theoretical physicist, Maurice Lévy, was also going to Tokyo then, he also asked Lévy to visit TIFR at the same time.

Speaking of his first visit in 1953, Marshak said:

> Bhabha was a severe taskmaster and requested Lévy and me to lecture during the religious holidays (of which there were many in India during August), except for the Parsi holiday which he observed; this gave us a long weekend respite to visit the three most famous caves in India: Elephanta, Ellora, and Ajanta, which Nehru exalts in his autobiography.[31]

There was one jarring incident, however, that showed how status-conscious Bhabha was. Marshak observed:

> I had a very pleasant time in Bombay despite the fact that Bhabha catered, to some extent to the caste system in India. For example, Maurice Lévy did not, at that time, have a professorship, so he was put in a different hotel. I was put in the Taj Mahal Hotel and he was put in a neighboring hotel, which was a little embarrassing to me, because scientifically Lévy had been doing very nice things and I considered him an equal. But in any case, he would come over and join me for dinner at the Taj Mahal Hotel.[32]

Marshak recollected the scientific conditions at the institute as being rather primitive, with most of the work being centred on cosmic ray nuclear emulsion work.

> For example, I remember entering that building, which was not completely covered by a roof—it wasn't that the roof was broken down but it was the type of building that was only partially covered, and you had pigeons all over the place, right in the labs—and I remember Bhabha issuing instructions one night to dispose of them. The next day they were all gone'.[33]

Bhabha, according to Marshak, ran TIFR as both the intellectual superior and the administrative dictator. He was the person who was unquestionably in charge and everything came to him. He said that Bhabha lived in a sort of royal splendour and, even in the old building, he was the only one to have an air-conditioned office. The rest of the people lived under rather unpleasant conditions.[34]

Bhabha was nothing if not hospitable. He took Marshak to some of Bombay's clubs for swimming and entertainment. 'He was very pleasant in trying to entertain me; he was very hospitable. We became quite good friends.'[35] In 1963, Marshak and his wife, Ruth, returned to India to spend six weeks at TIFR. During this time, the Marshaks also travelled to Madras, where they participated in the first anniversary symposium of the Institute of Mathematical Sciences (popularly known as Matscience) founded by Alladi Ramakrishnan, Bhabha's former student.

Paul Dirac's Visit to India

After Marshak, the most distinguished visitor at TIFR was Paul Dirac, who came to India in 1954. He was supposed to go to Princeton but was refused a visa due to his so-called Russian links. This was around the time of the Oppenheimer case when anti-Communist paranoia of the 1950s was at its height. Dirac's plans for a sabbatical in Russia also fell through, so he accepted Bhabha's long-standing invitation to visit India. Bhabha was thrilled and honoured to host the celebrated physicist, who, though well past his prime, was the Lucasian professor of mathematics at Cambridge.

An excited Bhabha wrote to J.R.D. Tata about the visit in July 1954. J.R.D. took great interest in meeting visiting scientists. He said he was delighted and wanted to know the arrangement with Dirac. At the end of September 1954, Dirac and his wife, Manci, set sail for Bombay, on the first stage of their round-the-world trip, which was scheduled to last almost a year. Bhabha made sure that the Diracs were treated like royalty from the moment they arrived on 13 October, though Bombay's unbearable heat and humidity were beyond his control. The visitors promptly escaped to the cooler climes of Mahabaleshwar, a hill station that is a few hours' drive from Bombay.

On Dirac's return, he delivered the usual lectures and seminars at TIFR. The Diracs then visited Madras, where he lectured at Madras University. When he spoke, the Senate Hall overflowed with listeners.

It is hard to picture this today, but people stood in the parking lot to listen to this lecture on loudspeakers.

After spending Christmas in Bangalore, they returned to Bombay and then travelled to the ISC in Baroda. On 5 January, Dirac delivered his talk in a special enclosure at Baroda's cricket ground to a large gathering of experts and non-experts alike. He spoke in simple, equation-free language and at the end of the 40-minute summary of the fundamentals of quantum mechanics, received warm applause.[36] Dirac met Nehru in New Delhi a week later and the Prime Minister asked him if he had any recommendations for the future of India. After pausing for reflection, Dirac replied: 'A common language, preferably English. Peace with Pakistan. The metric system.'[37] They apparently did not discuss nuclear weapons, though the subject must have been on their minds. The Diracs finally sailed from Bombay on 21 February 1955.

Bohr's Visit to India

Bhabha had invited Bohr to visit India several times previously, but he and his wife managed to come only on 3 January 1960. Apart from the usual lectures and seminars, the Bohrs found time to visit Elephanta, Ajanta and Ellora caves. On 5 January, Bohr delivered a popular lecture on 'Atoms and Human Knowledge'. In it, Bohr said that there was great scope for international cooperation and exchange of knowledge for the development of atomic science. He also delivered a convocation address at Bombay University and was conferred with a Doctor of Law degree.

Bohr offered to collaborate with India in its scientific development, cooperation that would take the form of exchange of scientific personnel. He was all praise for the facility at Trombay that Bhabha had created. When asked about whether India was in a position to make a nuclear weapon, he replied that any country that had powerful reactors could make the bomb, but fervently hoped that such a bomb would never be used.

Bohr visited Madras, Calcutta and New Delhi. While in Madras, he met Ramakrishnan. Upon his return to Delhi, Bohr was asked to comment on science in India. He said that two things impressed him the most—TIFR in Bombay and the small group of students trained by Ramakrishnan in Madras. For Ramakrishnan, the publicity that came with this statement from Bohr resulted in a meeting with Nehru, who asked him what he wanted. Ramakrishnan said that he was interested in setting up an institute for advanced fundamental research, like the IAS in Princeton. Nehru told C. Subramaniam, who was already a champion of the cause, to look into the matter. Apparently, Nehru consulted Bhabha who strangely said that he saw no need for such an institution in Madras. Perhaps, he did not want any competition on the mathematics front either. However, Subramaniam and Ramakrishnan persisted and Matscience became a reality on 3 January 1962. Ramakrishnan served as the director of Matscience for 21 years until his retirement in 1983. During his tenure, hundreds of important mathematicians and physicists visited the institute.[38]

In Delhi, Bohr was greatly impressed by Nehru. Bohr wrote a letter to him on 10 February 1960, calling his day in Delhi unforgettable, and saying that their conversation would be one of his most treasured remembrances. He offered Denmark's assistance in atomic affairs, however modest. 'In this great cause the closest possible international collaboration will surely be of a decisive importance, and here in Denmark we shall be happy, if in any field, even in a modest way, advice and assistance might be given from our small country.'[39]

In the same letter, Bohr expressed his admiration of Indian academic institutions and scientists and 'greatly admired the foresight with which the endeavours for promoting the welfare of the people on the basis of the progress of science and technology were planned under your leadership.' As far as the nuclear energy programme was concerned, he added: 'I talked with Dr Bhabha about the possibilities for such collaboration and we are looking forward to a visit from him in Denmark in the spring, when we will go thoroughly into technical matters.' Bohr said that he was eagerly looking forward to the visit of Bhabha and Dr M.S. Thacker, director general of CSIR, to Denmark soon.[40]

After Bohr's Bombay visit, he and Bhabha became quite close and a strong friendship developed between them. Bhabha even made a pencil sketch of Bohr, dated 25 January 1960, during this trip. The formality in their relationship seems to have disappeared as evidenced by the way they addressed each other in their letters. 'Dear Niels' replaced 'Dear Professor Bohr' and best wishes were sent in letters that closed with 'Yours ever'. Bohr sent letters addressed to 'Dear Homi' that ended with 'Yours ever'. There is evidence to show that Bohr misspelled Bhabha's name as 'Bhabba' in at least one of his letters written earlier. No doubt, he wrote it the way he pronounced it, his tongue being not nimble enough to deliver itself of two identical syllables consecutively. On his next visit to Denmark, Bhabha stayed at the Bohr residence, Carlsberg Palace.[41]

Bhabha went on what was virtually a world tour between 22 April 1960 and 12 June 1960. He visited Cairo, Vienna, Geneva, Copenhagen, Paris, Bonn, Zurich, London and Moscow. At the end of that year, he sent another invitation to Bohr to attend an inauguration ceremony at Trombay that Nehru was to attend. This was to be followed by lectures and a 12-day tour of India, culminating in the Republic Day celebrations on 26 January in Delhi. Bohr, who was 75 at the time, politely declined, citing the need to economize his time and strength. He promised to send a representative of the Danish Atomic Energy Commission in his stead.[42] Bohr, in a letter to Nehru, said that he would be sending H.H. Koch, chairman of the executive committee of the Danish Atomic Energy Commission to discuss issues with Bhabha and his collaborators.

THE ELUSIVE NOBEL PRIZE

Bhabha was both a physics Nobel Prize nominee as well as a nominator. He, of course, never won the prize despite being nominated five times by the French mathematician, Jacques Hadamard, for the prize in 1951, 1953, 1954, 1955 and 1956. Hadamard's proposal, however, was not backed up by convincing evidence and, as a result, was never taken seriously. For the 1951 prize, the Nobel committee noted that Hadamard had nominated Bhabha for his work on cosmic radiation without providing extensive reasons. While agreeing that Bhabha had done important work, the committee decided that it could not recommend him for the prize at that time.

Bhabha was nominated again for the 1953 prize but was rejected. The committee merely noted that it had investigated his nomination in the recent past and no new developments had occurred that could change their conclusions. Hadamard nominated him again in 1954, 1955 and 1956 but it had no impact and the reason given was the same—nothing new had happened to change their conclusions. One of the reasons why the eminent mathematician was not taken seriously was because his letters were too short and rudimentary and did not advocate Bhabha's candidature effectively. Hadamard was certainly aware of the process of nomination since, by then, he was something of a serial Nobel nominator, having sent 14 nominations for physics and 10 for chemistry in the first half of the twentieth century. The only explanation for his half-hearted proposals seems to be his age. The first time he nominated Bhabha, he was 85 years old. Presumably, he was too old and feeble to write a long proposal and enclose supporting evidence.[43]

Bhabha, unlike Raman, does not seem to have pushed very hard in this direction. Raman published numerous papers on the Raman Effect and sent copies to leading scientists, like Rutherford. In fact, it was Rutherford who, along with C.T.R. Wilson, jointly nominated Raman for the 1930 prize and enclosed a list containing 160 papers. Rutherford asked Bohr to nominate Raman for the prize. Raman also requested Bohr for a nomination.

After nominations for the prize are received, a list of valid candidates is drawn up. An experts group reports on the contributions of the candidates and a shortlist is made. The report of the Nobel Committee contains a short review of all proposed candidates and a detailed review of those who are potential candidates. Bhabha was never on the shortlist, and no detailed report of his achievements was ever written. It appears that he was never seen as a serious candidate for the prize.

It is surprising that given Bhabha's international contacts, none of his colleagues and friends in the scientific community ever thought of nominating him. The fact that British physicist Cecil Powell, also a Cavendish man and one of Rutherford's 'boys', had received the prize in 1950 for his development of the photographic method of studying nuclear processes and the resulting discovery of the pion made with this method may also have had something to do with it. The Nobel Committee presumably was not interested in work that they believed had already been recognized. Ironically, Powell was all praise for Bhabha's contribution.

> Homi Bhabha made decisive contributions to our understanding of how they (the showers) developed in terms of electromagnetic processes. He was also well known at this time for his attempts to account for those elementary particles then known to exist by a method using group theory. He was thus a very early exponent of those methods used many years later for a similar purpose by Gell-Mann and others.[44]

Bhabha was also a nominator for the physics Nobel, and in this he seems to have had more success. According to the rules, mainly Scandinavian scientists and Nobel laureates have a permanent right to nominate candidates. Academics from other universities are also invited. The committee also calls for nominations from other eminent persons who are thought worthy of the honour. They, however, only have an ad hoc status. Proposals are invited in September and these have to be submitted by 1 February of the next year. A nominator can propose one or more candidates.

Bhabha nominated John D. Cockcroft on 16 February 1951, presumably unaware of the 1 February deadline. Cockcroft did win that year, sharing the prize with Ernest Walton, but Bhabha's late recommendation was not taken into consideration. For the 1952 prize, he nominated Felix Bloch of Stanford University. Bloch won the same year. However, Bloch also had nominations from other heavyweights, like Enrico Fermi, James Franck and Gregor Wentzel.

Chapter 17

STUDYING THE EXTRATERRESTRIALS

> In cosmic rays nature has provided us with the biggest atom-smashing instrument in the world, and the whole surface of the earth is our laboratory.
>
> —Speech at the inaugural function of the TIFR, Bombay, 19 December 1945[1]

A TIFR stalwart, who joined the institute in 1955, was M.G.K. 'Goku' Menon. As we have seen, Bhabha had created a powerful group of researchers working on cosmic ray research to which Bernard Peters along with Yash Pal and Devendra Lal made a significant contribution. As noted, Bibha Chowdhuri had come to TIFR after working with Cecil Powell and Patrick Blackett in England. Others who joined the group included R.R. Daniel and Sukumar Biswas. B.V. Sreekantan had pioneered the KGF experiments in cosmic rays. Menon's arrival in 1955 completed the group.

M.G.K. MENON'S ARRIVAL

The Mangalore-born M.G.K. Menon completed his BSc in 1946 from Jaswant College, Jodhpur, Rajasthan, when he was just 18. Like his grandfather and father, he wanted to study law, but C.V. Raman's visit to his home changed all that. When the question of the young man's career options came up during a dinner conversation, Raman was most emphatic in his advice. He thumped the table and declared that the best thing Menon could do was to study physics and become a scientist!

He joined the Royal Institute of Science in Bombay, where he was introduced to photographic emulsions.[2] He was making significant progress in his research when his attention was drawn to the pioneering work being done at the H.H. Wills Laboratory at Bristol, under the leadership of Powell. This work had already led to the discovery of the pion and went on to win for him a Nobel Prize in 1950. Menon wrote directly to Powell, telling him of his own work and asking to work with the Bristol group. His confidence was not misplaced, for Powell responded with an immediate invitation. Menon sailed for Bristol in 1949. His own money, and a small grant from the AEC, helped him make ends meet in England.

Menon was warmly welcomed by Powell and his wife, who used to mother the entire group of young students and technicians working in the lab. By then, Powell, with his assistants 'Beppo' Occhialini, Hugh Muirhead and César Lattes, had already discovered the (charged) pion, in 1947, using the new boron-enhanced photographic emulsion invented by them. However, there were many other particles waiting to be discovered, and members of the Bristol laboratory were working at a furious pace. Menon also participated in the balloon flight experiments in Sardinia and the Po Valley.

As head of the laboratory and a celebrity to boot, Powell was a busy man. Menon worked with Muirhead and a young Swiss postdoctoral researcher, Olivier Rochat. Menon, Muirhead and Rochat published a paper in 1950 in the *Philosophical Magazine*, investigating nucleonic pion-capture reactions by studying thousands of plates exposed to cosmic rays in a long succession of balloon flights. However, this was not the major focus of his work at Bristol. In 1949, a little before Menon joined, the Bristol group had discovered another heavy meson, about a thousand times the mass of the electron. It was this heavy meson and its decay that became the focus of Menon's research.

This led Menon to what was called the tau-theta puzzle. The puzzle was that two heavy mesons, the tau-meson and the theta-meson, were almost identical—they had the same mass and the same lifetime. However, they had different parities[3]. Before we discuss the tau-theta puzzle, it is necessary to briefly touch upon the subject of beta decay pioneered by Enrico Fermi. Beta decay is a type of radioactive decay in which a beta particle[4] is emitted from an atomic nucleus. In beta minus decay (β^-), a neutron is converted to a proton, and the process creates an electron and an electron antineutrino. In beta plus decay (β^+), a proton is converted to a neutron and the process creates a positron and an electron neutrino. This is also known as positron emission. Fermi developed this theory, and introduced a new type of force, the weak interaction, completing the four basic forces known in nature: gravity and electromagnetism (which operate at long distances), and the strong force[5] and Fermi's weak force, which operate within nuclear dimensions.

When the theta particle decays (through the weak interaction), it disintegrates into two pions: one positive and the other neutral. When the tau particle decays, it disintegrates into three pions: two positive and one negative. The tau-meson has a parity of 1 and the theta-meson of -1. Since they have different parities (assuming no loss of parity in the decay process), it means that the tau and theta are different particles.[6] The notion of 'parity conservation' was a cherished belief of physicists and it implied that, for fundamental physical interactions, it is impossible to distinguish right from left and clockwise from counter-clockwise. If parity is conserved then it means that nature does not distinguish between left and right.

This was a riddle for particle physicists in the 1950s, who struggled to find a way to allow the tau and theta particles to be one and the same while preserving parity conservation. The paradox was eventually resolved in 1956 by Tsung-Dao Lee at Columbia University and Yang Chen-Ning of the IAS at Princeton by invoking parity violation in weak decays with tau and theta being identified as different decay modes of the same particle—the K-meson or kaon. Obviously, a single particle will have a single mass and lifetime. Lee and Chen-Ning showed that the weak interaction of radioactive decay could indeed violate parity conservation. This had never been experimentally tested, and the definitive experiment was carried out by Lee's Columbia colleague, Chien-Shiung Wu, who showed conclusively that parity is sometimes violated in weak interactions. Lee and Chen-Ning shared the 1957 Nobel Prize in physics, one of the quickest such awards ever made.[7]

Menon completed his PhD under Powell in early 1953. For the next two years, he held the Senior Award of the Exhibition of 1851, a prestigious fellowship, becoming only the second Indian after Bhabha to have been accorded this honour. Bhabha was aware of recent developments in the field, and Menon's work on kaon decays interested him greatly. When Menon delivered a talk as the spokesman for the Bristol group at the Rochester conference of 1955, Bhabha met him briefly during the tea break and invited him for dinner at his hotel.[8] The older man recognized a kindred spirit in the sharp, quick, decisive and competent Menon. The Anglophile in Bhabha also appreciated the many and close friendships Menon had with the English, and the fact that he could mix on easy and equal terms with them.[9] Like Bhabha, Menon, too, had been part of a network of scientists and was well-acquainted with Blackett and J.D. Bernal and had long conversations on cosmic rays with Peter Fowler, Ralph Fowler's son and Ernest Rutherford's grandson.

When Menon got to the hotel, he was taken aback because the guest list included the who's who of science, with Glenn Seaborg, J. Robert Oppenheimer and Robert Marshak in attendance. After dinner, Bhabha asked him to stay back. During the conversation, Bhabha offered him a position at TIFR. He told Menon that even though his achievements at Bristol merited an appointment to a professorial position right away, he was only going to give him a reader's position to avoid heartburn among other faculty who were older than him but were still at lower rungs of the hierarchy. The fact that Menon saw the point right away must have endeared him to Bhabha immediately. Menon said he would respond to the offer after consulting Powell.[10]

In an interview from April 2005, Menon recalled Bhabha saying, 'Once you are here and once, they recognise your merit and class and you know what you can deliver, there will be no objection to your going up. But they wouldn't want that to happen by coming in from the outside.'[11] Menon appreciated Bhabha's thinking and his acceptance showed that both he and Bhabha had similar views on what constituted fair processes in institution building. Both had a common vision for the institute where the struggle of those who persevered under difficult conditions was something to be valued and rewarded. As we shall see later, E.C. George Sudarshan was less understanding of Bhabha's offer for a position at TIFR, but he had his reasons.

Menon returned to India in the winter of 1955 and joined TIFR. In early 1956, Bhabha asked Menon to visit him at Mehrangir and quizzed the latter on recent research on elementary particles. This visit gave Menon an opportunity to meet Bhabha's mother and brother. He found Meherbai to be most caring and affectionate. Even after Bhabha passed away, Menon visited her regularly on Sunday evenings with his family and his two children grew very fond of her. Later that summer, Bhabha deputed Menon for a conference on high energy particle physics organized by the Soviet Academy of Sciences at Moscow.

One of the major projects that Menon was involved in during that period was the development of plastic balloons for cosmic ray research at stratospheric altitudes. In the past years, TIFR had undertaken an extensive series of balloon flights to study the penetrating components of cosmic rays using Geiger-Müller counter telescopes and nuclear emulsions as payloads. However, the balloons they used had severe limitations in terms of carrying heavier payloads and reaching higher altitudes.

Earlier, cosmic ray studies had been done with photographic plates exposed on high mountains, such as the Jungfrau and Pic-du-Midi in Europe and Sandakphu and Gulmarg in India. Towards the end of the Second World War, the Americans pioneered the use of rubber balloons filled with hydrogen or helium that carried the plates to higher altitudes. Since they could not afford to use expensive rubber balloons, the ever-resourceful Powell came up with the idea of using plastic balloons. His team fabricated small plastic balloons as clusters and flew them with varying degrees of success. Menon was a member of the group but played a peripheral role in these pioneering flights since his expertise lay in the payloads, rather than the balloons themselves. By the time Menon came to India, the Americans had already overcome the limitations of weather balloons by using large single-cell polyethylene balloons. However, since this plastic balloon technology had been developed under a military contract, it was unavailable to Indians.

When Menon joined TIFR, Peters put him in charge of a group that was responsible for development work in balloons. Peters suggested to Bhabha that TIFR should start a programme to develop large plastic balloons, and the latter immediately agreed. The group persevered in trying to develop plastic balloons at TIFR using the locally available polyethylene. In 1955, they started fabricating balloons using white-coloured commercial polyethylene, but failed. The balloons suddenly shattered like glass near the tropopause[12]. This was puzzling since atmospheric pressure goes down only gradually with height and the plastic had been tested and found to be strong enough to withstand much larger pressure differences than what could be expected during the whole flight.[13]

Investigations revealed that the balloon's fabric was freezing up, becoming brittle and then bursting due to the pressure of the gas. Menon's group solved this problem in an ingenious way by adding carbon black to the film before extrusion. By absorbing sunlight (which the white fabric had been reflecting away), it could maintain its temperature well above the brittle point while passing through the tropopause. They tested several black plastic balloons and all of them successfully passed the test. Thus, the group, in fact, invented the black balloon technique. The first successful flight was made from the grounds of Osmania University, Hyderabad, in February 1960.

The balloons were made on a special, 300-ft-long table at TIFR. The subsequent flights produced excellent results, and R.R. Daniel, S.S. Biswas, P.K. Malhotra and Menon all published papers. Hyderabad was selected as the launch site for a variety of reasons including low geomagnetic latitude and its location in the middle of the subcontinent. Osmania University provided all the early support for these flights. In 1970, the balloon operations were shifted to a new site at Maula Ali, Hyderabad.

At MIT, Sreekantan had worked with Bruno Rossi and perfected his skills in building equipment. Upon returning to India, he built one of the largest cloud chambers in the world at Ooty, where TIFR was developing new laboratories after hearing the news about the closure of the mines at Kolar. When this turned out to be a rumour, the scientists went back and set up their laboratories again, this time at a depth of 10,000 ft underground. It was the deepest laboratory in the world and was situated below an underground platform with offices and a cafeteria that sold coffee and dosa. The scientists had to descend in a bucket because the lifts did not go that far down.[14]

Bhabha and Menon both realized the great potential of the deep underground facilities at Kolar for conducting particle physics experiments. The group also had an international collaborator—Professor Saburo Miyake who was a cosmic ray physicist from Osaka City University, Japan. He was Menon's good friend and joined the group in 1960. The year before, Menon had seen images of air showers of cosmic radiation at Mount Norikura presented by Miyake at the International Cosmic Ray Conference in Moscow. He had also met Miyake in Japan and, perceiving a similarity of interests, invited him to TIFR. Miyake initially began experimentation at Ooty in the setup started by Sreekantan, but Menon soon persuaded him to move to the Kolar mines. Miyake's association from 1960 onwards was the start of a three-decade long collaboration with the TIFR group.

A new series of experiments was started in the early 1960s by Miyake, V.S. Narasimham and P.V. Ramanamurthy to methodically map the intensity of cosmic ray muons as a function of depth. After reaching the deepest site (2,700 m), they recorded no muons even after running the experiment continuously for two months. In a paper published in December 1963, Menon and his collaborators argued that such a deep underground site with almost no cosmic ray muon background would be an ideal place to study cosmic neutrinos. The paper resulted in the start of a series of important experiments conducted at Kolar over the next few years to study neutrinos.[15] The TIFR team that worked on cosmic ray neutrino detection in the early

1960s became one of the largest in the world.

The neutrino is a subatomic particle that was first proposed by Wolfgang Pauli in 1930, but remained an undetected particle till 1956. In June 1956, Reines, working with Clyde Cowan, confirmed the existence of the neutrino. Working in Hanford and Savannah River, Reines and Cowan developed the equipment and procedures to detect the supposedly undetectable particle. Since then, Reines dedicated the major part of his career to the study of the neutrino and his work influenced future research in the field. He won the 1995 Nobel Prize in physics for his work. Cowan, who died in 1974, missed out since the prize is not awarded posthumously.

Reines started out wanting to collaborate with the TIFR group but ended up becoming a competitor. He knew of the Kolar experiments from P.V. Ramanamurthy's thesis on the subject. The thesis had convinced Reines that the flux of muons was sufficiently attenuated to make the detection of neutrinos possible. Reines flew to India, met the scientists at TIFR and visited the Kolar mines. However, he wanted to bypass the TIFR team completely, taking all the credit for himself. He went to the IISc, met the director Satish Dhawan and then went to the KGF to work out an in-principle collaboration with IISc. The KGF signed the agreement with IISc. However, thanks to Bhabha, the Parliament had passed a rule that anything connected with nuclear science should first be cleared by the DAE. So, the KGF was obliged to refer it to him. Bhabha, then, told Sreekantan and his team that Reines had been going behind their backs and the so-called collaboration was called off. After being caught out, Reines tried to re-establish relations with the TIFR group, but it was too late. They had already established a connection with the Japanese and Durham groups, and Bhabha was in no mood to allow competition.[16]

Was it short-sighted on Bhabha's part to disallow Reines to lead the experiment since he had, after all, confirmed the existence of the neutrino? Sreekantan, who was very involved with the experiments at that time, said that Bhabha's reluctance came from Reines' already having contacted two local collaborators who were not part of TIFR, thereby giving the impression that the TIFR group was to be excluded completely from the experiments.

The TIFR group presented several papers at the international conference on cosmic rays held in Jaipur in 1963. Yash Pal and Peters presented a paper on meson production. Several TIFR group members, including Yash Pal, Ramanath Cowsik, and S.N. Tandon, presented their work on neutrinos. Menon delivered a talk on neutrino experiments underground; Miyake and his Indian collaborators presented their work on cosmic rays. The TIFR group had expanded its collaboration and was now partnering with Miyake's Osaka group and Arnold Wolfendale's group from Durham University, UK. The collaborators had recorded several clear examples of neutrino interactions and, in September 1965, presented their work at the international conference on cosmic rays in London. Reines became a formidable rival, choosing a deep gold mine in South Africa to conduct his experiments. His group detected the first neutrino tracks in February 1965, but the TIFR group seems to have beaten Reines' group in formal priority, if the date of submission for publication is considered. Their paper was received by the *Physics Letters* on 12 July while Reines' paper was received two weeks later on 26 July.[17]

In August 1964, Yash Pal attended an international conference on high energy physics at Dubna, which Reines was actively involved in. Bhabha realized that the names Reines had included were entirely American and questioned the international credentials of the conference. He asked Yash Pal to attend only as an observer and to not commit anything on behalf of TIFR. Clearly, Bhabha was not ready to trust Reines, highlighting the fact that while skills and resources are important in the formation of collaborations, so are human factors, such as trust.[18]

In the 1970s, the Kolar experiments were repurposed to look for proton decay, which had been first predicted by Jogesh C. Pati and Abdus Salam in 1974, followed by the Grand Unified Theory of Howard Georgi and Sheldon L. Glashow. No proton decays were seen and none have been till date.[19]

E.C. GEORGE SUDARSHAN

A bright young man who had joined Peters' group at TIFR was E.C. George Sudarshan[20]. He joined TIFR in the spring of 1952 after completing his postgraduation from Madras University. Sudarshan was initially asked to conduct experimental work in particle physics using the photographic emulsion technique. In this phase, he worked closely with Biswas, Daniel, and, of course, Peters. The three years, from 1952 to 1955, that Sudarshan spent at TIFR turned out to be enormously important in many ways.

Later, Sudarshan turned to theoretical work, and, when Dirac gave a course of lectures on quantum mechanics in 1954, the lecture notes were prepared by Sudarshan and Bhabha's student, K. K. Gupta. This brought Sudarshan and Dirac into close contact and provided the young man with a unique opportunity to learn the subject from the master himself. He greatly admired Dirac, and the two remained lifelong friends.

Robert Marshak's visit to TIFR in August 1953 had a major impact on Sudarshan's career. Marshak, a contemporary of Julian Schwinger and Richard Feynman, had created a vibrant group at Rochester University. He was also a good friend of Bhabha's. Marshak was so impressed by Sudarshan that he persuaded Sudarshan to come to Rochester and work with him for his PhD. Peters was hesitant to let him go, but Sudarshan was finally able to join in September 1955.

Sudarshan reached Rochester at an exciting time. The tau-theta puzzle was animating physicists the world over. He started working on his PhD in this extremely active field. The various particles, although ephemeral, are an essential ingredient of all modern theories of particle interaction, and their own properties affect the way in which those interactions, including the weak ones take place. Some of the properties they possess, related to their spin and parity, were called A, V, S, and T (shorthand for 'axial', 'vector', 'scalar' and 'tensor'). By early 1957, after an extensive study of all existing experimental results, and allowing for all theoretical possibilities, Sudarshan and Marshak had arrived at the V-A (vector minus axial vector) structure for the universal Fermi interaction. Others who were working on the same problem were Murray Gell-Mann and Feynman.

Sudarshan had spent a lot of time on the problem before the April 1957 Rochester conference, but since he was a student, he didn't get to make a presentation on the subject. Marshak did not refer to this subject at all, since he was busy presenting a major paper on another subject. Instead, he requested P.T. Mathews from England, then visiting Rochester, to present the V-A results as remarks during one of the discussions. Mathews agreed but never got around to doing it. Later, Marshak and Sudarshan met Gell-Mann in July that year, when they visited the Califorina Institute of Technology and the three of them discussed the implications of the V-A theory.

Soon after—as preparation for a talk Marshak was to give at the Padua–Venice conference in September 1957—Sudarshan and Marshak sent in an abstract on the V-A theory, and on 16 September 1957 (coincidentally Sudarshan's 26th birthday) they submitted their paper titled 'The Nature of the Four-Fermion Interaction'. As a matter of publication ethics, Marshak decided that the same results should not be submitted to two different places. Unfortunately, the publication of the Padua–Venice proceedings got delayed until May 1958. This delay became the cause for a lifetime of regret for Sudarshan.

On the very same day that Marshak sent his paper, Feynman and Gell-Mann submitted their paper titled 'Theory of the Fermi Interaction' to the *Physical Review*. It was published on 1 January 1958. Probably to make up for lost time, Marshak took the initiative to submit a very short paper with Sudarshan to the *Physical Review* on 10 January 1958 on the subject. It appeared in print on 1 March 1958. But the damage had been done. In most physicists' minds, the Feynman–Gell-Mann paper took precedence, and much to Marshak's and Sudarshan's chagrin, they got the credit for work that the Rochester duo had completed before them. Sudarshan could never overcome the bitterness he felt at the way credit was apportioned to Feynman and Gell-Mann, and understandably so. Feynman, after some reflection, decided that an injustice had been done. He gave Sudarshan and Marshak due credit when discussing the theory of the weak interaction. He described the theory as invented by Marshak and Sudarshan and published by Feynman and Gell-Mann.

After submitting his PhD thesis, Sudarshan spent two years (1957–1959) as a research fellow with Julian Schwinger at Harvard University. During this period, he continued his collaboration with Marshak. Sudarshan had apparently planned to return to India after the two years at Harvard. In a paper submitted to the *Physical Review* in July 1959, Sudarshan's name was accompanied by a footnote 'On leave of absence from the Tata Institute of Fundamental Research, Bombay, India'. In 1959, Sudarshan returned to Rochester and joined the physics faculty as assistant professor; two years later he became associate professor.

Bhabha was on the look out for a theoretical physicist who could build a group at TIFR. His own student, K.K. Gupta, had completed his PhD in 1954, receiving praise from Pauli, who had been the external examiner. Gupta joined Gregor Wentzel in Chicago for two years in 1957 (he had done the lecture notes for Wentzel when he visited TIFR). Upon returning, he started a group that researched the many-body problem[21], a subject that was one of the biggest challenges in quantum mechanics in the last century.

Bhabha thought that Sudarshan would be a good choice as a theoretical physicist and wrote to him in March 1962 with an offer to return to TIFR as an associate professor. He had met Sudarshan in Washington and discussed the possibility of returning to Bombay. Bhabha had written to Marshak before he made his offer, expressing the hope that Sudarshan would make up his mind to return soon. In his letter, Bhabha told Sudarshan that his promotion did not depend on existing vacancies and would take place as soon as he demonstrated his ability to work satisfactorily in the Indian environment. As associate professor, Sudarshan would be in charge of his own group and have full freedom. As far as the emoluments were concerned, Bhabha said that the revised pay scale of the institute made the pay of associate professor equivalent to the earlier pay scale of the professor.[22]

Marshak wrote to Bhabha in April 1962, attempting to persuade him to make Sudarshan a better offer. He argued that Sudarshan needed some inducement to leave Rochester. He asked Bhabha to consider the fact that Sudarshan was highly regarded at Rochester and had a bright future in what was a very active department. Sudarshan, Marshak said, felt that rank was important to work in the Indian environment.[23] It appears that Bhabha was unaware of Sudarshan's contribution to the V-A theory because, in his reply to Marshak, he said that Lal and K.G. Ramanathan had done work that was in no way inferior to what Sudarshan had done. He said that apart from Menon and Chandrasekharan, all the others were associate professors who ran their own sections with the greatest freedom.

Bhabha clearly felt very strongly about not granting a full professorship to newcomers because Marshak's letter advocating Sudarshan's case has a hand-written margin note next to the sentence: 'No!'[24] By the time Sudarshan responded in June, Marshak would have received Bhabha's letter and presumably discussed its contents with his protégé. Sudarshan, in his polite letter, closed the possibility of returning to India. He said that as the matter then stood, there was no possibility of him returning to TIFR. He did not entirely rule out a future reconsideration of the matter but felt that, given the substantial lack of agreement in their respective assessment of the situation, this would not perhaps be feasible.[25]

Many decades later, Sudarshan, in his oral history interview, clarified what he meant by the lack of agreement in their respective assessment of the situation. He said he found Bhabha's offer unacceptable because it did not conform to what they had discussed in Washington, the implication being that Bhabha had gone back on some promise he made. Sudarshan also realized that Bhabha was not aware of the work he had done on the V-A theory and perhaps felt a little undervalued in Bhabha's eyes.[26]

The ghost of the omission at Rochester, which deprived him of credit for the V-A theory, continued to haunt Sudarshan all his life. In an oral history interview with Charles Weiner for the American Institute of Physics in 1970, Marshak recalled Sudarshan's disappointment, saying that he had to suffer years of constant reference to the Feynman–Gell-Mann theory when his work was equally good and had been done earlier, and he was very disturbed about it.[27] At Sudarshan's 60th birthday celebrations in 1991, Marshak gave an after-dinner talk in which he spoke of the 'the pain and the joy of a major scientific discovery'.[28]

After recounting the full details of what had transpired with candour and humility, Marshak asked for Sudarshan's forgiveness for his three 'cardinal blunders', as he put it, which had cost his former student so dearly. Although Sudarshan spent most of his later life in the

US, he continued to have strong ties with Indian science. He set up the Centre for Theoretical Studies at IISc and was director at Matscience in Madras. A recipient of the Padma Bhushan (1974) and the Padma Vibhushan (2007), he was awarded the Dirac Medal in 2010.

Many years later Sudarshan, in his assessment of Bhabha's contribution to theoretical physics, said:

> Amongst people of our times, Bhabha's work on Cascade Theory is on par with Meghnad Saha's work on ionisation equilibrium. The only ones that surpassed it were Satyendranath Bose's discovery of Bose-Einstein statistics and C.V. Raman's of the Raman Effect. Yet it is a sad commentary that the momentum of Bhabha's discovery did not inspire comparable work in India—significant work on the frontiers...There was no close group of brilliant young colleagues and pupils who made up a critical mass to sustain his spirit.[29]

Chapter 18

A NEW HOME

> The building itself is only a shell to make possible the work that is done inside it. It is by the quality and volume of its scientific work that an Institute like this must be judged [...] and by its general impact on the scientific life of the country as a whole.
>
> —Speech at the inauguration of the new TIFR buildings, Bombay, 15 January 1962[1]

The Colaba premises of TIFR covered about 25,000 m^2 and consisted of a series of makeshift Second World War army barracks. Many of the experimental groups had moved there from the 1950s onwards, even though the library and the workshop continued to function from the OYC. Since there was no canteen at Colaba, everyone working there came to the OYC and patronized the many restaurants in the area. As already noted, Otto Koenigsberger had sketched the preliminary plans for the TIFR building, but by 1951 he had left India for London. The land for the new building was acquired only in 1953, nearly eight years after the institute had started functioning from Pedder Road. 'The Council,' Bhabha wrote in his Historical Note in 1954, 'took the view that the latest and the best ideas should be incorporated.'[2] Keeping this in mind, Bhabha appointed the Chicago-based firm, Messrs Holabird, Root and Burgee, as the designing architects.

DESIGNING THE NEW PREMISES

The NPL at New Delhi had been designed and constructed by the CSIR and the Central Public Works Department (CPWD), but Bhabha was too much of a perfectionist to settle for anything less than the best. Engaging the services of a foreign architect was not unprecedented at the time. Besides Koenigsberger, Joseph Allan Stein had designed the building of the India International Centre in New Delhi. Nehru had invited the firm, Mayer, Whittlesey and Glass, to design Chandigarh. The firm pulled out in 1950 and Maxwell Fry and Jane Drew were hired instead. They suggested that Le Corbusier, the Swiss architect and one of the founders of the International style, should be the master planner. Le Corbusier created Chandigarh, a city that signalled the arrival of Modernism in Indian architecture. (When Le Corbusier visited Bombay in May 1953, he stayed with J.R.D. Tata at his residence, 'The Cairn'. J.R.D. sent Bhabha a telegram in Bangalore inviting him to dine with them on 21 May[3]). Another example of the International style was Louis Kahn's design of the Indian Institute of Management, Ahmedabad.

One of the main reasons for Bhabha appointing Holabird, Root and Burgee was that they had the required expertise in laboratory design, having built a laboratory for the US AEC at Oak Ridge, and also had experience in building in tropical areas. Bhabha cited the firm's experience in building in South America as proof of the latter, even though the buildings designed by the firm in places like Bogota were International in style. The associate architects were Master, Sathe & Bhuta and later G.M. Bhuta & Associates.

Helmuth Bartsch, the designer, was at the time one of the partners in Holabird, Root and Burgee. Bartsch had been trained in Berlin and was a close friend of Ludwig Mies van der Rohe, the last director of Bauhaus, who had moved to the US in 1937, after the Nazis came to power. 'Less is more' is a dictum that owes its fame and popularity to van der Rohe. It was his favourite catchphrase, and though he didn't coin it, he was responsible for not only popularizing it but also making a practice of the paradox. His philosophy became a guiding principle for architects in the mid-twentieth century. Bartsch, too, was influenced by his way of thinking.

Bartsch was based in Chicago since 1926, and Bhabha met him in the course of his travels and chose him as the designer for the new TIFR buildings. This long-distance relationship came with its set of practical problems. Bartsch would send his drawings to Bombay, and a local architect, Achyut Kanvinde, hired from the CSIR would execute them. Kanvinde had worked under Walter Gropius at Harvard and his batchmates included famous names like I.M. Pei and Paul Rudolph. Later, with his partner Shaukat Rai, he set up the firm Kanvinde and Rai in Delhi.[4] He designed the ISKCON temple in Delhi and the buildings of Indian Institute of Technology, Kanpur. There were the inevitable delays and Bhabha noted with displeasure that even by March 1954, work on the new building had not begun. Bhabha upped the ante, and the finances for the project were sanctioned by the Ministry of Finance in April 1955.

Bhabha interacted with Bartsch on every aspect of the design of the new TIFR buildings. This unusual degree of interest, participation and constructive suggestions by Bhabha led Bartsch to comment that, in designing these buildings, he had 'worked with a client rather than for a client.' Bartsch found that Bhabha 'displayed unending interest and encouragement and constantly added intelligent suggestions and advice.' He hoped, in the end, that he had designed a building that would 'not only fulfil its function but should afford a great deal of enjoyment.'[5]

The principal contractor for this project was Shapoorji Pallonji & Co. Pvt. Ltd. Bombay, and the air conditioning was provided by Voltas. The rectilinear form of the TIFR building is horizontally orientated. The windows are metal louvred in horizontal bands with the metal-framed windows set flush with the exterior walls. Bartsch used floor-to-ceiling glass in the canteen and foyer. The glass wall is a feature of the International style, and its use in the TIFR building allowed Bartsch to show the Arabian Sea to advantage. The free-standing cantilever staircase that leads to the mezzanine library combines elements of the International style with modernist architecture.

The furniture in this centrally air-conditioned building also reflected the modern style and consisted of upholstered armchairs with metal legs that were manufactured in-house with designs from American catalogues. Bhabha was particularly pleased about the harmony between function and aesthetics. Bartsch observed that every item was manufactured in India. The aluminium frames in the windows and the glass walls required special effort. In his inaugural speech, Bhabha explained the processes of manufacturing and the extra effort required by Messrs Godrej & Boyce to fabricate the windows.

> This building has taken so long to construct because we had to do a lot of pioneering in the course of it, pioneering that has resulted in permanent benefit to the country. For example, the aluminium companies in India were induced to extrude special sections for windows and doors in a rustless alloy resistant to corrosion by the warm sea air. The well known firm, Messrs Godrej and Boyce, imported a costly machine in order to weld and fabricate these windows. There is a whole list of other items now readily available in the Indian market which were produced for the first time.[6]

Bartsch, too, emphasized the importance of the indigenous content in the buildings. 'Every item was manufactured in India with the exception of large aluminium sections on the ground floor. This was accomplished against all odds through the client's perseverance.'[7]

As we have already seen, Bhabha was passionate about gardens and plants. He had a wonderful terrace garden at home and he collected plants from all over the world for both his personal garden as well as for the ones in TIFR and AEET. He was an expert on trees, plants and flowers, and he experimented with exotic plants—cross-breeding bougainvillaea and roses on his terrace garden. J.R.D. Tata called him a 'master gardener' since whatever he built had to have gardens. Bhabha started designing the gardens at the institute almost as soon as the land in Colaba was acquired in December 1953. Unsurprisingly, his inspiration was European.

Bhabha and Shreepad Dwarkanath Vaidya, who joined AEET as the head of parks and gardens, combined elements of the gardens at Versailles with the English landscape garden and came up with a modified French model for the west lawn of the institute whose layout takes full advantage of the seaside promenade. The famous Amoeba Garden and the paved pathway combined the geometry of Versailles with the idea of a landscape garden. Bhabha personally supervised the landscaping of this feature; he would observe the Amoeba Garden from his fourth-floor office, and it kept changing shape till Bhabha was finally satisfied, becoming something of an iconic garden in his own lifetime. The east lawn, flanked by the artificial casuarina forest, was a sunken lawn with a single tree to one side and a clump of cacti on the other. The lounge on the ground floor of the institute offered a view of the Amoeba Garden and the Arabian Sea.

Bhabha's interest in plants made him source them from all over the world. Swiss-American physicist, Felix Bloch, who won the Nobel Prize in 1952, sent him begonia tubers and the Mexican politician, Emilio Porter Gils, sent him cacti for the garden at TIFR. Later, when Gils became the Mexican ambassador to India, he requested Bhabha for some of the same cacti for the Mexican Embassy in New Delhi. The barringtonia on the west lawn supposedly came from Africa, and the golden palm was from the West Indies. The international nature of the plants and trees reflected the internationalism of the building and could also be seen as an effective metaphor for the institute's internationalist aspirations.[8]

Bhabha, however, did not believe in inviting VIPs to participate in ceremonial tree planting, something that is still a customary feature of visits to government institutions in India. Nehru, even though he visited TIFR and AEET on more than one occasion, was never invited to plant trees. The only ceremonial tree planting Bhabha participated in was on the occasion of his 50th birthday at a function organized by the employees of AEC. His mother, Meherbai, once planted a conifer inside the premises of Kenilworth. Bhabha was also categorical that he wanted no photographs taken and no commemorative board hung to mark his birthplace.[9]

While afforestation was not unknown at the time, Bhabha perhaps pioneered the planned transplantation of full-grown trees. No one at the time thought that such large trees could be transplanted. In fact, the institute made place for its buildings by undertaking the transplantation of the many trees on the site. Individual trees in distress, or marked for the axe, received Bhabha's personal attention. Vaidya recollects that the very first tree transplant he undertook at Bhabha's request was the rain tree outside Kenilworth, which was to be felled to widen the road. Given the expense involved, Vaidya was hesitant at first, but Bhabha would not take no for an answer. It was ultimately lifted by crane and transplanted inside Kenilworth. This was possibly the first time that such a large tree had been transplanted in India. It soon became standard practice and, over the years, the TIFR complex and the housing colony became home to a number of full-grown trees that were transplanted on Bhabha's orders.

This practice continued even after Bhabha died; J.R.D. Tata, too, seems to have been a votary of the idea. In 1970, a massive baobab, about 18 ft in circumference was transplanted from Nepean Sea Road. Excavating this huge tree took nearly three days, with the TIFR gardening staff working in two shifts and J.R.D. Tata supervising the operation. The tree, which was moved with the help of three cranes and two trailers, took a full day to reach its destination inside the housing colony, where it exists to this day.[10] Mithoo Coorlawala recalls enlisting Bhabha's help to save two Gulmohar trees that had been knocked down by a monsoon storm in the compound of the building where she lived. Gulmohar trees are not very strong and Coorlawala was afraid that without intervention they would fail to take root again. She asked Bhabha to help, and he promptly sent someone to put the trees back.

According to M.G.K. Menon, it was sometimes remarked that the TIFR's buildings were more expensive to construct. But this was not the case.

> There is much else in India built at the same cost of construction per square foot. But what puts the Tata Institute in a class above these is the close personal attention to detail on the part of Homi Bhabha. It can truly be said that the buildings, the art collection, the gardens and, most important, the scientific programmes and standards bear the unmistakeable imprint of Homi's personality.[11]

The sea-facing land with its army barracks, which had once housed prisoners of war, was transformed into a modern building that overlooked landscaped gardens with exotic greenery collected from all over the world. Like the collection of paintings and the modernist architecture of the building, the gardens of the institute also attempted to create a congenial environment where scientists could discuss their work, or just relax after a busy day's work.

On the silver jubilee of the institute in 1970–1971, it was decided that Bhabha's birthday would be celebrated as Founder's Day. Bhabha had been dead for more than five years by then. On the occasion of the first Founder's Day, in October 1970, the institute presented several hundred trees to the city of Bombay for planting on Marine Drive. These trees, called *barringtonia speciosa,* which grow well near the sea, were grown in the gardens of the institute for four years before being gifted to the city to honour Bhabha's memory.[12]

THE FORMAL ENTRY

Nehru inaugurated the new TIFR buildings on 15 January 1962. The governor of Maharashtra, Sri Prakasa, in a speech laced with self-deprecating humour, narrated how inadequate he had felt when, as the minister for natural resources and scientific research, he had been invited to a seminar on scientific research by Bhabha in Poona. 'I must say that Dr Bhabha and other eminent scientists of the land were most kind and considerate and they thus enabled me to retain safely both my portfolio and my ignorance.'[13] He did not fail to express his admiration of the House of Tatas for all that they had done during the last three generations for the progress and advancement of the country in diverse spheres of its national life.

Speaking on the occasion, J.R.D. Tata, once again, referred to the allocation of land. V.K. Krishna Menon, the defence minister, who had come to campaign before the elections, was also present. J.R.D. thanked him for the land with his tongue firmly in his cheek.

> It gives me the opportunity to thank through him the Defence Ministry who were the original owners of this cosy plot which you might not have seen as yet because most of it is beyond

> the building. This magnificent plot used to be theirs and after fighting a friendly rear-guard action they agreed ultimately—thanks to the persistence of the PM to release it to this Institute. I am sure That Mr Krishna Menon and the Defence Ministry will agree that we have not harmed this plot, that we have perhaps embellished it.[14]

J.R.D. Tata, like his illustrious predecessor Jamsetji Tata, realized the increasing importance of scientific and technological advancement in national life. He said that the present civilization was being increasingly moulded by the advance of science and technology and that the abstract scientific theories of that day could well become practical realities of the next. India's vast programme of development would not succeed if it did not keep pace with the march of science and technology in other countries.[15] During the course of his speech, J.R.D. showered fulsome praise on Bhabha and his contributions. 'I do not wish to embarrass Dr Bhabha any further but it is right that on this occasion, we should pay tribute to him and recognise the immense contribution he has made and still continues to make to India's scientific and industrial progress.' In his conclusion, he did not forget to acknowledge the great contribution made by S.S. Bhatnagar, who, he said, was, 'a staunch supporter of the work of Dr Bhabha in this Institute and we honour his memory for the part he played in our development.'[16]

The chief minister of Maharashtra, Yashwantrao Chavan, in his address, said that there were not two but three institutions functioning there, since Bhabha himself was an institution. He spoke about the state government's support to the institute with pardonable pride. 'Thus, we have a little satisfaction at having made a pioneering effort in the right direction.'[17]

In his speech, Bhabha started by recapitulating the history of the institute and the manner in which it had grown. He also tried to justify the delay in construction. 'The Institute has taken so long to construct because we had to do a lot of pioneering which has resulted in permanent benefit to the country.' After the usual round of thanks, he said:

> The building itself is only a shell to make possible the work that is done inside it. It is by the quality and volume of its scientific work that an Institute like this must be judged, by the extent to which it has helped to explore and push back the frontiers of knowledge to open up new fields of knowledge, to provide the country with men highly trained in the newest fields of scientific endeavour and able to make their own contribution to the increase of scientific knowledge, and by its general impact on the scientific life of the country as a whole.[18]

Bhabha devoted the rest of his speech to the scientific work of the institute. Nehru, in his speech, equated scientists with the guardians of the sacred mysteries in the ancient Greek temples. 'We are now entering into an age when scientists begin to function like the high priests of old, who looked after the scared mysteries; we all bow down to them in reverence and awe and sometimes in little fear as to what they might be up to.'[19] He also thought that a scientist might be better at governing the country. 'But anyhow one cannot escape science and the scientist in this age [...] perhaps the scientist might—and I say "might"—make a better job of it, if he has a chance in the future, than the politician.'[20] Nehru also finally endorsed the name of the institute, when he recalled the pioneering role of Jamsetji Tata in encouraging heavy industry and fostering scientific and technical education. He said that Jamsetji's successors had followed his traditions and, therefore, this institute was appropriately named after the House of Tatas.[21]

Niels Bohr, who had been invited but who could not attend due to 'urgent obligations in Copenhagen', sent a message through Bernard Peters, which J.R.D. Tata read. At the end of September 1961, Bhabha had sent a letter to Bohr, inviting him to attend the inauguration ceremony of the new TIFR buildings, but Bohr sent his apologies, saying that it would not be possible for him to attend. H.H. Koch, who was supposed to represent him, could not make it either due to ill health. In his place, some of his colleagues, who were already in Bombay, had discussions with Bhabha. A little over a year later, on 18 November 1962, Bohr died of heart failure. On 7 October 1965, on what would have been his 80th birthday, the Institute for

Theoretical Physics at Copenhagen University was officially renamed the Niels Bohr Institute, its unofficial name for many years.

INTERNATIONAL NETWORKS

Bhabha had planned to replicate the atmosphere of Cambridge and Princeton at TIFR. However, he was aware that establishing science as a national activity in an underdeveloped country would be a slower process than in a country that had a mature pool of scientists to draw from. We have seen that Bhabha had put in place an international network of scientists, many of whom he knew from his time in Cambridge. They not only lectured at TIFR but also recommended students that had completed their PhDs under them. They provided a further service when they evaluated TIFR students as PhD examiners. For Bhabha, the international network was a source of information about Indian scientists trained abroad and also helped to confirm the level of expertise of the new recruits to the institute.

Bhabha's international contacts also served another purpose. They enabled students from TIFR to learn new technologies at the laboratories abroad. Perhaps, the best example was that of R.V.S. Sitaram, who had trained at the radar establishment in Malvern, Worcestershire, UK. Such international networking, apart from arranging the placement of students, ensured an information exchange that shaped the future course of study. This interaction allowed scientists to share findings, discuss problems, form partnerships and respond to new opportunities, something which Caroline Wagner has called the 'invisible college'.[22] She argued that the shift from big science to global networks created numerous opportunities for developing countries to tap the potential of science. The governments of such countries could leverage such networks by creating incentives for scientists to focus on research that addressed their concerns and by finding ways to tie knowledge to local problem solving.

Bhabha's model was unlike anything adopted by other institutions in India. He saw it as a unique experiment in institution building in the country. An example will make clear how Bhabha not only used his international contacts but also how he employed his own faculty in an international setting. Bhabha asked his younger faculty member, Kundan Singh Singwi, who was in Geneva then, to interview Lalit Kumar Pandit, who joined the institute in 1958. Pandit, who had done his PhD under Walter Heitler, had written to several institutions but only Bhabha responded. Later, Pandit also met B.M. Udgaonkar, who happened to be in Zurich at the time.[23]

There were times when Bhabha was not personally acquainted with a foreign scientist, but the fact that he had worked on a significant project was enough to satisfy him about his scientific credentials. Nuclear physicist, S.K. Bhattacharjee, had done his PhD under Bernard Waldman of the University of Notre Dame. Waldman had been a part of the Manhattan Project. Bhattacharjee wrote to Bhabha and the latter called him for an interview when he was in Chicago. Bhabha not only discussed Bhattacharjee's plans but also his own vision for building accelerators at TIFR.[24]

How current was Bhabha, given the unsettlingly fast pace of new developments in his field? He had always believed that science could grow satisfactorily only if it was administered and run by scientists. But he also felt that scientists who became too involved with administration would gradually become ineffective unless they had an abiding interest in research. Bhabha virtually stopped research after 1953–1954 and published nothing of consequence after 1954. Bhabha's work at TIFR had been mostly of a mathematical nature, but, in 1953, he returned to the phenomenological[25] work he had done in Cambridge many years ago. This concerned the production of mesons in collisions between nucleons. Bhabha wrote a paper on this topic, predicting the constraints on the meson production rate in nucleon collisions. According to Virendra Singh, Bhabha's model can be considered a precursor to Richard Feynman's 'parton model' of the structure of nucleons. This was his last scientific paper.[26]

Bhabha's increasing workload and responsibilities in the atomic energy programme made it impossible for him to remain at the centre of his field. Bhabha, the theoretical physicist par excellence in his younger days, started becoming increasingly peripheral to the field. He found it hard to keep up with the latest developments. This is hardly surprising because, by then, he was wearing three hats:

chairman of AEC, secretary of DAE and director of TIFR.

Bhabha's lack of contribution to new work drew criticism from Raman. He used to open up the latest periodicals and ask visitors, 'Do you find the name of Bhabha anywhere in the scientific literature nowadays? He has quit science.'[27] Although Raman had, all along, shared a cordial relationship with Bhabha, he hated bureaucracy and big organizations. Consequently, the two drifted apart. Bhabha did not visit Raman after 1954 and this upset the elder scientist very much. However, it appears that Bhabha had received diamonds from Raman's institute for research purposes. Raman had become something of an anachronism in the world of science. Having spent long years as a hobbyist at a time when science was increasingly professionalized, he tried his best to keep up with developments in quantum physics. He continued as a low-budget experimenter when experimental physics was making the transition to 'big science', with accelerators, cyclotrons and reactors, all requiring heavy capital outlays, being in vogue.[28]

Raman also resented Nehru's policy of concentrating research in specialized institutions, neglecting universities. He coined the phrase 'Nehru–Bhatnagar effect' to describe the mushrooming of CSIR laboratories, predicting that they would achieve very little. Bhatnagar had been on the committee whose report led to Raman's exit as director of IISc, and there was no love lost between the two. Raman quipped that like Shah Jahan, who had built the Taj Mahal to bury his wife, Bhatnagar had built laboratories to bury scientific instruments.[29] Raman, unlike Saha, did not wear his nationalist commitments on his sleeve and refused to join the National Planning Committee. Still, Nehru made him a national professor in 1947, ensuring that he would receive a full salary for the rest of his life. However, this warmth was never reciprocated. There is an anecdote about how Nehru, on a visit to Raman's laboratory in 1948, was tricked into identifying copper (glowing under UV rays) as gold. Raman was quick to pounce on the opportunity. 'Mr Prime Minister, everything that glitters is not gold.'[30]

A look at Bhabha's oeuvre during 1933–1967 reveals that, out of his 104 publications, 69 were journal articles, 21 were conference papers and 12 were speeches, with one meeting and one biographical piece each. Bhabha published his first paper 'Interaction of Radiation with Matter' in *Nature* in 1933 at the age of 24 for which he was awarded the Newton Studentship.[31] His last non-scientific paper, a biographical memoir on K.S. Krishnan, written in collaboration with Kathleen Lonsdale in 1967, was published a year after his tragic death.[32]

Bhattacharjee recalled that although Bhabha's knowledge was a little out of date, 'ultimately the man was so sharp that he would hardly miss anything.'[33] Notwithstanding Bhabha's curiosity, the fact remained that he was out of touch. Bhabha knew it too, and from 1956 onwards, gave instructions that he should be addressed as Dr Bhabha and not Professor Bhabha. By then he was hardly doing any teaching and he said that he didn't want to be called professor when he was no longer doing that job.[34]

Pandit recalled that while Bhabha's vision kept TIFR from becoming 'provincial', he felt that he and the others were constantly struggling to keep up with the terrifying pace of change in a discipline like physics.[35] Though the institute had done much good work that received international recognition, many scientists felt that they were far removed from where the action really was. Scientific thinking was changing at an unsettling speed, and it is no wonder that TIFR scientists felt disconnected and left out.

P. Babu, a cosmic ray physicist who later turned to molecular biology, spoke about the yearly ritual where the work of the physicist who had received that year's Nobel Prize was discussed. 'Even in TIFR we were all peripheral players. It was like a football match where we sitting and cheering.'[36] Although Indian scientists were not isolated from scholarly discourse, they did feel that they were marginal to the science in the US or Europe, despite occasional publications in important journals. Bhabha was an excellent facilitator, enabler and outstanding fundraiser but as a teacher who led others to greatness, he was not in the league of other great scientists. He was certainly no Rutherford or Bohr; closer home, J.C. Bose, P.C. Ray and Raman exerted more influence as teachers.[37]

The Summer Schools in Theoretical Physics—the first of which was organized in 1961 in Bangalore—proved to be very important both for the institute and the development of theoretical physics in the country.

Richard Dalitz and Murray Gell-Mann lectured there, and they came at a critical time for particle physics—one in which new ideas on symmetries and phenomenology were being developed. Writing about it, G. Rajasekaran, at the time a student at TIFR, recalled how the TIFR group missed out on what could have been a major contribution—the discovery of quarks. All the TIFR heavyweights, along with Bhabha attended. Bhabha, who was always late for meetings, was there on time at the front desk taking notes and inscribing his own comments.

Dalitz and Gell-Mann lectured on SU (3) and the Eightfold Way[38] even before they had been published. Had the scientists pursued this, quarks could perhaps have been discovered at TIFR. Rajsekaran felt that since most of the TIFR scientists subscribed to the 'bootstrap'[39] model of Geoffrey Chew, who was at Berkeley, this chance was missed. This missed opportunity, according to Indira Chowdhury, pointed to the unrealistic expectations scientists at TIFR had about their ability to keep pace with rapid changes in their fields of study.[40]

This was not the only chance Bhabha missed. With his nationalist focus, he was unable to appreciate other models of training. A case in point is his refusal to support Abdus Salam's International Centre for Theoretical Physics (ICTP). Bhabha wrote to Arthur Lall, India's ambassador in Vienna, expressing his disapproval of Salam's scheme. He called it 'basically unsound' and not in the interest of the underdeveloped countries, since it would draw away talent from these countries, leaving them even poorer. He said that its main purpose was to find an international job for Salam or an institute that he could direct. Bhabha felt that Salam should have done this in his own country and it could only be done with a lot of hard work and no short cuts.[41]

This was nothing if not arrogant since it completely ignored the circumstances that had compelled Salam to leave Pakistan and establish his institute in Trieste, Italy. Bhabha had Nehru's full backing in everything he did: government support was continuous and something that he could always take for granted. Salam, on the other hand, had to leave Pakistan because he belonged to the Ahmediya sect of Shia Islam, which the Sunnis considered heretical. In 1953, there were widespread anti-Ahmediya riots, and Salam, fearing that he would also be a target, was obliged to return to Cambridge in 1954. Bhabha also does not seem to have given much thought to the role that TIFR could have played in training scientists for the rest of the developing world.

M.G.K. Menon, who succeeded Bhabha as director in 1966, saw the merit in Salam's endeavour. Both Salam and Menon had been admitted into the Pontifical Academy of Sciences in 1981. While attending the meeting at the Vatican in 1981, they discussed the idea of creating an international science academy that would create opportunities for scientists from the developing countries to exchange ideas. At the instance of Salam and Menon, scientists from, or originating from developing countries, formed the Third World Academy of Sciences (TWAS), based in the ICTP at Trieste. Salam was the first president and Menon became vice president. Today, TWAS plays a crucial role in promoting scientific research in many developing countries.[42] Salam's institute went on to become the most significant training institute for scientists from the developing countries. Bhabha, of course, did not live to see the success of Salam's institute. Nor did he witness the success of Mudumbai Seshachalu Narasimhan, who, after his time at TIFR, went on to head the mathematics group at ICTP. Scientists from TIFR have regularly won awards instituted by Salam's institute.

Chapter 19

BUILDING A SCHOOL OF MATHEMATICS

> The Institute has a school of mathematics which is perhaps the largest and most active in India today [...] Judged by the quality and extent of research work that is carried out, it is second to none in this country.
>
> —Historical note on TIFR to Nehru on 1 January 1954[1]

By the time K. Chandrasekharan arrived at TIFR from Princeton in July 1949, he was famous for his work on mathematical analysis and analytic number theory. The most well known and influential persons in mathematics in India at the time were from Madras. There was R. Ananda Rau, Chandrasekharan's teacher who had been a student of G.H. Hardy at Cambridge; R. Vaidyanathaswamy, at the newly created mathematics department in Madras University; and Father Charles Racine, who had received his doctorate under Élie Cartan in Paris, at Loyola College. Besides Chandrasekharan, there were other highly qualified mathematicians produced by the 'Madras School', such as S.S. Pillai, S. Minakshisundaram, V. Ganapathy Iyer, P. Kesava Menon and K.G. Ramanathan.[2]

Chandrasekharan went to the IAS at Princeton in 1946 and was

Hermann Weyl's assistant. He was also close to John von Neumann. Chandrasekharan collaborated extensively with Salomon Bochner, who was at Princeton. He worked closely with some of the best mathematical brains in the world, but his exposure to the working of the School of Mathematics at Princeton was perhaps more important from the viewpoint of institution building. Oswald Veblen had worked tirelessly to build this school, which attracted the best mathematical minds in the world. Veblen was responsible for attracting James Alexander, Albert Einstein, Marston Morse, von Neumann, Weyl and Kurt Gödel to the institute. As J. Robert Oppenheimer, who was the director of the institute remarked, Veblen had provided a real clearing house at the institute for mathematicians all over the world.

ENTER K. CHANDRASEKHARAN

Although D.D. Kosambi and F.W. Levi had joined TIFR before Chandrasekharan, there had been little or no progress towards establishing a school of mathematics. Kosambi was a solitary genius, happy to plough a lonely furrow, and Levi knew that his stay was never going to be permanent. Chandrasekharan had to start building his group from scratch, and he systematically set about doing so. He seems to have been influenced by what he had seen at Princeton in the way he handled this task. His idea was to build expertise in many important disciplines in mathematics and to build something similar to a graduate school of an American university, where students have to study basic disciplines before working towards a doctoral degree. While Chandrasekharan's mathematical achievements are of the highest rank, his talents as an extraordinarily gifted organizer and administrator are even more important.

Chandrasekharan remained closely linked with the Madras University. The presence of Father Charles Racine—the French Jesuit from Paris who taught at Loyola College and whose contributions to the mathematical milieu influenced a whole generation of mathematicians—was of great importance. Racine had worked with Élie Cartan and Jacques Hadamard—both legendary figures in mathematics. His friends included André Weil and Henri Cartan (Élie Cartan's son) who was a reputed mathematician. With this background, Racine began weaning Indian mathematicians away from traditional Cambridge-inspired areas and was responsible for encouraging and guiding a veritable galaxy of Indian mathematicians. Apart from his influence on Minakshisundaram, other young mathematicians he influenced included K.G. Ramanathan, M.S. Narasimhan, C.S. Seshadri, Raghavan Narasimhan and C.P. Ramanujan. Madabusi Santanam Raghunathan, who was a student of M.S. Narasimhan, recalled with pleasure, the one meeting he had with Racine in the company of his teacher. 'He was an excellent example—by no means unique—of the coexistence of the cassock with a lively disposition.'[3] On Racine's advice, Narasimhan and Conjeevaram Srirangachari Seshadri, who later became famous for the Narasimhan–Seshadri theorem, joined TIFR.

Apart from doing and encouraging research, Chandrasekharan also devoted his time to publishing it. He was a member of the Indian Mathematical Society, which published India's oldest mathematics journal—*The Journal of the Indian Mathematical Society*. To ensure its publication, Chandrasekharan requested Bhabha for funds in 1950. Bhabha sent a recommendation to K.S. Krishnan and informed him with some pride that Chandrasekharan had been invited to represent India at the International Mathematical Congress at Harvard that year.

Chandrasekharan's visit to Harvard in 1950 coincided with the conference of the International Mathematical Union (IMU) that was held in Columbia at the end of August that year. This was the beginning of a long and fruitful association that saw Chandrasekharan become its president in 1970. Within four years of the conference, he was elected to its executive committee and remained on it for the next 25 years, first, becoming secretary and then vice president before his promotion as president. He undertook the task of compiling the *World Directory of Mathematicians*, the first edition being published in 1958. It was a mammoth task, and one for which he will always be remembered. The directory listed mathematicians from 36 countries who were members of the IMU and included the names of mathematicians from 27 non-member countries.

BUILDING A SCHOOL OF MATHEMATICS

K.G. Ramanathan assisted Chandrasekharan in the task of building a school of mathematics. Ramanathan had done his PhD from Princeton under Emil Artin, and he had also worked as an assistant to Weyl. He was also deeply influenced by the mathematics of Carl Siegel. There is a story about how Ramanathan, whose office was only a few rooms away from Einstein's at Princeton, had managed to get the senior scientist interested in Carnatic music. Ramanathan joined TIFR in 1951 and played an important role in inspiring young mathematicians by sharing stories about great mathematicians with them.

Along with Ramanathan, Chandrasekharan initiated a graduate programme at TIFR where a few talented students were selected every year on the basis of interviews. Chandrasekharan chose his students carefully and then made available the resources with which to train them. They were assigned readings and after a year of basic training in algebra, topology and analysis, they attended courses in different fields taught by distinguished mathematicians who were invited to spend a few months at TIFR. The notes of each course were written by students attending them. These lecture notes were written up and published as the 'TIFR lecture notes series', and became quite famous over the years. They enjoy a good reputation in the mathematical community to this day. Chandrasekharan's own course on zeta functions was also published in the lecture notes series. The writing up of the lecture notes not only clarified the intricacies of a subject but, as was the case with physics, also gave the students a chance to interact with important mathematicians they held in awe. This helped clear the way for a less formal relationship and perhaps even collaboration.

Initially, Chandrasekharan and Ramanathan taught basic courses, in which students also gave seminars. Chandrasekharan was a great judge of excellence in areas outside his expertise. He managed to instil a commitment to hard work without detracting from the romance of mathematics. He gave his students plenty of freedom to work on whatever they fancied, and, given their exposure to foreign mathematicians and different mathematical areas, it was but natural that these areas of interest were often very different from his own areas of specialization.[4]

Although TIFR had been inviting some of the leading mathematicians of the world from 1946 onwards, it was only later that the visiting programme had the specific purpose of training young mathematicians. Bhabha and Chandrasekharan, initially, at least, shared the same vision for the institute. Chandrasekharan wanted the mathematicians at TIFR to study current problems and be abreast of the recent developments in their field. He also ensured that the library was well-stocked with books.

The programme of foreign visitors started in 1953. That year, the American mathematician, Warren Ambrose, considered one of the fathers of modern geometry, lectured on several topics. It was all about the atmosphere at TIFR: Narasimhan recalled that Chandrasekharan, Ramanathan and K. Balagangadharan greatly contributed to the creation of a vibrant academic environment. As Narasimhan observed tellingly, 'Research is not just your own intelligence, it's not just the opportunities that are made available—it's the atmosphere.'[5] The visitors to TIFR from 1954 onwards included some of the biggest names in mathematics: Norbert Wiener, Carl Siegel, Hans Maass, Hans Rademacher, Oscar Zariski, Martin Eichler and key Bourbakis, like André Weil, Samuel Eilenberg, Jean Dieudonné and Laurent Schwartz.

Schwartz's visit was of particular importance, since he made French mathematicians like Henri Cartan aware of the talent at TIFR. Another visitor who greatly influenced the way mathematics developed at TIFR was Siegel. Schwartz persuaded many of his colleagues and students to visit TIFR. There is little doubt that it was Chandrasekharan's and Ramanathan's association with eminent mathematicians in Princeton that brought them to TIFR.[6] Weil played an important role in keeping them abreast of the latest developments in France.

Ramanathan's role was no less important. Through him, many at TIFR were exposed to some of Siegel's work and this was reflected in the kind of research that was pursued in the institute during the 1960s and 1970s. Ramanathan, with Chandrasekharan's encouragement, was also responsible for publishing Srinivasa Ramanujan's notebooks. Ramanujan had recorded his discoveries from his late teens, and these notebooks, after several years of hibernation in the archives of the Madras University, were published with financial assistance from the Sir Dorabji Tata Trust. The *Notebooks of Srinivasa Ramanujan* was published in two volumes in 1957. In 1960, Bhabha presented a special leather-bound set to the Royal Society of London. They were also shown to Nehru when he came to inaugurate the new buildings of the TIFR in January 1962.

Chandrasekharan also initiated a four-yearly international colloquium on various topics in mathematics, which had the support of the IMU. The first colloquium was on zeta functions, which was held in 1956. The Norwegian mathematician Atle Selberg delivered four lectures in this colloquium and the famous Selberg trace formula figured in these talks.[7]

In 1957, M.S. Narasimhan and Seshadri were deputed to work in Paris. Their interaction with visiting mathematicians at TIFR had laid the foundation for a period of rapid learning in Paris. They were there for three academic years and returned to TIFR in 1960. Narasimhan worked with Schwartz in Paris and Seshadri worked with Claude Chevalley, another Bourbaki. Seshadri notes that that was a great period for mathematics in France. The Grothendieck Revolution in algebraic geometry was the rage in Paris at the time, and, since Seshadri was studying algebraic geometry with Chevalley, he was able to understand Grothendieck's work. Alexander Grothendieck was the leading figure in the creation of modern algebraic geometry and is considered one of the greatest mathematicians of the twentieth century. He was awarded the Fields medal in 1966, generally considered to be the equivalent of the Nobel Prize in mathematics. Grothendieck visited TIFR during the International Colloquium on Algebraic Geometry in 1968.

By the mid-1960s, the work done by the School of Mathematics covered a wide range of topics. Much of this work was of a very high calibre and TIFR gained the reputation of being a world-class centre for mathematics. Chandrasekharan not only wanted to improve the quality of mathematics but also help the teaching community in India. He had a very good rapport with mathematics teachers at various colleges in Bombay. He was also responsible for creating a training programme for mathematics teachers from colleges all over India. He used to organize annual summer schools at TIFR on various topics in mathematics.

One of Chandrasekharan's important achievements was the setting up of the department of mathematics at Bombay University and appointing the renowned Sharadchandra Shankar Shrikhande as the head of the department. In 1959, Shrikhande and R.C. Bose, along with their collaborator E.T. Parker, disproved Leonhard Euler's famous conjecture concerning Latin squares[8]. At the time of the discovery, Shrikhande and Bose were in the department of mathematical statistics at the University of North Carolina, Chapel Hill. Chandrasekharan was also responsible for making the University Grants Commission (UGC) recognize the department of mathematics of Bombay University as a UGC centre for promoting research. He deputed two members of the School of Mathematics to work with Shrikhande for the development of the centre.

Seshadri recalls how, when he met the Australian-born Cambridge mathematician John Coates for the first time at a metro station in Paris, the first question Coates asked him was 'How did you manage to do it?'[9] The query referred to the fact that they had been able to build a first-rate mathematics centre at TIFR, when, at that time, there was no comparable institution for mathematics in Australia. Nearly two decades later, when Seshadri was at the IAS in 1975–1976, S. Chandrasekhar, who was also visiting the institute, told him that mathematicians, like Andre Weil, had expressed a very high opinion of TIFR's School of Mathematics.[10]

KOSAMBI'S EXIT

As Kosambi's interests in historical analysis increased in the 1950s, his involvement in mathematics suffered. He travelled to the USSR and China during this period and wrote on a variety of social issues. All these activities were at variance with the TIFR ethos. His many-sided curiosity and multifaceted genius came in the way of a more conventional (and specialized) role that was expected of him. Bhabha had little patience with Kosambi's other pursuits, which were peripheral to the work at hand and led to Bhabha doubting his usefulness to the institute.

Kosambi's visit to 'Iron Curtain' countries and connections with

Marxists in Europe was a cause for concern to the TIFR authorities. In 1951, he visited Paris to meet Frédéric Joliot-Curie in connection with the World Peace Conference, a gathering identified with communists, and something that the members of the TIFR council viewed with suspicion. Rustum Choksi wrote to Bhabha, who was in England at the time, reporting how he and the others had warned Kosambi about participating in political activities. He said that when he met Kosambi at the airport[11], the latter had agreed to desist from such activities after his Paris visit. Choksi was upset enough at this unseemly blending of politics and science, to call Kosambi 'incorrigible'.[12]

Bhabha's response was muted but showed his displeasure with Kosambi's contribution to TIFR. He said he would not mind Kosambi taking part in the peace conference provided his academic contributions were what they should be. However, he hoped no drastic action would be taken immediately.[13] In the same letter, Bhabha spoke about the agitation among people about the possibility of his accepting an invitation to the USSR that year. While Bhabha said that he had decided to decline since it did not fit in with his plans, he believed that such invitations should be accepted in the interest of good relations. Bhabha believed that intellectual freedom had to be maintained at all costs in the institute and strongly defended an individual's right (in this case Kosambi's) to associate with Left ideology. He also made the point that such intellectual freedom had to be maintained if the institute was to attract men of calibre.[14]

As we have seen Bhabha had offered a job to Bernard Peters, who had been accused of communist sympathies in 1950. In the light of the Kosambi episode, hardline anti-Communists in India began to suspect, wholly without foundation, that Bhabha was a 'fellow traveller'. Bhabha's pragmatism and idealism allowed him to walk a fine line between representing the Indian state, the scientific community and the interests of the Tatas. Kosambi's outspoken Marxist views and his public condemnation of atomic energy posed a challenge to this balancing act, and he was ultimately sacked in 1962.[15] Kosambi may have guessed he was living dangerously, since, as early as December 1950, he wrote to his friend, Agnes Conklin, wife of Professor R.J.

Conklin at Harvard about the possibility of his dismissal, a letter in which he called Bhabha a 'rank opportunist'.

> In this country, a considerable portion of the business community is all for war, though they pay faint lip service to peace. I have decided to take prominent part, in the peace movement [...] As one result I face dismissal—not immediately, I believe, but eventually. The warning has been given not only by the director of my Institute (a rank opportunist) but by the real power behind him, one of India's biggest leaders of industry and finance.[16]

Although Kosambi was prescient enough to anticipate the fallout of his political activities, his loss of standing in the academic community contributed in no small measure to his downfall. Towards the end of the 1950s, Kosambi started working on the Riemann Hypothesis, a conjecture made by Bernhard Riemann that exactly predicts the pattern of the distribution of prime numbers. Such is the complexity of this problem that it is still considered the 'holy grail of mathematics'. He published two papers offering a proof of this problem, in the *Journal of the Indian Society of Agricultural Statistics*. His approach, which used probability, did not flow from his earlier work and his motivation for attempting this proof, which has still eluded mathematicians (and is infamous for the number of flawed proofs it has attracted), remains a mystery. His choice of the journal was inexplicable and his outrageous claim exposed him to ridicule.

Mathematicians speak of this phase of his life with an air of embarrassment. The scientific fraternity at TIFR was also greatly embarrassed. Chandrasekharan, in a handwritten note, told Bhabha about his own feeling of humiliation when the subject came up in Zurich. He called it an international scandal and compared Kosambi's claim with Goebbels's claims about victory in the Second World War. He pointed out that the proof of the Riemann Hypothesis was something that had eluded mathematicians of the calibre of Hardy, J.E. Littlewood, Siegel, Weil and Ivan Vinogradov. Chandrasekharan said that he had to hang his head down in shame at a meeting in Zurich when Professor Peter Hilton of the University of Birmingham, with his tongue firmly in cheek said, that it seemed Indian Mathematics was very good since they had a colleague who had proved the Riemann Hypothesis in his own way.[17]

Chandrasekharan was obviously worried about the effect such bogus claims would have on the credibility of his own mathematics group at TIFR. As he told Bhabha in the same letter, Kosambi had worked independently without discussing his claims with either the mathematics group at TIFR, or with the visiting mathematician, Siegel, who was an expert on Riemann's work. Chandrasekharan said that if Kosambi had confided in the mathematics group, it would have taken them very little time to 'explode the nonsense'. He asked Bhabha to take some notice of 'scientific dishonesty' and deal with it in his own way, quietly and effectively.[18]

In another joint communication addressed to Bhabha and the TIFR council in August 1960, Chandrasekharan, along with Seshadri, Narasimhan and Ramanathan, had denounced Kosambi for his total lack of interest or participation in the activities of the group, saying that Kosambi was only interested in advertising his high claims abroad, which had a demoralizing influence on the students.[19] The group was quite openly sceptical about Kosambi's credentials for undertaking such a task, given the fact that he was not familiar with the subject. Kosambi's proof had been examined by famous experts, who had rejected it. They questioned Kosambi's professional ethics, saying that it was against all cannons of rectitude to claim any result without a published or verifiable proof of it, and to persist in that claim, even after his efforts had been rejected by experts in the field.[20]

They then described their embarrassment at having to publicly refute the unsubstantiated claims of a colleague, and the adverse impact on the reputation of the institute. The mathematicians felt that Kosambi's claim was setting a dangerous precedent and was jeopardizing the professional standing of the mathematics group in the eyes of the world. The group was working on different aspects of the Riemann problem at TIFR and it was a matter of sorrow to them that the institute should be despoiled of the seriousness of these pursuits, and that a member of the institute should want to perform on the flying trapeze.[21]

Perhaps the most damaging was their comparison of Kosambi's efforts with those of other 'cranks or maniacs'. They also conveyed to Bhabha that Kosambi was making similar claims for Fermat's Last Theorem. The mathematicians said that they had been assisting Bombay University in the disposal of such cases and cited the recent example of a Pakistani Air Force officer who had tried to become famous. They said it was a strange and compelling irony that a professor of TIFR should want to join that category.[22]

After receiving this letter, Bhabha wrote to Kosambi in September 1960, asking him for a copy of the article containing the purported proof. Kosambi's response was understandably sharp, and one in which he revealed his unhappiness at Bhabha not allowing him to attend conferences abroad. Kosambi wrote to Bhabha wanting to know the name of the person who had communicated the 'claim' to him. He unhesitatingly added that this had been done without his authorization. He hoped that Bhabha would not take it amiss if he pointed out that the referee's task would have been materially shortened and the problem would have been settled through personal discussion had Bhabha been a little more reasonable in the matter of deputing him for conferences and meetings abroad.[23]

After some difficulty, Chandrasekharan procured a copy of Kosambi's 'proof' and sent it to Siegel on 1 November 1960, who had also taken the trouble to show it to Selberg, who recalled seeing an earlier version that Kosambi had sent the Croatian–American probabilist William Fellner, at Princeton. Fellner, Selberg said, had rubbished the proof, saying that it used the same erroneous ideas and was completely wrong. The matter was discussed at a meeting of the council in 1962, where letters from Selberg and Fellner were read out by Chandrasekharan. Kosambi had drawn a blank from all possible sources. His proof had been pronounced as erroneous by all experts. The council noted that Kosambi would turn 55 on 31 July 1962 and decided against renewing his contract, making him one of the very few people to have been fired from TIFR.

It was a sudden and precipitous descent into mathematical ignominy for someone who had at one point been considered worthy to adjudicate on the candidature of mathematicians for the Fields medal. By the early 1960s, Kosambi had become marginal to mathematical research at TIFR. His relationship with Bhabha soured to such an extent that the two avoided each other. Between 1962 and 1964, Kosambi was without a formal position, although he published papers both in and outside mathematics. In 1964, Kosambi was appointed a scientist emeritus by the CSIR for five years. This grant was routed through the Maharashtra Association for the Cultivation of Science, Poona, a position he held until his death in 1966. A few days after completing his first book on mathematics (the typescript of which has mysteriously disappeared), Kosambi died in his sleep.

What went wrong? Kosambi is generally seen as a victim in this whole affair. Those who see his departure in this light, point to his clash with Bhabha about India's atomic policy. Kosambi, in a talk at the Rotary Club in Poona on 25 July 1960, had pointed to the cost of setting up atomic power plants and argued that the prestige of having nuclear energy was not compensated by the extra expenditure or the dangers involved. His lecture shows that he was, indeed, a man who saw tomorrow when he advocated the cause of solar energy. With solar energy, you can supply power locally, with or without a grid. 'Solar power would be the best available source of energy for dispersed small industry and local use in India.'[24]

Kosambi argued that the biggest advantage of solar energy would be decentralization. He was also perspicacious enough to point out that the problem of reforestation could only be solved if the population had cheap fuel for cooking so that they did not need to cut down trees. In an essay written four years later, Kosambi bemoaned the fact that questions were asked in Parliament about his speech. The answer given was that such projects were designed to keep India in the bullock-cart age. 'A question of science, technology and economics was reduced to one of ostentation and prestige. However, the sun has not yet been abolished by decree, so the matter may be taken up at some future date when common sense gets a chance.'[25]

Kosambi's Poona speech made in 1960 could not have pleased Bhabha very much, but it was not as if he himself had not thought about solar energy. More than a decade ago, in October 1950, Bhabha had written to Bhatnagar about his suggestion to Nehru regarding the appointment of a committee to study the problem of utilizing solar energy for generation of power on a small scale, for refrigeration and air conditioning. He had suggested the names of K.S. Krishnan, S.S. Bhatnagar, D.S. Kothari, William McBain and himself as members of the committee.

The outspoken Kosambi was also guilty of another faux pas. According to a famous anecdote, at a social gathering in TIFR, numerous speakers in their speeches praised the Tatas for their benevolence in contributing their hard earned money for the noble cause of setting up TIFR. Kosambi disagreed with the description of the Tata money as 'hard earned' explaining that capital often starts off as rogue capital before turning respectable. He illustrated this by pointing out that the Tatas had earned big money from trading in opium.[26] These remarks must have annoyed Bhabha very much.

Another view sees Kosambi as a victim of scientific management, which eliminated knowledgeable and creative persons. As Kosambi put it: 'The greatest obstacles to research in any backward, underdeveloped country are often those needlessly created by the scientist's or scholar's fellow citizens [...] The meretricious ability to please the right people, a convincing pose, masterly charlatanism, and a clever press agent are indispensable for success.'[27] However, these are extreme views. Kosambi's departure had to do primarily with mathematics and its institutionalization at TIFR. Even when Bhabha's relationship with Kosambi was not strained, there was a sense of impatience with his statistical work. In the early days, Kosambi had worked out a statistical analysis of cosmic ray experimental data. In 1952, he wanted Bhabha to extend the appointments of six casual workers, who were collecting data of births and deaths, by six months. Bhabha agreed so that the work on hand could be completed, but he firmly refused to make them permanent, telling Kosambi that the institute had been set up to do research in the basic sciences and not in applied sciences.

After Kosambi's departure, he became something of a pariah as far as the institute was concerned. His presence was no longer acknowledged by the institute in any of its reports or publications. The Silver Jubilee Report did not mention any of his publications, excising his memory from the history of the institute, the focus being on Chandrasekharan and his group of mathematicians. While the section on mathematics gave credit to Chandrasekharan for his foundational role in setting up the School of Mathematics, there was absolutely no mention of Kosambi's role in the early part of TIFR's existence. Like all official histories that tend to ignore uncomfortable details, the report does not narrate the circumstances surrounding Kosambi's exit. All the publications that have been mentioned in the report begin from 1949 and focus on the mathematics done by Chandrasekharan and his group.[28] The report also did not mention the work done by Levi. Thus, 1949 became the acknowledged date of founding of mathematics at the institute.

CHANDRASEKHARAN LEAVES FOR ZURICH

Chandrasekharan left TIFR towards the end of 1965 to join the ETH, Zurich, and remained there till his death in 2017. Seshadri observes that although Chandrasekharan admired Bhabha, he also found faults with many things that the latter did, and their relationship grew strained. After he moved to Zurich, Chandrasekharan became more active in mathematics. In fact, his collaboration with Narasimhan, which began at TIFR, continued even after he went to Zurich. Seshadri moved to Chennai from TIFR due to personal reasons and was instrumental in setting up the Chennai Mathematical Institute. He met Chandrasekharan a few times in Zurich. He said that Chandrasekharan was always nostalgic about his TIFR days and the institute was always at the centre of the conversation.[29]

Why did Chandrasekharan leave? We have already noted that Bhabha and Chandrasekharan's relationship was strained. The mathematician Jal Choksi recalled that Chandrasekharan had talked about leaving from time to time. Jal and Narasimhan were quite close to Chandrasekharan, and, while he did talk of going away, they didn't think that he would really do it. Jal, when asked the reasons for his exit, attributed it to a clash of personalities. Bhabha was a very dominating personality and was used to having his way; Chandrasekharan was no pushover either. Seshadri describes Chandrasekharan as an 'intimidating personality', who could sometimes even be rude. He believed in calling a spade a spade. It seems to have been a case of diamond cut diamond. Jal put Chandrasekharan's departure down to irritation and dissatisfaction with Bhabha. As he put it, 'KC was very down to earth and matter of fact and said exactly what he felt. I think Bhabha was more intuitive and maybe more original. But also, KC was very quick and extremely to the point. And the relationship between the two was amazingly complex. There was a deep personal bond but there was something else too.'[30]

Perhaps, Bhabha's dominating personality made him somewhat difficult to get on with, especially so for those who were his intellectual equals. Sreerup Raychaudhuri, a theoretical physicist at TIFR, in a talk on the history of particle physics in India, narrated how Bernard Peters had first escaped from the Nazis and then, escaped from the US where he had been branded a communist. Referring to his departure from TIFR, he mentioned in a semi-jocular vein (and almost sotto voce) that people said, Peters, was 'escaping from Bhabha'.[31] While too much should not be read into this innuendo, it does point to the fact that Bhabha's autocratic style may not have gelled with many. However, there is also no denying the fact that Peters had already spent eight years at TIFR and may have been looking for, newer, and perhaps greener pastures, when he left to join Bohr's institute in Copenhagen.

Another reason for Chandrasekharan's departure was that he used to spend the summer every year at the ETH and was a good friend of the mathematician Beno Eckmann who wanted him to settle in Zurich. According to the mathematician, R. Sridharan, who joined TIFR in 1955, Jal and Narasimhan left TIFR after Chandrasekharan departed. Many felt that Chandrasekharan should have been asked to return to head TIFR after Bhabha's unfortunate demise, but that did not happen. Rather, attempts seem to have been made to erase his memory from the TIFR's institutional legacy. Even when Bhabha was director, there existed two cultures that could be identified with the two groups—physics and mathematics.[32]

In 1967, a year after Bhabha's death, an exhibition was put together, which was first displayed at the Royal Society in London and later installed at the foyer of the Homi Bhabha auditorium. The exhibit aimed to honour the founder, but the overwhelming emphasis on Bhabha, discarded a part of the institutional legacy. In two key photographs with Nehru, Chandrasekharan's image was excised from the originals. The first photograph was of J.R.D. Tata, Bhabha and Nehru in the lounge of the institute, and the second was of Bhabha and Nehru looking at the 1957 facsimile edition of Ramanujan's *Notebooks*. Both had been taken on 15 January 1962, at the inauguration of TIFR's new buildings. As Indira Chowdhury has observed, 'In removing Chandrasekharan from the photos presented at the exhibition of 1967, the institute was choosing to identify completely with the founder. In the 1967 exhibition, physics and its applications reigned supreme; the narratives about the pre-eminence of mathematics were held in abeyance.'[33]

Chapter 20

COMPUTERS AND ELECTRONICS

> Electronic digital computers are becoming more and more important in modern scientific research and there are many fields today in which progress would be impossible without their help.

> —Speech at the inauguration of the new TIFR buildings, Bombay, 15 January 1962[1]

Bhabha not only saw the need for India to acquire both theoretical and practical knowledge of atomic and nuclear physics but also realized the importance of the emerging fields of electronics and computers very early on. He wanted to develop a digital computing group in the early 1950s. It was an audacious idea at the time, considering that John von Neumann was still laying the foundations of computer science in the US, and only a handful of computers had been built there. There was no dearth of resources in the US, but in stark contrast, postcolonial India, which was emerging from the ravages of colonial rule and a bloody partition, was struggling to stabilize its polity and taking the first hesitant steps towards building a modern industrial base. The Indian expertise in electronics was limited to a few radio engineers in All India Radio and some manufacturers, who assembled valve radios.

EARLY ATTEMPTS AT BUILDING COMPUTERS: MAHALANOBIS AND BHABHA

Bhabha managed to accomplish as much as he did in a variety of fields because of his rare abilities to spot and nurture exceptional talent and recognize new fields whose importance would grow later. We have seen that his model of institution building depended on identifying exceptional talent—people came first, buildings later. The growth of computer science, and as we shall see later, molecular biology, had much to do with Bhabha's identification of exceptional talent. The stalwarts in computer science were D.Y. Phadke and Rangaswamy Narasimhan. Work on the first digital computer in India began after R. Narasimhan joined TIFR in 1954.

For his experiments in cosmic rays, Bhabha had to build his own equipment, which consisted of cosmic ray detector systems, such as counters and other related electronics. When he moved to Bombay to set up TIFR, he pioneered the further development of such equipment. However, Bhabha had many irons in the fire and needed an experienced person to assist him. He found one in Phadke, who had taught at the Xavier's Technical Institute and joined TIFR as an associate professor in 1949, becoming the head of the new instrumentation group. He was deeply interested in a range of technical subjects but had no interest in publishing papers or taking out patents. However, he was a visionary, who understood the significance of innovation. M.G.K. Menon recalled having several conversations with him on subjects, like vacuum technologies, microwave engineering, accelerator technologies and mechanical engineering.[2]

It was also in 1949, that the Electronic Delay Storage Automatic Computer (EDSAC) started functioning in Cambridge. The brainchild of Sir Maurice Wilkes, who was the director of the Mathematical Laboratory at Cambridge University, the EDSAC was the world's second practical stored program computer. Bhabha, who knew Wilkes from his Cambridge days, discussed the subject with him.

After getting Phadke onboard, Bhabha persuaded R. Narasimhan to join TIFR in 1954. A telecommunications engineer from the College of Engineering, Guindy, he had gone to the California Institute of

Technology to do his MS on a Tata Endowment scholarship. Later, he completed his PhD in mathematics from Indiana University. In an interview, Bhabha asked R. Narasimhan if he was aware of the von Neumann report, which Bhabha had procured on his visit to Princeton. When R. Narasimhan answered in the affirmative, Bhabha hired him and asked him to design a digital computer. Work on the pilot machine started in 1954 with Narasimhan's arrival. While TIFR eventually became the first institute in India to build its own digital computer, it was not the first Indian institute to possess one. P.C. Mahalanobis's ISI in Calcutta achieved that distinction when it acquired a British-made digital computer Hollerith Electronic Digital Computer—2M (HEC-2M) in 1956.

The HEC-2M was not very suitable for large data processing work, but it was a start. Mahalanobis's idea was to familiarize his technicians and scientists with a small computer, experience that he hoped would prove useful in building a bigger one. In 1959, a larger computer, the URAL, was obtained as aid from the USSR through the UN. At that time, ISI became something of a national computer centre, catering to the needs of the defence ministry, meteorology department, the AEC and TIFR. It is worth noting that, until TIFR had its own computer, the secret calculations for the AEC were done in Calcutta by ISI.

Mahalanobis's original training had been in physics, but a chance encounter with statistical journals in Cambridge led him to statistics. In 1932, he established ISI in Calcutta, which started out from a tiny room in Presidency College. The ISI introduced mechanical desk calculators for the first time in the country, and over the next two decades, it helped disseminate the use of mechanical, electrical and electronic calculating machines across the country. In 1953, Mahalanobis developed an analogue computer at ISI with the help of two young graduates: Samarendra Kumar Mitra and S.M. Bose.

By the 1950s, ISI moved to a large campus in north Calcutta and evolved into a premier centre for statistical research and application. Mahalanobis, like Bhabha, was a highly networked and mobile scientist who maintained close links with scientists across the world, including the Soviet bloc. Like Bhabha, he also made frequent trips abroad (in fact, much more than Bhabha). The story goes that once, in 1947,

Nehru pulled his leg by asking him, 'Are you coming back to India or going out.'[3] His biographer, Ashok Rudra, likened him to the *Flying Dutchman*, the legendary ship that was supposed to be under a curse which doomed it to sail continually on the high seas with no port of refuge. However, Mahalanobis never gave any sign of regretting his predicament; on the contrary, he seemed to enjoy it.[4]

He criss-crossed continents, visiting Europe, the US and the USSR, constantly networking. W. Edwards Deming recalled, 'Probably no statistician ever attended so many meetings the world over.'[5] The British-born Indian geneticist John Burdon Sanderson Haldane, who worked as a regular staff member of the ISI for nearly four years, thought that the institute suffered because of Mahalanobis's peregrinations. He observed, 'The journeying of our Director, define a novel random vector.'[6] Haldane left the ISI after falling out with Mahalanobis, who ran the ISI like a personal fiefdom, generous in his favours but expecting in return to be permitted his autocratic ways.[7] Bhabha, too, was accused of being an autocrat, but he was not a control freak like Mahalanobis who interfered in virtually everything, including the travel arrangements for visitors.

Mahalanobis was respected globally for his contributions to the development of statistics through new concepts including the 'Mahalanobis Distance' in multivariate space. Under him, the ISI blossomed into a multidisciplinary centre of research. 'A physicist by training, a statistician by instinct and a planner by conviction',[8] Mahalanobis combined scientific talent with a tremendous drive to engage in practical matters. Consequently, he became intrinsically involved not just with independent India's first forays into economic planning and development but also with the development and establishment of statistics, most notably, sample survey techniques as a principal tool for knowing the world.

Mahalanobis's desire to use statistics to solve real-world problems led to his involvement in economic planning and institution building. He had met Nehru several times before the Second World War, usually when the young Congress leader called on Rabindranath Tagore. Their conversations on planning began in 1940 at Anand Bhavan, Nehru's ancestral home in Allahabad. Mahalanobis recalled late night discussions on the future of economic planning in India. When Nehru visited Calcutta in 1946 for the ISC he practically invited himself to the Mahalanobis residence so that they could spend more time together. Mahalanobis became a de facto member of the Planning Commission in 1955 and was charged with drawing up India's Second Five Year Plan (1955–1959). It was only in 1960 that he became a de jure member of the Planning Commission. In preparation for this work, Mahalanobis spent the summer of 1954 travelling through the US, the UK, Eastern Europe and the USSR, conferring with economists, statisticians and planners. The two-sector growth model that he created in 1953, similar to G.A. Feldman's 1928 model, became the basis for India's Second Five Year Plan.

Bhabha and Mahalanobis shared a fierce professional rivalry. They were too alike to eventually not get in each other's way. 'Facing off may have felt like looking in a mirror',[9] even though, physically the cleft-chinned Bhabha, portly in middle-age, had little in common with the rigid angularities of the tall Mahalanobis. They had both initiated efforts in computer development and the centres they created served as nuclei for the development of early hardware and program writing skills. Both these favourites of Nehru began a race to build the first computer in India. However, there was no acrimony involved and the two were always on cordial terms. In early 1946, Bhabha invited Mahalanobis to deliver a lecture on 'India and the Russian Experience'.

In his letter to Bhabha in January, Mahalanobis thanked him for his hospitality. He said it was very kind of Bhabha to have asked him and his wife to the Rotary dinner. He said that they were very glad to have had the opportunity of meeting his mother and it was very charming of her to have taken them to the Bhabha house straight from the lecture and to have given them dinner before they went to the station. Mahalanobis responded with an invitation of his own. He told Bhabha that in case he ever came to Calcutta in the near future he should let them know and if it suited his convenience even stay with them.[10]

Like Bhabha, Mahalanobis also had a wide circle of contacts abroad. He kept abreast of the latest developments in digital computers through his friends, including Norbert Weiner of MIT and J.D. Bernal of Birkbeck College, London. Bernal also knew Bhabha well. Bernal had personal relations with the Indian scientific elite and had participated in the planning process of CSIR. Mahalanobis also had close contacts with the USSR, which resulted in the large computer URAL being installed at the ISI in 1958. There was also a collaborative effort between ISI and Jadavpur University to build a second-generation transistor-based computer called ISIJU-1. It was commissioned in 1966 but was not very successful.[11]

Both Mahalanobis and Bhabha were known for inviting distinguished scientists from abroad to visit their institutes. Wilkes came from Cambridge to TIFR, as did Wiener, who was present at the foundation-laying ceremony of the new TIFR building in 1954. Weiner met K. Chandrasekharan and Pesi Masani (a mathematician at TIFR), and of course Bhabha, who he mistakenly calls a member of the Tata family in his autobiography. He called his Bombay stay one of the most profitable parts of his India visit. Weiner's comment on the Parsis shows his remarkable perspicacity:

> I found the Parsis an extremely interesting group, accepting the new India very thoroughly. Yet they were partly divided in soul between their patriotism and the quasi-European position, almost as westerners, which their small minority of a few hundred thousand souls had accepted under the British.[12]

Weiner also renewed his acquaintance with Kosambi and made a telling remark. Did he realize even at that time that Kosambi didn't really fit in?

> My special crony during this episode was Professor Kosambi, who had been a boy at the Cambridge High School during a period when his father, a refugee from British India, was working over the rich Sanskrit material of the Harvard Library. The son, perhaps owing to his American early training, has been a bit more of a fighter and a bit less of a serene Indian scholar than most of his fellow countrymen.[13]

Wiener indicated his willingness to return the following year, and Mahalanobis extended his second invitation without arranging for the finances. S.S. Bhatnagar's death in January 1955 proved to be inconvenient in this regard because he had been banking on the CSIR to provide most of the funds. Since the ISI could provide only part of the funds, he wrote to Meghnad Saha, C.V. Raman, K.S. Krishnan, M.S. Thacker (of the IISc) and Bhabha. He drew a blank everywhere. Bhabha was travelling at the time and his personal assistant acknowledged the letter. After he returned, he wrote to Mahalanobis that since TIFR's programme for a visiting scientist had already been finalized, he was unable to contribute anything to the expenses for Weiner's visit. In the end, Mahalanobis decided to bear all the expenses himself and wrote to Weiner confirming his invitation.[14]

TIFRAC

The decade of the mid-1940s to the mid-1950s saw the evolution of first-generation computers built out of vacuum tubes, semiconductor diodes and ferrite core memories. The basic logical scheme for these computers was described by John von Neumann in a now classic report in 1946, and he started a project at the IAS, Princeton, to build the computer he described. This was completed in 1952 and spawned a number of copies with universities and atomic energy establishments all over the world making computers based on the Princeton machine.

One such machine the Illinois Automatic Computer (ILLIAC I), completed in 1952 at the University of Illinois, Urbana-Champaign, was the first computer in the US to be built and owned entirely by an educational institution. The TIFR computer's control logic design benefitted greatly from the design details provided by the ILLIAC I team. Bhabha had recruited R. Narasimhan to TIFR with the express mandate that he would build a digital computing group as part of a low-profile instrumentation group. After some preliminary efforts at building digital logic sub-assemblies, a decision was taken, in late 1954, to design and build a full-scale general-purpose electronic digital computer using contemporary technology. The group consisted of six people, besides R. Narasimhan: Minoo M. Dosabhai, S.P. Srivastava, M.M. Farookhi, D.S. Kamat, B.K. Basu and K.S. Kane. Most of them had an MS in physics with a specialization in electronics. None of them, except R. Narasimhan, had travelled abroad. Moreover, none of them had ever used a computer, much less designed or built one![15]

Apart from Phadke and R. Narasimhan, Bhabha had recruited Ayyagari Sambasiva Rao, a Tata scholar and an electrical engineer from Stanford, who joined TIFR as a reader. A.S. Rao helped develop instruments for the balloon experiments at TIFR. Later, he was appointed head of the electronics unit at AEET. In 1958, the electronics group under him developed a general-purpose computer called EAC-62, 10 units of which were delivered to different research institutes by 1965. This computer calculated the necessary mathematical information for designing control systems of nuclear reactors. A.S. Rao later served as the member secretary of the Electronics Committee, which was chaired by Bhabha, with Vikram Sarabhai and Suri Bhagvantam as its members. The team, led by R. Narasimhan, went about their task in two phases. The first phase involved the design and construction of a pilot machine, which served as a proving ground for ideas in circuit and logic system design. It also provided them a chance to experiment with engineering design and fabrication techniques. With the experience gained in building the pilot machine, they then went on to design a full-scale computer in the second phase.

The design work on the pilot machine started in early 1955, and, by October 1956, it was ready. It was set up in the cramped space of the OYC. Narasimhan's machine was based on the ILLIAC I—the machine built at the Illinois University, where he had spent some time. This connection saved the day for the TIFR team—when R. Narasimhan was stymied, it was the computer group at Illinois that made their control logic scheme available to him.

Work on the full-scale machine started in early 1957, and it was completed in early 1959. Lack of suitable air conditioning delayed its testing by nearly a year and it was only in February 1960 that it was commissioned. In 1961, R. Narasimhan built the subroutine library and trained programers who would be able to assist the users. When Nehru came to inaugurate the new buildings in January 1962 it was named Tata Institute of Fundamental Research Automatic Computer (TIFRAC). While it was not a technological breakthrough, it helped Indians gain capability in various fields of computer design, operation and maintenance. It also helped to spread computer awareness among scientists beyond TIFR.

That the team at TIFR was at par with the best in the world became clear when the TIFR team was evaluating the CDC 3600 computer of the Control Data Corporation (discussed later in this chapter) and had the opportunity of witnessing the ongoing development of a later model, CDC 6600. The new model also had a cathode-ray tube display with a line segment display for text characters. The TIFR team was pleased to know that its design was world-class.

The pilot machine, except for its size, was state of the art for 1954, and the full-scale TIFRAC, too, was not very much behind the attempts elsewhere in 1957. However, by the time it was commissioned, in 1960, computer technology had surged ahead to the second generation. Bhabha, although he was not involved in the day-to-day affairs of R. Narasimhan's team, was familiar with von Neumann's work and was deeply interested in his abilities in applied mathematics. According to M.G.K. Menon, Bhabha spent a lot of time reading books on computers. In 1959, Bhabha visited Wilkes, the director of the University Mathematical Laboratory at Cambridge, and invited him to TIFR while also proposing training for TIFR scientists in Wilkes' laboratory. Bhabha had similar interactions with the electrical engineering laboratories of Manchester University, National Research Development Corporation and the British Computer Society.

TIFRAC ran from 1960 to 1964. It was used for data analysis by scientists not only from TIFR and AEET but was also made available to any organization that wished to use it. The TIFR teams ran special computer programming courses in a number of centres for a variety of user groups. The AEET had its first introduction to computers through TIFRAC. Packages developed at TIFR were used in the oil companies and the naval dockyard laboratories, in addition to town planning. By 1964, the machine was in great demand and was operated in two shifts every day, as scientists from all over India used it for their computational needs.

This universal use was what distinguished it from other institutions, notably in Calcutta, which also had computers in the early 1960s: scientific equipment was meant to be shared and associated processes disseminated widely. The role played by TIFRAC in disseminating the importance of computer literacy in India can hardly be underestimated. Its success established a lasting culture of self-confidence and innovation, which qualitatively influenced the progress of information technology in India in the following decades.

RIVALRY

The demand for TIFRAC grew exponentially and it became obvious that in a few years' time it would be inadequate both in terms of speed and volume. It became clear to Bhabha and his team that TIFR would need to acquire one of the large, commercially available computers. Mahalanobis, too, was of the opinion that he needed more powerful, standard, commercially available computers for his research. Both of them were on the lookout for state-of-the-art computers and were in touch with leading computer laboratories in the West.

This soon led to a battle for supremacy between Bhabha and Mahalanobis, both of whom wanted their institutes to house a national computing centre. Mahalanobis had already acquired two large first-generation computers—one from the UK and the other from the USSR. His attempts to get government support for a larger machine failed as did his attempts to get American funding for a Universal Automatic Computer (UNIVAC). Ultimately, he was forced to hire an IBM 1401 and, later, got a Honeywell 400 for his institute. Since the ISI had been a pioneer in the field, he wanted it to be declared a national computing centre.[16]

Bhabha, too, had similar ambitions for TIFR. He discussed the idea with Emanuel R. Piore, IBM's director of research, whom he met by chance on a Paris–Zurich flight in June 1959. Piore advised starting with a powerful machine but not the most powerful one, since the programming of such a machine required considerable experience. He felt such experience could be gained even on a machine with lesser computing power. IBM sent Richard Lawrence Garwin, a professor of physics at Columbia University (who was a consultant to IBM), to evaluate TIFRAC and advise on furthering indigenous efforts in the field of computers in India. Garwin supported the idea of a powerful computer that would serve as the nucleus of a national centre at TIFR. He thought it worthwhile for atomic scientists to use computers to design and run nuclear plants. Since TIFRAC was a non-standard computer with limited speed, he suggested that Bhabha purchase or hire an IBM 704 and wrote a blueprint for the proposed computer centre.[17] IBM also had a faster machine, the 7090, which was more expensive and would take longer to import.

When Bhabha went to attend the centenary celebrations of MIT in early 1961 he discussed the proposal with Jerome Wisener, who advised President John F. Kennedy on scientific matters, and who had earlier been a professor at MIT's Research Laboratory of Electronics. Wisener suggested the 7090, which was 10 times faster and better suited to scientific applications. Bhabha sent M.G.K. Menon to IBM and the UNIVAC division of the Sperry Rand Corporation in the US. Since IBM didn't want to miss an opportunity to consolidate their position in India, they offered the 7090 at an education discount of $518,560. M.G.K. Menon also met Wisener, who reiterated his recommendation of the 7090. Having identified the computer he wanted, Bhabha started lobbying for external aid. He met the US ambassador John Kenneth Galbraith and handed over a note on the computer activity at TIFR, which also included parts of Garwin's report. Galbraith promised to help and even visited TIFR.[18]

In February 1962, Mahalanobis suggested setting up a large computer in New Delhi as a national facility under the Department of Statistics. Bhabha opposed the move. Phadke wrote to the Department of Statistics on Bhabha's behalf saying that if only one high-speed computer was to be set up in India, it had to be at TIFR, but there could be justification for setting up more than one centre in India. He said that the TIFRAC had a speed close to that of the IBM 704 and compared favourably with other computers developed in the USSR and Japan. Bhabha also raised the issue with Mahalanobis, opining that they should press for two computers, which he thought could become available with US aid. He also told Mahalanobis that the UNIVAC, which the ISI wanted to procure, was less powerful than both the IBM 704 and the 7090.

In a letter written in August 1961 to Ajudhia Nath Khosla of the

Planning Commission, Bhabha made a case for more than one large computer in the country. 'In a big country like India I think there would be a legitimate case for having two computing centres, and getting two computers, one of the types that we have asked for it at Bombay, and the other of the type Prof. Mahalanobis desires.'[19] This observation, if taken out of context, would be equivalent to the remark attributed to Bill Gates about 640k being sufficient memory. Gates has, however, denied saying this. A quote may sound amusing later, but it would do well to remember the context in which it was made.

Mahalanobis didn't agree with Bhabha. He said that in case, for some reason, two computers could not be procured, priority should be given to the data processing for the National Sample Survey, which needed a powerful computer for the speedy processing of large volumes of data. As Dinesh C. Sharma has observed, 'The clash of two of Nehru's close advisors on a crucial high technology issue is intriguing. The computer projects of both Mahalanobis and Bhabha had the blessings of the prime minister [...] Both scientists held key positions in the government. Yet Nehru chose not to interfere on the issue of a national computer centre.'[20]

THE CDC 3600

Undeterred by the tussle with Mahalanobis, Bhabha decided to purchase a powerful computer. Since word had spread that Bhabha was out shopping, computer companies in the US and the UK sent their proposals to TIFR. Bhabha set up a four-member committee of engineers in the US, which was headed by R. Narasimhan, who was then at the Digital Computer Laboratory of Illinois University. The committee, after making a detailed technical and financial analysis of four large computers, settled for the CDC 3600, manufactured by Control Data Corporation. Its chief designer was Seymour Cray, who later became famous for the supercomputers bearing his name. P.V.S. Rao, who was part of the TIFRAC team, observed that when the time came to procure a commercially produced machine, the group took the decisive step of selecting a machine that was still on the drawing boards (over an existing machine that was already obsolete).

The TIFR team took on the responsibility of maintenance rather than leaving it to the manufacturer. Rao noted that problems that baffled the manufacturer were solved by the TIFR team.[21]

The CDC 3600, which was installed at TIFR in May 1964, was acquired for $1.5 million and was financed through a loan to the Government of India by the United States Agency for International Development (USAID), which Bhabha had arranged. R. Narasimhan coordinated the training activities and TIFR's engineers were present when the machine was being assembled and tested. Bhabha had also arranged for all existing FORTRAN-63 library subroutines to be sent to Bombay. From October 1964, the CDC system started functioning as a national computation facility.

This purchase gave Bhabha a chance to befriend Jim Miles, senior vice president of CDC, whom he had visited in Minneapolis. From his conversation with Miles, Bhabha learned that the memory systems for the CDC computer were being made in Hong Kong using low-cost labour for manufacturing ferrite core memory stacks. Bhabha believed that the same could be done in India, which would kick-start the manufacture of sub-systems for large computers in the country. He visited Cray, and later, told M.G.K. Menon that he thought it possible for India to build and design a supercomputer on its own. 'His mind was always moving ahead and fast,' observed M.G.K. Menon.[22]

After visiting the CDC facilities in the US, Bhabha set up a small group to work on printed circuit cards and approached the Tatas to fund research on monolithic and hybrid circuits. The Tatas, who were in talks with Fairchild Semiconductor for microelectronics production in India, showed little interest in research. Talks started between the Tatas and CDC, but, despite three years of discussions, the TIFR–CDC–Tata Sons venture did not take off because of Bhabha's sudden demise and CDC's own apprehensions about bureaucratic delays and the investment climate in India.

BHABHA AND ELECTRONICS

We have already seen how Bhabha set up the electronics group at TIFR, which developed most of the electronic equipment required for cosmic ray and other experiments. The electronics division at AEET grew out of this group at TIFR, and a large part of the reactor control systems and other equipment was manufactured in-house. Notwithstanding all these achievements, the electronics industry was still at a nascent stage in India and much remained to be done. This became painfully apparent during the Indo–China conflict of 1962, when Bhabha realized India's great dependence on foreign sources in a field that was vital for both defence and national development. Nehru also realized the need for reorganizing scientific research so that it catered to the defence needs of the country. Even though the Defence Research and Development Organisation (DRDO) already existed, Nehru wanted to involve other scientific groups in such strategic research. He made Bhabha a member of the National Defence Council after the war with China had ended.

Soon after the hostilities ended, the AEC recommended to the government that an electronics committee be set up to recommend measures for the planned development of electronics in the country. Apart from Bhabha, Sarabhai, Bhagvantam and A.S. Rao were its members. Rao was the member secretary. A large number of representatives from user ministries, research organizations and industry assisted the committee's work.

M.G.K. Menon recalled Bhabha telling him how he regarded electronics as an all-pervasive and empowering technology.[23] Bhabha and the committee took a comprehensive view of the field. The committee submitted a series of 21 interim reports from February 1964 to September 1965, which were consolidated in a final report in December 1965. Bhabha saw electronics as an essential component of a technologically advanced society, noting that the existing industry was at a primitive level and mainly involved the entertainment sector. He considered electronics to be an enabling technology for national development. By then, electronics had become the fifth largest industry in the US even though it was just two decades old in that country.

Bhabha laid great emphasis on computers, realizing the important role they would play in fields ranging from pure research to industry. He felt that the use of computers would become inevitable in railways and communications, missile guidance and meteorology and even in such a complex field as national planning. Bhabha observed that computers were giving rise to a new world view and a new scientific culture in society.[24] The committee recognized that civilian and military production would have to be dovetailed along with the private and public sector. It examined the field from the viewpoint of materials, components and equipment, with standardization and self-reliance as key elements. The committee recognized that this was a rapidly developing area, and the rate of change was so great that they did not think it appropriate to undertake market surveys.[25]

The committee laid down the blueprint for a 10-year plan for the development of the electronics industry in India. One of its most significant recommendations was that India should leapfrog rather than take the step-by-step route of developed countries. It wanted the government to take policy decisions to ensure that the industry grew on the scale recommended by it and wanted those important executive decisions to be in the hands of scientists and technologists rather than bureaucrats to avoid delays. Bhabha worked very hard on the report, and had he lived longer, it would have been one of the areas he would have concentrated on. The final report was ready in December 1965, but Bhabha perished in an air crash on 24 January 1966. Before his demise, the report had been circulated for review among the committee members, and Bhabha had told his colleagues that he would sign it on his return from Vienna.

In the event, the report was formally presented to Prime Minister Indira Gandhi by Sarabhai, who stepped into Bhabha's shoes. He called the electronics committee report one of Bhabha's most memorable contributions. The message at the beginning of the committee's report by Indira Gandhi acknowledged Bhabha's role and the importance he attached to electronics in national development. '...Every page of the Report bears the imprint of his genius, skill and thoroughness and attention to detail,' she observed.[26]

However, since the subject of electronics was under the administrative control of the Department of Defence Supplies, a turf war ensued between the atomic scientists and the defence wing. When the question of starting manufacturing facilities arose, the electronics division of AEET became an independent public enterprise in Hyderabad called the Electronics Corporation of India Limited, headed by A.S. Rao, something that M.G.K. Menon and R. Narasimhan called a perfect example of 'government entrepreneurship'.[27] Later, an electronics commission and the Department of Electronics (DOE) was established by the government.

In 1971, Menon was appointed as the secretary of the newly constituted DOE, a position he held concurrently with the directorship of TIFR. The department, under M.G.K. Menon, founded the National Centre for Software Development, National Informatics Centre, Computer Maintenance Corporation and several state electronic development corporations to promote the manufacture of electronics components, instruments and consumer products. The fruits of these efforts are abundantly visible in India today. A lesser man would have faltered under this burden of responsibilities, but not Menon. Sarabhai's sudden demise in December 1971 added further responsibilities. Menon became the interim chairman of the Indian Space Research Organization and the director of the PRL in January 1972, additional responsibilities that he shouldered with customary aplomb.

Chapter 21

SCIENCE, ART AND CULTURE

> ...I love the consciousness of life and I want as much of it as I can get. But the span of one's life is limited [...] I will increase it by increasing its intensity. Art, music, poetry and everything else that I do have this one purpose—increasing the intensity of my consciousness of life.
>
> —Letter from Bhabha to his friend Jessie Maver, 8 December 1934[1]

Several European intellectuals and artists who had fled the Continent after the outbreak of the Second World War found themselves in India, playing active roles in supporting a bourgeoning art community. We have already encountered Rudolf von Leyden in this regard. Many came to Bombay, a city which was known to reshape narrow regional and parochial identities into a cosmopolitan sense of belonging. The city witnessed the arrival of a group of Jewish refugees from Austria and Germany over a period of time. Bombay had vibrant artistic, musical and film milieux that welcomed German-speaking exiles into their networks. The European émigrés enriched and stimulated the local cultural scene. These transcultural dialogues, which were a reflection of Bombay's cosmopolitan openness, resulted in the emergence of rejuvenated institutions and plural networks. Nothing better signified the spirit of the times than the city itself. The intellectual and cultural effervescence of the city saw many writers, painters and theatre and cinema directors come together to address the needs of a nation in the making.

THE INFLUENCE OF ÉMIGRÉS

The émigré presence decisively moulded groundbreaking movements, the most important of which was the Progressive Artists' Group (PAG). They mixed with influential figures in colonial Bombay to build an art scene that benefitted from the contributions of minorities, like the Parsis and the Jews. Bombay reaped legendary aesthetes from what the artist Krishen Khanna called Hitler's bounty to India. In a letter to von Leyden, Khanna wrote that the events which brought him to Bombay, were unfortunate but their gain was incalculable. The timing seems to have been just right.

Two particularly interesting personalities from this period were the art critic, Rudolf von Leyden, and the first art director of *The Times of India*, Walter Langhammer. It is not entirely clear why von Leyden moved to India in 1933. Of course, as a Jew, it would have been dangerous for him to have stayed in Germany after the rise of Nazism, but it is not known for sure if he was fleeing persecution at the time he left for India. His choice of India was probably dictated by the fact that his elder brother, Albrecht, had been living and working in Bombay since 1927. Rudolf von Leyden had just finished his studies (he received his PhD in geology from the Göttingen University in 1932) and would soon start his own career in an entirely different field. Whatever the reasons for his leaving Germany, von Leyden arrived in Bombay in 1933.[2]

In Bombay, von Leyden began working in publicity for a textiles firm, but he was also interested in visual art. He set up the Leyden Commercial Art Studio, produced watercolour scenes on his travels around India and, more importantly, began working on his series of political cartoons. Starting as a layout artist at *The Times of India*, he soon became the newspaper's art critic. In addition, von Leyden also worked at the cotton-trading firm Volkart Brothers and sketched cartoons for *The Illustrated Weekly of India* under the pseudonym Denley.[3]

Rudolf von Leyden favoured experimental artists, for instance, the painter, Krishnaji Howlaji Ara. He supported Ara in his writings, introduced him to the art dealer, Kekoo Gandhy, and even supported him financially. He became a central figure in the art scene in Bombay. A collector of Indian art, he organized exhibitions and actively promoted contemporary artists. He was also a contributing editor of the leading art review journal *Marg* since its inception in 1946. He advised Bhabha on which paintings to buy for TIFR, and it ended up owning one of the most important collections of post-independence Indian art. Rudolf von Leyden was also an authority on antique board games and Indian playing cards. However, it is as an art critic that he is best remembered. His reviews in *The Times of India* had a decisive impact on artists; and coupled with his lectures and other art activities, he virtually steered the formative years of modern art in India.

The other significant émigré was von Leyden's friend, the Austrian expatriate, Walter Langhammer, who came to India in 1938 after the death of his Jewish father-in-law. He was an artist, critic and patron, and, like von Leyden, his importance lies not so much in his own paintings but as a promoter of the modern art movement in Bombay during the 1940s. Langhammer, who was also a Jew, fled to India with his wife in 1938, shortly before the Second World War. He was an art student at the time the famous painter Oskar Kokoschka was teaching at the Viennese Academy. Langhammer had made friends with a Parsi student, Silloo Vakil, who subsequently returned to India. When she became aware of his travails, she managed to persuade Sir Francis Low, editor of *The Times of India*, to appoint Langhammer as the first art director of the paper.[4]

An accomplished painter, Langhammer exhibited regularly at the Bombay Art Society, winning prizes. He encouraged artists to visit his studio on Nepean Sea Road, and young Indian artists, who later developed into the great names of the Indian progressive art movement, would meet at his studio on Sundays to learn about the European art scene. He introduced these young Indian painters to foreign art periodicals and catalogues. The artists at his atelier included Francis Newton Souza, Ara, M.F. Husain, Sadanand Bakre, Sayed Haider Raza and Hari Ambadas Gade, who founded the Progressive Artists' Group (PAG) in 1947. Apart from this group, other important artists associated with the PAG, included V.S. Gaitonde, Akbar Padamsee, Khanna, Tyeb Mehta, Ram Kumar and Bal Chhabda. The impetus to move away from the realism of the British academic school of art and a revival of the Ajanta fresco style had already commenced when the PAG, led by Souza, began forming.

In 1939, Langhammer became chairman of the Bombay Art Society, and von Leyden a board member. With financial support from Sir Cowasji Jehangir, the society's president, the Bombay Art Society Salon was created. Apart from the PAG, its shows featured artists from beyond Bombay and included George Keyt from Ceylon, Jamini Roy from Calcutta and Sam Tata from Shanghai.[5] Langhammer exposed the group to different schools of art, like Expressionism, at the weekly Sunday salons held at his home and provided the members with informed feedback on their work. The creation of the PAG is entwined with India's independence. A culturally and religiously diverse group of artists formed an informal group in cosmopolitan Bombay of the 1940s and ended up creating a modern art movement for the newly emergent nation state.

The PAG operated from a small room in the Artists' Aid Centre (later renamed Artists' Centre) on Rampart Row in South Bombay. Novelist and art critic Mulk Raj Anand and von Leyden spoke there frequently. Bhabha attended these meetings accompanied by his mother. The artist, H.A. Gade, who became secretary of the PAG in later years, recalled Bhabha delivering a lecture that presented a broad view of the art of painting. Bhabha would drop in regularly from the nearby OYC to see the paintings on display. K. Chandrasekaran who accompanied Bhabha ended up becoming an amateur painter himself.[6]

Similarly, another émigré, the Austrian, Emmanuel Schlesinger, was also crucial in exposing Indian artists to European developments. A pharmacist by profession, he had set up Indo-Pharma, a successful pharmaceutical company in Bombay. He was an avid collector of modern Indian art, and his patronage was also important to the burgeoning art movement in Bombay at the time. It is said that he had been an avid art collector even before his departure from Austria and had to leave most of his collection behind but had managed to bring with him significant artworks including, it is said, paintings by Kokoschka and Egon Schiele.[7]

Raza became close to Langhammer, who allowed him to work in his studio. He would paint during the day and show Langhammer his work in the evening. When Raza left for Paris, Langhammer gave him his overcoat and a pair of shoes. Gandhy, who later founded the long-running Chemould Gallery of Modern Art in Bombay, recalled that Langhammer would tell them what made a good painting. He would share his experiences of Europe and tell them about events in the art world that he had seen first-hand. Langhammer was commissioned by the Tatas to do a series of artworks on the Jamshedpur steel plant in the late 1940s. Along with Langhammer, there was also an influx of other Jewish professionals, who valued art, provided patronage and, in fact, pioneered the buying of art.

Like Bhabha, the war had also prevented Gandhy from returning to England to pursue economics at Pembroke College, Cambridge. A fortuitous meeting with a Belgian by the name of Roger van Damme (whose father had been a framer and restorer of Flemish masters), inspired Gandhy to set up a factory that made frames, called the Chemical Moulding Manufacturing Company, which was later abbreviated to Chemould. Gandhy's framing shop on Princess Street hosted numerous exhibitions of the PAG and others. Gandhy soon began selling paintings that artists had brought for framing. Notably, the artists did not have to incur any expenses for framing, as the costs were reimbursed by the buyer after purchase. The first display of paintings by Langhammer and others was inaugurated by Bhabha in what became Bombay's first commercial art gallery.[8]

In the 1940s, Chemould was not just a framing studio but also an art salon—something Bombay had not seen before. Artists, curators, architects and authors would meet at Chemould. Gandhy, who was the patron-in-chief, facilitator and catalyst had no formal training or education in art. Rudolf von Leyden introduced Ara to Gandhy; and Schlesinger developed an important friendship with the artist Mohan

Samant. In 1963, Gandhy found a gallery space on the first floor of Jehangir Art Gallery, a public space dedicated to the display of modern art, which was founded by Sir Cowasji Jehangir at the urging of Gandhy, von Leyden, Hebbar, Langhammer and others. Bhabha, too, played a role in its founding. Built in 1952, on land given by the Prince of Wales Museum (now Chhatrapati Shivaji Vastu Sanghrahlaya), the gallery named after Sir Coswaji's late son, Jehangir, became a regular haunt of many aspiring and struggling artists soon after its inauguration. Since then, Chemould has been a cradle of modern and contemporary art.

Langhammer and von Leyden, Europeans in a colonial space, were deeply interested and invested in Indian culture and this interest and involvement in modern art was prompted by a strong belief in Indian contemporary art. Their talent was first spotted by Anand, who, at the opening of their first exhibition in 1948, spoke about the urgency of providing a platform for new voices. Langhammer's dinner parties and salons were opportunities for aspiring artists to rub shoulders and exchange ideas with Bombay's cultured and moneyed elites, among whom the Parsi community was particularly significant.

Unsurprisingly, Bhabha, who was a connoisseur of the European modern masters, was attracted to this new group of artists, which also exemplified internationalism and a spirit of freedom.[9] Bhabha's friendship with Langhammer, von Leyden and Gandhy enabled him to develop an art collection at TIFR, which is still considered one of the best collections of contemporary Indian art of that period. Two other key figures involved in Bhabha's quest for art were the art collector, scholar and lawyer Karl Khandalavala and Pipsy Wadia, Bhabha's companion.

For Bhabha, the arts were not just a form of recreation but a serious pursuit to which he attached great importance and devoted as much attention as his scientific work. Bhabha's eye was precise and sure, and he knew exactly what he wanted. The paintings he selected are among the most significant paintings of the TIFR collection. Bhabha was always given a preview when Chemould exhibited, and if, for some reason, he could not attend, they would not sell till he had exercised his right of first refusal.

Bhabha paid considerable personal attention to the way a painting was framed and hung after it was bought. He would drop in at the shop and ask for frames to be held against the paintings. Sometimes he would forget that a painting was being held for a long time.[10] Gandhy observed, 'The contagion spread among our set from Langhammer and Leyden to Homi Bhabha and Farookh Mulla, PRO of the Tatas. They were all patrons of the Chemould shop.'[11] Langhammer and the Bohemian composer and musician, Walter Kaufmann, both lectured at the Rotary Club's regular luncheon meetings at Green's Hotel, next to the Taj Mahal Hotel, as did Bhabha and Anand. Gandhy curated exhibitions at the Taj and he also showed art in a 'window' at the hotel, with von Leyden reporting on these shows.

TIFR's ART COLLECTION

Bhabha's deep interest in the cosmopolitan aesthetics nurtured by the Bombay Progressives resulted in TIFR spending government money to buy paintings and other works of art. With Nehru's permission, he began spending one per cent of the institute's annual budget on purchasing works of art.[12] Bhabha started the TIFR collection in the early 1950s, and, within a decade, had amassed the most significant collection of Indian modern art anywhere in the world. Most of the art collection consisted of contemporary pieces by practising artists and included women artists like Pilloo Pochkhanawala, Prabha Agge and Jyoti Shah. When Sir Maurice Wilkes who visited TIFR in 1964 remarked on the artworks displayed, Bhabha told him that 'the government of India was one of those enlightened governments that allowed a small percentage of the total funds spent on new government buildings to be spent on artistic enrichment.' Bhabha said that, unlike many of his colleagues, he took advantage of this provision 'up to the hilt'.[13]

The first painting acquired by TIFR was *Window* by G.M. Hazarnis in February 1952. The second was *Window Light*, a water colour by Ara, a purchase that marked the beginning of Ara's special relationship with the institute, one which continued long after Bhabha passed away in 1966. In the 1950s, Bhabha acquired works by almost all the Progressive artists from Bombay—Souza, Husain, Bakre, Raza,

Gade and Gaitonde. The third piece purchased for the collection was a decorative panel by Langhammer in 1952, followed by a landscape in 1955. Langhammer's portrait of von Leyden titled, *The Critic,* also found its way into the collection.

Other modern Indian artists, whose works were purchased for TIFR, include Kattingeri Krishna Hebbar, Badri Narayan, Ardeshir Davierwalla, Shiavax Chavda and Laxman Pai. Bhabha also bought the early works of Ganesh Pyne and K.G. Subramanyan. His patronage and informed criticism of the plastic, pictorial and performing arts assumed an added dimension since he was an accomplished artist himself.

Anand, in his letter that he never posted due to Bhabha's demise, said that Bhabha's frequent appearances in a casual non-official manner at the Artist's Aid Centre confirmed for many of those young talents that in him they would have a discriminating patron who would talk from within the experience of the pictorial situation rather than in the generalized manner of art critics.[14]

Bhabha's institutional patronage was driven by his own artistic inclinations, and his desire to refine the aesthetic sensibilities of the scientists at the institute.

His patronage of the American–British sculptor, Sir Jacob Epstein, attracted much criticism from government circles. In 1954, Bhabha purchased one of Epstein's bronze heads of Einstein[15] for what was then considered the extravagant sum of ₹3,818.72. Six months after Bhabha's death in January 1966, a letter from Marlborough Fine Arts Limited, London, was discussed at the 50th meeting of the TIFR council. This letter informed the council that a bronze head of Einstein had been sold at Sotheby's for £2,100 in July 1964. The institute had paid £200 for it in 1954.[16] It is not known what the government thought of the more than tenfold appreciation.

However, Bhabha was hardly going to be affected by such criticism. Apart from his own strong convictions about the importance of aesthetics to science, he also had Nehru firmly by his side. A wall in the entrance hall of the new TIFR building was 45 ft long and nine feet high, and Bhabha wanted to cover it with a mural. In keeping with his desire for excellence, he set in motion an ambitious plan to invite

Pablo Picasso to India to paint the mural for the institute. Bhabha was interested in encouraging a 'different pedagogy through which the sensibilities of the citizen-scientist could be refined'. Scientists at TIFR were being 'envisaged not only as leaders in an expert-led world but also as connoisseurs capable of holding conversations with the world'.[17] It could also be seen as an attempt to bring together the modernity of Indian art into the modern International style building designed by Helmuth Bartsch. Bhabha did not want art, culture and science to be in silos.

In November 1962, he wrote to J.D. Bernal, the famous crystallographer and biologist, who he had befriended during his stay in England. Bernal had established Birkbeck's Biomolecular Research Laboratory in two Georgian houses in Torrington Square. Picasso was close to Bernal, and, when he came to England in November 1950 to attend a world peace conference, he made a drawing on the wall of the sitting room of Bernal's flat, which was just above his laboratory at Birkbeck College. During a dinner party held at Bernal's home, Picasso, using pencil and crayon, drew a portrait of a man and woman with wings on the wall of Bernal's sitting room. This was the only mural Picasso ever made in England. When Bernal's flat was demolished to make way for a new college building, that section of the wall was removed for preservation and is now part of the Wellcome Collection. This original section of the wall called *Bernal's Picasso* is currently installed in the Tate Liverpool gallery.

Referring to a previous conversation on the subject in which Bhabha had asked Bernal to sound out Picasso and ascertain his interest in the project, Bhabha wrote to Bernal to ask if he had met with any success. Citing the foreign exchange crunch, he said it would be impossible to pay him anything in foreign exchange, leave aside a price appropriate for Picasso.[18] Aware of his financial limitations, Bhabha offered Picasso first class return airfare, a stay at the Taj and trips to places of historic interest. Bhabha was obviously relying a great deal on Bernal's friendship with Picasso for an affirmative answer since what he was offering was very modest, and as he admitted, certainly not in Picasso's league. We do not know if Bernal replied; there is no evidence on record to suggest that he did.

Bhabha, who must have realized that the chances of Picasso responding favourably were slim, had already set the ball rolling for identifying a suitable artist in India. In September 1962, two months before he wrote his letter to Bernal, he had invited several artists from Bombay to attend a meeting to discuss the mural. A telegram was also sent to Narayan Shridhar Bendre in Baroda to attend the same meeting. In October 1962, the institute invited 12 modern Indian artists to submit preliminary designs for the mural. In what would be termed a 'call out' in modern parlance, N.R. Puthran, the registrar of the institute, invited the artists emphasizing that the mural was intended to be a tribute to Indian art, and a stimulus to the aesthetic sensibility of the many young scientists who passed through the building. Puthran's letter declared TIFR's intention to survey a wide range of artistic possibilities which were in harmony with the aesthetic concept of the building.

The institute had allotted ₹15,000 for the finished mural. The preliminary sketch was to be 9ft 8 in. long and 2 ft wide, and each artist was to receive a remuneration of ₹800 for it.[19] Ara, Bendre, Gaitonde, Satish Gujral, Hebbar, Husain, Padamsee, B. Prabha, R.D. Raval, Raza, Baburao Sadwelkar and G.M. Solegaonkar were requested to submit entries. In addition, the names of Shanti Dave, Ram Kumar and Jamini Roy formed part of a separate list in case any of those in the first list were not able to participate. An advertisement was published, calling for applications. From the initial list of invited artists, Raza, Padamsee and Gaitonde did not submit entries.

Nine artists responded: Ara, Bendre, Gujral, Hebbar, Husain, B. Prabha, Badri Narayan, Solegaonkar and Raval. Bhabha also invited the senior-most among contemporary Indian artists, Roy, to contribute. His painting titled *Krishna and Balarama* was acquired by the institute in 1963. Since this work is similar in size to the preliminary sketches for the mural, it was thought that Bhabha had rejected this piece for the mural, but this is not true.[20] Roy was one of the most significant artists of the twentieth century and Bhabha wrote to him in October 1962, calling him the doyen of contemporary artists in India and telling him that his design would not be considered for competition with the rest, but would be treasured as a work of art as part of the art collection of the institute.

Bhabha wrote about his wish to capture the entire range of Indian modern art, including experiments in Bengal as well as Bombay.

> I may mention in this connection that the Institute has one of the best collections of contemporary Indian art in the country, and the absence of any painting by you is a serious lacuna which we wish to fill [...] May I take this opportunity to express to you my admiration for your pioneering and original contributions to contemporary Indian painting, and I would like to call on you when I next visit Calcutta.[21]

Roy seemed rather touched by Bhabha's letter. He wrote about the similarity between both of them, Bhabha 'pioneering in scientific researches' and he 'throughout my life researching in art'. He promised to send the sketch by the end of December even though 'time left is short'. He ended with, 'I am really very happy for your letter and particularly for the spirit with which it has been written.'[22] In keeping with his status as the senior-most among the invited artists, Roy was paid ₹1,000 for *Krishna and Balarama*.[23]

The four-member committee that was appointed to assist Bhabha in selecting the mural consisted of Khandalavala, von Leyden, K. Chandrasekharan (who was an amateur painter himself), and Pipsy Wadia, who was an art connoisseur. Khandalavala was an accomplished lawyer and art collector, who was also an authority on Indian miniatures. He published landmark studies of miniatures and other books on Indian sculpture and paintings. He went on to chair the board of trustees of both the Chhatrapati Shivaji Vastu Sangrahlaya and the Lalit Kala Akademi. He bequeathed his priceless art collection to the museum. As a lawyer, he was the defence counsel in the Kawas Nanavati cause célèbre[24]. Khandalavala lauded Bhabha's artistic achievements, which he believed to be praiseworthy even when judged by sound professional standards.

Bhabha was on tour abroad when the committee shortlisted the entries of Raval and Husain. Upon Bhabha's return, they were reviewed and, in the end, Husain's *Bharat Bhagya Vidhata*, which echoes the second line of the Indian national anthem, found favour. After summarizing the views of the committee, Bhabha wrote: 'The consensus of opinion is that the Husain is to be preferred.' Bhabha gave his own reasons for choosing Husain. He considered the Raval to be more elegant but thought that it would wear less well with time. Husain's offering, he thought, would be more substantial and enduring, and he expected the full-scale mural to be richer in texture and detail than the sketch.[25]

Husain was no stranger to either Bhabha or the institute. Two of his water colours, *Yellow Face Tempera* and *Mother and Child* had been acquired by Bhabha in February 1954 at Husain's first exhibition, which Bhabha had opened at the Chemould gallery. Two oils, *The Lamp* and *Naika (Nritya)*, were added in 1959, and, in 1961, the institute added two more oils *Ragamala* and *Horses*. In a reversal of the patron–artist relationship, in January 1961, Bhabha sketched Husain, who sat for him. The TIFR building was being constructed at the time and Husain recalled visiting Bhabha at his home once or twice a month.

Husain, it appears, had not been very keen about applying to paint the mural. On the morning of the last day for submission, he made a half-hearted sketch with crayons and submitted it. One of the reasons for his hesitation was that von Leyden's favourite painter, Ara, who also happened to be a favourite of high society at the time, had also applied. Husain thought that, with Ara's promoter on the selection committee, his chances were minimal. As it turned out, he couldn't have been more wrong. When a few months had passed and TIFR had yet to declare the results of the competition, Husain met Pipsy Wadia and asked her what was happening. She seemed surprised that Husain had not been intimated yet. She went inside and phoned Bhabha. At the end of the conversation, she came out and said, 'You got the mural.'[26]

Husain not only enjoyed painting the mural but also the atmosphere at the institute. Like the scientists, he had a room with his name on it and interacted with them over lunch at the canteen. He liked being at TIFR so much that he took two years to finish the mural, even though he could have finished it in a few months. Bhabha would come to see him paint. Husain's mural is celebratory in mood, with a mingling of Indian and western influences. Husain, at that time, was in the middle of a series of paintings based on Rajasthan and so unsurprisingly, a Rajasthani landscape serves as the setting for the action. While there is no leading narrative, a large number of vignettes play out on their own, where animals, architecture and people compete for the viewer's attention. The repeated use of the elephant as a motif adds to the impression of monumentality that a mural naturally evokes. A decade later, Husain painted a series on art and science, which he made available to the TIFR scientists for their outreach programme publication for school children.

Husain thought that Bhabha was the first corporate patron of contemporary Indian art. According to him 'society' (by which he meant art critics and rich patrons), did not give Indian artists the recognition they deserved. He felt that in the beginning, only Rabindranath Tagore and Bhabha recognized the significance of contemporary Indian art.[27]

Husain was also instrumental in selling Khanna's painting to Bhabha. Khanna had a covenanted job in Grindlays Bank in Madras and had been introduced to the Progressives by Husain. He did not know Bhabha but was thrilled to know that one of his small oil paintings, *Spring Nude*, had been sold to Bhabha for ₹200. Khanna gratefully remembers that the initial momentum created by Bhabha was in some way responsible for the many fortunate events which came his way. He bore a sense of gratitude to someone who instigated events of which he was never aware.[28]

Jehangir Sabavala's *Seagulls and Sails* was acquired for the collection in 1962. There also seems to have been the practice of borrowing works and displaying them with the permanent collection at the institute. Sabavala recalled that one of his paintings, a still life from his Paris days, was in the personal collection of Pipsy Wadia, which she loaned to TIFR since Bhabha decided that it would hang well alongside the rest of the collection. When Bhabha died, he left a personal collection of fine works of art (including a Picasso), paintings by some of India's finest artists whom he had patronized, along with an impressive collection of his own paintings and sketches.

Pipsy Wadia, Bhabha's long-time companion and partner, was Rustum and Jehangir Choksi's sister. She was married to a businessman called A.B. Wadia. The couple had three daughters, but the marriage was an unhappy one. The tall, sophisticated, cultured and elegant Pipsy had nothing much in common with her husband, who it is said, preferred to spend most of his time at the Mahalaxmi race course. The Choksis came from a humble background and the story goes that Pipsy married her husband on the understanding that the Wadia family would provide the funds for the higher education of her two brothers in England. When they reneged on that promise, the marriage is said to have come under strain almost immediately.[29]

That Bhabha and Pipsy were an item was common knowledge, and the couple appeared in public freely, with no stigma attached to their relationship. They were even jointly invited to events. Bhabha, in fact, insisted on Pipsy being invited and as an incident narrated by Mithoo Coorlawala recounted later shows, such requests sometimes resulted in a slightly embarrassing position for the hosts. When one of Pipsy's daughters was diagnosed with cancer, Bhabha arranged for the best Harley Street doctor to examine her. In a letter to the doctor, Bhabha said that the girl should be treated as part of his family.

Coorlawala, who went to Newnham College, Cambridge, recalls being present at the performance of Mozart's *Idomeneo* at Cambridge, for which Bhabha had designed the sets.[30] At Cambridge, the hierarchies of academic life had stood in the way of their interaction. 'I was a first-year student—and he was an exalted Fellow', she said in an interview more than seven decades later.[31] After they had both returned to Bombay, she got to meet him socially, at dinner parties at Petit Hall and in the home of Rodabeh, J.R.D. Tata's sister, and at Scylla Petit's home. The Cambridge Society performed a Shakespeare play annually to raise money for their scholarship fund, and Bhabha was very involved with the fundraising. Before the play, Coorlawala hosted a dinner at her home and Bhabha would attend, if he was in Bombay.

Pipsy, who was the vice president of Bombay's Time and Talents Club, organized a Christian Dior Fashion Show at TIFR in 1962, which saw Parisian models walking the ramp at the west lawn. Coorlawala was the secretary and a close friend of Pipsy. Obviously, it was Pipsy who got Bhabha to agree to host the event at TIFR since, without his support, there would have been no show. Coorlawala mentioned how she and Rodabeh had corresponded with Dior and organized the show. Pipsy had set up the venue. A ramp was erected from behind the west canteen, and it ran into the lawn. The models and Marc Bohan's Autumn Collection of that year came from France.

Uttara Parikh, a former deputy commercial director of Air India, was then a young student at the Government Law College in Bombay, recalls being one of the sari models. 'The Dior girls wearing the western wear collection were deeply impressed by the colour, pattern, and texture of the silk and cotton fabrics the Indians wore. Their professionalism and efficiency impressed us.'[32] Parikh also recalled how Bhabha attended nearly every weekly exhibition when she worked as manager at the Taj Art Gallery before joining Air India. Shortly before he left for Geneva on the ill-fated Air India flight, he had inquired about when a certain artist was going to exhibit. On being told it was after two weeks, he said, 'Good, I'm going to Geneva tonight but will be back in a few days. See you then.' Parikh says that she was devastated when she received the news of the crash.[33] Air India was one of the first corporate art collectors in India and some of Bhabha's pencil drawings are part of the Air India Collection.

The fashion show aimed to raise money for charity. Bhabha hosted an elegant dinner after the show. It was remarkable that an institution famous for its scientific research should have hosted a fashion show on its premises. Perhaps, it was only Bhabha's international exposure and breadth of vision that allowed these two dissimilar cultures to coexist in the same space, even if it was only for one evening.[34] Pipsy also used her influence with Bhabha in other matters. In the 1960s and 1970s, tucked between the OYC (now a sea-facing, restored part of the DAE) and Pier No. 1 at the Gateway of India, was a small café called Victory Stall. It was run by the Time and Talents Club, and it is said that Pipsy was instrumental in getting the space for the café from Bhabha. The café closed in Homi Sethna's time, but the Time and Talents Club continued to have an office in the atomic energy building, even if it was just a few tables.

Coorlawala also spoke about the dinner she organized as part of the Cambridge Society to honour Bhabha after he was admitted to the Royal Society. On the day of the dinner, Bhabha had called Coorlawala to announce that even though they had not invited her, he would be bringing Pipsy to the Cambridge Society dinner. Sir Cowasji was the president of the society and Bishop Lash, an Englishman who belonged to the Church of England, was its chairman. The bishop, who had strong scriptural concerns about matrimony, felt that the 'learned doctor had put them in a predicament', as theirs was an inappropriate relationship since they were not married.[35]

Coorlawala responded with some indignation that it did not matter to anybody in Bombay, as they were both well accepted and respected. She recalls that Bhabha and Pipsy '...were very well respected in high society and they were always thought of being together—and were invited together'.[36] Bhabha and Pipsy did attend the dinner in the end, mainly because of Coorlawala's persistence in the matter. The problem was solved by seating Pipsy between Sir Cowasji and Lady Jehangir, who was well-married, with the prudish bishop being seated on Lady Jehangir's other side.[37] Obviously, public opinion mattered little to Bhabha or Pipsy. It also reflects the liberal free-spirited culture of Bombay at the time, which was characterized by a cosmopolitan sophistication. Bombay was just another link in the chain that connected New York to London to Paris to Beirut to Cairo and, then, to Hong Kong. The decades between 1940 and 1960 are remembered as the city's golden era—a time when the city was carefree, leisurely and, most of all, liberal.

Coorlawala's narrative, in many ways, countered official and more deferential narratives about Bhabha the scientist. She affectionately recounted her relationship with Pipsy. 'She looked amazing. She was tall and she had platinum blond hair. She wore a sari very elegantly and moved very elegantly. She didn't bother about her appearance. She was just beautiful. We liked each other—so we did things together. We went to plays and we used to read poetry together. T.S. Eliot at that time was a favourite.'[38]

Journalist Coomi Kapoor, in her book on the Emergency, recalls how letters for her sister Roxna, who is married to the politician Subramanian Swamy, were sent to their friends when Swamy was on the run. Pipsy, who was a good friend of Coomi's mother, also received some letters. The police were talking to all these friends and paid Pipsy a visit. Kapoor observes:

> The only person unfazed by the police raids was Pipsy Wadia, my mother's friend in Bombay and fellow member of the Time and Talents Club. This grand dame, long-time companion of atomic scientist Homi Bhabha, was cool as a cucumber and told the agitated police officer who had come to question her to calm down. She had no particular liking for Indira Gandhi, who she felt had set her cap at Homi Bhabha at one stage.[39]

And what did Bhabha think of Indira Gandhi? If Sethna is to be believed, he had a poor opinion of her intellectual capabilities. Sethna quoted Bhabha as saying, 'Homi she is not very bright, unlike her father.'[40]

Jamshed Bhabha, in an interview, was candid enough to admit that Bhabha had many lady friends, but 'Even they [the lady friends] realised that mother came first', he said. [41] When asked if Bhabha had met with any failures in his life, he said that the one failure he could recall was of such an intimate nature that he could not talk about it. It involved Bhabha's failure in his relationship with a lady.[42] It is impossible to know who Jamshed was referring to.

Chapter 22

'WITHOUT HIM ART IS ORPHANED'

> Art like Science knows no frontiers [...] We all hope that with its newly achieved freedom, India will become the leading country of Asia and one of the leaders in cultural matters, and it can achieve this in the artistic sphere not by a mere repetition of its ancient forms but by the creation of new art forms, possibly through synthesis of the ancient Indian and European traditions in art.
>
> —Letter to Abul Kalam Azad, 24 July 1947[1]

Bhabha's pedagogy included creating an artistic taste among Indian scientists by training their aesthetic sensibilities. TIFR was supposed to create a scientific culture, but it was also expected to become a part of the sociocultural milieu in which it existed. Bhabha's vision showed that he was sensitive to the humanistic values within which a scientific research culture could grow.[2] His attempt to recreate the world he knew at Cambridge was supported, in no small measure, by the international cosmopolitanism unique to the Bombay of those times. As Indira Chowdhury has observed, 'the cultural ethos of Bombay of the 1950s resonated with Bhabha's institutional design'.[3]

AN ART COLLECTION

Any discussion of great institutional art in India has to contend with the paintings and sculptures that form the TIFR collection. The collection is not very large, with less than 300 paintings and sculptures, but, among these, the number of master works is very high. This is in itself remarkable since it would do well to remember that Bhabha was putting his faith in artists who had no track record at all. From the outset, Bhabha was interested in housing his scientists in an environment that was sensitized to beauty. Between 1961 and 1968, more than a dozen artworks were purchased every year, and they were housed in a modern building, set among beautifully landscaped gardens.[4] The Indian art at the institute was contemporaneous with its science, except for the Rajasthan temple pillar, several sculptures of the Chola period, wood carvings from Gujarat and South India and Sir Jacob Epstein's bronze head of Einstein.

The TIFR collection also provides new insights for a student of modern Indian art since it has numerous works by artists, who are little known today. Perhaps the best example is the work of artist Homi Patel, who is all but absent from present scholarship. Luckily, M.G.K. Menon, who succeeded Bhabha, not only maintained the collection but also added to it substantially. In 2010, public interest in the collection was rekindled with a large-scale conservation programme, the publication of a book on the collection, and a major exhibition mounted at the National Gallery of Modern Art, Mumbai. Works from the collection also featured in major international exhibitions. Before his death in 2016, Menon donated his own collection of paintings, which he had acquired at auctions while accompanying Bhabha. They include a portrait of Bhabha by K.K. Hebbar. He also donated a little antique statue of a Bharatanatyam dancer, which once adorned Bhabha's study desk and had been gifted to Menon by the grieving Meherbai Bhabha, after he became director of TIFR.[5]

V.S. Gaitonde was Bhabha's favourite painter and K.H. Ara was a close friend who gave painting lessons to the children of the staff of the institute. When the new buildings of TIFR were inaugurated in January 1962, artists were special invitees seated in the front row.

Later, Bhabha introduced them to Nehru with comments about their work.[6] Ara remained connected with the institute till his death in 1985.

Among the other artists Bhabha patronized was Ardeshir Davierwalla, one of the first generation of modernist sculptors. Davierwalla was a self-taught sculptor who was greatly influenced by Epstein and Henry Moore. He had trained as a chemist before turning to wood and metal sculpting. He knew Bhabha very well and recounted that Bhabha would visit his exhibitions and purchase his work. Bhabha purchased his series of four pieces, *Quartette* in 1962. He also appears to have asked Chandrasekharan to visit one of Davierwalla's exhibitions. After Bhabha's sudden demise, Davierwalla wrote a letter to the editor of *The Times of India*, which bears reproducing in some detail since it encapsulates Bhabha's approach to art in a nutshell:

> He was never cursory; in fact, he often visited an exhibition a second time. His grasp of art forms, old and new, was intuitive, his judgment clear and unbiased, his purchases magnanimous. Over the past decades he used the funds at his disposal to build a comprehensive collection of contemporary painting and sculpture which today ranks among the finest in the country. Dr Bhabha bought each art work after personal scrutiny, chose each frame and pedestal after considering various alternatives and finally sought a position where the work would be seen to the greatest advantage. Once an art work was included in this collection, it was not the end of it. The paintings and sculptures were maintained in perfect condition, copyrights of reproductions secured against prior payment being made to the artist. The collection was always open to the public. The message came through clear—the artist and his work was loved and cared for and the artist's self-respect never overlooked. This fine collection is bereft of its patron but stands today as the legacy of a genius who was a prince among men! Without him art is orphaned.[7]

In 2017, Davierwalla's metal sculpture *Galaxy* sold for $187,500 at a Christie's auction in New York. The sculpture, which is made from scrap metal, is an Atlas-like figure holding up two globes and has 'Bhabha' inscribed on its body as a homage to the late scientist.[8]

Bhabha also found other scientists who were interested in art. Like Bhabha, Robert Marshak, too, was fond of the arts and humanities throughout his life. The two bonded over shared interests, not only in scientific matters but also in the arts. Ruth, Marshak's wife, was equally enthusiastic about art and culture. Consequently, during their travels, the Marshaks collected objets d'art and paintings whenever possible. They had a limited budget but good taste and took the advice of their hosts.

During their 1963 visit, the Marshaks first saw and learned about contemporary Indian art from Bhabha. At TIFR, they saw the works of Gaitonde and Davierwalla, which were still very affordable at that time. Both artists were promoted by Bhabha and feature extensively in the TIFR collection. Davierwalla '[was] bought and commissioned voraciously by Bhabha and the TIFR, the Atomic Energy Establishment, and his own personal collection'.[9] The painting by Gaitonde acquired by Marshak resembles some of the paintings at TIFR. Bhabha owned a Davierwalla work titled *Dryad* that closely resembles the piece purchased by the Marshaks. In his oral history interview, Marshak recalled how two of the paintings in his collection were by Gaitonde along with some of the sculptures by Davierwalla. He expressed his happiness at the chance to 'acquire some of the first original art that we ever purchased anywhere [...] I never did turn over the Bombay works to the University of Rochester because it was not that expensive and I decided I might as well keep them for my family's own pleasure'.[10]

It was just as well that Marshak chose to keep his Bombay acquisitions—he left his heirs a small fortune in the form of this art. Perhaps, not so small after all! In March 2021, a Gaitonde from the Marshaks' collection was auctioned for $1.9 million. In the same month, another painting by Gaitonde fetched a record-breaking ₹39.98 crore at an auction in Mumbai.

THE MAGDA NACHMAN ACHARYA EPISODE

Bhabha believed that science was international and knew no frontiers. He also believed in a more inclusive idea of citizenship, as is evident from his taking up cudgels with Sarojini Naidu and Abul Kalam Azad on behalf of Magda Nachman Acharya, a Russian painter who had married an Indian and settled in Bombay after fleeing Hitler's Germany.

In 1920, Nachman Acharya met and fell in love with the Indian communist, M.P.T. Acharya, in Moscow. Acharya had arrived in Bolshevik Russia with a group of Indian comrades looking for support in the struggle against the British. The couple married in 1921 and immigrated to Berlin. With the rise of Hitler, the Acharyas fled to Switzerland. Subsequently, they visited Paris to say goodbye to Madame Bhikaiji Cama, who Acharya had previously worked with, before he sailed for Bombay in 1935. Nachman Acharya followed a year later. She managed to sail to India only because of her husband's nine-year struggle with the British authorities during the couple's years in Berlin, which finally got them British passports and an assurance that Acharya would not, as a former freedom fighter, be arrested on his arrival in India.[11]

In Bombay, Nachman Acharya began to exhibit and became a member of the Bombay Art Society. She made the acquaintance of the artists who would later be a part of the PAG and became their teacher and friend. She also worked with the artist Shiavax Chavda, who painted the Tantric murals that adorn the walls of the Tata theatre in Mumbai and created costumes for the Viennese expressionist dancer, Hilde Holger. In the Indian press, she was called the 'Great little lady of the art world of Bombay'. She painted a portrait of artist Jehangir Sabavala in 1942. Sabavala's wife, Shireen, recalled sitting for a portrait, too. According to Lina Bernstein who has written a biography of Nachman Acharya, Shireen maintained that Nachman Acharya had made several portraits of Pipsy Wadia, but her quest for discovering these paintings met with limited success.

Nachman Acharya lived and worked at house No. 63, opposite Ara's modest living quarters in Walkeshwar. Artist Akbar Padamsee recalled how a furious Nachman Acharya barged into Ara's studio one day, railing against the fact that her works had been rejected by the committee selecting paintings for the exhibition of Indian art to be held in London in the winter of 1947–1948. She had applied to be included in the 'Exhibition of Art Chiefly from the Dominions of India and Pakistan, 2400 BC to 1947 AD', which was to be mounted at the Royal Academy of Arts at Burlington House. She was rejected on the grounds of her European roots and ethnicity.

Bhabha referred to this rejection in his letter to the education minister, Maulana Abul Kalam Azad, raising questions about the definition of Indian citizenship and its connection with Indian cultural life. Acharya had written to Kekoo Gandhy asking him to intervene on his wife's behalf. Gandhy asked him to write to Bhabha. Acharya sent Bhabha a copy of the letter he had written to Gandhy, which was, presumably, the first time Bhabha heard about the case. Bhabha was in Simla when he received the letter and had also met Maulana Azad, the man with whom he would correspond on this issue.

Acharya's communication resulted in Bhabha writing a series of letters in defence of Nachman Acharya. Bhabha's own painting *Dove sono I belli momenti* (1938), inspired by the countess' aria from Mozart's opera *The Marriage of Figaro,* was chosen to be part of the exhibition. Bhabha who headed the regional committee wrote to Naidu, who headed the all-India committee for the London exhibition about the rejection.

In his letter to Maulana Azad dated 24 July 1947, Bhabha argued that the rejection of Nachman Acharya's paintings had more to do with her European origins and less to do with her style of painting. Advocating a more liberal cultural nationalism, he argued for respecting plurality and individuality. Bhabha used Nachman Acharya's case to elucidate the question of who should be considered Indian at the dawn of Independence. The question was how far removed from Europe did one have to be in order to represent India. 'Art like Science knows no frontiers', he said, explaining how he believed that the committee had taken a very narrow and parochial view of the identity of an Indian painter.[12] He gave the examples of Pablo Picasso, who was included in French exhibitions, Amrita Sher-Gil who was both Hungarian and Indian, and Simkie, the French dancer who was an indispensable part of Uday Shankar's dance troupe. Bhabha observed that half a dozen of the paintings chosen by the committee were painted in purely European style while practically all the others showed the influence of European art.[13]

In his letter to Maulana Azad, Bhabha argued that India's cultural life was not made up only of the work of Indian artists but foreign artists domiciled in India also contributed to it. He said that the great strides the US had made in the last 20 years was due to the liberal policy it had been following in cultural and scientific matters, and it would set a poor precedent for the future if the wrong done to the Russian émigré painter was not righted. 'I trust you will make some statement which will lay down the policy of our national government on such questions for the future guidance of official and semi-official bodies, and establish a valuable principle in line with the broad ideals for which the Congress has always fought.'[14]

Bhabha wanted Azad to make a statement that would help in laying down the policy for the disposal of such cases in the future. He probably realized that such a situation could recur and wanted to have a precedent to cite so that future cases could be dealt with in the same way. In what is now a famous quotation in Indian art circles, Bhabha said that they all hoped that with its newly achieved freedom, India would become the leading country of Asia and one of the leaders in cultural matters. This, he said, could be achieved in the artistic sphere not by a mere repetition of its ancient forms but by the creation of new art forms, possibly through synthesis of the ancient Indian and European traditions in art.[15]

It is not known if Maulana Azad replied to Bhabha's letter. If he did, there is no record of his reply. Nachman Acharya wrote to Bhabha, expressing her pain at being considered an alien, even after she had married an Indian and established herself not only as an artist but also as a mentor to Indian artists. She thanked Bhabha and Pipsy Wadia (even though she misspelled her name as Thiroja instead of Piroja) 'for having the courage to stand up against the prejudice of the great majority and to fight out of idealism even a losing battle'.[16] Nachman Acharya died on 12 February 1951, hours before an exhibition of her paintings was to open at the hall of the Foreign Languages Institute in Bombay.

MARG

Bhabha also had a hand, albeit indirectly, in the establishment of the magazine, *Marg*. Its founding editor was the novelist Mulk Raj

Anand, who had met Bhabha at Cambridge. He was also a friend of J.D. Bernal and recalled the latter telling him that if Bhabha devoted more time to his cosmic ray research, he would one day win the Nobel Prize for physics. In 1946, Anand gathered a group of artists, architects and intellectuals around him to launch an art journal that was truly world-class. The name of the group of devotees of modern architecture responsible for founding the journal was the Modern Architectural Research Group (MARG) and the acronym could not have been a more appropriate one, for 'marg' means 'path' or 'pathway' in Hindi. *Marg*, which still exists, surveys classical and contemporary Indian art, architecture, clothing, crafts, dance and photography. It also reports on contemporary foreign art and architecture, especially that of the US and Europe.

Anand was widely regarded as India's pre-eminent English novelist of his generation. He studied at University College, London, where he came into contact with the Bloomsbury Group and T.S. Eliot. His first book on art history, *Persian Painting*, appeared in response to an exhibition held in London in 1929–1930. This was followed by *The Hindu View of Art* in 1933. Anand's first novel *The Untouchable* which was published in 1935, had a preface written by E.M. Forster. The same year he co-founded the All-India Progressive Writers' Association. He assumed the editorship of *Marg* shortly after returning to India from London, where he had spent most of the previous two decades.[17]

Among the other important figures involved in the founding of *Marg* was Otto Koenigsberger. The founding of *Marg* was also a way of keeping abreast of new developments in art and architecture in the world. In addition to Koenigsberger, the other architects involved with *Marg* included the Sri Lankan architect Minnette de Silva, who had worked for Koenigsberger in Bangalore in 1944. An alumna of Bombay's J.J. School of Architecture, she was a founding member and the architectural editor of the publication. Her elder sister, Anil Marcia de Silva—who Bhabha used to meet socially with Vikram Sarabhai in Bangalore—was the assistant editor responsible for the magazine's innovative layout.

In Bombay, Anil came into her own; she was photographed by Henri Cartier-Bresson and was sketched by Bhabha. In the 1940s, the sisters resided at 25, Cuffe Parade, Jassim House, also known as Taraporewala Mansion. They were later joined by Anand in 1945. Their shared apartment became a cultural hub: Bhabha and Gandhy were regular guests, as were Rudolf von Leyden, Walter Langhammer and Emmanuel Schlesinger. The Hamied family also lived there from 1933 to 1970. K.A. Hamied, who set up the pharmaceutical major Cipla, and his Polish wife, Luba, fostered exchange among Hindus, Jews and Muslims alike. The concept for an Indian modernity was shaped, through *Marg*, in this space.[18] After a live-in relationship with Anand, Anil inexplicably left him to marry a Frenchman, settling in Paris in 1950. There, she ran salons, one for Americans who were refugees from McCarthyism in the 1950s, and the other for Indian artists, like S.H. Raza, F.N. Souza and Padamsee. Later, Anand married the Indian classical dancer Shirin Vajifdar, whose sister, Khurshid, was married to Chavda.

While von Leyden contributed articles to *Marg*, Anand's 'at-home-soirees', much like Langhammer's dinner parties and salons provided a forum for aspiring young artists and architects to meet and exchange ideas with Bombay's elites, the most significant of whom belonged to the Parsi community. The Parsis were crucial contributors to the development of Bombay. The most prominent philanthropists in nineteenth century Bombay were Parsis and it has been commented that it was they, rather than the British, who built Bombay.

When Anand and his group were founding *Marg*, he approached Bhabha for financial help. Bhabha introduced Anand to J.R.D. Tata, and the Tata chairman, who was never known to decline a request for help, duly obliged. The Tatas provided *Marg* an office space (two rooms) in the Army and Navy Building opposite which the Jehangir Art Gallery was built seven years later. The publication was financed in the form of guaranteed revenue from seven advertising spaces booked by the Tatas. While the Tatas had a record of philanthropy in technical and scientific education, *Marg* was perhaps their first commitment to education in architecture and the arts. In return for the office space and financing, *Marg* was to be a low-cost, educational, non-profit magazine that could be accessible to a larger group of people. It is said that Nehru wanted each state that was depicted by *Marg* in its issues on various Indian states to buy a thousand copies.

'JUDICIOUS AND LEARNED NATURALISM'

What of Bhabha's own style of painting and sketching? Bhabha, in his travels around Europe, had witnessed first-hand, the works of the masters he had seen in books as a boy which profoundly affected his painting. He abandoned the traditional academic style of painting in favour of a more individualistic, expressive style. In his abstract canvases, he infused inanimate objects with dynamism and motion, inspired by the Baroque and Rococo architecture and sculpture he had encountered in Europe. One of his famous abstract paintings in oil shows the interrelationship between a vase-like form and a cross. He adopted this theme and its development from music.

Michelangelo, Leonardo da Vinci and El Greco seemed to have the greatest influence on him. Indeed the 'precision and swiftness of line', which he attributed to the drawings of da Vinci in an article he wrote for *Marg* for the artist's quincentenary could also be a commentary on his own drawings. The influence of El Greco is seen in the surreal and elongated figures and shadowy background of his *Dove sono I belli momenti*, regarded as his most famous painting. Bhabha executed a large number of figure and portrait studies on paper and favoured graphic mediums over oil paints. His pencil and charcoal sketches were particularly good. Bhabha even sketched during scientific conferences and doodled while attending lectures.

Rudolf von Leyden, in an essay for *Marg* in June 1953, commented on Bhabha's drawings. Bhabha was very involved with the layout of *Marg* and supervised the blocks with his drawings before the article went to press. In analysing Bhabha's ability as an artist, von Leyden remarked that while he would have to be considered an amateur since he was not a professional in the accepted sense of the term, the use of this term for Bhabha was 'peculiarly inappropriate'.

> His work may suffer occasionally from weaknesses which, considering the absence of methodical training and continuous practice are only natural: but it does not suffer from the superficiality and the lack

of direction which characterizes and amateur's work. His drawings in particular show that he penetrates beyond effect and detail to a total integrated appreciation of his subjects and thus creates, not studies or portraits but works of art in their own right.[19]

He went on to observe:

> Dr Bhabha's style is best described by 'judicious and learned naturalism' in which the usual maligned term 'naturalism' is meant to convey an understanding and mastering of nature and not its servile imitation. This naturalism has romantic overtones caused by the artist's personal attitude to his subject. What makes his drawings so appealing is that his personal attitude is one of modesty and humility, a kind of wooing of his subject, a penetrating struggle to understand and reveal. And thus we find the work of this man, whose mind can visualize the working of the Cosmos in the austerity of a mathematical equation, is flooded with a gentle and tender humanity. And that in many ways is gratifying and re-assuring.[20]

When Bhabha perished in the air crash over Mont Blanc, in January 1966, von Leyden said, 'India's art world is orphaned.'[21]

In 1952, to mark the quincentenary of the birth of Leonardo da Vinci, the Royal Asiatic Society invited Bhabha for a talk on da Vinci, the scientist in April. However, Bhabha was travelling at that time, and his speech was read out by Dr George Moraes. In it, Bhabha asserted that da Vinci not only had great powers of observation 'but a hand which could put down, seemingly effortlessly, whatever he wanted.' He observed:

> His best drawings have a precision and swiftness of line and a sensitiveness which was unequalled by any of his great contemporaries, and has probably hardly been equalled in any age. And so, even his botanical and anatomical studies, or his records of the flow of water usually have the aesthetic quality of the finest drawing.[22]

Anand published Bhabha's talk on da Vinci in *Marg*. They argued about the cover of the issue. Anand favoured a Kalighat painting and

Bhabha insisted on a da Vinci drawing, which, according to him, would raise the prestige of India's artists in the world. In the end, the editor bowed to the wishes of the artist–scientist. Karl Khandalavala thought that Bhabha's appreciation of art was not only marked by unusual insight but also the courage to express his views, even when they were contrary to general opinion. When Sher-Gil was neglected, even scoffed at, Bhabha saw her potential and said that she would be hailed as a great painter, a prophecy that was fulfilled after her death. The biracial, bicultural and bisexual artist, Sher-Gil, spearheaded the path of modernity in Indian art by imbuing her work with aspects of both western and eastern traditions, famously declaring: 'Europe belongs to Picasso, Matisse and many others, India belongs only to me.'[23]

Bhabha, who encouraged several artists, was well-aware of the importance of a patron and, in 1934, wrote about Lorenzo de Medici's face at the entrance of the Uffizi Gallery in Florence. Lorenzo de Medici had set the Renaissance in motion, and Bhabha felt that while historians and art critics talked of geniuses in art, they failed to acknowledge the genius that made the flowering of these artists possible. Bhabha, perhaps, saw himself as a de Medici on a smaller scale when he patronized the art of the PAG. *Marg* brought out a fine publication in 1968, with an introduction and an essay on 'Music and Literature in the Life of Homi Bhabha' by Jamshed Bhabha. These essays provide glimpses of Bhabha's life and career along with an insight into Jamshed's own personality through his analysis of his brother.

Dr K.N. Vyas, the chairman of the AEC in 2022, recounted how Bhabha's involvement with the pictorial arts included an interest in photography. He recalled that a cousin from his mother's side by the name of Bharat Upadhyay worked as a photographer at TIFR. When Bhabha saw his camera[24], he wanted a similar one for himself. Bhabha did manage to procure one, and, according to Vyas, took it with him on the ill-fated flight to Geneva.[25]

◆

Bhabha was equally fond of the performing arts and used to frequent dance performances since his Bangalore days. He would discuss music and dance with Vikram Sarabhai and his wife Mrinalini, who was trained in Bharatanatyam, Kathakali and Mohiniyattam. Educated at Shantiniketan under Rabindranath Tagore, she was also a poet, writer and environmentalist. In 1949, she started her dance school Darpana in Ahmedabad. Bhabha made two sketches of Mrinalini, including one when he and Vikram had gone to Mexico to attend a conference there. Mrinalini had asked Bhabha who he would bring for her dance performance, and, to her astonishment, he brought the actress Dolores del Río.[26]

The nuclear scientist Malur Ramasamy Srinivasan recalls that the International Council of Scientific Unions had met in Bombay in January 1966, just two weeks before Bhabha's tragic demise. One evening at the Raj Bhavan, Bhabha asked him if there was a Bharatanatyam programme in town that evening to which he could invite his international guests. At that time, Rukmini Devi Arundale was staying at Srinivasan's home, and her group was to stage a performance at the Kishanchand Chellaram College near Churchgate station.

He told Bhabha that Rukmini Devi would be delighted to welcome him and his guests. During the interval, Bhabha had to make a telephone call, so he walked to the nearby Ritz Hotel with Srinivasan. Bhabha was in a reflective mood and confided in Srinivasan.

> People tell me that I am too fussy and want perfection in everything. I know I can move ahead faster if I relax the standards. The trouble in India is that we have lost our quest of excellence. Look at our dance and music and this evening's programme—here we saw perfection. I want to prove that we are once again capable of achieving that perfection.[27]

Anand also recollected that Bhabha was interested in the nuances, the undertones and the finest points of each technique, be it a music recital or a dance programme. He and Bhabha exhorted a young dancer, after her Bharatnatyam recital, to go to Martha Graham[28] and use the inspirations of the traditional dance form for the development of a more contemporary Indian style.

Chapter 23

AN EMERGING INSTITUTIONAL CULTURE

> The administration of scientific research and development is an even more subtle matter than the administration of industrial enterprises, and I am convinced that it cannot be done on the basis of borrowed knowledge. It must necessarily be done, as in the technologically advanced countries, by scientists and technologists themselves.
>
> —Lecture delivered at the International Council of Scientific Unions, Bombay, 7 January 1966[1]

While science had flourished in the West, certain elements had to be introduced for the development of a scientific culture in India. This was all the more relevant since systematic scientific thinking in the modern sense was not a part of traditional indigenous cultures. Also, colonial modernity had introduced an educational system that emphasized humanistic, administrative and legal studies, which favoured entry into the civil services over scientific work. Even after Independence, the fact that science was a state-funded activity meant that the deeply entrenched bureaucratic culture could not be overcome. Aware of all these shortcomings, Bhabha wanted to create a culture of science that was free of this baggage from the past.

We have seen how Bhabha created an expansive intellectual culture, which included art and music and served to broaden the intellectual base and widen the range of interests of the staff. Scientifically, too, he had created an enabling environment that encouraged experimentation and free discussions, even though, in the early days, the infrastructure was poor and TIFR was engaged in a narrow area of research. Having laid the ground for a new scientific culture, he turned his attention to nurturing a new administrative culture, something in which he was only partially successful.

A NEW CULTURE

A culture consists of shared beliefs and values established by leaders and communicated and reinforced through various methods, ultimately shaping employee perceptions and behaviour. An institution's culture defines the proper way to behave within the institution and sets the context for everything it does. Leaders in successful institutions and organizations live their cultures every day and go out of the way to communicate and reinforce their cultural identities to employees. Bhabha was one such leader. He was trying, as we have seen in some detail, to establish a distinctive institutional culture that included not just science and its accoutrements but also buildings, gardens and even paintings. There are documents in the archives that speak of appropriate clothing and demeanour, rules for cleanliness, conventions of scientific publication and much else. While some are in the form of direct instructions, other conventions that came to define institutional culture can be found in memos, correspondence, notes and minutes.

Bhabha wanted the colonial administrative machinery to be adapted to suit the requirements of scientific work. He, however, did respect the abilities of administrators since he appointed E.C. Allardice, a retired British civil servant who had stayed back in India after 1947, as deputy director of administration at TIFR. It was not uncommon at the time for industrial houses and other institutions to employ retired administrators. J.R.D. Tata had been put through his paces by a retired ICS officer, John Peterson, when he joined the Tatas as a young man in 1932.

The scientists were given 70 days annual leave, like university staff. The policymaking body was called the 'Common Room', a term borrowed from Cambridge. It consisted of senior scientific leaders who could discuss and initiate new programmes, invite foreign scientists, interview new candidates and even review the budget. The library and laboratories were open day and night. Academic staff that preferred to stay in the hostel was provided accommodation, and discussions and lectures were frequent and productive. In fact, the administration was so unique that many wondered how Bhabha got away with doing whatever he wanted when he should have followed central government rules. The recruitment policy was based on long interviews to enable the authorities to judge the applicants' potential.

The assessment and promotion of scientific staff was completely revolutionized at TIFR. To attract the best talent, the starting grade of scientists was made equal to that of Indian Administrative Service officers. A confidential reporting system was introduced, which described very clearly the work of the candidate. The nature of assessments involved more than one assessor so that a proper record of the scientists' careers was maintained. A periodic review of the confidential reports decided promotions irrespective of seniority. Bhabha introduced a fast-track scale so that the candidates who did outstanding work could move up faster and take up leadership positions. At the interviews for promotion, outside experts were invited to ensure that no biases were involved.

Allardice didn't move his office to the new buildings and, in the words of Rangaswamy Narasimhan, was not very helpful or forthcoming. N.R. Puthran was Allardice's second-in-command and ought to have been in constant touch with Bhabha, but the gulf separating them was so large that it was difficult to imagine Puthran having the courage to approach Bhabha directly.

There was also the politics between the different personalities to consider. There were differences of opinion between Bhabha and K. Chandrasekharan, and between Chandrasekharan and M.G.K. Menon. Bhabha tried to get around these problems by recruiting more and more people, but the crucial need to train administrators who understood science and its needs did not get the attention it deserved.

Bhabha seems to have been aware of these administrative shortcomings. In an interview with *International Science and Technology* in 1963, he said that getting the right administrative setup was the most serious problem he faced. He said that the administration and administrators were not ready to administer laboratories, plants and factories in the manner required. The TIFR way of doing things was in sharp contrast with the rigidity of the bureaucratized university system. As we have seen, the focus was on nurturing individual talent and no organizational chart of the future development of the institute was submitted when it was founded, or even later.

The growth of the institute had been based entirely on individuals who entered either as students or teachers. The TIFR was modelled on western institutions and it remained western in its culture. While Bhabha managed to create an atmosphere conducive to research, he was less successful in creating the administrative structure essential to the sustenance of a scientific culture. Regardless of his dislike for a rigid hierarchical structure, Bhabha could not prevent bureaucracy from prevailing to some extent, even when he was alive. After his death, things deteriorated further.

NEW MORES

The new buildings and premises of TIFR demanded modern behaviour from the staff. They were asked to give up their old habits and adopt new ones, especially when it came to preserving the quality of the buildings and their cleanliness. There were very obvious differences in the class and culture of the westernized Bhabha and his colleagues and students. Bhabha, who generally remained aloof, realized this and tried his best to ensure that the transition to the kind of behaviour and comportment he desired was a smooth one.

Physicist Ramanath Cowsik, who began his career at the Atomic Energy Training School, spoke about Bhabha's tact in this regard. He narrated an incident in which Bhabha gently corrected one of Cowsik's friends who had not tucked his shirt into his trousers. After asking him what he did at the institute, Bhabha politely told him that he should tuck his shirt in since it was the norm, and that it would be nice if everyone adhered to the discipline and decorum of the institute. Bhabha did this without making the student feel bad about it and, more importantly, had the sensitivity to do it out of earshot of his friends.[2] Puthran quoted Bhabha as saying that the answer to a shirt was a tie. He added that the only time Bhabha saw K.G. Ramanathan without a tie, the latter immediately offered an explanation for it.

Bhabha was always perfectly turned out in his trademark double-breasted suits and thought nothing of changing his attire a few times in the day. Cowsik remembers how, on Wednesday, a day Bhabha would spend entirely at the institute, he could be seen wearing a bush-coat with pockets early in the morning. Later, he would appear for the 4.00 p.m. colloquium wearing a day suit—dark brown or grey. And if one stayed on at the campus late into the evening and night, one could perhaps catch a freshly shaven Bhabha wearing a dark evening suit.

Giving Bhabha good competition in sartorial elegance was Chandrasekharan. The mathematician M.S. Raghunathan has an interesting anecdote to narrate in this regard. Raghunathan and the mathematician, Raghavan Narasimhan, were visiting members at the IAS in Princeton in 1966–1967. On one occasion, the two of them went to a very expensive clothing shop in Palmer Square. Raghavan Narasimhan ordered a suit for himself while Raghunathan bought a scarf. The shopkeeper struck up a conversation with them and, at one point, asked if they knew John von Neumann. When they said that they had only heard of him, he asked them with a typical American nasal twang if they knew 'Chandrasekharaaan'. When they answered in the affirmative, he said: 'They were the only gentlemen from that Institute who knew how to dress!'[3] John von Neumann wore a three-piece suit at almost all times, including, it is said, on a donkey ride down the Grand Canyon. Even as a student, his clothes stood out; he was so well dressed that after meeting him for the first time, the mathematician, David Hilbert, could ask only one question: 'Who is his tailor'?[4]

Chandrasekharan dressed formally—he would always turn up in a suit at TIFR, even on the hottest summer days. One could not assume that he was always available, and colleagues who wanted to meet him had to take an appointment. He always spoke in English in a deliberate measured accent, which was neither British nor American but not Indian either. Like Bhabha, his personality did not allow for intimacy, and many students at TIFR were inclined to believe that no one outside his immediate family ever got really close to him. As Raghunathan observed: 'The way he dressed and the way he conducted himself at the rare official parties suggested to us that he was a stickler for (western) etiquette who could give a lesson or two to Emily Post!'[5]

ARROGANT OR JUST RESERVED?

Bhabha was an intimidating personality for most, and many saw him as haughty and arrogant. 'Homi was a distant Alp,' observed George Greenstein, professor of astronomy at Amherst College. 'One did not presume to climb it.'[6] While there was always an element of appreciation and respect whenever people spoke of him, Greenstein did not encounter friendship. Bhabha had no personal friends at TIFR and, throughout his career, there was always an element of social distance between him and the people he worked with. 'He was not one to hang around corridors chatting—most people there hardly ever saw him. An air of fastidiousness about him, a formality, effectively forbade casual contact.'[7] Pabitra Kumar Maitra, who joined the molecular biology group in 1963, said that he was afraid of Bhabha because he was not a man he identified with in any way.

S. Chandrasekhar had another story to tell, which he said was narrated to him by the geophysicist, Chaim Perkeris, when he visited India. When Bhabha and Perkeris had gone to attend a public lecture, presumably on the TIFR campus, they found that not a single seat was vacant. Bhabha waved his hand and everyone in the front row got up and just disappeared.[8] Though this anecdote has not been corroborated by other sources and could well be false like the Pauli episode, Bhabha does seem to have been the sort of person about whom such tales accumulate.

His manner towards his subordinates was often described as disdainful, and he took great delight in behaving imperiously with westerners. As a senior American academician observed, 'He liked to make the white man wait.'[9] That may have been true, but the fact was that Bhabha made everyone wait. Punctuality was not his strong suit. Visitors, even those who had appointments, would wait for hours to meet him. It was, as Chandrasekhar put it, a 'common disease', especially among the powerful, but one which Bhabha seems to have been afflicted with right from the beginning, and in a virulent form.

When Raja Ramanna returned from England in 1949, he and one of his colleagues from King's College, London, R.R. Roy, had both been offered positions at TIFR. R.R. Roy had a higher opinion of himself than his abilities justified, and was critical of anything Indian. He also thought that Bhabha was not giving him the treatment he deserved. They had both been called for an interview at 11.00 a.m., but when the duo reached there on time, there was no sign of Bhabha. The 'great man', as Ramanna put it, arrived eight hours later, at 7.00 p.m. Things did not go well because R.R. Roy stormed out of the room in an agitated state and took a ship back to Europe the very next day.[10] Bhabha's office tried to contact him but to no avail.

When he went to the airport, Bhabha always took a retinue of aides, associates and secretaries with him, and he would give last minute instructions to them before boarding the plane. Not one to do things by halves, it is said that once, while dining overseas, he brought along an aide whose sole function was to pay the bill. 'It would not have been seemly for a gentleman to be seen dipping his hand into his pocket, and above all else Bhabha was a gentleman.'[11] In this, however, he was not alone. According to Mahalanobis's biographer, Ashok Rudra, Mahalanobis never carried any cash with him. It was his personal assistants who spent money even for his personal needs and who maintained accounts of the expenditure.[12]

Bhabha was also conscious of his own importance. He once told Sir Maurice Wilkes that, apart from scientists, he hardly had any time for meeting people from other countries, unless they were heads of state, 'although he would stretch a point for the Prime Ministers of the more important countries'. Wilkes observed:

> I said that I hoped that he included the United Kingdom among the latter. He replied with one of his engaging smiles, that he certainly did, adding that it would be a poor day if this were

ever no longer the case. I make no comment on the exchange of banter.[13]

On the flipside, there are several anecdotes about Bhabha's kindness and consideration. Some thought that while he appeared to be an extrovert, he was, in reality, a reserved person, and this led people to think that he was haughty. One of those who subscribed to this view was Vasudev K. Iya who narrated how, in October 1960, when Bhabha hosted a reception at the house of the Indian ambassador in Paris, he asked Iya to bring his wife, in case she was still with him. Iya commented:

> Bhabha was a shy and sensitive person, and endowed with a sharp sense of humour. Being reserved by nature, he had a tendency to remain aloof, and this was sometime mistaken for arrogance [...] He was in fact acutely sensitive to the sufferings of others, of even the very poor with whom there was no social contact...[14]

Devendra Lal has his own story to tell about Bhabha's kindness. Once, he had an appointment with a doctor to remove a cyst. Bhabha sent him a message asking him to cancel the appointment and go to the Tata Memorial Hospital instead, where he had arranged a meeting with a doctor. Lal says that he was surprised that Bhabha knew about his doctor's appointment and wondered why the latter cared so much about what he was doing.[15]

Similarly, in 1960, when H.L.N. Murthy, the glassblower, needed urgent medical attention due to a congenital heart complaint, Bhabha sent him to the Boston General Hospital, where he was operated upon free of charge. Bhabha also lined up 18 blood donors for Murthy's open-heart surgery. After Murthy had recovered, Bhabha sent him to the Corning Glass Works and to Philips in Eindhoven. Murthy worked for 13 more years at TIFR. Bhabha also shifted Anand Panchal, a TIFR carpenter, from a public hospital to a private nursing home with a full-time nurse in attendance. Bhabha said that he was not a carpenter but an artist. Unfortunately, Panchal did not survive. When Dr Das, who was attending an international conference at Bonn, was diagnosed with galloping leukaemia, Bhabha ensured that his wife was rushed to his bedside in his last days (he died within ten days).[16] As R.M. Lala noted, beyond the culture of science was a culture of care, where people mattered as much as projects.[17]

In another anecdote, S. Ramanan, who joined TIFR as a mathematics researcher in 1956, recounted a nice gesture by Bhabha when he was struggling to use cutlery. There was a tradition that new students had to treat seniors, including the faculty. Ramanan took them all out for tea to a restaurant called Rendezvous, where the new Taj Intercontinental now stands. Bhabha and Chandrasekharan were also invited. The problem arose when Ramanan tried to cut his samosa with a fork and much to his embarrassment found the pieces flying in all directions. Bhabha who was sitting next to him realized what was happening and picked up his own samosa with his hand and bit into it. Ramanan was both pleased and relieved.

Bhabha made it a point to show his appreciation when any of the lower-level staff did good work. Patrick 'Shiner' Braganza, who joined the institute in 1965, polished the brass panels of the elevators. He was thrilled when Bhabha praised him with the words, 'Good work, Patrick.'[18] As Gustav Born, Max Born's son, recalled about Bhabha: 'I remember his round open face, his gracious manner; his (left?) shoulder tic. I remember his quick responses; openness; bright comments; personal tolerances and friendliness up and down the Indian social divisions. He was much too big a person to invalidate the servants who adored him.'[19]

The key to understanding Bhabha's behaviour can perhaps be found in a letter he wrote to his friend, Homi Seervai, in March 1928. Bhabha was aware that he could be moody. 'By nature, I am very sensitive to outside influence, but by reasoning and philosophy (?) I have trained myself to be, as far as possible, independent of outside circumstances [...] Being very subject to mood, my being influenced by outside circumstances depends upon which mood predominates, the philosophic or otherwise.'[20] In the same letter he wrote that he preferred his 'own company, and nature'.

THE INDIAN–WESTERN DICHOTOMY

The members of TIFR were expected to conform to western mores and styles. We have already seen how Bhabha disapproved of shirts that were not tucked into trousers. Similarly, saucers were meant for holding the cup, not for drinking tea. While Bhabha showed an appreciation of Indian cultural heritage by encouraging Indian artists and investing in their art, he showed little tolerance or understanding of local value systems and regional cultural usages and forms. He ordered the staff to wear ties when the Soviet leaders Nikita Khrushchev and Nikolai Bulganin visited TIFR in 1956.

Not all fell in line with such sartorial diktats though. The most notable nonconformist was Yash Pal, who steadfastly stuck to his pyjama–kurta and Nehru jacket. He did so even at MIT and other international conferences, including the international conference on elementary particles in Bombay in 1950 and the Rochester conference in 1954, where he presented the work of the TIFR group. Bhabha was present at the meeting and expressed his disapproval at the way he was dressed. Yash Pal was disappointed that Bhabha thought it important to remark on his clothes rather than the paper he had presented. In an interview in August 2008, he told Indira Chowdhury, 'Was that the only thing he could say?'[21]

Greenstein believes that Bhabha's wealth and aristocratic background set him apart and goes as far as saying that in some fundamental sense he did not understand the life of the ordinary Indian. His outlook was international and most people he regarded as his equals lived in the West.[22] If Bhabha was cosmopolitan in his approach, he was also, in a sense, limitedly metropolitan in his outlook.

When dignitaries made formal visits to the institute, Bhabha circulated instructions about the expected dress code. Thus, in April 1964, before Admiral Karmarkar's visit, he issued instructions about the way he expected his staff to dress:

> It is desirable that in these Faculty lunches a certain minimum formality should be observed in the dress, even when no guests are expected, as there may be occasion to bring in a guest unexpectedly. It seems to be that an appropriate dress for the occasion would be lounge suit, or bundgalla, or bush coat and trousers. If a shirt is worn, so should a tie invariably, even if the coat has been dispensed with on less formal occasions. On more formal occasions the dress should invariably be lounge suit or bundgalla.[23]

The institute's buildings corresponded to international standards, and Bhabha expected each member of the staff to take personal pride in them and ensure their maintenance. In February 1961, even before the buildings were inaugurated, he observed:

> These buildings are as fine as any corresponding buildings to be found anywhere in the world [...] The buildings have been paid for by generous grants from the Sir Dorabji Tata Trust and the Government of Bombay, and above all very substantial grants made by the Government of India. These buildings, have therefore, been made available to the staff of the Institute by the people of India through the taxes they have paid.
>
> Each member of the staff should have a sense of personal pride in these buildings, which have been given for his use, and it is his duty to take personal interest in their proper maintenance and to see that he himself uses them in such a way as to maintain their quality and cleanliness so as not to cause inconvenience to others. Certain normal and elementary good manners with regard to the use of the buildings and fixtures and furniture in them must be observed. A member who sees another not using the buildings properly should draw his attention to the proper conduct in such matters. If any person continues to misbehave, the matter should be reported to his superior.[24]

In the same office order, he issued orders that are still as relevant, nearly six decades later. 'Feet should be wiped on a mat before entering the building, and marks from dirty hands should not be left on the walls. Lavatories should be used properly and kept spotlessly clean.'[25]

Apart from this, it is said that Bhabha would often instruct people to not use the WC like an Indian-style toilet. He would keep a check on whether the WC was being used correctly, with footprints on the seat being reported. The toilets were equipped only with toilet paper—using water was not an option. Four years later, in 1965, he expressed his unhappiness with the way the facilities were being used and said that 'those who continue to behave in an unhygienic and unsocial manner will have to leave the Institute'.[26] These are, as Greenstein observes, not memos to grown men and women but memos from a father to his children.[27]

The Indian–western dichotomy was also reflected in the two canteens at TIFR. The west canteen served western food and the use of forks and knives was compulsory. It also attempted 'a variety of hybrid culinary experiments', with locally available vegetables. The east canteen, on the other hand, served Indian food from the beginning, and people were free to use their hands. While Bhabha's diktat normally prevailed, it did, on occasion, provoke a sharp response. Chandrasekharan's response to the institute's decision to stop free tea before a colloquium was one such instance. Chandrasekharan said that while he didn't care for tea, he did care for academic tradition. Bhabha cited the case of the Cavendish, where tea was paid for, but it cut no ice with his deputy director. Chandrasekharan's sharp response is an indication that he was, perhaps, one of the very few people who had no hesitation in speaking his mind freely to Bhabha. He said that Rutherford had been dead for many years, and the description of the Cambridge tradition given by Bhabha was rather incomplete, since even food was free for Fellows at Cambridge. He said that they were not a university, and the comparison with Cambridge did not seem quite right. The IAS at Princeton, and several similar Institutes in the US served tea in their common rooms every day of the week free of charge. This helped to build an atmosphere, which, he said, even Dirac would recognize. If this was impossible to secure, in a building of their size, he would give it all up.[28]

The mathematician R. Sridharan feels that Chandrasekharan may have left TIFR because of differences of opinion with Bhabha. When the furniture in Sridharan's cabin was being changed, his old table was replaced with a bigger one. While transferring some of his old correspondence and other papers to the empty drawers of the new table, he found two small pieces of paper with some notes. They turned out to be parts of an exchange between Chandrasekharan and Bhabha, which must have been left there when Chandrasekharan's table was being cleared. One of them was a note to Bhabha, signed by Chandrasekharan, wherein some faculty members had complained about the TIFR canteen being shut on a certain afternoon to accommodate a meeting of the DAE. Bhabha told Chandrasekharan that he should bring it to the notice of the faculty that TIFR was funded by the DAE, and, therefore, such a complaint was quite pointless. Chandrasekharan seems to have started a retort but gave up halfway.[29]

If Chandrasekhar is to be believed, there was a smouldering resentment against Bhabha even before the new buildings were inaugurated. He told Wali about his experience when he visited TIFR in 1961. He was disappointed by the fact that almost everyone, from senior colleagues to students, were highly critical of Bhabha. Chandrasekhar wanted them to be more tolerant and understanding of Bhabha. 'Why don't you be a little tolerant? he asked them. 'After all he has built this institute. And so far as I know, this is the only institute where admissions are made on the basis of merit without the considerations of region or caste or creed. If he has certain weaknesses, why don't you overlook them?'[30] Physicist Basanti Dulal Nagchaudhuri echoed a similar view when he said that Bhabha was 'singularly free from any sectarian bias or traditional inhibitions.'[31] Cecil Powell, perhaps, put it best when he observed, 'Without a certain ruthless arrogance he would never have achieved what he did, and he did much.'[32]

The new social mores promoted by Bhabha at the institute can be best viewed as a kind of training to prepare its members for participation in global scientific networks. Bhabha wanted his scientists to be adequate not only in their work but also in the social and cultural side of international interaction. As Indira Chowdhury observes, 'The laboratory thus encapsulated its own double—the invisible cultural workshop which performed an equally significant pedagogic function'.[33]

However, at times, Bhabha's approach to social problems took little account of social realities. He had infinite faith in the power of science as an instrument of social change, and the problem of national development was for him solely a problem of the development of modern technology. His approach to India's population is indicative of this approach. In February 1959, at an address before the Sixth International Conference on Planned Parenthood, Bhabha proposed a solution to India's population problem. He called for a research programme aimed at developing a substance that, when mixed with rice, grain, or salt, would reduce fertility by about 30 per cent.[34] It is difficult to believe that Bhabha would not have been aware of the opposition, if not outrage, his proposal would arouse. Possibly, he viewed the problem with clinical objectivity, merely a stumbling block in the way of his programme's implementation. He ignored the fact that children, at least at that time, and to a large extent even today, were the only form of social security in rural India: sons who would care for them in their old age were all that stood between ageing villagers and starvation.

Chapter 24

THE OUTLIERS

> No organisational chart of the future development of the Institute was submitted either when it was founded or later, and the philosophy has always been to support ability whenever it was found in the fields of work covered by the Institute.
>
> —Homi Bhabha's Lecture on 'Science and the Problems of Development'[1]

Bhabha did not care much for conventional jurisdictional authority or territorial boundaries within the domain of Indian science. Like all men of stature, Bhabha's influence transcended his own field of specialization. He used the DAE's umbrella to support pure and applied research, not limiting it to strictly atomic-energy-related development. His strategy was twofold: first, if there was an able researcher capable of building a group, he would support him to the hilt, giving him the necessary independence to function effectively. Second, he would take any institution that he felt was languishing without sufficient support under the umbrella of the DAE. The initiation of molecular biology and radio astronomy at TIFR are examples of the first kind of support. His support to Vikram Sarabhai's space programme and the taking over of the Tata Memorial Hospital are examples of the latter.

The great scientific advance of the first half of the twentieth century was, undoubtedly, nuclear physics. Relativity and quantum theory began unlocking the secrets of the atom, and nuclear physics became the cutting edge of human knowledge. However, the discovery of the double helix structure of the DNA molecule by Francis Crick and James Watson in March 1953 created the entirely new scientific field of molecular biology, which began unlocking the secrets of life itself. Crick and Watson's discovery of the ultimate building blocks of life allowed scientists to understand how the making of life forms was regulated. Molecular biology became the nuclear physics of the second half of the twentieth century.

The movement towards the unification of modern biology began in the early 1940s. The realization of the centrality of the nature of the gene to all of biology was at the heart of this movement. Microbial genetics, in particular, the genetics of enteric bacteria and the viruses called bacteriophage, became the vehicle of a great intellectual revolution that led to the confluence of genetics and biochemistry.

The mid-twentieth century revolution in biology traces its roots to the Cavendish, presided over by William Lawrence Bragg, one of the founders of crystallography, who shared the Nobel Prize for physics in 1915 with his father, William Henry Bragg. He was only 25 at the time, and remains the youngest ever Nobel laureate in physics to this day. The origins of molecular biology can be traced to the pioneering work of Linus Pauling and Max Delbrück at the California Institute of Technology. Pauling was an early practitioner of quantum mechanics and crystallography. Delbrück began his career with Niels Bohr; Francis Crick, too, started out as a physicist. The drift into biology coincided with a feeling that physics, as perceived by the romantics of the 1930s, was reaching an end: the future, it was thought, lay in biology, which might be able to uncover new laws of nature.

OBAID SIDDIQI AND MOLECULAR BIOLOGY

When A.V. Hill visited India in 1944, he advised Bhabha to include 'biophysics' in his institute since, apart from the Bose Institute in

Calcutta, the subject seemed non-existent in India. A biologist himself, Hill thought that many of the most important future applications of physics would be in biology. Bhabha internalized that advice but waited for the arrival of the right man to finally take the plunge. That man was Obaid Siddiqi, who joined TIFR on Bhabha's invitation in 1962. The physicist Leo Szilard, who had turned to biology in 1947 after significant contributions to physics and who knew Bhabha from his Cambridge days, recommended Siddiqi to Bhabha. Szilard had joined Delbrück's phage biology course at Cold Spring Harbor Laboratory, New York, in 1947, and 14 years later, in 1961, met Siddiqi during a course there.

Siddiqi had been introduced to plant genetics and embryology at AMU, his alma mater. From 1954, he taught there for three years before joining as a research scholar at the Indian Agricultural Research Institute, New Delhi, which focussed on applied research in genetics to increase agricultural production. It is said that a hailstorm destroyed his wheat crop. So, rather than waiting for the next crop season, Siddiqi wrote to Guido Pontecorvo, at the University of Glasgow to accept him as a PhD student. Pontecorvo was a British–Italian geneticist who discovered the process of genetic recombination in the fungus *Aspergillus*. He also studied the evolutionary divergence in the common fruit fly *Drosophilia*. The conviction that research in microbial genetics could lead to increased production of the drug penicillin—much needed in the Second World War—led him in 1943 to the genetics of fungi.

Pontecorvo asked Siddiqi to come for an interview. When he reached Glasgow in 1958, he was taken directly to the laboratory. Puzzled, he asked Pontecorvo when the interview would be held, only to be told that he had passed the test by merely showing up in Glasgow. The senior geneticist wanted to gauge Siddiqi's interest, and the very fact that he had come all the way from India despite the risk of rejection had been enough for him. Siddiqi completed his PhD under Pontecorvo and, in 1961, moved to Alan Garen's laboratory at the University of Pennsylvania for his postdoctoral work. Siddiqi had lost his job at AMU because the university authorities had refused to extend his leave, just as he was completing his thesis in Glasgow. While he was at Aligarh, he had tried, unsuccessfully, to introduce modern genetics at the postgraduate level a—particular case of a general resistance to change within university departments characterized by a refusal to restructure curricula.

While in Philadelphia, Siddiqi was introduced to Maurice Fox, who was, at the time, an associate professor at the Rockefeller University. The two soon became fast friends. Siddiqi told Fox that he wanted to go back to India to introduce the new discipline of molecular biology there but was worried about the way that would be received, visualizing resistance from establishment biologists. Fox suggested that he return to India and join a physics department. He said that physicists would appreciate his work and would also not interfere with it. Fox, who knew Szilard well, seemed to remember that the latter had been in touch with Indian physicists during his time in Cambridge (the obvious reference here is to Bhabha). Fox asked Szilard to help. Siddiqi met Szilard, and the latter wrote to Bhabha on 3 March 1962 after talking to Garen.[2] He enclosed two letters of recommendation, one from Pontecorvo and the other from Garen, asking Bhabha to follow up with whatever action he thought appropriate.[3]

Bhabha wrote to Siddiqi on 12 March 1962, expressing support for research in molecular biology and genetics and promising an appropriate offer, either at TIFR or at the biology division at AEET. He also wanted Siddiqi to send his curriculum vitae.[4] It is interesting to note that the letters of recommendation came in first, and all Bhabha wanted was Siddiqi's biodata for the record. Bhabha trusted Szilard's judgement, who in turn, trusted the judgements of Garen and Pontecorvo.

Pontecorvo, in his letter to Bhabha about Siddiqi, made a very cogent observation about Indian science administration. He thought it very important for the progress of biology in India that Siddiqi should go back to a job in which his abilities would be fully expressed. In fact, he said he was really baffled as to why India seemed to promote mediocre scientist–politicians and did nothing to retain their really good scientists. Pontecorvo's perspicacious comment struck a chord with Bhabha, who wrote to Nehru about it while informing him about Siddiqi's appointment.[5] He said that Pontecorvo's impression was not entirely ill-founded; they had not yet got over the inheritance of the pre-Independence days in which whoever spent his time on administration was considered to hold a more important position than someone who was working in a laboratory, although the latter was perhaps of greater significance from a global point of view.[6]

Siddiqi wrote to Bhabha from Philadelphia, expressing his desire to work in a physics institute, displaying, perhaps with good reason, a lack of confidence with traditional biological institutions. Part of this optimism about physics institutes stemmed from the fact that there was a precedent for a molecular biology laboratory at the Cavendish.

Bhabha presented Siddiqi's application to the TIFR faculty, asking for their opinion. The physicists at the institute were not particularly keen on welcoming the newcomer. Apart from raising questions about logistics, they also demurred on a more fundamental level, questioning the compatibility of the new discipline with the ongoing agenda of the institute. B.V. Sreekantan and Raghavan Narasimhan were in favour, but S.S. Dharmatti, B.V. Thosar and Devendra Lal opposed it. D.Y. Phadke gave it lukewarm support. M.G.K. Menon, who had already been taken into confidence by Bhabha, said nothing. It would do well to remember that in the 1950s, and even later, physics was the pre-eminent area of inquiry in science, the discipline at the top of the intellectual pyramid. Physicists looked down on chemists who turned their noses up at biologists.

J.V. Kotwal, who was the secretary of the physics faculty, recalled Bhabha's response at the meeting. After hearing what the others had to say, Bhabha ended all debate when he said, 'Gentlemen, I have decided to take Siddiqi and start research in molecular biology.'[7] Thanks to Bhabha's unflinching support, Siddiqi was made an offer, which he accepted and joined TIFR in 1962.

This decision was in consonance with Bhabha's philosophy of supporting ability wherever it was found. In a speech in early January 1966, he observed:

> Indeed, the philosophy underlying the founding of the institute was the same as that underlying the Max Planck Institute in Germany: The Kaiser Wilhelm Society shall not first build an institute for research and then seek out the suitable man but shall

first pick up an outstanding man, and then build an institute for him.[8]

It may not have been the most democratic way of deciding the matter, but Bhabha's judgement and vision in such matters could never be faulted. In any case, there was no doubt in anyone's mind as to whose institute it was—it was Bhabha's, and he did not hesitate to run it the way he wanted. His decision was to have far reaching consequences for biology in India.

The TIFR faculty, faced with a fait accompli, suggested that Siddiqi work in the biology division of AEET and spend as much time as he liked at the institute. The lukewarm response is understandable given that molecular biology was not a subject that had captured the imagination of the scientists nor were they fully aware of its potential. In any case, it was a discipline perceived to be outside physics, and it redounds to Bhabha's great credit that he saw the potential of the relatively new field at the time he did.

Siddiqi, when he joined TIFR, faced the inevitable teething problems since the mechanisms for supporting interdisciplinary research were not yet in place. The institute had been built to study physics, and there were practical difficulties to be overcome. They had to do with equipment and facilities. Initially, Siddiqi and his student, B.N. Apte, worked in the AEET laboratories at Byculla. Things improved gradually, largely due to Bhabha's control over budgetary allocations and the power to reappropriate funds into a discipline that had not found a place in TIFR's original plans. The ability to define its own research priorities and control budgetary allocations was a privilege the institute had under Bhabha. In 1966, it had a budget of ₹1.5 crore, a large sum in those days. However, foreign exchange continued to be a major constraint.

Soon after Siddiqi joined TIFR, India was at war with China. Bhabha announced that only those new projects that were absolutely essential in the national interest would be funded. When it was Siddiqi's turn to state how much money he needed, Bhabha said, 'I know you cannot say what you are doing is indispensable. Nevertheless, how much do you need?'[9] When Siddiqi named his figure, Bhabha asked N.R. Puthran if it was possible to meet Siddiqi's demand. Puthran thought that the savings from electricity and water would suffice. Siddiqi observed many years later, 'I find it both amusing and reassuring that under the difficult circumstances of an emergency, molecular biology at TIFR could be started with savings under electricity and water.'[10]

Foreign exchange was also a perpetual problem. Aware of this stringency, Bhabha told Siddiqi about the visit of Sir Michael Perrin, the creator of polythene and chairman of the Wellcome Trust[11], to Bombay. Bhabha saw this as an opportunity for Siddiqi to apply for a grant, and E.C. Allardice facilitated a presentation. Siddiqi's vison for molecular biology was presented as an article in *Nuclear India*. The grant of £40,000, which Siddiqi got from the Wellcome Trust, ensured the survival of the molecular biology group.[12]

Bhabha was very proud of the TIFR buildings and he was a stickler for aesthetics when it came to alterations in buildings. The buildings at TIFR had no provision for wet labs, which require sinks, drains and fume hoods with ducts. Since Bhabha was reluctant to allow ducts to protrude from glass windows, he decided to construct new laboratories adding two floors to the 'B' block. He made several changes in Siddiqi's detailed floor plan for the wet labs till he was satisfied. Bhabha also paid the same attention to the furniture. Later, finding the walls of Siddiqi's room bare, he offered him paintings from the TIFR strong room.[13] Siddiqi recalled that two such paintings adorned his room for four decades.

At TIFR, Siddiqi established a small but strong bacterial genetics group. Their work on delinking DNA transfer and replication and recombination in bacteria was widely recognized. In 1963, P.K. Maitra, a PhD in biochemistry from Calcutta who had done postdoctoral work at the Johnson Foundation in Philadelphia, joined Siddiqi. Working under Siddiqi, Maitra introduced yeast genetics to the institute. Maitra and his wife, Zita Lobo, became leaders in the genetics of sugar metabolism. A combination of institutional vision and collegial concern allowed Siddiqi to recruit a number of brilliant scientists to TIFR. By 1968, the group grew to include seven members, some of whom left within two years of joining.

Siddiqi organized several training schools at the institute, where faculty members taught alongside international scientists. These courses, which started in 1967 after Bhabha had passed away, marked a coming of age of molecular biology at the institute. The first winter school was also organized that year. As the group grew, its international networks were further strengthened in the 1970s with visits by highly distinguished international scientists. The famous Mahabaleshwar Seminars in Modern Biology, which trained over 800 scientists in the first 25 years of TIFR, were also started during this time. Thus, all members of the molecular group found themselves included in an 'exclusive freemasonry of outstanding scientists'[14] with the TIFR molecular biology scientists gaining access to the brightest minds in the field. Just as the physicists had found with Bhabha, the molecular biologists discovered that their association with Siddiqi opened many doors and greatly expanded their scientific worlds.

By the early 1970s, Siddiqi's focus had shifted to neurobiology. Influenced by Seymour Benzer in the 1970s, he changed his scientific orientation to using genetics to understand the nervous system. Siddiqi studied temperature-sensitive paralytic mutants in the genus of fruit flies, called *Drosophila*. This was pioneering work that changed our understanding of how nerve signals are generated and transmitted. Siddiqi and his students and collaborators focussed increasingly on *Drosophila*, with R.N. Singh and K. VijayRaghavan working on this subject. In 1976, Siddiqi, with his student Veronica Rodrigues, used the neurogenetic approach to understand the processes of smell and taste perception in *Drosophila*. A breakthrough came in 1978 when they isolated genetic mutants with defects in smell or taste. While their work led to an improved understanding of how olfactory information is encoded in the brain of the fly, their study of chemosensory genetics inspired others to address this challenging field.

By the 1980s, this once peripheral field was not only an established discipline but the molecular biology group, which had received a hesitant welcome, was also at the vanguard of the latest research in the field. As Venkatraman Ramakrishnan, the Nobel laureate from Cambridge put it: 'It is not so difficult to do first-rate work at the LMB [Laboratory of Molecular Biology in Cambridge], but to do it in India while building up institutions with a good tradition is truly remarkable.'[15] Bhabha knew this all too well. He had written to

B.M. Udgaonkar in 1963, conveying a similar sentiment. 'For every thousand scientists who can do reasonably good work in a good scientific atmosphere, there is only one who can create the atmosphere for himself in a place where it does not exist, and this alone is a test of the outstanding scientist.'[16]

The National Centre for Biological Sciences (NCBS) came into being under the TIFR umbrella in 1991, in Bangalore. Siddiqi, taking a cue from Bhabha's insistence on aesthetics, formed a committee with molecular biologist, Jayant B. Udgaonkar, B.M. Udgaonkar's son, as its head. J.B. Udgaonkar was also put in charge of overseeing the construction of the new buildings and, once, wrote a strong letter of complaint to the AEC officials about the poor quality of construction.[17] J.B. Udgaonkar, whose research interests include protein folding and protein aggregation, is presently the director of the Indian Institute of Science Education and Research in Pune.

RADIO ASTRONOMY

At about the same time that Bhabha invited Siddiqi to work on molecular biology, he also invited a group of young radio astronomers to join TIFR. Like molecular biology, radio astronomy, too, had not been planned as a discipline at TIFR, and the simultaneous beginnings of two not very congruent subjects at the institute attracted comment in scientific circles. But radio astronomy faced lesser resistance from the scientists at TIFR since they considered it a part of physics. Moreover, the physicists at TIFR had, over a period of time, achieved a certain level of technological expertise that gave Bhabha the confidence to begin such a field. Many prominent astronomers and physicists played direct and indirect roles in the establishment of radio astronomy at TIFR in 1963. Apart from Bhabha and Joseph L. Pawsey, who were both from Cambridge, other scientists contributed to it over a period of several decades, including, Wilbur Norman Christiansen, Ronald N. Bracewell, K.S. Krishnan, Govind Swarup, T.K. Menon, Mukul R. Kundu, T. Krishnan, Mark Oliphant and Bart Bok.

Pawsey, who had done his MSc in physics from the University of Melbourne in 1931, completed his PhD from the Cavendish on the properties of the 'E' layer of the ionosphere. He worked on early television problems from 1934 to 1939 before leaving for Australia following the outbreak of the Second World War. He began work at the laboratories of the Commonwealth Scientific and Industrial Research Organisation (CSIRO) in February 1940, working on radar projects until 1945. Pawsey's main achievement was establishing and leading the Australian radio astronomy research group in October 1945. During the war and immediately afterwards, he recruited and mentored many of the pioneers of early radio astronomy, including Bernard Mills, Bracewell, Paul Wild, Frank Kerr, Christiansen, Ruby Payne-Scott and John Bolton.

In 1952, the International Union of Radio Science held its 10th general assembly at the University of Sydney. The congress was a vehicle to showcase the remarkable achievements of radio science, particularly radio astronomy in Australia seven years after the war. The three Indian delegates at the congress were K.S. Krishnan, director of NPL, S.K. Mitra of the Calcutta University and A.P. Mitra, who went on to become director of NPL and the director general of CSIR.

Impressed by what he saw in Australia, Krishnan, on his return, described the developments at a colloquium at NPL. He described the remarkable discoveries being made in the field by Pawsey's team with the radio telescopes that they had constructed to investigate radio emission from the sun and distant cosmic sources. Govind Swarup, a student of Krishnan's from Allahabad University, was greatly inspired by his account. Swarup had completed his MSc degree from Allahabad University in 1950 and was working on problems in paramagnetic resonance with Krishnan at the NPL. Swarup went to the library and after reading the scientific papers written by the Australians, was entranced by this new field. Swarup had met Bhabha during the latter's visit to the NPL in New Delhi in 1951 and recalled Bhabha's drive and attention to detail, which he called awesome and infectious.

Krishnan, who was interested in starting research on radio astronomy at NPL, recommended Swarup's name for a two-year fellowship to Sydney under the Colombo Plan. In March 1953, Swarup arrived at CSIRO to work with Pawsey. During his stay of two years and four months, he also worked with Christiansen, Wild, Mills and

Bolton. Bhabha visited Sydney while Swarup was there and was given a tour of the Potts Hill station near Sydney. Six years later, when Swarup and the others approached him to start radio astronomy at TIFR, Bhabha knew exactly what they were talking about. Work on the array at Potts Hill proved to be of great value to the young scientists who learnt to work independently on hardware construction, observations, data reduction and the interpretation of the results.

Before Swarup left Sydney, he asked Pawsey if it was possible for CSIRO to donate the 32 six-foot dishes to NPL when they were no longer needed. Krishnan had agreed to pay the small shipping costs (£700). The NPL seemed all set to start a modest radio astronomy programme. Swarup had planned a grating interferometer with a baseline of 640 m for solar observations using the 32 dishes. The proposal calculated to cost ₹10,000 was sent to Krishnan in January 1955 from Sydney. Swarup returned to NPL in July 1955 and commenced work on building a 500 MHz receiver for the new grating array. However, since foreign exchange was a huge problem at the time, Krishnan failed to provide the funds and, in the end, even the shipping costs were paid by CSIRO.

During the mid-1950s, several other young scientists were also associated with NPL and the nascent radio astronomy programme. They were all to play significant roles in the early years of radio astronomy in India. T. Krishnan joined NPL in 1956, after completing a Tripos at Cambridge and spending a year there working with the radio astronomer, Martin Ryle. In late 1958, he went to Sydney to work with Pawsey and Christiansen. Kundu joined NPL in 1958 after completing his PhD in solar radio astronomy from France. M.N. Joshi and N.V.G. Sarma had already joined NPL by then, after completing their MSc degrees in India in 1956.

T.K. Menon graduated from the IISc in 1950 and went to Harvard University in 1952, where he completed his doctorate. He was on the faculty of the astronomy department from 1956 to 1958, before joining the (US) National Radio Astronomy Observatory in 1959 as a senior scientist. In 1958, M.R. Kundu joined the radioastronomy group led by Fred Haddock at the University of Michigan, and he soon became an internationally recognized leader in the field of solar radio astronomy.

When there was little progress in India, in late 1955, Swarup decided to go to the US. He spent a year at the Harvard Radio Astronomy Station at Fort Davis, Texas, and five-and-a-half years at Stanford from where he did his PhD under Bracewell.[18] In September 1960, a few months before completing his PhD, Swarup wrote to Pawsey. He explained that he would like to remain at Stanford for a year or two and then return to India. By then, discussions had started about a new Indian radio astronomy group. Bracewell suggested that the four Indian radio astronomers working in the US could meet at the next American Astronomical Society meeting to discuss this joint effort. The three of them already working in the US had discussed the possibility of returning to India and forming a major radio astronomy group on several occasions. In 1960–1961, Swarup also corresponded with Christiansen, Kerr and Pawsey in this regard. They recommended T. Krishnan, who was then at the Sydney University but was planning to leave.[19]

Swarup wrote to T.K. Menon in an undated letter, (most likely in mid-1961), explaining the conditions surrounding the possible formation of the new group. T. Krishnan, Menon, Swarup and Kundu met during the IAU meet, in August 1961, and discussed the plan to return to India after the formation of a radio astronomy group. A plan called 'Proposal for the Formation of a Radio Astronomy Group in India' was completed on 23 September 1961. Their initial project was to be a modest solar telescope using the 32 dishes that the NPL had received a decade earlier from CSIRO. The proposal was sent to AEC, TIFR, CSIR, UGC and PRL. Five distinguished astronomers outside India were requested to give their observations: Jan Oort, Bok, Pawsey, J.F. Denisse and Harlow Shapley. Bhabha was away in Geneva when the proposal reached TIFR, and it was forwarded to him by M.G.K. Menon.

T.K. Menon recalled that a chance meeting with the astronomer M.K. Vainu Bappu drew their attention to Bhabha. Initially, they had planned on sending their proposal only to CSIR and UGC. T.K. Menon was walking up the Berkeley Hill with Vainu Bappu, when he told the latter about their plans. He said that they should send a copy to Bhabha, if their proposal was to have any chance of success. Till then, nobody had thought of him. T.K. Menon admitted that they had no idea of the power equations in Indian science and were unaware of who could get things done.[20]

Pawsey wrote to Bhabha on 6 October 1961, praising Swarup and T. Krishnan's abilities and asked him to 'try to assist them in their efforts to work out something worthwhile'.[21] Bok was also just as upbeat. He wrote to Bhabha on 23 October 1961, asking him to act upon the offer 'which comes only rarely in the history of scientific development of a nation which, scientifically is obviously coming of age'.[22] Oort was less gushing in his response but fully supported the proposal, as did Oliphant.[23]

T.K. Menon met Bhabha for two hours in Washington, D.C. on 10 November 1961 to discuss various aspects of the proposal. An optimistic T.K. Menon wrote to Swarup about Bhabha's enthusiasm and positive response, including large final outlays of ₹50 lakh to ₹1 crore. 'In general, he had very little good [*sic*] to say about UGC and CSIR and particularly NPL [...] As you can see, I am much encouraged by the whole discussion and I hope you are too [...] Bhabha suggested the major project could be put anywhere in India suitable for them. Suggested places Bombay, Ootacamund, Kodaikanal, Poona or anywhere else we want.'[24]

Bhabha took a few months to respond. In mid-January 1962, Bok visited TIFR for a two-week lecture tour on Bhabha's invitation. Bhabha was evaluating the proposal at the time and discussed it with Bok. The offer was sent as a telegram a few days later, on 20 January 1962, to Swarup and, presumably, the others: 'Reference proposal from Menon, Kundu, Krishnan and yourself forwarded by M.G.K. Menon. We have decided to establish radio astronomy group. Letter follows with offer. Bhabha.'[25]

Bok wrote two letters, one to Swarup and the other to T.K. Menon, which was to be shared with the others. The first, handwritten letter to Swarup was dated 21 January, in which he said that the group would be wise to accept Bhabha's offer. The longer letter to T.K. Menon (and the other three) expanded on this theme. He said that he had been impressed with TIFR and Bhabha, especially his ability 'to get anything he asks for from the government [...] My advice to you and the others is to make your headquarters here at the Institute in Bombay [...] then set up your field station away from disturbances [...] I do not advise your going from the start to either Kodaikanal or Ootie'.[26]

During the course of that year, there were deliberations between Bhabha and the four radio astronomers on the offer. These complex negotiations continued for over a year. The offer consisted of the positions of reader for T.K. Menon, fellow for Swarup and Kundu and research fellow for T. Krishnan. The group of four had previously agreed among themselves that maintaining a small difference in rank and salaries was important to promote efficient cooperation in the beginning. Swarup and T.K. Menon accepted the offer from Bhabha while Kundu, who was initially quite displeased with the disparity, finally accepted the offer in mid-April 1962. However, T.K. Menon arrived only in early 1965.

In early 1962, when Bhabha made the first offers to the radio astronomers, the expectation was that T.K. Menon would be the most senior member of the radio astronomy group. Throughout most of the year, he, Swarup and Kundu wrote several versions of the plans for the interim solar observatory at Kalyan as a phase one project. However, he became a member of the TIFR radio astronomy group only in 1970.

T. Krishnan considered the offer to be unacceptable, and his negotiations with Bhabha broke down. Swarup returned to India in March 1963, and the 32 Potts Hill dishes were transferred to TIFR by mid-1963. In the meantime, he developed a design involving 20-foot dishes to complement the 32 smaller dishes for operating at a longer wavelength. V.K. Kapahi and J.D. Isloor of AEET joined the group in August 1963. Ramesh P. Sinha and D.S. Bagri followed in August 1964. In the same year, N.V.G. Sarma and M.N. Joshi left the NPL to join the TIFR radio astronomy group. As a first step, a grating type of radio interferometer was set up at Kalyan, near Bombay, to observe the sun at 610 MHz. The array consisted of 32 Potts Hill parabolas. The Kalyan Radio Telescope was completed by April 1965.

Swarup wanted to construct a large cylindrical radio telescope on a suitably inclined hill in South India to make its axis parallel to the earth's axis, to take advantage of India's close proximity to the equator. He discussed this idea with M.G.K. Menon, who responded enthusiastically. He also had a long discussion with Bhabha, who grilled him for over two hours. When Swarup asked Bhabha whether he should write a detailed project report, Bhabha replied: 'Young man, do not waste your time writing a project report; your main problem would be to collect a team; when you have managed that, you can submit a project report and proceed with its design and construction.'[27] The unique design of the Ooty Radio Telescope (ORT) called for a special site on a hill slope parallel to the spin axis of the earth. The cost of earthwork needed for creating such a slope would have been prohibitive so they had to look for an existing hill slope that would meet their needs. R.P. Sinha has narrated how the final site, Fairlawn, was selected.

Swarup and R.P. Sinha set off on a month-long 3,000-km search in the Nilgiri, Kodaikanal and Anamalai hills. Armed with three maps, a theodolite and an experienced multilingual driver, they scoured the picturesque countryside in search of a suitable site. They presented their detailed findings to Bhabha, who narrowed down the choice to four sites. The one near Kodaikanal found favour with Bernard Peters because it would have the advantage of sharing library and workshop facilities with the Kodaikanal Observatory. The other three were in the Nilgiri hills, and Bhabha, who vacationed there regularly, was familiar with the terrain. These sites had the advantage of easy access to both Coimbatore and Bangalore.

Bhabha, accompanied by M.G.K. Menon and Peters, arrived at the Ooty Raj Bhavan in a convoy of cars provided by the Karnataka and Tamil Nadu state governments. Sinha and Sreekantan were already there, as was the chief secretary of Tamil Nadu. The Pykara Lake site in Wenlock Downs was the first site they saw. On one side of the site was a prominent Toda village and Sinha and Swarup had been warned about acquiring land there by the villagers. Bhabha rejected it, as there was not enough 'elbow room'. At the second site, the hill slope was deemed too small. The third site, Fairlawn, which was a small forest reservation, met all their requirements.

Bhabha visited the site once more before it was finalized. On that visit, he was confronted by a very colourful delegation of Toda elders, who came to demonstrate against the selection of the Wenlock Downs site, not realizing that it had been rejected. Bhabha received the delegation with great respect and defused the situation by patiently listening to them and promising due consideration.[28] In late 1965, Bhabha approved the setting up of the ORT at Fairlawn as part of an Inter University Centre (IUC). After corresponding with Nehru, Bhabha obtained funding for the project and a 600-acre plot of land was earmarked by the Tamil Nadu government for the IUC. Unfortunately, Bhabha's demise soon after made it a nonstarter.

The structural and mechanical design of the ORT was done by Tata Consulting Engineers (TCE). Bhabha had decided not to hire consultants to save money, but, later, decided to approach TCE, which collaborated with the American design company, Ebasco Services. When the hesitant consultants asked Bhabha if he could give them $50,000 to get the design checked in the US, Bhabha told them that if he had that kind of money at his disposal, he wouldn't have needed them in the first place. When the consultants didn't know what to say, Bhabha told them that he would take the responsibility for any failures if TCE promised to put their best engineers on the job.[29] The ORT was completed in December 1969. The radio astronomy group continued to receive support from M.G.K. Menon, who succeeded Bhabha as director of TIFR.

During 1984 to 1996, Swarup also conceived and directed the design and construction of the Giant Metrewave Radio Telescope (GMRT), which consists of 30 large fully steerable antennas, each 45 metres in diameter spread over 25 km, about 80 km from Poona. It occupies pride of place as one of the biggest basic science projects in the country. Swarup became the project director of GMRT in 1987. In 1993, he became the first director of the National Centre for Radio Astrophysics of TIFR. With Swarup's death on 7 September 2020, the scientific community in India lost one of its most pioneering scientists.

PIONEERING SPACE RESEARCH

Vikram Sarabhai had set up the PRL at Ahmedabad in 1947 for atmospheric and geomagnetic research. The foundation stone was laid in February 1952 by C.V. Raman in the presence of Bhatnagar, Bhabha and the industrialist Kasturbhai Lalbhai. The main building was inaugurated in April 1954 by Sarabhai's family friend, Nehru. Over the years, Sarabhai's interests had moved away from cosmic rays to planetary magnetic fields and solar physics. He became interested in satellite-based observations, which eventually led to the launch of Indian satellites.

Ever since the first man-made satellite, Sputnik I, was launched by the USSR in October 1957, Sarabhai had been thinking of setting up a rocket-launching programme. It was an audacious idea and even K.R. Ramanathan, who had retired from the Indian Meteorological Department and was Sarabhai's co-director at PRL, was openly sceptical. Sarabhai was interested in space research and its potential applications, visualizing the use of satellite communications for broadcasting educational programmes to the rural hinterland. He also thought that remote sensing technology could help the progress of the country. While most dismissed his ambitions as a pipe dream, there was one man who did not—the one man who really counted in Indian science administration—Homi Bhabha.

Bhabha took his friend's vision to heart. In August 1961, at Bhabha's behest, the central government identified an area known as 'space research and the peaceful uses of outer space' and placed it within the jurisdiction of the DAE. Sarabhai was co-opted onto the board of AEC, and PRL was recognized as the appropriate centre for research in the space sciences. In February 1962, the DAE created the Indian National Committee for Space Research (INCOSPAR), with Sarabhai as its chairman, to oversee all aspects of space research. Thanks to Bhabha, Sarabhai overcame the first hurdle.

His next challenge was identifying a suitable location from where a sounding rocket[30] programme could be run. More specifically, Sarabhai was interested in studying the magnetic equator using small sounding rockets. Scientists from all over the world wanted to launch rocket payloads from this region because they wished to study a unique phenomenon called the 'electrojet'—a strong belt of charged particles that runs above the magnetic equator and greatly impacts radio communications. This phenomenon could only be studied from under the magnetic equator by launching instruments carried by sounding rockets.

One of Sarabhai's first students, Ekanth Vasant Chitnis, was given the task of finding a suitable site. Chitnis had joined Sarabhai at PRL in 1953 on a ₹100-per-month scholarship, after rejecting a better-paid job as an engineer at All India Radio. Udupi Ramachandra Rao, who went on to become chairman of ISRO, also joined around this time, with an MSc from Banaras. By the end of the 1950s, many of Sarabhai's students had completed their PhDs from American universities. The launch of Sputnik had set off a space race and these bright Indians were welcomed by American universities.

Chitnis first met Sarabhai and Bhabha as a student of physics in Poona in 1950 when the two had come to deliver lectures at ISC. Chitnis went to MIT in 1957 and worked with Bruno Rossi on cosmic rays as part of his PhD programme. He returned after three years to become a faculty member at PRL and joined the newly set up Space Application Centre. In 1961 he set up the Satellite Telemetry Station at PRL. When the INCOSPAR was constituted in 1962, he became its member-secretary. Chitnis was virtually Sarabhai's second-in-command being closely involved with all aspects of the space programme including identifying a suitable site for launching sounding rockets.[31]

Others in Sarabhai's group included his student, Praful Bhavsar, a cosmic ray specialist, and S.R. Thakore, who had studied cosmic rays at PRL before going to Rochester to study computers. U.R. Rao, who was also Sarabhai's student at PRL, had studied cosmic rays using American spacecrafts during a stint at MIT. Sarabhai had selected him for setting up the satellite centre for the space programme.[32] Chitnis set off on the first of many trips across the southern coast of India in July 1962. In view of the falling debris, a site for a sounding rocket needs to be isolated. He made some 200 flights in a Dakota aircraft before identifying three options. In November 1962, Bhabha and Sarabhai went to take a look. There is a photograph of Bhabha (rounding but not yet rotund) and Sarabhai taken on this mission, which allows us to visualize the moment. 'There is Bhabha, a trifle portly with thick-framed glasses in a short-sleeved safari suit pointing into the distance. Vikram, in a short with sleeves rolled up and light trousers, follows Bhabha's pointed finger. Behind them the tall, proud palm trees rise into a clear sky.'[33] Chitnis remembers Bhabha as being a no-nonsense person who was totally professional. 'He knew how to get things done,' he recalled.[34]

The plane carrying Bhabha, Sarabhai and Chitnis flew low over the coastline between Trivandrum and Alleppey. A fleet of cars waited at Trivandrum airport to take them to the shortlisted sites. Vellanathuruthu, 25 km outside Quilon (now Kollam), was rejected. Vellanathuruthu means 'white elephant island' and, perhaps, this also contributed to its prompt rejection. 'We steered clear of it for fear of it becoming a national joke,' Sarabhai is reported to have commented.[35]

Though the magnetic equator passed south of Quilon, in the end, the consensus was in favour of Thumba, a picturesque fishing village on the outskirts of Trivandrum. It had the advantage of being situated close to the airport and was also sparsely populated. A small church was located on the land they wanted to acquire. Thanks to the bishop's efforts, permission was granted. The bishop's home was quickly converted into an office, the church became the workshop and cattle sheds served as storage houses and laboratories. Legend has it that the altar served as the design table and the bishop's house the living quarters! Sarabhai's giant leap of faith in the field of space exploration began in a church!

Since they had planned for the site to be a research and development centre and not just a rocket launching site, the area included not just the long stretch of beach but also a part of the hill behind it. Sarabhai, it seems, had initially planned for the main building to only accommodate 200 people, but Bhabha, realizing that as the programme grew it would need more people, advised him to plan for 400. Aware of how humid it was, Bhabha announced that the setup would have to be air conditioned as well.[36]

Sarabhai had to put together a team for the upcoming programme. An advertisement put out in an Indian embassy publication in Washington D.C. had lured many Indians from American universities including some of his former students from PRL. Some were poached from the atomic energy programme. Around this time, Sarabhai hired one of his brightest recruits—a shy, slim Tamil Muslim boy from the

Aeronautical Development Establishment in Bangalore. A.P.J. Abdul Kalam, for that was the young man's name, was called to Bombay to meet Sarabhai, and the rest, to use a cliché, is history. Chitnis narrates that when H.G.S. Murthy, the first test director of Thumba was expanding his team, he had one post but two very good candidates, and he wanted to hire both. They were all sitting in a hotel in Boston, and Sarabhai handed the biodata of the second candidate—Kalam—to Chitnis and asked him for his opinion. Chitnis recommended that he be taken.[37] Kalam went on to head the country's missile development programme and ended his career as the president of India.

Sarabhai put together a small group, which included B. Ramakrishna Rao, Pramod P. Kale, A.S. Prakasa Rao and R. Aravamudan, to visit NASA's (National Aeronautics and Space Administration) sounding rocket branch at Goddard Space Flight Centre at Beltsville, Maryland, in December 1962. Aravamudan had resigned from the DAE to join the rocket programme. The others were PRL employees, which, at the time, was still a private organization. In March 1963, Kalam arrived in the US, along with two others, H.G.S. Murthy and D. Easwaradas. After spending a few days in Maryland, they were sent to NASA's sounding rocket launch station on Wallops Island in Virginia to be trained in rocket assembly, launching and explosives safety.

By mid-1963, the basic range had started taking shape, under the leadership of a civil engineer R.D. John. By the end of 1963, the preparations for the launch were at their peak. On 21 November 1963, Thumba was ready to stage its very first blast-off. Some big names gathered for the occasion: Jacques E. Blamont of the French space agency was there with his assistants, and there was a team from NASA, which included Arnold Frutkin and Ed Bissell. The India team included Bhabha, A.P. Mitra, and P.R. Pisharoty, founder director of the Indian Institute of Tropical Meteorology at Poona. The administration was represented by the governor V.V. Giri and the district collector. The bishop who had convinced his flock to relocate was also present.

The launch was not without its problems. The Nike–Apache rocket supplied by NASA had been flown to Delhi by the US, but the truck that was bringing it to Cochin had broken down along the way. When this was sorted out, it was discovered that the French payload could not be fitted into the American rocket. Ratilal Panchal, the PRL's expert mechanic was flown in from Ahmedabad to tackle the problem. Bhavsar and Kalam supervised him as he scraped the payload by hand. The rocket had been assembled in the church by Easwaradas while Kalam was in charge of safety. Back then, even rocket parts and payloads were transported by bullock carts and bicycles to the launch pad.

However, there were still some more glitches to be overcome. Just as the rocket was about to be placed on the launch pad, it started tilting because there was a leak in the hydraulic system. The technicians had to physically hold the rocket and place it correctly on the launcher. When the rocket was in position, the remote system to raise the launcher to the correct position malfunctioned. A man was sent to operate the controls on the launcher itself. Just when all seemed ready, P.P. Kale, then a new student at PRL, noticed a worker still fiddling with the launcher controls and rushed to pull him to safety.[38]

The Nike–Apache rocket was finally launched into the evening sky at 6.25 p.m., and minutes later, a vertical line of sodium vapour, tinged orange by the setting sun, was formed. All those present in the oval canteen at Thumba jumped for joy, including Bhabha. Sarabhai was modest about the achievement. He sent home a telegram: 'Gee whiz wonderful rocket shot.'[39] The two-stage rocket was believed to have reached an altitude of 180 km. The first stage Nike rocket burnt out after three-and-a-half seconds and dropped into the sea. The second stage Apache rocket sank into the sea at 105 nautical miles from the coast, 26 seconds after take off. Immediately after the launch, Bhabha told the press that India would launch its own rocket within 18 months. This marked the beginning of India's long and ongoing journey into space exploration. Chitnis narrated that Bhabha was frustrated by the bureaucratic processes he was facing at AEC and had promised a more convenient and flexible structure for ISRO. Unfortunately, Bhabha perished prematurely in January 1966 and ISRO ended up with a structure similar to AEC.[40]

It is not generally known that V.K. Krishna Menon had opposed this launch. The strained relations between him and Bhabha were no secret. On 11 September 1963, Krishna Menon wrote to Nehru with a sense of outrage and indignation that he had read in the press that the US intended to 'fire her Nikes in India' and 'the Nike as you know is a nuclear weapon'. He added:

> You may remember that when I was in Government, I protested against our agreeing to set up [satellite] tracking stations, over which and the data collected we have no control. I also raised my ineffective voice against placing that programme on our side, in the hands of those who are not greatly concerned about our non-alignment or independence and are also outside normal governmental control.[41]

Clearly, Krishna Menon was targeting Bhabha, whose special position in the government had been a source of irritation and annoyance to many, including Krishna Menon. Nehru's former defence minister used his arrogance as a defence mechanism to conceal lifelong insecurities, marrying emotional insecurity to political petulance. Nehru disabused Krishna Menon of his biased and baseless notions that same day.

> The experiments intended are entirely meant for weather research [...] These rockets are sounding rockets, and are not military rockets [...] Nike itself is a booster and is used in the first stage of both the sounding and military rockets. I do not see any harm in our associating ourselves with this. It is essentially a UN scheme and we are being helped in it not only by the US, but also the USSR, France etc.[42]

We do not know, if and how, Krishna Menon responded.

SCIENTIST AS ADMINISTRATOR

The development of nuclear energy was a truly international process, emerging from widely scattered research laboratories in the US and Europe. The ideas and work of scientists in one country stimulated the minds of their colleagues in others. In the 1920s and 1930s, the leading physicists and chemists of Europe and the US started unravelling the secrets of the atom.

FISSION

In 1934, Irène and Frédéric Joliot-Curie created radioactive nitrogen and phosphorous by bombarding boron and aluminium with alpha particles. Their experiment showed that radioactivity could be created. While natural radioactivity had been studied for three decades, artificial radioactivity was a novelty. A little after the Joliot-Curies' discovery, in 1934, Enrico Fermi and his team showed that the same phenomenon could be achieved more efficiently through bombardment with neutrons, especially those that had been slowed down. This introduction of slow neutrons to probe nuclear structure was one that would have extraordinary ramifications in the years to come.

Towards the end of 1938, in Berlin, Otto Hahn, a German chemist, and his younger associate, Fritz Strassmann, who had been bombarding uranium with slow neutrons, discovered a radioactive isotope of barium as an end product of their experiment. Since the barium nucleus has only 56 protons, while uranium has 92, this seemed impossible. The target uranium nucleus had moved an astonishing 36 places down the periodic table. Bombarding uranium with neutrons had caused the uranium nucleus to split virtually in half. This was an extraordinary result and was initially received with expected incredulity by all those who came to know about it.

On 19 December 1938, Hahn wrote to Lise Meitner about the anomalous product of their experiment. Meitner had just arrived in Sweden from Berlin as a Jewish refugee. She was part of the vast diaspora seeking refuge after the rise of Nazism. Meitner, who was born in Vienna, received a doctorate in physics from Vienna University, becoming only the second woman to achieve that distinction. In July 1938, she had been in Sweden for only a few months after having escaped from Germany, where anti-Semitism, a nasty undercurrent for a long time, had become a public danger.

When Meitner, then in her twenties, moved to Berlin, she met Hahn who was also interested in radioactivity. The two collaborated for the next 30 years at the Kaiser Wilhelm Institute, where Meitner was asked to set up a physics department. Within a decade, she had achieved such success that she became one of the first women to receive the title of professor in Germany. The physicist, Otto Frisch, her nephew, had also studied at Vienna University and had worked in Germany, but emigrated when Hitler became chancellor of Germany. He had also worked with Niels Bohr in Copenhagen for four years.

When Frisch came to meet his aunt on the morning of 23 December 1938, he found her reading a letter from Hahn in which he described the results of his experiments with uranium. They discussed Hahn's results, and Frisch thought that perhaps there had been some mistake. They found their answer the next day while walking in the woods as they discussed the incongruity. They worked out the calculations of the phenomenon and realized that the uranium nucleus had split into two smaller nuclei.* Their calculations showed that if the uranium atom did split, as they thought it might, it could produce barium (which has 56 protons) and krypton (which has 36 protons), adding up to 92 protons.

However, energy had to be conserved in this process since it

---

*Segrè, Gino and Bettina Hoerlin, *The Pope of Physics: Enrico Fermi and the Birth of the Atomic Age*, Henry Holt and Company, New York, 2016, p. 129.

was a fundamental law of physics that could not be broken. Meitner recalled attending a lecture by Albert Einstein in Salzburg in 1909, where he had derived his elegant equivalency, $E=mc^2$. The idea that mass could be converted into energy had left a deep impression on her. She also realized that the nuclear masses of the fragments created by splitting a uranium nucleus would not add up to the mass of the original nucleus. After doing the math, she was certain that a neutron caused the uranium nucleus to split in two, converting a tiny amount of mass into energy along the way.*

Meitner urged Hahn to try to further purify the uranium bombardment samples and assess whether they were, in fact, made up of radium or its chemical cousin, barium. Hahn complied and found Meitner's conjecture to be correct: the element in the sample was barium not radium. Hahn's finding suggested that the uranium nucleus had split into two, becoming two different elements with smaller nuclei, just as Meitner had suspected. On 1 January 1939, Meitner wrote to Hahn, telling him that both she and Frisch thought that a nucleus splitting was *perhaps* possible. Two days later, Frisch, who was back in Copenhagen, gave Bohr the news. Frisch recalled that he had barely started speaking when Bohr struck his head with his fist and said, 'Oh, what idiots we have all been! Oh, but this is wonderful! This is just as it must be.'†

By 6 January, Frisch and Meitner, working over the phone, drafted a paper that they intended to send to *Nature*. Frisch gave Bohr, who was leaving for the US the next day, two typed pages. Bohr assured him that he would not mention fission till Frisch's paper was received by *Nature*.‡ To be doubly sure, Frisch repeated the experiment, and his finding suggested that the uranium nucleus had split into two—becoming two different elements with smaller

*Ibid. 130.
†Ibid. 131.
‡Ibid. 131.

nuclei—just as Meitner had suspected, releasing nuclear energy in the process. He wrote a paper on his experiments and sent both papers to *Nature* on 16 January 1939.

Even after she had taken a position in Stockholm, Meitner continued to correspond on the subject with Hahn and Strassmann. The discovery of barium was a fruit of that collaboration. Yet, when it was time to publish, Hahn knew that including the name of a Jewish woman as a co-author would ruin his career. He published without her, falsely claiming credit only for himself. In fact, Hahn had trouble explaining his own findings. In his paper, he put forth no plausible mechanism explaining how the uranium atoms split into barium atoms. However, Meitner had the explanation.*

In a few weeks, on 11 February 1939, Meitner wrote her now famous letter to the editor of *Nature,* in which she explained 'Hahn's discovery'. In that letter, Meitner, along with Frisch, provided a physical explanation of how nuclear fission took place. It was in this letter that the word 'fission' was first used.† Bohr brought news of the splitting of the atom and its possibilities to scientists in the US. He also grasped the fact that uranium 238, the common isotope, is very stable, but uranium 235, the less common form, was susceptible to fission.‡

For years, Einstein had been asked whether his equation could unleash the energy locked inside the atom. He would always say that the energy released by one atom was too small to be of any practical use. The key insight into how to exploit Einstein's equation for practical use came from the Hungarian–American physicist Leo

---

*Jorgensen, Timothy J., 'Lise Meitner—the Forgotten Woman of Nuclear Physics Who Deserved a Nobel Prize', *The Wire*, 19 February 2019, https://tinyurl.com/yn9da8wd. Accessed on 11 August 2022.

†Rhodes, Richard, *The Making of the Atomic Bomb*, Simon & Schuster, UK, 2012, p. 263.

‡Watson, Peter, *A Terrible Beauty: A History of the People and Ideas that Shaped the Modern Mind*, Weidenfeld & Nicolson, 2000, pp. 395–6.

Szilard, who was working in New York. He showed that in the uranium fission process about two neutrons are emitted whenever a neutron released by this process collides with the nucleus of another uranium 235 atom. While the energy released by splitting a single uranium atom is minuscule, Szilard realized that a self-sustaining chain reaction is possible, which would cause an exponential rise in the number of split uranium atoms, eventually creating enough energy to level a city. In May 1939, Irène Joliot-Curie, the Austro-Hungarian scientist, Hans Halban, and the Polish scientist, Lew Kowarski, who were refugees in France, repeated Szilard's experiment and patented the production of nuclear energy as well as nuclear explosives. The potential applications of nuclear energy, both military and civilian, were beginning to unfold at the same time.

Like many other Jewish scientists, Einstein had immigrated to the US to escape Nazi Germany. On 2 August 1939, at the urging of Szilard and, fellow Hungarian, Edward Teller, Einstein wrote to warn President Franklin D. Roosevelt that Germany was trying to produce enriched uranium and urged him to make sure that the US armed itself with nuclear weapons before Germany did. Roosevelt took heed. He set in motion what eventually became the Manhattan Project, the vast American undertaking that led to the production of the atomic bomb. The project was named after the Manhattan Engineer District of the US Army Corps of Engineers because much of the early research was done in New York City. British scientists, helped by refugees from Hitler's Germany, had been ahead of the US on the path to developing the bomb in 1940 and 1941, and they made a decisive contribution to an enterprise that changed the world forever.

Another element, plutonium, was to prove as powerful as enriched uranium when used as a nuclear explosive or a source of energy. However, plutonium essentially does not exist on earth and is a man-made element. In late 1940, the American chemist,

Glenn T. Seaborg, and his colleagues at the University of California, produced a trace of one of plutonium's isotopic forms. Seaborg named the new element after the solar system's outermost planet, Pluto, the ancient Greek god of the underworld and wealth.

Unknown to all but a very small and select group of scientists and government officials, Fermi was building an atomic pile in Chicago. On 2 December 1942, James B. Conant, the long-time president of Harvard University and member of the wartime National Defence and Research Committee, received a telephone call from Chicago, 'The Italian navigator has just landed in the New World.' When Conant asked, 'And how did he find the natives?', he was told, 'Everyone landed safe and happy.'* On the other end of the telephone was Arthur Compton, a key figure in the Manhattan Project. In 1942, Compton became the head of the Metallurgical Laboratory responsible for building nuclear reactors to convert uranium into plutonium, finding ways to separate plutonium from uranium and designing an atomic bomb.

The 'Italian navigator' was Fermi, and the coded cryptic message sought to convey that Fermi had experimentally confirmed the possibility of a controlled nuclear chain reaction, something that had been established theoretically three years ago. The world's first atomic pile had been built in an abandoned squash court at Chicago University, and it is these humble beginnings that led to the building of the world's first atomic bombs, which were used to such devastating effect against Japan in 1945. Compton oversaw Fermi's creation of Chicago Pile-1, the first nuclear reactor. Furthermore, the Metallurgical Laboratory was also responsible for designing and operating the X-10 Graphite Reactor at Oak Ridge, Tennessee.

On 3 July 1944, Bohr sent a memorandum to the ailing Roosevelt,

---

*Segrè, Gino and Bettina Hoerlin, *The Pope of Physics: Enrico Fermi and the Birth of the Atomic Age*, Henry Holt and Company, New York, 2016, p. 197.

urging that the US and the UK should take the USSR into confidence about the progress they were making towards manufacturing a nuclear weapon. The idea was rejected particularly sharply by Churchill, who recommended that steps be taken to ensure that Bohr passed no information to Moscow. Luckily, the Nazi bomb project, led by Heisenberg, was considerably behind the effort of the Allies. Heisenberg was yet to calculate the amount of enriched uranium required to create a chain reaction. The Nazi programme was underfunded, and its base was also being bombed by the Allies.

Considering the fact that Einstein was an overt pacifist, it is ironic that the destructive use of the matter-to-energy conversion ($E=mc^2$) provided some of the strongest evidence for his Special Theory of Relativity.

'LITTLE BOY' AND 'FAT MAN'

The world was rudely introduced to the atomic age in August 1945 with the bombing of the Japanese cities of Hiroshima and Nagasaki. The first nuclear test was conducted on 16 July 1945 in the vast wasteland of Alamogordo in New Mexico. Code-named Trinity, the test would radically change the world and the entire outlook to mankind. As a mighty ball of fire growing larger and larger rose heavenwards, only Trinity's author, the scientific director of the super-secret Manhattan Project, J. Robert Oppenheimer really comprehended what had happened. He later quoted from the Bhagavad Gita to describe the momentous occasion: 'I am become Death, the destroyer of worlds.'*

At exactly 45 seconds past 5.29 a.m., early risers more than 150 miles away heard the roar of the bomb. The world's first atomic bomb called simply the 'Gadget' was detonated 120 miles south

---

*Strathern, Paul, *Oppenheimer and the Bomb*, Arrow Books, London, 1998, p. 74.

of Albuquerque at Alamogordo in the New Mexico desert. It was a plutonium bomb and was detonated atop a 100-ft steel tower. The darkness of the early morning sky was suddenly rent by an intense blinding flash. This was followed by a silent blast of heat. About 9 km from this point zero, Oppenheimer and his scientists monitored the explosion from a protected bunker. Dignitaries and less essential staff viewed the show from the base camp 20 miles away. Moments after the explosion, the roar of the shock wave tore over the bunker, echoing and re-echoing as the earth rumbled with its power.

Oppenheimer and his team watched with awe as a vast molten fireball emerged from the horizon and rose into the sky. A huge mushroom cloud formed, gradually rising 40,000 ft into the upper atmosphere.* The 100-ft steel tower simply vaporized down to the twisted wreckage of its footing, and the heat scorched the desert sand for a radius of 800 yards into a sheet of glass. When the explosion was measured, it was found to be the equivalent of 18,600 tonnes of TNT, nearly four times the general estimate of the Los Alamos scientists.

The bomb required an incredible amount of money and effort to make. In 1942, the Manhattan District of the Corps of Engineers was formed to construct three secret 'cities' for major portions of atomic bomb development. Bohr, who had argued that building an atomic bomb 'can never be done unless you turn the United States into one huge factory', told his colleague Teller, 'I told you it couldn't be done without turning the whole country into a factory. You have done just that.'†

The Manhattan Project was under the administrative charge of Major General Leslie Groves of the Army Corps of Engineers,

---

*Ibid. 75.

† Rhodes, Richard, *The Making of the Atomic Bomb*, Simon & Schuster, UK, 2012, p. 500.

who had supervised the construction of the Pentagon. Secret atomic energy communities were created almost overnight in Oak Ridge, Tennessee, at Los Alamos, New Mexico, and in Hanford, Washington, to house the workers and the gigantic new machinery needed to produce the bomb. The weapon itself was to be built at the Los Alamos laboratory, under the direction of Oppenheimer.

For obvious reasons, the Manhattan Project was carried out in extreme secrecy. At its height, the project was the size of the entire American automobile industry, employing 130,000 people and costing a total of $2 billion. Among these employees were some of the greatest scientists that have ever lived, many of them émigrés: Niels Bohr, Enrico Fermi, Hans Bethe, Richard Feynman, David Bohm, Leo Szilard, Eugene Wigner, Felix Bloch, Edward Teller, Klaus Fuchs and Harold Urey. Bohr had escaped from Nazi-occupied Copenhagen in September 1942 with the help of the Danish resistance and the British Secret Intelligence Service and arrived at Los Alamos in January 1943. Revered by most at Los Alamos, he became a kind of father figure to some of the younger physicists, particularly the American physicist, Richard Feynman.

During the next three years, thousands of scientists and technicians worked on building the bomb. In the end, they produced three: 'Gadget,' 'Little Boy,' and 'Fat Man'. Gadget, the plutonium device exploded in Alamogordo, proved to the scientists that their design for Fat Man (a plutonium-based bomb) would work. Exactly 21 days after the Trinity test, the first uranium bomb, Little Boy, was delivered untested to the army and was dropped on Hiroshima on 6 August 1945, killing at least 70,000 inhabitants. On 9 August 1945, the plutonium bomb Fat Man, virtually identical to the Trinity device, was dropped on Nagasaki, killing at least 35,000 inhabitants. By the time the atom bomb was ready to be used, Germany had, in fact, already surrendered. Szilard and James Franck drafted a petition signed by nearly 70 scientists who were opposed to the use of the bomb on moral grounds. On 10 August, the Japanese surrendered. The Second World War ended.

Was the bombing of Japan justified? The vast literature on the subject proves that Japan would have surrendered soon enough without the use of nuclear weapons. Patrick Blackett declared pertinently and pithily that the dropping of the bomb was not the final military operation of World War II but the first diplomatic move in the Cold War. After the bombing of Japan, there was a wave of uneasiness and disapproval in the world. Einstein was deeply disturbed about how things had turned out. He told his secretary: 'Had I known that the Germans would not succeed in producing an atom bomb, I would never have lifted a finger. Not a single finger!'*

When Nehru voiced his concern about this event to Bhabha, the latter urged Nehru to take note of the terrific power that had been put in men's hands through the fundamental research that had been done in physics during the last 15 years, without any practical application in mind. Bhabha argued that it was indicative of fundamental research on a big scale always paying in the long run. Four months after this response, Bhabha started TIFR in Bombay.

*Bodanis, David, *E = MC²: A Biography of the World's Most Famous Equation*, Pan Books, London, 2000, p. 218.

Chapter 25

THE TWO FACES OF JANUS

> In a broad view of human history, it is possible to discern three great epochs. The first is marked by the emergence of the early civilizations in the valleys of the Euphrates, the Indus, and the Nile, the second by the Industrial Revolution, leading to the civilization in which we live, and the third by the discovery of atomic energy and the dawn of the Atomic Age, which we are just entering. Each epoch marks a change in the energy pattern of society.
>
> —Address at the Geneva Conference in 1955[1]

INDIAN SCIENTISTS AND NUCLEAR PHYSICS

Nuclear physics was the frontier of scientific inquiry in the early 1930s and the scientists in India were eager to participate in the new area. Radioactivity as a phenomenon had attracted the attention of some early Indian scientists, and among the first to engage with research in this field was Ruchi Ram Sahni (father of the famous Birbal Sahni) of Forman Christian College, Lahore, who had studied under Ernest Rutherford at Manchester University from 1912 to 1914. With the growth of nuclear physics as a field of research, it is hardly surprising that Indian scientists showed an interest in it. In the following decade, the others who engaged in research on radioactivity and nuclear physics included S.N. Bose, then at the University of Dhaka who had worked with Marie Curie in Paris between 1924 and 1925, and Rajendra Lal De (also from Dhaka) who had worked with both Curie and with Otto Hahn in Berlin. D.M. Bose first worked at the Cavendish and then with Erich Regener in Berlin in 1919.

No one would have thought a third world nuclear power a possibility in 1948, but among the newly decolonized countries, India alone had a number of physicists of outstanding ability who not only understood the significance of the discovery of nuclear fission, but felt that developing an atomic energy programme should be a state priority. Bhabha, Meghnad Saha and S.S. Bhatnagar all believed that an Indian atomic energy programme should become a national imperative. They realized that the first priority lay in establishing the capabilities for pursuing nuclear physics in India. The other was to participate in research in leading laboratories in Europe and the US. The Cavendish had an established tradition of training Indian students and the Radiation Laboratory at the University of California, Berkeley, with its successful implementation of the cyclotron, was also assuming global prominence.

Saha's interest in nuclear physics was not new. As far back as 1933, he had attempted to posit a theory of beta decay with his student Daulat Singh Kothari, before Enrico Fermi had put forward his own now famous theory on the same subject. In 1936, towards the end of his stay in Allahabad, Saha was awarded a grant by the Carnegie Foundation, which allowed him to visit both Europe and the US. He visited the Radiation Laboratory at the University of California, Berkeley, where he met Ernest Lawrence and was greatly impressed with his cyclotron, convinced of its use for both nuclear research and medical treatment. In June 1936, he was in Copenhagen to attend Niels Bohr's conference on nuclear physics. Bhabha was also attending this conference and perhaps it is here that the two met for the first time. Bohr's institute in Copenhagen was in the initial stages of building a cyclotron with funds from the Rockefeller Foundation. On returning, Saha reported on the conference in his magazine *Science and Culture*, which served as a forum for his writings on atomic energy and national problems.

Saha also visited Princeton and met Milton White who was in charge of the cyclotron. His interest in the cyclotron soon became an obsession. He wanted to build a cyclotron in his laboratory because it would not only build infrastructure for science but also elevate the status of his laboratory. Around this time, Saha came to know that Arthur Compton of Chicago University, supported by Paul Langevin at the College de France, was nominating him for the Nobel Prize. He wrote to Bohr asking for support for his candidature. Saha did not win the prize, but his stock in the international physics community had never been higher.[2]

The discovery of nuclear fission gave a boost to the study of nuclear physics worldwide. In a talk before the Indian Physical Society in March 1941, Saha noted '[...] the discovery opens up a new vision of energy-production which is extremely exciting'.[3] He was aware of the possibility of a chain reaction but said that the practicability of the idea could only be gauged once experiments were conducted. He was also alert to the idea of explosions in fissile material. In 1941, he wrote in *Science and Culture* about the destructive potential of the new discovery. 'It is quite possible that a process may be discovered which renders the [chain] reactions to proceed with explosive violence [...] the idea that a tablet of U^{235} [...] may blow off a mighty Super Dreadnought [battleship] cannot but be an exciting one.'[4] This was indeed prescient because he wrote this a full four years before his visit to North America and his tour of the sites associated with the Manhattan Project. It was also long before he suspected that work on a major project involving uranium 235 was underway.

In 1939, by the time the news of fission had spread across the scientific world, Saha was back in Calcutta from Allahabad, and Bhatnagar was in the process of moving from Lahore to Delhi. Bhabha, as we know, had also returned to India. All three probably read about nuclear fission in an article in *Nature* and would have understood its theory and experimental requirements. The three were in regular touch, though it is impossible to tell if they discussed fission. Also, from 1940 onwards, there was a freeze on publications on fission brought about, in part, by Patrick Blackett in London. The race to build the first nuclear weapon was just beginning, and the Allies did not want scientific details on the new discovery to benefit Germany. However, Saha and Bhabha, in their own way, knew that something very secret was afoot.

Early in 1942, Georgii Flerov, a Russian physicist who had worked on fission, found that neither were there articles on fission in any of the American journals he read nor was there any response to his work with a colleague on the same subject. Flerov's, 'dog that didn't bark', found a parallel in India. Govind Swarup remembers Bhabha telling him that by 1944, he was convinced that the Americans had started a nuclear weapons programme. Bhabha's reasons for believing this were similar to Flerov's. He had been corresponding with a number of physicists around the world through the war years. Letters from colleagues in the US, which always took long to reach him because of the distance—a situation made worse due to the war—had practically dried up by 1943. Bhabha did not give it much thought at the time, assuming that distance and the war had slowed his mail down. In 1944, not having heard from his colleagues despite writing several letters, Bhabha sat down and made a list of the people who would be likely candidates for a nuclear programme. When he made a list of his silent correspondents, he found that the two lists were almost identical.[5]

After the bombing of Japan all the suspicions and piecemeal information of the Indian scientists finally fell into place. A month after the bombing, Saha, along with B.D. Nagchaudhuri, wrote an account 'The Story of the Atomic Bomb' for *Science and Culture*. Saha was also aware of the moral issues arising out of atomic bombs. In an editorial in *Science and Culture*, in September 1945, he rubbished the arguments of Allied statesmen who justified the use of the atomic bomb because it had saved lives by cutting short the war. He compared it with a similar humanitarian plea by Genghis Khan when he slaughtered the inhabitants of Balkh. Saha ended on a note of optimism, hoping that science would solve problems of reconstruction and 'centuries of neglect can be compressed into decades'.[6]

Later, Saha also summarized Henry DeWolfe Smyth's account

of the American participation in the Manhattan Project, which was made available only three days after Fat Man was dropped on Nagasaki. The report had been commissioned by Major General Leslie Groves. Realizing that the public would want to know what happened, the report was a way of explaining key concepts to the public. The report by Smyth, a Princeton University physicist, provided a background and a brief history of what had transpired. To everyone's astonishment, it went on to become a *New York Times* bestseller, remaining on the list for many months.[7] Saha published his own article 'The Atom Bomb' giving details about how an atomic bomb was made and methods of uranium separation, which appeared in the June 1946 edition of *Science and Culture*.

PLANNING AND SCIENCE

In the decades leading up to Independence, planning emerged as the language through which Indian aspirations were expressed. Japan's development after the Meiji Restoration of 1868 showed what a central authority could accomplish; later the USSR's rapid advance through its first five-year plan raised the hopes of many in India who thought that planning was the way forward.

In October 1938, the Congress's National Planning Committee (NPC) was set up for a complete survey of the Indian economy. Well before Independence, India's technological and industrial trajectory had been a matter of intense debate between politicians and businessmen. There were those, like Madan Mohan Malviya and Sir Mokshagundam Visvesvaraya, who advocated the widespread use of machinery in all sectors and believed that technology was critical to the country's advancement, citing pre-war Japan and Germany as models for emulation.

Their confidence in technology was shared by Saha, who was a votary of the path Bolshevik Russia took in its thrust to industrialization. Saha was also closely associated with Subash Chandra Bose, then president of the Indian National Congress, and this connection was crucial in setting up this body that brought together the country's scientists, engineers, politicians and industrialists. In February 1938,

S.C. Bose, in a meeting in Haripura, Gujarat, spoke of the need for a socialistic solution to India's problems through a planning commission that would begin a comprehensive scheme of industrial development under state-ownership and control.[8]

A meeting that included politicians and other prominent men of India was called in October 1938 with the intention of forming the NPC. Initially, Visvesvaraya, the engineer–technocrat, who was the diwan of Mysore, became chairman of the committee. Visvesvaraya had been writing on Indian reconstruction and was famous as the author of a book, *Planned Economy for India* was published in 1934. Saha wrote to Nehru on behalf of the scientists of India asking him to accept the chairmanship of the NPC. S.C. Bose also wrote to Nehru, offering him the job in a telegram. Nehru, who was in London at the time, accepted the offer when he returned in November 1938 and Visvesvaraya stepped down. Saha believed that Nehru was the best person to head the NPC since he wanted the committee to be led by a political leader who would give it the visibility it needed. More importantly, Gandhi had been promoting rural development as a strategy of reconstruction and a section of the nationalistic elite were on the lookout for an antidote to Gandhi's trajectory, which they believed to be catastrophic.

If Nehru accepted leadership of the NPC, it would signify wider acceptance of planning for industrialization, ensure its implementation and pave the way for science-based industrialization. Tagore wrote to Nehru to say that, after a long discussion with Saha about the importance of planning, he was now convinced of its importance.[9] Led by Nehru and managed by the economist K.T. Shah, the 15 member committee included industrialists, scientists, economists and a sceptical Gandhian, J.C. Kumarappa. The NPC's office was housed in a number of places in Bombay, including the Bombay University building. After the Second World War, J.R.D Tata provided room for it in the New India Assurance Building.[10]

S.C. Bose, like Saha, believed that economic planning in India should mean planning for the industrialization of India. Saha was critical of Gandhi's policy of khadi and agreed with S.C. Bose that Indian industrialization could only proceed at a brisk pace if it was supervised by a strong state. His promotion of modernity and complete industrialization and his backing of S.C. Bose exposed Saha to criticism and suspicion. Nehru, at the time, held a more moderate position, although he also believed in a planned economy with the planning component being dominated by science and technology. However, unlike Saha, who wanted to introduce machine-led industrialization, Nehru viewed centralized planning as an effective way to manage India's post-war reconstruction.

One objective of the NPC was to get active scientists to play a greater role in decision-making and to guide the government toward 'liberating' the scientific and industrial side of the economy. With the declaration of war in 1939, the Indian leadership realized that Britain would need all the help it could get, and the perception of this need further accelerated the demand for Independence. With a growing awareness of the unorganized state of scientific and industrial research in the country, there was clamour for the constitution of a body on the model of the Department of Scientific and Industrial Research (DSIR) in England.

In April 1940, the Government of India established the Board of Scientific and Industrial Research (BSIR) and Bhatnagar was made its director. Saha had written to Sir Ramaswamy Mudaliar twice in early 1940 asking him to set up BSIR and to make a distinction between scientific and industrial research, noting that setting up industries was different from promoting them.[11] Saha, J.C. Ghosh and Nazir Ahmed, a Cambridge trained physicist who had turned to cotton and textile research, were its members. It was first proposed that the governing body of BSIR should not contain any other scientists when all the three scientists and Bhatnagar threatened to resign, the decision was reversed.

Sir Ardeshir Dalal, who possessed considerable theoretical knowledge of planning, was an important member of BSIR. There were few people as familiar with the reports of the NPC as Dalal. He was also a close associate of Visvesvaraya. In 1944, Dalal would be one of the authors of the Bombay Plan. Saha, apart from lobbying Mudaliar, had also been in personal touch with Dalal on the formation of the BSIR. Dalal's time at Tata Steel had given him wide experience in industrial management, and from 1941 onwards, he was closely associated with the government. As the liaison man for the Ministry of Supply, he was the channel through which information regarding India's capacity to produce steel products for war was passed to and from the government's supply department. From 1938 onwards, he took a wider interest in scientific and technological education, the cement and textile industries, and insurance and banking. Technological and scientific developments and education claimed Dalal's special attention, and he was on the board of many scientific bodies in the country.[12]

There was a need for a legal entity to disburse funds, and the CSIR came into existence in September 1942 with Mudaliar at the helm. 'Steamship Bhatnagar' had arrived in Delhi in the summer of 1940 as the director of BSIR, only to discover that there was no place for him to build his laboratories. It was decided that his labs and researchers would move to spacious buildings in Calcutta, and he would supervise them once a month. Since the Japanese had, by then, occupied Burma in 1942, the testing laboratory was moved to Delhi from Calcutta, which was considered too close to the theatre of conflict. Along the way, Bhatnagar picked up a knighthood in January 1941.

Bhatnagar was in regular contact with the three most powerful men responsible for industry within the government: Mudaliar, Dalal and Sir Azizul Haque. He started projects that substituted Indian raw materials and skills for those that had previously been imported, such as textiles, steel, glass and oil. The imports were to be made good through indigenous production and this, according to Bhatnagar, presented a vast field of investigations for the technical men and universities. While both BSIR, and its later avatar, CSIR, were the first experiments in science-based industrialization in India, they had little to do with national development—they were both meant to serve the war effort. Jahnavi Phalkey has observed that the major goal of the BSIR and the CSIR was to defend an amorphous territory called British India.[13]

NEHRU AND SCIENTIFIC TEMPER

Nehruvian science was basically state science, science conducted for the people but at the direction and discretion of the State. Nehru believed that science would help in a transformation based on rationality and logic. He believed that it would help solve the problems of hunger and poverty. Nehru's personal engagement with science helps explain how it received so much public validation. His interest lay as much in the culture of science as in its material achievements, and he repeatedly invoked the term 'spirit of science'. Nehru was the first to introduce the term 'scientific temper' in the country. Nehru's philosophical understanding of science enabled him to present his own social ideas as being grounded in rationality while condemning his opponents, especially Hindu and Muslim communalists, as 'woefully unmodern'.[14]

Nehru believed that while the applications of science were inevitable—the scientific approach, the refusal to accept anything without testing and trial—was equally important. He was preoccupied with what he, at different times, called the 'scientific method', the 'scientific approach', and the 'scientific temper'. The phrase scientific temper has come to define Nehru, much like the equally memorable phrase 'tryst with destiny'.[15] Nehru's use of the term is not a straightforward one; he used it in different ways. The first was as critique and exploration but the second was something more.

Science dealt with the domain of positive knowledge but the temper that it was to produce went beyond that domain. For Nehru, such a temper was an attribute of freedom—not only one embodied in democratic institutions but also in the ecstasy of creativity. In this sense, he used the expression 'scientific temper' for artistic creativity. What lay within the bounds of reason, he called science and the scientific method; what extended on the far side was pure creativity. Science, by itself, marked stagnation on the plateau of reason; but the scientific temper was creative, inventive and spiritual.[16]

Nehru's unflinching faith in the benefits of science and technology led him to introduce a Science Policy Resolution in Parliament in March 1958. Bhabha played an important role in this, and the policy approved by Parliament has his distinct imprint. Bhabha believed that science and technology, as developed in the West, could solve the economic problems of any society, irrespective of its history and culture. Science and technology, considered to be unambiguously modern, were central because the practical realization of modernity came about primarily through technological transformation. According to M.G.K. Menon, it was not Bhabha's intention to plant science and technology as an imported item but to grow science indigenously, a science that would bear the imprint of the tradition and culture of Indian people.

The opening statement of the resolution that, in all probability Bhabha drafted was as follows:

> The key to national prosperity, apart from the spirit of the people, lies in the modern age, in the effective combination of three factors, technology, raw materials and capital, of which the first is perhaps the most important, since the creation and adoption of new scientific techniques can, in fact, make up for a deficiency in natural resources, and reduce the demands on capital. But technology can only grow out of the study of science and its applications.[17]

The atom became the primary symbol of the new era—the atomic age. If the atom held terror and death, as the explosions of Little Boy and Fat Man had so dramatically proven, many also saw in it a major opportunity, a permanent resolution of the ever-increasing need for electrical energy at little cost. That was the paradox of the age—extermination or great material prosperity. The symbol for either consummation was the same—the atom. Proponents of nuclear power believed that electricity from atomic reactors was a key factor for economic development. It was believed that economic development was a function of energy consumption, which, in turn, depended on energy supply. It was, therefore, thought that a leap in energy consumption would translate to a major leap in economic development. Bhabha, like Nehru, believed that mastery over the atomic nucleus represented the apogee of science.[18] The British had deliberately retarded industrial development in India, but Indian science would propel the country to modernity.

India's tryst with atomic energy followed her tryst with freedom. Atomic research at the time was at the vanguard of scientific research and provided a means for a rapid march towards modernity.[19] Also, since it was a new field, the developing nations would start on a more or less even footing. Nehru thought that the application of science for evil ends (read nuclear weapons) was unconscionable and that India, with its restraining influence and long background of culture, would come up trumps in this area.

Nehru considered modernity as synonymous with the advancement of science and technology. He sharply disagreed with his political mentor, Gandhi, on the role of science and technology in India's future as a modern state. While Gandhi saw the real origins of colonialism in modern science, Nehru argued that it was the lack of scientific progress that had allowed India to be colonized in the first place. Nehru was also highly impressed by the prestige associated with this field of scientific inquiry. He saw it as a way of making India a leading scientific power among developing countries and, at the same time, raising India's prestige in the eyes of the technologically advanced countries of the West.

The possibilities for both progress and destruction inherent in atomic energy were not lost on anyone. It could be used with equal ease to invoke economic progress as well as military supremacy. If the promise of atomic energy inspired Nehru, its destructive potential imbued in him a moral aversion to nuclear weapons. Yet, Nehru could not be blind to the reality of a nuclear world, where nuclear weapons provided the ultimate military capability. This ambiguity was to be a defining feature of India's nuclear policy for decades to come.[20] Nehru continued to exhibit an ambivalence that could not distinguish between the twin goals of state action—development and security. Even as he hoped to harness the power of science in the service of the nation, the association between war and science was always on his mind. His realpolitik instincts made him realize that defence was a primary need, and he saw, in the development of atomic energy, a nuclear option for India's security.

On 26 June 1946, Nehru declared:

> As long as the world is constituted as it is, every country will have to devise and use the latest scientific devices for its protection. I have no doubt that India will develop its scientific researches and I hope the Indian scientists will use the atomic force for constructive purposes. But if India is threatened, she will inevitably try to defend herself by all means at her disposal. I hope India in common with other countries will prevent the use of atomic bombs.[21]

In February 1947, explaining India's defence policy, he said that the use of atomic energy in warfare was likely to revolutionize all known concepts of war and defence. While India could, for the moment, disregard atomic energy, it was absolutely essential to develop the methods of using it for both civil and military purposes. Thus, moral aversion notwithstanding, Nehru, with Bhabha by his side, invested heavily in research and development of atomic energy. Nuclear capability would help with both material prosperity, and if the need arose, India's defence. Nehru said, '... Science has two faces like Janus: science has its destructive side and a constructive, creative side. Both have gone on side by side and both still go on.'[22]

In August 1947, in a constitutional committee debate, Nehru argued about the conflict between the atom bomb and what it represented and the spirit of humanity. But there was also a more rational reason for this aversion. A nuclear war would affect the whole world, including India. A nuclear arms race and heightened tensions would complicate India's policy of non-alignment. A stable international system would allow India to benefit from the assistance of both superpowers. Pragmatic reasons apart, there was also the aspect of moral ascendancy, which seems to have been important to Nehru: nuclear disarmament provided him with a heft in international politics.[23]

Bhabha and Nehru both came from patrician backgrounds, both were Cantabrigians (Nehru had a Natural Sciences Tripos) and shared an interest in arts and music. Over the years, they developed a deep friendship, recognizing in each other, the intellectual curiosity, cosmopolitanism, intelligence and breadth of knowledge that they both prized. M.R. Srinivasan, who became chairman of the AEC in 1987, said that while they all were aware of the close relationship Bhabha and Nehru shared, he became aware of just how close they were only after he read the correspondence between the two. 'I doubt very much if anyone barring Mahatma Gandhi, Indira Gandhi, and other close members of the Nehru family, had such a close relationship with Panditji. Krishna Menon may be the only other exception.'[24]

Bhabha, over the years, also got to be on familiar terms with Indira Gandhi, and Nehru's sisters, Vijay Lakshmi Pandit and Krishna Hutheesing. Nehru spent his time in jail reading books on diverse subjects, including science. He read books by James Jeans, H.G. Wells, Aldous Huxley, Bertrand Russell and J.B.S. Haldane. He once observed, 'We have to keep in mind that the present-day world and our lives are yoked to science. I have been interested in relativity and quantum theory, biology, sociology, psychology and psychoanalysis.'[25] He tapped into various sources to get books. He wanted Hutheesing to ask Bhabha for names of recent books on relativity and quantum theory. He also approached K.S. Krishnan, then at Allahabad University, for books on science.

Indira Gandhi recalled many years later, that the life of a politician lacked many of those warm moments of sensitivity that other people take for granted in their everyday life. She knew that Bhabha opened one such 'window' for her father. He always found time for him, not only because the problems Bhabha brought were important and he wanted to give them urgent attention, but because he found at the same time it was relaxing and it was an entirely new world.[26]

One of the reasons the two got on so well together is that both were a curious amalgam of East and West. Nehru observed in his autobiography, '...I have become a queer mixture of the East and the West, out of place everywhere, at home nowhere.' He also said, 'I am a stranger and alien in the West... But in my own country also, sometimes I have an exile's feeling.'[27] This doubleness of identity across two cultures made both Bhabha and Nehru exemplary figures. They both epitomized the modern traveller who brings ideas from elsewhere. The double home of East and West created a tension, both aesthetic and spiritual. To belong and not to belong was their cultural predicament: both their identities mirrored an impressive heterogeneity.

Nehru and Bhabha were already modern. India was an old idea lagging behind in time. Despite being ushered in by a colonial power, modernity gave the colonized a chance to be a part of universal history. As Nehru confessed: 'India was in my blood and there was much in her that instinctively thrilled me...To some extent I came to her via the West, and looked at her as a friendly westerner might have done. I was eager and anxious to change her outlook and appearance and give her the garb of modernity.'[28] Nehru could well have been speaking on Bhabha's behalf as well.

Atomic energy was also a source of international prestige and Nehru was trying to emerge as a world leader. It is, therefore, no surprise that among all scientific endeavours, the atomic energy programme received the maximum patronage in independent India. Nehru wanted science and technology to create what he called 'temples of modern India', and atomic energy establishments became the holiest of these shrines.[29] Helping him was Bhabha, who, with his impeccable scientific credentials, shaped India's nuclear policies according to his vision. Both chafed at the gradualism inherent in the development process and wanted to jumpstart India's economic growth.

At a press conference soon after the first atom bomb was dropped on Hiroshima in August 1945, Nehru said that he was hopeful that free India would use the atomic force for constructive purposes but, if threatened, it would try to defend itself by all means at its disposal.[30] 'I have a scientist's outlook', he said at the same press conference and hoped that the new invention would be used for constructive purposes. Little wonder then that Bhabha became a hero in a popular culture which valorized scientists, experts and development itself. Nehru counted a number of scientists as friends: Patrick Blackett, Arthur Compton, Niels Bohr and Albert Einstein, among others. He met Einstein in 1949, when he visited the US and gave him a copy of his book *The Discovery of India*. Einstein wrote back saying that he had read the marvellous book with extreme interest although the first half was not easy reading for a westerner. However, he added that it gave an understanding of the glorious intellectual and spiritual tradition of India.

Nehru's faith in science and technology was not a blind one. While he saw technology as being crucial to improving standards of living in the country, he also believed that too much emphasis on technology had led to too great a power in human hands without the moral capacity to use it appropriately.[31] Nor did he believe that scientists should have a monopoly over the truth. They needed the vision of an intellectual and the pragmatism of a politician. He believed that India's problems could not be solved by politicians or by scientists operating alone.[32]

The NPC discontinued work when the Second World War broke and the Congress leaders resigned from the provincial government protesting entry into the war. Many, including Nehru, were jailed. The NPC held 78 meetings between 1938 and 1946. Nehru attended and chaired all but one meeting. He was almost single-handedly the arbiter of India's technological trajectory. As noted, for Nehru, the nuclear programme represented the pinnacle of scientific advancement. The nuclear programme must, therefore, be seen in this context—as the zenith of science, it would herald India's return to the stage of technical progress; as a source of power, it would represent a return to the golden ages of India's past; and finally, it would allow India to demonstrate its restraint and establish moral ascendancy over the West. Bhabha and Nehru believed that India had been held back temporarily by its lack of technological advancement, and it was in this context that they started the nuclear programme. As Jahnavi Phalkey has observed, 'Modern India was being forged as at once a scientific and political project.'[33]

Chapter 26

AN INSTITUTIONAL FRAMEWORK

> ...the development of atomic energy should be entrusted to a very small and high-powered body composed of say three people with executive power, and answerable directly to the Prime Minister without any intervening link. For brevity, this body may be referred to as the Atomic Energy Commission.
>
> —Homi Bhabha's Note on the Organisation of Atomic Research in India, 26 April 1948[1]

The Nehruvian idea of science as an essential accoutrement of a modern society was one of the markers of independent India. Nuclear power was attractive for leaders like Nehru because of its multiple uses—it promised material upliftment by providing cheap electricity and could also be used to build weapons of the most destructive kind. It was also advanced science and held the promise of the latest modern technologies.

Bhabha was appointed the chairman of the Board of Research on Atomic Energy, which was set up under the umbrella of CSIR. But he was far too independent and ambitious to be tied down by such an institutional structure. As M.V. Ramana has observed, 'Bhabha was unique not only in the audacity of his vision for nuclear energy in

India but in his ability to successfully meld science with politics, capital with government.'[2] He wanted to preside over his own organization, and to this end, he sent a note to Nehru on 26 April 1948 titled 'Note on the Organisation of Atomic Research in India'.[3] Despite the economy of words, Bhabha managed to masterfully touch upon all the important issues concerning the organization of atomic energy in the country. As Itty Abraham has observed, 'In this note Bhabha strikes themes that would characterize the growth of atomic energy programmes worldwide: secrecy, centralization, and unaccountable executive power.'[4]

Bhabha wrote the note after returning from a foreign trip, which included visits to France and Canada. Soon after Independence, Bhabha began to look abroad for resources and expertise. He approached France, UK and Canada, and, in fact, had even asked Nehru for an open-ended statement confirming his position as the official representative of the Indian government for this trip. Nehru was a little reluctant since he did not want to officially approach foreign governments at that stage but he did give Bhabha a guarded letter of reference. In the past, too, Bhabha had volunteered to represent the government while in the US and Europe. Before he went to Mexico to attend the UNESCO conference at the end of 1947, he wrote to Nehru, asking him if he wanted to entrust any commission to him when he was in Europe and US.[5] Bhabha's trip was a success with the Canadians promising a tonne of uranium oxide and the French agreeing to train scientists and share scientific information. The French would also help build a factory for the extraction of fissile materials in Travancore. His visit also helped Bhabha get a sense of the kind of progress made by these countries and he concluded that India was not very far behind.

On his return, Bhabha wrote a long report to Nehru and also had discussions with Defence Minister Baldev Singh. It was apparent that defence considerations were on the minds of all concerned. Nehru asked Baldev Singh to set up a scientific advisory committee and appoint Bhabha on it. The Scientific Advisory Committee to the Ministry of Defence was established on 16 July 1948. Its members included the usual suspects: Bhabha, S.S. Bhatnagar and K.S. Krishnan.

Patrick Blackett also accepted membership, and, a few months, later D.S. Kothari was also appointed as the scientific advisor to the Ministry of Defence. The committee was later renamed the Defence Science Advisory Board.

THE ATOMIC ENERGY COMMISSION

The Atomic Energy Bill, which Nehru introduced before the Constituent Assembly in early April 1948, aimed to implement an Atomic Energy Act to create an Atomic Energy Commission and the legal framework for its operation. Secrecy was the main point of debate. Though the Act was modelled on Britain's Atomic Energy Act, it was far more restrictive than either the British or the American legislation. Research was to be completely secret, and the government would control all fissile material relevant to atomic energy, especially uranium and thorium.

Introducing the Bill, Nehru felt obliged to explain the secrecy: 'The advantage of our research would go to others before we even reaped it, and it would become impossible for us to cooperate with any country which is prepared to cooperate with us in this matter because it will not be prepared for the results of researches to become public.'[6] According to him, such great secrecy arose from the need to protect Indian materials and prospective knowledge from exploitation by 'industrialised countries in a colonial manner' and assure foreign powers that were inclined to collaborate with India that their secrets were safe with India.[7]

The Bill was criticized for its secrecy provisions, especially since they went so far as to not permit any legislative oversight at all. While most members praised the government for starting an atomic energy programme, S.V. Krishnamurthy Rao, the delegate from Mysore, raised objections about the illiberal nature of the Bill due to the excessive secrecy involved. He said that though the stated objective of the Bill was developing and controlling atomic energy, in its current form, it only seemed to deal with controls. Krishnamurthy Rao argued that while secrecy was understandable for military purposes, it was difficult to comprehend why such secrecy should also extend to peaceful uses, especially when countries that were far more advanced in the field did not have such restrictions.[8]

Nehru had to admit that the secrecy he was advocating could not be separated from the defence related ends of the atomic energy project. Unable to sustain the dichotomy, he was forced to concede: 'I do not know how to distinguish between the two [peaceful and defence purposes].'[9] The connection between developing nuclear energy and building nuclear weapons was certainly clear to Nehru. In the late 1940s the distinction between the two was blurred: it was assumed implicitly that the facilities and expertise used for generating electricity could also be used for making weapons. The Bill was passed with minor amendments, with Nehru all but acknowledging that atomic technology was essentially dual-purpose. The association of atomic energy with national defence had served to support the provisions of the Bill, not undermine them. Since the atomic energy programme would bolster India's defence, the need for secrecy and control became obvious, making Krishnamurthy Rao's concerns about secrecy redundant.

Nehru's dilemma is obvious from his statements when the Bill was being debated. He observed, 'I think we must develop it for peaceful purposes', but later added, 'Of course, if we are compelled as a nation to use it for other purposes, possibly no pious sentiments will stop the nation from using it that way.'[10] Nehru merely acknowledged that, like the UK and the US, the nuclear project in India, too, was a potential source of both military and economic power. Soon after the debates on the Bill ended, the Board of Research on Atomic Energy passed a resolution urging the government to set up an atomic pile (a nuclear reactor) and allocate ₹80 lakh for capital equipment and research. It also wanted the government to set up heavy water capabilities as soon as possible. Considering the need for secrecy, the results of all the surveys for atomic materials were to be centralized under the board.[11]

In his note to Nehru on 26 April 1948, Bhabha opined that two basic facts had to be kept in mind while deciding the structure of the new organization that would be in charge of atomic energy in the country. First, there was to be absolute secrecy. Second, given the paucity of scientific and technical personnel, some of the 'top people' would have to do more than one job at a time. The first condition required 'that the development of atomic energy should be entrusted to a very small and high-powered body composed of say three people with executive power, and answerable directly to the prime minister without any intervening link. For brevity, this body may be referred to as the Atomic Energy Commission'.[12]

Commissions, albeit with a wide variety, were familiar instruments of governance in the Anglo-Indian administrative tradition. What they all had in common was the provision of independence and protection from short-term and narrowly defined political interests. However, as Itty Abraham observes, Bhabha broke new ground when he adopted this structure for the development of atomic energy because commissions had never before been used as instruments for developing new technologies.[13]

Bhabha convinced Nehru that due to its strategic significance the AEC required an unprecedentedly higher degree of autonomy from normal governmental procedures, including financial and recruitment regulations. He wanted negotiations with international partners to take place outside the external affairs ministry channels with accountability restricted solely to the prime minister, bypassing existing rules and norms. This created a strategic enclave, which not only segregated the atomic energy programme from existing regulatory practices but public scrutiny as well. The commission model was very successful in creating advanced technology sectors where none had existed before, where a State monopoly faced no competition and where private-sector investment was unlikely due to the large-scale investment needed and the low likelihood of commercial returns.[14]

When Bhatnagar sent the organizational chart for the proposed DSIR to Nehru, he added the Board of Research on Atomic Energy as a part of the department. Bhabha objected to this on the grounds of secrecy and security. He wanted the AEC to report directly to the prime minister. If DSIR was ever put under a minister without a portfolio, it would be necessary to separate the AEC from the department if it was ever made a part of it. Bhabha noted, 'The Atomic Energy Commission whose work has important international implications must always be attached directly to the Prime Minister, and I am, therefore, strongly of the opinion that it should from the start be organised directly under the Prime Minister and not as a part of the Department of Scientific and Industrial Research.'[15]

Bhabha also pointed out that the existing Board of Research on Atomic Energy could not be entrusted with this work, since it was only an advisory body that reported to the governing council of the CSIR, which comprised 28 members. He realized that the structure of the CSIR's atomic energy committee was unwieldy and impractical, composed as it was of not only scientists and bureaucrats but also businessmen with vastly varying degrees of scientific expertise. The committee members had little in common and he convinced Nehru to create an entirely different structure to organize atomic energy. Bhabha also wanted the AEC to have its own independent secretariat. He wanted ₹1 crore to be made available to the commission over the next four years. He also asked for permission to continue his negotiations with France, UK and Norway, under conditions of absolute secrecy. He wanted to prepare each bilateral agreement in its entirety, which would then be submitted to the government for approval.[16]

A number of factors explain Bhabha's remarkable success in creating this new structure. First, there were the close personal ties between him and Nehru. The Prime Minister realized that if anyone could translate his vision for the technological transformation of the country into a reality, it was Bhabha. He also had the support of the Tatas, and Bhatnagar who was the most powerful technocrat. Bhatnagar repeatedly enabled Bhabha's plans by blocking resistance from other Indian scientists (mostly Saha), who questioned the power and resources being given to the AEC. Bhabha argued that to successfully develop atomic power in India, he would need to completely control the entire enterprise. It was not difficult to marginalize other Indian scientists given Bhabha's professional standing combined with Bhatnagar's institutional clout.[17]

On 28 April 1948, two days after his note to Nehru, Bhabha apprised Bhatnagar of the suggestions he had made.

> In fact I have said [...] that the Department of Scientific and Industrial Research should be directly under the Prime Minister. He appears fairly keen on this himself. As I have told you, I have suggested that there should be an Atomic Energy Commission

> composed of three people with you as the chief administrative officer so as to ensure close liaison with the Department of Scientific and Industrial Research of which you would be the Secretary.[18]

Bhatnagar responded on 3 May, enclosing a letter he had written to the Prime Minister. The tone of the letter suggests that Bhatnagar was too tired to cross Bhabha in any way. 'The whole of the work should now be taken up by you in the country's good. I wish to remain outside this strife as you can fight much better as a young man with definitely more vigour than myself [...] I don't care what role I am given.'[19]

In his letter to Nehru, Bhatnagar said: 'I agree with him on all points excepting my own position. I should trust the whole of this work to Dr Bhabha. I have no hesitation to be of service where I can but Bhabha may be entrusted with executive duties for atom work. As a matter of fact, Bhabha should be given a government post right away.'[20]

On 5 May, Bhabha wrote to John Cockcroft for details on the administrative setup of the atomic energy establishment in the UK. He wanted to know if the Telecommunications Research Establishment at Malvern was an integral part of the setup or merely developed suitable instruments for the atomic energy establishment. In the second part of the letter, he wanted designs for the magnet for the Wilson Chamber. He said that he would understand if Cockcroft could not send him the designs but he would be disappointed nonetheless.[21]

On 22 May 1948, Nehru sent a note to the cabinet to authorize the funds Bhabha had requested. He agreed with Bhabha on secrecy and asked for ₹1 crore to be disbursed over a period of four years. He also wanted a heavy water plant to be set up and recommended that the AEC—which was to be a small, high-powered body—should report directly to the prime minister.[22] Nehru was doing exactly as Bhabha was telling him, an indication of both his trust in Bhabha's abilities, and the excellent rapport the two shared. The two had dinner or tea together almost every fortnight. As Robert S. Anderson has observed, 'Their mutual attraction enabled them to speak the same language.'[23]

The Atomic Energy Act was passed on 10 August 1948, and the AEC was legislated into existence. Although the Act did not explicitly mention the development of a nuclear weapons structure as one of the programme's objectives, its careful wording avoided any provisions that would exclude it in the future. If one reads between the lines, Bhabha's note makes it clear that he was in no doubt that a national nuclear programme would eventually acquire a military objective. India's policymakers perceived the nuclear programme as a national endeavour within global competition and, therefore, the decision-making structure was designed to meet the central requirements of efficiency and secrecy. Bhabha and Nehru, the only two who really mattered, led the nuclear programme in an autocratic, informal and opaque manner.[24] Itty Abraham, among others, has argued that the institutional structure guaranteed the secrecy necessary for maintaining the myth of nuclear technology as the symbol of India's path to modernity.

The AEC was a mirror image of the scientific advisory committee to the prime minister. Bhabha became chairman with Bhatnagar and K.S. Krishnan (soon to be appointed director of the NPL) being the other two members. The continuance of Bhatnagar as a key scientific administrator, even in independent India, and Bhabha's rise as the leader of the national nuclear establishment became apparent in 1946 and was cemented in 1948 with the formation of the AEC.

A little before the passing of the Atomic Energy Act, Bhatnagar had met Saha and S.N. Bose at a friend's house, where the conversation seems to have taken an ugly turn. He later wrote to Saha about his unhappiness with the 'anti-Nehru' conversation. Saha tried to cover up his indiscretion by saying that his complaints were against the AEC, and he had brought Nehru into the picture only to emphasize that without Nehru's continued help he could not have made any progress. Saha's defence shows that he was aware that he could not afford to lose Nehru as an ally. He reminded Bhatnagar that he had made Nehru's acquaintance well before both Bhatnagar and Bhabha, and this contact had been established at a time when such associations were certainly not profitable if not downright dangerous. In a not-so-subtle dig at Bhatnagar, he said, 'Witness that I was never knighted by the British Government.'[25]

Saha's grievances may have been valid, but it was Bhabha's star that was in the ascendant. In a few short months, Nehru and Bhabha had put in place the mechanisms with which they completely dominated atomic energy under their joint command. Bhabha was given a free hand to negotiate with foreign powers in the name of the State while Nehru managed atomic affairs at the level of the cabinet. Bhabha had managed to delink the atomic energy programme from oversight by scientific or any other committees. Unlike most policy matters where the cabinet was the ultimate authority, its role in overseeing developments in atomic energy was limited, if not non-existent. This was because the AEC reported directly to the prime minister's office. Bhabha's establishment managed to avoid standard checks and balances from the very beginning.

Nehru was also very much a part of the secrecy. In 1952, Finance Minister Chintaman Dwarakanath Deshmukh, frustrated with the lack of information about the working of the AEC and the high costs associated with funding it, had asked Nehru for more information. The latter had replied: 'The work of the AEC is shrouded in secrecy. I try to keep in touch with it and get reports from time to time. I do not know how else we can proceed in this matter.' This was, at best,

disingenuous since Nehru was being regularly briefed by both Bhabha and Bhatnagar on the subject. He summarized the AEC's report to Deshmukh but asked him not to share it with anyone else. Nehru also marked a copy of this letter to Home Minister Gulzarilal Nanda. He did not want anyone else to see it.[26]

In the same letter, he provided Deshmukh with budget estimates for the next four years, observing that the Planning Commission would have to form its own opinion about the AEC's programme, giving the impression that it would not count for much. As Robert S. Anderson has pithily observed, 'This is the basic agreement imposed on the cabinet about the whole AEC operation: Nehru is saying, read our summary version of the atomic program plan, return the document, don't ask further questions in cabinet, and approve the money.'[27]

Nehru was to address the question of secrecy once more when he laid the foundation stone of TIFR in January 1954. While admitting that science does not flourish in secrecy, he said that if India was to continue dealing with countries that had advanced nuclear programmes, secrecy would be paramount because those countries desired it. The secrecy was not imposed simply because he and Bhabha wanted it: it was also the result of conditions imposed by foreign agencies on which India would necessarily depend for the negotiation of nuclear power. Bhabha had already been told that agencies of other governments could not communicate with him or the AEC unless legally enforceable secrecy governed his commission.

It was more difficult, however, to neutralize the civil service, and here again Nehru's support was critical. Bhabha kept the influence of bureaucrats to a minimum, ensuring that scientist–technocrats would always hold a majority of positions on the commission. More importantly, he insisted that the chairman of the AEC be given a simultaneous appointment as secretary to the Government of India, effectively joining technical expertise with State power. This would ensure that disagreements over policy would now be between equals in the government's administrative hierarchy, giving the technically qualified AEC scientists an edge.

Finally, after consolidating the formal powers of the commission and his own, Bhabha wanted the headquarters of the AEC to be located in Bombay rather than New Delhi. This was extraordinary to say the least, but Nehru agreed, and this, perhaps more than anything else, reflected the extent of Bhabha's power. Bombay was home for Bhabha. The TIFR was there, as were the headquarters of the Tata group. In Bombay, far from the interfering eyes of the bureaucracy, Bhabha could organize and run the atomic energy project as he saw fit.

SAHA'S EXCLUSION

Where did Saha figure in all this? His snubbing and exclusion from the AEC, and his marginalization in atomic energy matters has been the dominant narrative, which has had a long life in Indian science historiography. The truth, however, is different. What is generally not so well known is the fact that Saha, despite his severe criticism of Bhabha's plans, was thought necessary to the AEC. Nehru invited Saha to be a member of the commission on not one but two occasions. After the first refusal, Nehru spoke to Saha over the telephone to get him to change his mind. He failed. When Nehru persisted, making another offer, Saha spoke to Nehru directly, expressing his inability to be part of the government's atomic energy setup.[28] It was only then that K.S. Krishnan was appointed as the third member.

Saha had opposed the idea of concentrating nuclear research under one institution or umbrella organization from the start. He had also expressed his reservations about the choice of Bombay as the centre of nuclear research, given its coastal location. In a letter to S.P. Mookerjee written in April 1948, he wrote: '[...] there is a very dangerous clause in the speech of Pandit Nehru in introducing the Atomic Energy Bill that all research would be more or less concentrated. I believe that he has done so under the influence of Bhabha and Bhatnagar.'[29]

Saha was opposed to many of Bhabha's ideas on how to initiate a full nuclear programme in the country. In a letter he wrote to Saha in January 1947, Bhabha had said that India needed to import one or two really first-class men from abroad, for a limited period, say, two years to give a fillip to experimental work in cosmic rays and nuclear physics. Saha, on the other hand, wanted the country to first address the dearth of good mechanics and laboratory men, as well as the lack of engineering and manufacturing firms for the production of electrical goods, machinery and scientific instruments. In his reply to Bhabha in February 1947, Saha bemoaned the fact that he had been unable to get the equipment he desired from abroad, even after he had visited the firms personally and made his request.[30]

Saha had also opposed the creation of the AEC on the grounds that India did not have enough trained scientists and engineers for the development of nuclear energy. He felt that India first needed to grow industrially and train personnel in nuclear physics in the universities. Saha emphasized a university position, something that both Bhabha and Bhatnagar lacked, which he viewed as a serious deficiency. He wanted India to follow the French model since France, too, had faced a similar shortage of raw materials, instruments and trained personnel. Saha was speaking from first-hand knowledge, having discussed this subject with Frédéric Joliot-Curie in 1948 at the 10th anniversary memorial conference honouring Ernest Rutherford in Paris and later when he came to lay the foundation stone of the Saha Institute.

Saha advocated a more regionally distributed system of scientific research based primarily in universities and not institutes directly administered by the State. He argued for technological self-sufficiency so that India could have its own equipment and institutions to carry out original physics research. His idea was to learn from foreigners without becoming dependent on them. He called for the 'Indianization' of science—Indians should eventually contribute the specialized knowledge and advanced techniques required for all Indian industrial needs. Yet, even as he argued for the long-term technical autonomy of the postcolonial Indian state, he was in no way advocating outright rejection of international dependence. Instead, he wanted a more strategic use of international scholarly alliances for the benefit of the country.[31]

Saha's refusal to join the AEC must have come as a relief to Bhabha and the others who were kept on their toes by his constant criticism. Saha perhaps thought that his refusal would send the most critical signal possible. And since he did not hold an official position, it would be easier for him to criticize Bhabha and his programme. But, he would eventually feel the humiliation of being financially dependent on CSIR, and later, AEC. Though Saha's rejection of the AEC membership deprived him of important information, his interest in the subject was unceasing. He used his network of former students and friends to keep abreast of developments at CSIR and AEC.

Saha's old friend, J.C. Ghosh, his former student, D.S. Kothari, along with the geologist, D.N. Wadia, all kept him informed of developments at AEC. It is certain that fewer questions would have been raised publicly had Saha not persisted in his criticism of what Bhabha was doing. His editorials in *Science and Culture* were highly influential and often proved to be the starting point for journalists to pursue stories on India's nuclear programme. There is also little doubt that much of the early information on the atomic energy programme was made public largely due to his efforts.[32]

BHABHA, BLACKETT AND NEHRU

The AEC met for the first time on 15 August 1948, exactly a year after Independence. Nehru and Patrick Blackett were present at this meeting. Blackett had played an important role in India as a military consultant and 'scientific intervener', ever since he had visited India for the first time in 1947 on Nehru's invitation. At the 1946 Empire Scientific Conference in London, Mahalanobis and Bhatnagar proposed a scheme called 'Short Visits of Scientists from Abroad'. Mahalanobis drew up a list for this proposal, which included J. Robert Oppenheimer, Norbert Wiener, Niels Bohr, Patrick Blackett, Henry Tizard and J.B.S. Haldane. Nehru agreed to sign the invitations to give these visits prominence.

Blackett may have also met some Indian scientists in London, including Bhabha. He was invited in late 1946 and accepted the invitation. When he came to India to attend the meetings of the Indian Science Congress in early 1947, he found himself on the same aircraft as Nehru and they managed to talk. Later, he lunched with Nehru at his home in Delhi. Much of Blackett's engagement in India resulted from these conversations with Nehru. From 1948 onwards, he usually stayed as Nehru's guest; scientists realized that one way to Nehru, and to Bhabha and Bhatnagar, was through Blackett.[33] Blackett also believed that the affection Nehru felt for Bhabha filled a deeper need. 'He liked intellectual company, and he did not get it except in Homi and people like that.'[34]

By 1948, the idea of science as an instrument of national security had been explicitly defined within the state apparatus, and the Blackett Report of 1948 positioned science and defence as inextricably related to each other through the common goal of self-reliance. His report—essentially a long-term plan for the defence of the state—crucially depended on the growth of an independent scientific branch. Blackett had been closely associated with the invention of operations research techniques for wartime logistics and planning and was 'the epitome of the figure of the scientist allied to the state for common purpose'.[35]

Blackett also became very friendly with both Bhabha and Bhatnagar. As we have seen, he already knew the former from his Cambridge days. When Bhabha was setting up TIFR, he confided in Blackett. In 1941, he asked Blackett to lend him the big magnet he had built at Cambridge and Manchester for cosmic ray research but nothing came of it. Blackett used to attend parties at Bhabha's house and Bhabha even sketched his portrait. Blackett, in turn, hosted Bhabha at his own home in Manchester in 1946.

Whenever Blackett was in India, Bhabha would 'command' part of his schedule and arrange meetings for him, including introductions to captains of industry, many of whom were Parsis. Blackett called Bhabha 'my best personal friend' at the time of Bhabha's death in 1966. Blackett also shared a personal relationship with Bhatnagar and the two met frequently in Delhi and London.[36] Blackett also helped shape the career of D.S. Kothari, who headed the Defence Science Organisation when it was established in June 1949. Its structure was modelled on the one Blackett had prescribed for the UK in 1946. That Blackett had a powerful set of friends, who were part of the establishment and wanted changes in it, only enhanced his influence with Nehru.

As a confidant of Nehru, Bhabha and Bhatnagar, Blackett was kept abreast with Indian developments on atomic research from the beginning. The minutes of the meeting held on 15 August 1948 are not publicly available. However, Blackett's views on them are available from a letter Bhabha wrote to him to verify the recording of the proceedings. Blackett had said that nuclear physics research was carried out through big machines and cosmic rays. By big machines, he meant particle accelerators and, possibly, reactors. Blackett advised Bhabha against big machines since India did not possess the engineering capacity for the job. Smaller machines were no longer useful for pioneering work. The cyclotron in Calcutta could be used for training technical manpower, which India lacked. Like Bhabha, Blackett also showed a clear preference for cosmic ray research. He wanted cosmic ray research to be encouraged, opining that this made sense for India since the geomagnetic equator ran through the southern part of the country. If India wanted to make fundamental advances in scientific research, cosmic ray research would have to be encouraged.[37]

Blackett also wanted India to build a small atomic pile that would be useful as a neutron source, not arouse suspicion and boost the self-confidence of the people working on it. A small pile would be more politic and also scientifically appropriate since India was not in a position to make a bomb. It must be remembered that in those days, it was not possible to sustain a distinction between nuclear weapons research and atomic energy research because both required the construction and operation of nuclear piles. Blackett felt that cosmic ray physics and medical research were two areas of fundamental research that India could gainfully pursue. Regarding the teaching of nuclear physics, he wanted it taught at the postgraduate level, in line with Bhabha's and Saha's thinking on the subject.[38]

Blackett also helped Bhabha later by giving his expert views on the proper organization of atomic research, drawing on his knowledge of the setup in Britain. In 1951, he met the scientific attaché to the Indian High Commission who was studying the organization of institutions where atomic research had started. Although the UK Atomic Energy Authority was cautious about it, Blackett knew it was being done for Bhabha and Bhatnagar and gave his views in a candid manner.[39] Bhabha confided in Blackett a great deal, something which kept the latter updated about the current strategic thinking in Delhi and in atomic energy circles elsewhere. Like Bhabha, Blackett was passionate about the role of science and technology in developing countries. Both received unstinting support from Nehru who had assigned an important place to scientists in national development and had even subordinated civil servants to them. The Prime Minister also got carried away on occasion. At the inauguration of NPL in Delhi in January 1950, he declared: 'I am so fascinated [...] that I feel how much better it would have been for me to be the Director of this Laboratory, if I had the competence than to be the Prime Minister.'[40]

The Atomic Energy Research Committee met again on 1 September 1948, possibly one of its last meetings before it made way for the newly established AEC. Saha was absent, claiming he could not procure timely passage to Bombay. He had attended only two meetings until then, and his absence was doing little to further his ambitions of shaping policy on atomic energy. His absence from the meeting that had decided on the proposal to start nuclear physics research in Bangalore had been crucial. Supporting this proposal would have strengthened his position and that of advanced studies in nuclear physics in Indian universities. Saha was also involved with multiple projects at the time. They included river and flood management and refugee housing in Calcutta. Since his energy was spread over diverse fields, Saha lacked a singular focus upon this aspect of organizing for science. While his laboratory in Calcutta continued research in nuclear physics, it was constrained by its origins in the university system and struggled to achieve the status of a national institute.

Chapter 27

FISSILE MATERIALS AND THE FRENCH CONNECTION

> ...it was pointed out that the quickest and most desirable way of developing atomic energy in India would be to come to an agreement with the Governments or atomic energy agencies of one or more countries such as Great Britain, France and Norway. Such agreements would be on mutually advantageous terms involving the exchange of raw materials used... and the pooling of scientific and technical information.
>
> —Homi Bhabha's Note on the 'Organisation of Atomic Research in India', 26 April 1948[1]

Before uranium was discovered in the Jaduguda mines in Bihar, the monazite sand found in Travancore, a principality on the southwest coast of India, was considered the primary fissile resource in India. In 1945–1946, the episode with the Maharaja and the Diwan of Travancore came to the fore. The sand, which contained thorium, had been sold to European countries primarily for use in the gas lighting and mantle industry. Four firms—of which a British firm, Hopkins and Williams Ltd. was the largest—had been awarded concessions for the extraction of minerals from the sand.

TRAVANCORE AND MONAZITE

With the outbreak of the Second World War, the British became aware of the strategic importance of monazite. So did the canny Diwan of Travancore, C.P. Ramaswami Iyer, who permitted the American geologist Andrew Corry (who later became the US attaché) to survey the area, hoping to attract other firms to bid for concessions. In fact, he had been made aware of the value of his reserves in 1937, when the Nobel Prize-winning English chemist, Frederick Soddy, an authority on radioactivity, visited Travancore to study its natural mineral reserves. Soddy told Iyer that the scientific extraction of monazite would bring in ₹2–3 crores annually to the state. This made the Diwan order the first survey of Travancore's mineral reserves.

In August of 1945, Iyer told Corry that he was going to place an embargo on all shipments of monazite, except those needed for the war. Since the war was almost over, it meant all exports would cease.[2] Travancore wanted to end the sale of raw sands and desired access to the capital and expertise needed to set up domestic processing plants. When the newly formed CSIR announced that they would survey Travancore's thorium deposits, Iyer refused permission to do so. He said that Travancore was the sole owner of the sands and would not surrender control to any agency, including the British government.

Monazite sand had by then been transformed from a commercial mineral to a strategic resource. The Diwan wrote to the Maharaja that atomic power would revolutionize all industry and power production. Aware of the special position Travancore enjoyed by virtue of its mineral resources, Iyer developed ambitions for his state.[3] A staunch monarchist who distrusted the Congress leadership, the Diwan made a strong case for Travancore's right to regain its independence after the British left, believing that Travancore's economic resources—especially its valuable export commodities that included tea, coffee and rubber—allowed for a viable independent existence.[4]

In September 1946, Travancore began secret negotiations with monazite extracting companies in the UK. The news soon reached the tabloid press and an exposé appeared in the *Blitz*. Nehru soon came to know about the deal as well as the fact that Travancore was flirting with the idea of independence. When questioned on the subject in the Legislative Assembly, Iyer prevaricated saying that no foreign company had been given a contract, when in fact secret negotiations with a British firm were already in an advanced stage. The negotiations were concluded in early 1947, and the British informed the interim Government of India as soon as the agreement was signed. Bhabha was also told about it and expressed his satisfaction at having being informed in advance. At that time, at least, he does not seem to have thought that Travancore's action conflicted with India's interests.[5]

However, when rumours began to circulate about the Travancore arrangement, there was consternation among the scientists at the ISC in January 1947. Nehru got a resolution passed that the State should own and control all such minerals, especially those required for generating atomic energy. In a note written on the subject in February 1947, Nehru observed: 'I agree with Dr Bhatnagar's suggestion that the Central Government should not allow surplus monazite or thorium nitrate to be exported from Travancore except through the Government of India, who should purchase the whole of the produce.'[6]

Speaking to the Indian Cabinet in April 1947, Nehru reportedly said that he would be willing to use air power against Travancore to make it toe the line. He was in no doubt that the sale of India's mineral resources should involve the Government of India. He asked Bhatnagar to obtain first-hand information about Travancore's deals with Britain. Bhatnagar was furious that he had been blindsided despite the *Blitz* exposé. He took Bhabha with him and the two met the Diwan. Surprisingly, Iyer capitulated without a fight and agreed to the setting up of the [Travancore–India] Joint Commission on Atomic Energy in early June 1947.

On his return, Bhabha convinced Nehru about the monopoly of the State over all rare minerals. He no longer thought that it was fine to allow Travancore to export monazite, which was a national asset that could be used to obtain foreign technology and expertise. It was far too valuable a resource to be controlled by anyone other than the State.[7]

After India's independence, Travancore was merged with Cochin to become part of the Travancore–Cochin union and, later, the state of Kerala. This episode proved to be decisive in the larger political battle of establishing territorial sovereignty over constituent parts of India and was an object lesson for others who harboured thoughts of independence.

The Atomic Energy Act of 1948 gave the government sole right over substances like uranium, thorium, beryllium and others. Accordingly, a Rare Minerals Survey Unit was set up, which subsequently became the Atomic Minerals Division under the Atomic Energy Commission in October 1950. The distinguished geologist D.N. Wadia (who became an FRS in 1957) was entrusted with the task of organizing the activities of this division. His classic textbook, *Geology of India,* written in 1919, is used even today.

In a related development, Indian Rare Earths Ltd. (IREL) was incorporated in August 1950 as a private limited company that was jointly owned by the Government of India and the then government of Travancore–Cochin. The immediate objective of IREL was the processing of monazite for recovering uranium and thorium values in the form of a uranium–thorium concentrate and for producing rare earths compounds. The company's operations started with a plant at Udyogmandal, Kerala, in 1952, which was subsequently called the Rare Earths Division. It is not generally known that Nehru, at the prompting of Bhabha and Bhatnagar, had invited J.R.D. Tata to be the chairman of IREL in March 1949, telling him that the actual work of the chairman would be minimal. J.R.D. declined saying that unfortunately he was one of those who could not undertake a job without taking a deep and detailed interest in it. His deputy, J.D. Choksi, Rustum's elder brother became chairman.[8]

URANIUM MINING

The main task of the chemistry programme initiated by Bhabha when TIFR was still located at Kenilworth was to analyse uranium ore and beryl samples. When space became a constraint, the chemistry laboratories moved to abandoned sheds in the Bombay Dyeing compound in central Bombay in 1953. It was around this time that the Atomic Minerals Division identified a relatively low-grade uranium deposit at Jaduguda in the Singhbhum district of Bihar. In 1954, Bhabha became chairman of a committee called the Uranium Survey Development and Beneficiation Committee set up to periodically review the position of uranium production. Its other members included Homi Sethna, Brahm Prakash, D.N. Wadia and Jagdish Shankar.

It was decided to set up a uranium metal plant at Trombay and a large uranium mill project at Jaduguda. Sethna, in association with S. Fareeduddin, was entrusted with the task of setting up both. Fareeduddin who joined AEC in March 1953 had an MS in chemical engineering from the US. The first uranium ingot was produced on 30 January 1959, about a year after the construction of the plant had commenced. Development work on the extraction of uranium from the Jaduguda ore started in the late 1950s at Trombay. Pioneering work was done during the project phase at Jaduguda, which lasted from 1962 to 1967. Not only was the technology new, even the scale of operation was large. As Fareeduddin observed, 'We had decided to take on this programme without any outside help. Both Bhabha and Sethna were fully supportive on this aspect.'[9]

COOPERATION WITH THE FRENCH

Given India's foreign exchange crunch after independence, Bhabha was compelled to use the AEC's monopoly on fissile materials for profit. He was aware that a strategic interest had to be developed in the minds of foreign collaborators. Instead of using sterling balances for purchasing materials and equipment, he shrewdly used thorium and beryl as bargaining chips in negotiations for reactors and heavy water. Neither did India have a strategic resource, like oil, nor did it manufacture anything that appealed to foreign powers. Thus, it was thorium and beryl, coupled with the rivalry of the US, Canada and Britain with France and Russia that strengthened India's bargaining power to some extent. Even with limited resources, Bhabha and Bhatnagar were given a free hand in negotiations, access to scarce hard currency and the assistance of India's diplomatic machinery abroad. Bhabha was on friendly terms with Braj Kumar Nehru, Pandit Nehru's cousin, whom he had met in Bombay in 1947–1948. B.K. Nehru had occupied many diplomatic posts, including being India's ambassador to the US and was privy to the best intelligence at the highest levels, something that Bhabha found useful.

Bhabha and Bhatnagar traded thorium and beryl not for cash but for equipment, designs, drawings and training. They were travelling to London and other foreign capitals about thrice a year and to North America once or twice a year. Occasionally, Nehru was also directly involved. He travelled to Ottawa in 1948 to discuss nuclear cooperation and to Geneva and Brussels in 1949 to procure centrifuges for uranium separation.

India's first cooperation in the nuclear field was with the French. When an agreement was signed between Indian Rare Earths and a French firm to construct a monazite processing plant in India, Bhabha wrote to Frédéric Joliot-Curie in February 1949 expressing his delight and the hope that the agreement would further promote cooperation in scientific matters between the two countries.

France remained outside the Manhattan Project, as it had been occupied by the Nazis. However, five French scientists had participated in the effort: Bertrand Goldschmidt, Jules Guéron, Pierre Auger, Hans Halban and Lew Kowarski. France viewed the Manhattan Project as an essentially Anglo-Saxon endeavour. Charles de Gaulle, the leader of Free France during the Second World War, was never officially informed of the progress made on the bomb. It was left to these scientists to keep him posted of progress in this field when he visited Ottawa in July 1944. The post-war feeling that the French contribution to nuclear research had gone unappreciated, coupled with feelings of hurt national pride, resulted in liberated France pursuing nuclear objectives with great enthusiasm. Adding to this feeling was the Anglo-American information censorship, which the French viewed as an attempt to block their resurgence during the Cold War. It has been argued that the nuclear programmes of both India and France had much to do with restoring national pride through technological prowess—atomic energy involved technology that only a few countries could access, making it a desirable symbol for national progress and uniqueness.

In the post-war years, the US attempted to preserve the censorship of information and control strategic minerals around the world. This included the monazite in Travancore. However, Washington met with stiff resistance from New Delhi. Bhabha was quick to identify the potential of this strategic resource and got the government to impose an embargo on the export of monaztie. The Indian ban on monazite export also affected the British. Initially, there was some ambiguity about the applicability of the Atomic Energy Act of 1948 to Travancore since it was a princely state (and not a part of the Indian Union). However, after it joined the Indian Union in July 1949 to form the province of Travancore–Cochin, this ambiguity no longer existed. In 1950, a corporation was set up for the joint exploitation of monazite in the state, ending British hopes of future access to this resource. Given this background, France provided a good alternative for India.

In January 1950, Frédéric Joliot-Curie came to India to attend the ISC. Nehru invited him to attend a special meeting of AEC on 16 January 1950 with Bhabha, Bhatnagar and K.S. Krishnan. Joliot-Curie had almost single-handedly guided the organization of nuclear energy research in France without support from the British or the Americans, something that did not make him very popular in those circles. His belief that science and technology were forces for the reconstruction and independence of France resonated with Nehru's science-based industrialization. His Left-leaning sympathies made him identify closely with the Indian situation. When Nehru made clear India's political commitment to the peaceful uses of atomic energy on matters of principle, it struck a chord with Joliot-Curie, who was also a proponent of harnessing the power of the atom for peaceful uses.[10]

When Nehru asked him how long it would take for nuclear energy to be available, Joliot-Curie thought the prospects were promising in a decade or so. More importantly his views on how to proceed on the matter must have given Bhabha's goals a shot in the arm. He suggested that, given India's lack of trained manpower and resources it was necessary to establish only one centre for atomic energy, combining both scientific and industrial characteristics.[11]Joliot-Curie offered a bilateral agreement concerning strategic minerals, such as beryllium and thorium, which was eventually finalized in 1951.

A day after the meeting with Nehru, Joliot-Curie met Bhabha, K.S. Krishnan and Bhatnagar at Bhatnagar's residence for a secret meeting. Joliot-Curie offered to share technical information on the purification of uranium, graphite reprocessing and designs of a low-power reactor in exchange for exports of thorium, beryllium and mineral oil for manufacturing graphite. The offer also included the sale of uranium, should it be discovered in large quantities at a later date. There is no doubt that this was an unusual offer, given that all the nuclear countries at the time were keen on maintaining control over nuclear technology and information.

Joliot-Curie visited TIFR and Saha's institute in Calcutta. He came away quite impressed with TIFR, convinced that it had the necessary qualities for successful work in nuclear energy. Saha's institute in Calcutta left him less than impressed. He commented adversely on the lack of cleanliness and the thoroughness of the procedures in the laboratory. He also criticized the lack of an expert who could guide the young researchers. Joliot-Curie had no reason to be biased against Saha, so his views can be taken at face value even if there are no other contemporary accounts to corroborate his observations about the institute in Calcutta. His assessment would certainly have enhanced Bhabha's stature with Nehru and perhaps taken the sheen off Saha's claims about his own institute.[12]

Like Blackett, Joliot-Curie also advised Bhabha to build a small pile. Bhabha wrote, among others, to Hans Halban (Joliot-Curie's former assistant) at Oxford, asking if he was interested in spending a few months or even a year at TIFR. They were procuring a radium–beryllium source, Bhabha said, but anything else he would have to bring with him. Bhabha also told him that they had no accelerators at TIFR. Halban declined. He said that he had spent four years getting his equipment established, and what was more, graduate students needed his constant support and supervision.

In March 1950, a contract was drawn between the Commissariat à l'énergie atomique et aux énergies alternatives (CEA) and the AEC on beryllium reprocessing. A month later, Joliot-Curie was relieved of his duties as head of CEA, but progress on the agreement continued. In July 1950, Jules Guéron met Bhabha in London to negotiate the details of the agreement. Francis Perrin, who succeeded Joliot-Curie at CEA, visited Bombay in December 1950 and met Bhabha for further negotiations. The agreement was finalized in August 1951 and was the first bilateral agreement that CEA signed with the atomic energy commission of another country. That same year, the French began constructing the monazite processing plant in Alwaye, Kerala.[13] Goldschmidt observed that they had been very isolated due to obvious political reasons and the prevailing Anglo-American policy of atomic secrecy. He added that Bhabha had been the first leader to propose a collaboration with France.

The Indo–French agreement called for two stages of bilateral cooperation. In the first stage, preliminary theoretical studies were to be conducted on the construction of a beryllium-moderated nuclear reactor of moderate power, which would use natural uranium as its fuel. This first stage would take place in France. In the second stage, the reactor would be constructed in India over a period of five years. At the end of the five-year period, it was decided that CEA would transfer the control of the reactor to AEC.

During the same five-year period, AEC would supply beryllium to CEA for constructing a similar reactor in France. Since the original reactor in India would become the property of AEC, the construction costs would either be paid directly or eventually reimbursed by AEC. Each party would be responsible for the expenses related to their personnel involved and for the material costs for the studies undertaken in their own countries. The agreement required CEA and AEC to cooperate fully in all appropriate scientific and technological matters.[14]

Under the agreement, France was responsible for supplying all the uranium required for research and construction of the reactor while India was to supply the beryllium. The AEC granted CEA priority access to Indian beryllium. In the face of the US monopoly over strategic minerals, notably uranium, in Belgian Congo, Joliot-Curie sought out international partners. Finding himself at the receiving end of American and British attempts at controlling their own strategic minerals, Bhabha also gravitated towards bilateral cooperation with the French.

However, at the same time, Bhabha was in the midst of negotiating a similar deal with the British. When John Cockcroft found out about the arrangements Bhabha had made with the French, he informed the British Cabinet Committee on Atomic Energy that there was no need for the UK to negotiate a beryllium agreement with India. The British and the Americans were a little apprehensive that the Indians should pursue their atomic energy activities under French tutelage and wanted to wean India away. When Cockcroft informed the government about the Franco–Indian deal, he expressed the view that it was perhaps a delayed response from the UK that had made the Indians partner with the French.

F.C. How, of the Ministry of Supply, wrote a letter to Sir Roger Makins of the Foreign Office, which shows that the British knew of Bhabha being in talks with the French.

> Then at the Official Committee, as you will remember, Perrin said that it had been a guiding principle in the negotiations with Bhabha that we would not become entangled with the French. In these circumstances, I do not think it would be safe to make any move until we have got round a table. It should be possible to do this about the end of the month, and an Oriental should not regard a delay of less than three months in answering a letter as being abnormal! I do not think it was our delay in answering his earlier approaches which caused him to turn to the French because it appears from Cockcroft's paper that Bhabha had made his arrangements with the French at a time when he was still amicably discussing possible similar arrangements with us.[15]

In the same letter, How observed: 'The proposals set out in Bhabha's letter seem to me to be extremely one-sided and to involve our giving far more than we are likely to receive.'[16]

By May 1951, information regarding low-power reactors was declassified by the US, but personnel who had worked at the US AEC were prevented from conducting research in any other country on security grounds. Furthermore, the British were also prevented from exchanges of information and personnel without prior approval of the US. No wonder then that Cockcroft observed with some regret that the American policy of information censorship had left the field free for the French to step in.[17]

Joliot-Curie's visit resulted in a slightly curious newspaper report in January 1951. Nehru, according to the report, was supposed to have offered the French scientist a leading position in India's atomic energy setup. This was immediately denied by Joliot-Curie, and Nehru told Bhabha in a subsequent conversation that no such thing had been discussed in his brief meeting with the Frenchman. Bhabha wrote to J.R.D. Tata that Joliot-Curie had hinted two years ago that he might be prepared to come to India and work if the situation in France became too difficult for him. However, Bhabha said he had never taken the hint and said that he was unaware of any offer of even a post under AEC, much less as its head.[18] If Joliot-Curie had indeed voiced such a wish, it is not difficult to see why he changed his mind later on. His moving to India in a position subordinate to Bhabha is difficult to imagine; similarly, it is inconceivable that Bhabha would have agreed to work as Joliot-Curie's subordinate.

In February 1960, France exploded its first nuclear bomb in the French Sahara. Two more tests followed in April and December of that year. Differences intensified between the French and the Americans when the latter refused to readily admit France into the exclusive nuclear club. In 1961–1962, Bhabha began to explore the possibility of importing an unsafeguarded natural uranium reactor from France. However, this agreement failed to materialize by 1963. Owing to the attractive Exim Bank loan terms arranged by Washington, Bhabha was lured into accepting the US's offer instead, which, as we shall see later, involved the construction of two boiling water reactors in Tarapur as part of a turnkey project.

Chapter 28

ATOMIC ENCLAVE

A booster in the form of a foreign collaboration can give a plane an assisted take off, but it will be incapable of independent flight unless it is powered by engines of its own. If Indian industry is to take off and be capable of independent flight it must be powered by science and technology based in the country.

—Lecture on 'Science and the Problems of Development' at the International Council of Scientific Unions, Bombay, 7 January 1966[1]

Meghnad Saha and Bhabha had different notions about the goals of science and technology and the means to achieve them. Scientist-turned-historian of science, Abha Sur, has argued that Saha emphasized judicious and equitable distribution and advocated participatory democracy even in engineering projects that involved highly technical information.[2] Policymaking in the 1950s was dominated, if not monopolized, by economists and technocrats. The emphasis on technical and other expertise meant that the public discussion about India's progress had little room for debate. Most important decisions were taken by this small cohort of experts and nowhere was this more apparent than in the working of the

Planning Commission, an exclusive enclave removed from Parliament and the Cabinet, which was in charge of policy formulation. Despite Saha's deep political roots in the nationalist movement, Bhabha's more elitist approach prevailed.

SAHA'S CRITICISM

Saha's influence in national science policymaking declined greatly after Bhabha's return to India and the consolidation of the latter's close ties with Nehru. Once Saha realized that he would be excluded from taking major policy decisions for science, he became a very vocal critic of Bhabha and the AEC. Perhaps, it was this marginalization that made him contest the country's first general election as a Lok Sabha candidate in 1952. He contested the election as an independent candidate from the constituency of northwest Calcutta and won by a handsome majority with the backing of the Left parties. Saha said that his interest in planning made him join politics. Whatever his reasons, ignoring Saha was not going to be easy.

P.C. Mahalanobis had been appointed chief advisor to Nehru and he dominated planning. Saha was in general accord with planning based on statistical analysis but intended to enforce those plans while criticizing the way the government implemented them. Saha was convinced that his presence in Parliament would make a difference in plan implementation. That was also the year in which Saha persuaded the Senate of Calcutta University to declare the Institute of Nuclear Physics an all-India institution, thereby making it eligible for central government funding. S.N. Bose was nominated to the Rajya Sabha in 1952, but there is nothing to suggest that the two had anything more than a casual interaction. They certainly did not act as a team. Given the relations between the two, this is hardly surprising.

Once in the Lok Sabha, Saha used his position to become an influential critic of the government's policies, especially on atomic energy policy. He was one of the very few who challenged Nehru, particularly in matters pertaining to atomic energy. He needled Nehru regularly with some success. Saha was aware of Nehru's intellectual limits on the subject and the Prime Minister himself was not always sufficiently prepared to give a forceful reply to Saha's criticism, which was a result of meticulous preparation. An example of Nehru's weak grasp of the technicalities involved was his confusion regarding a reactor and an atomic pile. Writing to the Finance Minister C.D. Deshmukh, he said that a reactor was the basis for atomic energy work. It was not, he said, an atomic pile that could be used to make a bomb. His questionable technical knowledge notwithstanding, Nehru took pains to answer most of Saha's questions, replies which were often mingled with sarcasm and irony.[3]

In 1953, Saha wrote to Nehru saying there could be no development of atomic energy without the development of an atomic instrument industry and associated chemical and metallurgical industries.[4] Saha also deplored what he called the 'meddling' of foreign scientists in Indian affairs, a not so oblique reference to Joliot-Curie.[5] He drew attention to the huge gap between what Bhabha and AEC had promised, and what they had, in fact, delivered. In the parliamentary debates in May 1954, he repeatedly accused AEC of incompetence.

> India has got an Atomic Energy Commission and five years ago [1948] it had announced that we were going to set up a nuclear reactor within five years. This is 1954 and a nuclear reactor has not been set up [...] I do not know how long it will take us to make good this proposal. It may take us years to set up a nuclear reactor.[6]

In a letter to Nehru written on 23 May 1953, Saha was more explicit. He said that the only justification for the style of functioning and secrecy of AEC was that it was building a bomb. Saha was also critical of the fact that none of the AEC members gave atomic energy their exclusive attention. He said that in every country that had taken atomic energy seriously, the chief men to whom the work was entrusted were full-time workers. He cited the example of the US Atomic Energy Act, which did not permit any of its members to engage in any other business, vocation or employment.

Saha, even though he was not part of AEC, had collected enormous data on the work done in other countries and regularly passed it on to the government directly or indirectly through the pages of *Science and Culture*. When, even five years after the formation of AEC, there had been no tangible progress, Saha thought it fit to write to Nehru, criticizing the tardy progress and the extreme secrecy surrounding the activities of AEC.[7] Saha sent Nehru 17 typed pages that also included an assessment of the AEC's achievements contrasted with its announced objectives. He bemoaned the fact that AEC gave no details of its working—no budget or progress report and no information about its activities. 'When the Indian AEC was started, it began to out-Herod Herod in secrecy,' he observed.[8]

Saha said that secrecy was doing great damage and creating wrong impressions. There was no reason to exclude Indian scientists, particularly when foreign nationals seemed to have been taken into confidence. Otherwise, it would have been impossible for Gordon Dean, a former chairman of the US AEC, to comment favourably on the work done in India. Secrecy, Saha said, was being imposed by the AEC 'to hide its do-nothingness and concentration of all power and distribution of patronage in a few hands…'[9] As far as exploration for minerals was concerned, Saha said that the fact that the monazite sand in Travancore contained thorium had been known for the last three decades, even though AEC had only recently announced that it had discovered monazite sand containing thorium in Travancore.[10]

Saha also attacked the AEC's expenditure on cosmic ray research. Most of the grants for fundamental research in the laboratories, he said, were being spent on cosmic ray research. He challenged Bhabha to prove that such research had played any role in atomic energy developments in the country.[11] At the end of 1954, Saha wrote to Nehru informing him that AEC was exporting beryl to the US and making huge profits in the bargain. Beryl was one of the minerals whose sale was banned by the Atomic Energy Act. The official reply was a model of obfuscation. Saha was told that most of the beryl was used for stockpiling and some was used in nuclear reactors (never mind that none existed at the time) and only a small fraction was sold to the US government with the knowledge of the Prime Minister.[12]

In July 1955, Saha wrote to Nehru just after returning from Moscow. He noted that the 'latest administrative measures' (Saha is in all probability referring to the creation of DAE under Bhabha), had made Bhabha a führer in atomic energy development in India, just like Heisenberg had been in 1944 in Germany. 'The analogy is almost complete and I have no doubt that it will lead to the same disastrous results.'[13] This was certainly unfair. Saha was unable to keep his personal biases and sense of isolation out of his criticism of Bhabha. Much of his annoyance stemmed from the fact that he was totally dependent on Bhabha for the disbursement of funds for his projects. Bhabha controlled the purse strings and nothing moved without his permission. Even Bhatnagar, who was Saha's old friend, deferred to Bhabha in matters concerning atomic energy and never did anything without consulting his junior, but now more powerful, colleague.

Saha's opposition was not to atomic energy as a means of development but to Bhabha's strategy for developing atomic energy, and the special status given to him and his clique of scientists. India's nuclear institutions were largely, if not exclusively, in the hands of Bhabha and Nehru, who led in an autocratic and opaque manner, ensuring that their decisions were kept out of parliamentary or public debate. The ad libitum shaping of India's nuclear policy by its scientists also meant that, in 1966, when both Bhabha and Lal Bahadur Shastri died within two weeks of each other, much of the basic knowledge of India's planning vanished with them.

In a session of Parliament in May 1954, Saha said: 'The Atomic Energy Commission, which you have now got, has to be scrapped and we must start our work on a broad basis. First of all, there should be no secrecy. If you read our Atomic Energy Act you find that it does not tell us what to do, but it simply tells us what is not to be done...' He did not believe that atomic scientists were in a special class by themselves. He said that the atomic scientists had been ordinary chemists, physicists and others and was 'not a new race that has come into existence'.[14] Saha also pointed to what he called 'the sterilizing influence of secrecy', explanations for which lacked conviction. He complained that the government had failed to set up an atomic energy establishment, like Harwell in UK and Saclay in France. Warning against overdependence on any one country, he wanted India to aim for atomic autonomy, like France. Saha proposed a Central Atomic Energy Laboratory, which would organize all aspects of atomic energy production. He also wanted the government to set up a central service that would implement the atomic energy programme.

Saha was equally critical of the existing administrative machinery, which he felt did not have 'the spirit nor the stature of the machineries which have been found suitable in other countries' and, therefore, was unable to make any progress. He wanted it to be replaced by new machinery that would secure the cooperation of the general body of scientists in the work of policymaking and research and development. In conclusion, he said, 'It is now an urgent necessity that the policy of hesitancy be given up, and the old Atomic Energy Act be repealed, and a new Act be brought into existence, embodying the experience of other countries.'[15] As Itty Abraham has observed, it was the first time that the political leadership was hearing of a sustained attack on one of Nehru's vanguard projects from a source whose nationalist and scientific credentials were impeccable.[16]

THE DEPARTMENT OF ATOMIC ENERGY

Saha's concern was heightened by two developments in 1954, which threatened to marginalize him and his institute even more. On 3 January 1954, the Government of India created the Atomic Energy Establishment at Trombay, with Bhabha as the founding director. It was established to consolidate all the research and development activities for nuclear reactors and technology under AEC. All scientists and engineers engaged in the fields of reactor design and development, instrumentation, metallurgy and material science were transferred with their respective programmes from TIFR to AEET, with TIFR retaining its original focus for fundamental research in the sciences.

Though Bhabha had been chairman of the AEC since 1948, as a commission, it enjoyed only consulting powers to the government. It was essentially a policymaking body and such a body could obviously not carry out a full-fledged atomic energy programme. This required a regular department of the government, which would fund, create, and operate all the facilities needed to run the atomic energy programme. Seven months later, on 3 August 1954, the Department of Atomic Energy was created, with Bhabha as a secretary to the Government of India—the highest bureaucratic office in the system. He was now well and truly part of the government machinery.

Bhabha's position at DAE made him even more powerful than other secretaries because the DAE reported directly to the prime minister, who became the minister in charge of atomic energy. Thus, DAE, like AEC, managed to bypass the Cabinet. Bhabha overcame the challenge from the bureaucracy by becoming a high-ranking bureaucrat himself. Clothed with representative authority, he could now do as he pleased. As already noted, in a distinct break from the past, Bhabha insisted that DAE, like AEC, have its secretariat not in Delhi but in Bombay. This made ample sense because Bhabha was director of both TIFR and AEET.

More importantly, he had the clout to push it through. Such an arrangement also had the added benefit of being far away from the bureaucratic environment that Bhabha found so stifling. Most scientific work was centred in Bombay. Bhabha used this to justify the location of the department since he did not want to be divorced from his work. He was aware of his unorthodox request and gave full credit to Nehru for acquiescing to it. It was a revolutionary step and a few decades later, the director of IISc, Satish Dhawan, followed this example when he requested that the Department of Space be located in Bangalore.[17] Bhabha also ensured complete autonomy in matters of staffing, construction and supplies. The new DAE was vested with powers to make its own appointments without reference to the UPSC; carry out all civil engineering work through its own engineering division; and independently purchase its materials without going through the Director General of Supplies and Disposal.[18]

The TIFR was already a training ground for nuclear physicists required for India's future nuclear programme until a separate research centre was established. Thus, A.S. Rao, who had built the instruments for the cosmic ray experiments, was given the responsibility of training electronics engineers and H.J. Taylor also helped in training nuclear physicists who had returned from foreign universities. In fact, till 1957, all of AEET's work was carried on under the umbrella of TIFR. E.C. Allardice, who was the deputy director at TIFR, helped Bhabha with the administration of AEET, and for a time, TIFR managed all administrative matters, including purchases. Later, Allardice became the controller of administration at AEET.

Scientists shared the facilities, including the library and the workshop, and, in those days, there was little difference between a member of TIFR and of AEET. From a modest beginning at Kenilworth, the research activities at TIFR expanded rapidly and fresh space had to be found. In 1953, the metallurgy programme was shifted to the OYC, and a place was found for chemistry, spectroscopy and allied activities in an old mill at Cadell Road (now Swatantrya Veer Savarkar Marg). The physics and electronics sections were housed in the barracks at Colaba.

A.S. Rao held a dual position, as did Raja Ramanna, who headed the physics part of the atomic energy programme. As mentioned earlier, Bhabha had met the young Ramanna at Mysore in 1944 and had been impressed by the earnest young man with the bristly moustache who was a skilled pianist. Bhabha was instrumental in getting Ramanna a J.N. Tata scholarship to study at King's College, London, where his PhD guide was the infamous Alan Nunn May who was arrested for leaking atomic secrets to the Russians. Ramanna obtained his doctorate in nuclear fission in 1948 and Bhabha offered him a job at TIFR immediately, but asked him to stay back in England for a year to familiarize himself with latest developments in the field.

Ramanna recalled that while Bhabha had offered many of them jobs, he had made no commitment about paying for their passage back to India. He met V.K. Krishna Menon, the Indian High Commissioner, who, after making a few disparaging remarks about Bhabha, suggested that he meet Parmeshwar Narayan Haksar, the Deputy High Commissioner. Ramanna observed that all these worthies were so concerned about the reactions of the Finance Ministry on the issue of paying their travel expenses that he learnt early in his career, that the clerks and not members at the top rung of government called the shots.[19]

Haksar arranged for Ramanna's passage to India. He took a liking to him and, years later, would rely on him to become India's *bombmeister*. The feeling, it appears, was mutual. Ramanna noted, 'A brilliant man Haksar made a deep impression on me, and would be one of those I'd always remember with respect.'[20] Ramanna went on to become the director of AEET, chairman of AEC, a Member of Parliament and a minister of state in what was a glorious career dedicated to science.

In 1949, Bhabha recruited another young man who also went on to become a star: Homi Sethna. Stocky, with wavy hair, Sethna had done his BSc (Tech) from the University Department of Chemical Technology, Bombay University, in 1944 and completed his MS from the University of Ann Arbor, Michigan in 1946. In 1947, he joined the Imperial Chemical Industries (ICI) as a trainee under the Tata–ICI scheme. He returned to India in 1949 and is said to have met Bhabha for the first time while he was swimming at the Willingdon Sports Club.

Bhabha called his namesake for an interview the very next day, and was impressed enough to send him to work as manager at the Indian Rare Earths, Alwaye, Kerala, where rare earths were separated from monazite. In a recorded meeting with Professor A.K. Grover of TIFR in 2009, Sethna said his contact with Bhabha at TIFR had been peripheral. 'For instance, he would ask me, what I thought of Mr. A or Professor B or whatever it is and I had to tell him frankly, what it was.'[21]

As we shall see later, Sethna was project manager of the 40 MW research reactor, CIRUS, during its formative years between 1956 and 1958. In 1959, he joined AEET as chief scientific officer. Later, he became director of the engineering group responsible for designing and constructing the plutonium plant at Trombay, and concurrently, of the Uranium Mill at Jaduguda in Bihar. He was the director of BARC from 1966 to 1972 and succeeded Vikram Sarabhai as the chairman of AEC and the secretary of DAE, in 1972. He was the second-longest serving chairman of AEC, after Bhabha, when he retired in 1983.

Sethna was responsible for the development of mixed oxide fuel as an alternative for the enriched uranium fuel for the American reactors at Tarapur when the US reneged on its commitment to supply enriched uranium in the aftermath of Pokhran in 1974. He pioneered reprocessing technology in India, and the plutonium that went into making the nuclear device that was tested on 18 May 1974 was the result of his early efforts in this field. When Pokhran took place, he was the chairman of AEC. He continued to work well into his 80s, and, in an interview in 2006, said that he belonged to Bhabha's school of thought, and Bhabha had taught them to be ageless.

A CONFERENCE

In September 1954, Saha returned from a trip abroad, and soon after succeeded in convening a meeting of leading politicians to discuss his proposals regarding the future of atomic energy in India. The meeting was held at Teen Murti House, Nehru's residence. After a long introduction, Saha made some concrete suggestions. He wanted India to follow the French model, which pursued self-reliance but eschewed secrecy. Saha also wanted a change in the Atomic Energy Act and a complete revamp of AEC. He recommended advisory boards of scientists, who would advise the prime minister. In what was a direct attack on Bhabha, he said that under no circumstances should an atomic energy administrator be permitted to wear more than one hat. However, Saha had lost the battle when AEC was formed with Bhabha at its head. Furthermore, Saha's rejection of the membership of AEC didn't help matters. According to Anderson, 'Bhabha and Bhatnagar represented political orientations to which Saha was strongly opposed, specifically private industry working in collaboration with foreign firms.'[22]

Faced with constant criticism from Saha, Nehru agreed to hold a special conference, inviting scientists from all over the country to assess the progress of the atomic energy programme. The conference on 'The Development of Atomic Energy for Peaceful Purposes in India' was held at NPL in New Delhi on 26 and 27 November 1954. Nehru and his Cabinet colleagues were present. The agenda of the conference, the people invited, and those selected to speak, made it apparent that the purpose of the conference was not to question AEC but to allow it to assert its legitimacy in a very public way.[23] The choice of NPL as the venue for the first semi-public discussion on atomic energy made criticism of AEC difficult. Bhabha aimed to avoid criticism at all costs and the way the agenda was framed reflected this. Rather than a discussion on the appropriateness of the AEC's strategy in developing atomic energy, it became a means to share information about achievements in the field and, more importantly, to justify what AEC was doing.

In his inaugural address, Nehru said that atomic energy had two components: scientific and technological. The former could be the domain of the universities, but the latter had to be restricted to the State alone due to the costs involved. Bhabha described how nuclear energy would eventually meet the major part of India's energy needs. He also touched upon the complex technological problems they were facing and his plans for resolving them. The veil of secrecy had been partially lifted as the progress was described in some detail.

It was at this conference that Bhabha outlined his famous three-stage nuclear programme aimed at utilizing the country's limited reserves of relatively high-grade uranium ore to pave the way for exploiting India's much larger resources of thorium. Pressurized heavy water reactors fuelled by natural uranium would produce chain-reacting plutonium, which would, then, be separated from the spent fuel. The resulting plutonium stockpile would be used to provide the start-up fuel for fast breeder reactors. The cores of these reactors would be surrounded by a blanket of either depleted or natural uranium to produce more plutonium to fuel more breeder reactors. Eventually, thorium would be substituted for uranium in the blankets to produce fissile uranium 233, the start-up fuel for the third stage, consisting of breeder reactors using uranium 233 in their cores and thorium in their blankets.[24] The primary goal was to base the growth of nuclear power on the plentifully available thorium rather than the relatively scarce uranium. Thus, the nuclear programme, as envisaged by Bhabha, was based on a 'closed fuel cycle', wherein the spent fuel of one stage would be reprocessed to produce fuel for the next one.

Practically all the presentations at this conference were made by people associated with the establishment. The majority of the speakers were from TIFR and AEET. The text of the papers avoided specific details about the Indian programme, choosing instead to speak in 'vague generalities' about the value of atomic energy to development. Saha presented a paper on 'The Design of Reactors' but his intervention had only a very limited impact. Sethna read a paper on 'Uranium and

Thorium Extraction and Purification', which elicited a sarcastic remark from S.N. Bose, a person not generally known to have a sharp tongue: 'What waste have you to dispose of except night soil?'[25]

The host, K.S. Krishnan's humour and gentle manner helped keep tempers under control. He was chosen to summarize the proceedings of the conference, and, expectedly, his summary favoured AEC. Raja Ramanna, who was present at the conference, said that Krishnan 'saved the day for us'. He exhorted the scientists to get on with their work instead of indulging in endless unproductive discussions. He eased the tension and the symposium ended on a happy note. Ramanna overheard him saying at lunch: 'I hope I have not stepped on the toes of too many people.'[26]

Regardless of Saha's views on the conference, at the end of the year, he was, once again, involved in a not so pleasant exchange with Nehru. Saha wrote to Nehru that he had been told by Maulana Abul Kalam Azad that he (Saha) had boasted to some persons 'that now that Dr D.S. Kothari was becoming Secretary of the Ministry of N.R. & S.R., and Director of Scientific & Industrial Research, I was going to control all the national laboratories and scientific institutions of the country.'[27] He was given to understand that Nehru had believed this and he wanted to set the record straight.

> May I assure you that this story is a pure and simple fabrication and that you should give no credence to it? My request to you is that you do not smother your Desdemonas on the report of men like this particular Yago [*sic*]. I sometimes believe that there are too many Yagos [*sic*] about you, as there have been in history about every person of power, and prestige.[28]

The allusion to Bhabha, and perhaps also to Bhatnagar, is not difficult to see. Whatever the justification for Saha's hypercritical attitude, it did little to endear him to people. Even someone as mild and even-tempered as S. Chandrasekhar wrote about it to his father back in June 1936. Having talked to Saha, Chandrasekhar wrote to his father of 'his scheming, sneering attitude—an attitude which desires to destroy all things productive—an attitude which blurts out when in inviting me to come back to India he asks me to "join in the politics"'.[29]

The challenge to AEC was warded off at least for the time being—the conference managed to defuse the most public challenge to its monopoly over atomic energy activities in the country. Despite Saha's standing as a scientist, and the fact that he was also a Member of Parliament, he was unable to shift the boundaries of atomic energy, either as an institution or in strategic terms.[30] This was the last serious discussion on changing the institutional framework of AEC or of subjecting its activities to a degree of public scrutiny. The conference served only to confirm the dominance of atomic scientists over other scientists and state agencies. The most tangible challenge Bhabha's critics could mount was to suggest that AEC would not be able to build a working nuclear reactor for many years to come.[31]

Bhabha set about trying to remedy this lack of tangible achievement, starting with the small research reactor, Apsara. However, to acquire a reactor, Bhabha had to request foreign assistance, effectively admitting that, despite his tall claims, he had nothing much to show for the huge investment in his department. The conditions of secrecy that surrounded atomic energy would now be invaluable in order to shield Bhabha and his scientists from domestic scrutiny.

Chapter 29

BHABHA'S EMPIRE

> The problem of developing the underdeveloped countries is therefore the problem of establishing modern science in them and transforming their economy to one based on modern science and technology.

> —Lecture delivered at the International Council of Scientific Unions, Bombay, 7 January 1966[1]

Bhabha acquired 5 sq. km of land (approximately 1,200 acres) in Trombay, an industrial area along Bombay's south-east harbour to house the new atomic energy establishment that was planned on lines identical to that of the Atomic Energy Establishment, Harwell, UK. This scrubland adjoining an oil refinery was an unlikely site for a nuclear establishment but Bhabha had envisioned it as a scientific Versailles. Bhabha, apart from trying to achieve his ambition of establishing himself and India as sources of world-class science at TIFR, was also laying the foundation of India's nuclear programme. He often referred to TIFR as the cradle of the Indian atomic energy programme, as indeed it was. Till 1957, all the work of AEET was carried out under the umbrella of TIFR. The Trombay scientists shared facilities, like the workshop and the library with the institute. As noted before, the TIFR managed administrative matters, including purchases. E.C. Allardice, the deputy director of TIFR, also assisted Bhabha with administrative matters at AEET; in those days, the only difference between members of TIFR and AEET was that their salaries came from different budgets.[2]

The various divisions, such as physics, chemistry, engineering and reactor physics, controlled research activities at AEET and the head of each division was given a free hand to choose areas of research. Starting with Apsara, a small research reactor, the complex would have an entire family of reactors, a large engineering laboratory and even a plutonium processing plant.

In a note to Nehru, 'Architects for AEE', dated 18 June 1956, Bhabha informed the Prime Minister that the main site was virgin with no buildings. He said that it would have 23 large and small buildings, and that the estimated budget was between ₹2–3 crores, and a provision for the same had been made in the second Five Year Plan. Bhabha thought that this was an opportunity for developing a project that was not only functional and efficient but also of architectural significance. 'There is indeed an obligation on the Department to do the best it can and see that the opportunity is not missed.'[3]

Aware of the limitations of the engineers of the CPWD, Bhabha established his own engineering division to ensure control over the project. He chose Rustum B.J. Patell, an understudy of the great Frank Lloyd Wright, as his chief architect. This selection came after Bhabha had shortlisted five firms and architects that included D.S. Bajpai, Messrs Bhuta and Associates, Messrs Kanvinde and Rai and Messrs Polk and Stein. But it was not easy. He had to deal with opposition from the Ministry of Finance, which did not agree with his proposal of awarding a prize of ₹10,000 to the architect whose proposal was selected. He had also suggested an honorarium of ₹3,000 to each of the other four firms for the time and labour spent in making the proposal. The finance ministry wanted it to be treated like a normal tender and was of the view that since a tenderer was not paid any money nor should the architects. The CPWD, the ministry opined, never made any such payment, and Bhabha's proposal was turned down.

Bhabha wrote to Nehru on 18 June 1956 seeking his intervention.

> The comparison between asking a number of architects to submit designs with a calling of tenders by contractors is clearly misconceived […] Moreover, anyone who has any experience of intellectual activity knows that the conception of an architectural project is the most important part of it in which the architect's real merit is shown, whereas the working up and the execution of the designs can be left to juniors and subordinates […] It is said in the Finance Ministry's note that no such honorarium is paid by CPWD. This may be one of the reasons why CPWD work is architecturally so insignificant. The general tendency in India today to employ civil engineers instead of architects […] is also probably one of the causes why architecture, which was one of the glories of Indian civilization in the past has now sunk into complete insignificance…[4]

Bhabha also said that, given the magnitude of the project, it was possible that each of the five firms would be entrusted with the designing of at least one of the buildings. He wanted the Prime Minister to approve his proposal, after which the papers would be sent back to the finance ministry for ex post facto sanction. Expectedly, Nehru sided with Bhabha. The very next day, he cleared the proposal.

> If we are to have any improvement in our present low level of architecture, we shall have to adopt other methods than those of the Central PWD […] I do not think that the proposal of Dr Bhabha involves any marked additional expenditure considering the size of the project. It is likely to result in much better building, I would, therefore, approve of it.[5]

MAHARAJA OF TROMBAY

Despite his efforts to hire a chief architect for this project, Bhabha remained the true master builder of Trombay. He kept a site plan of Trombay in his office at home and would constantly fine-tune it. People saw Bhabha's car, bearing the registration number BMY 3333, in Trombay regardless of the time of day. In 1959, he wrote to Nehru that he imagined both Trombay and TIFR to be architecturally and botanically beautiful when they were completed.[6] A deep sense of beauty underlay the design of everything that Bhabha built. J.R.D. Tata noted his admiration for Bhabha's sense of natural value, sense of nature and the fact that whatever he built had to be beautiful.

Bhabha believed that buildings, too, were an expression of creativity. While travelling in Europe, he had carefully observed the buildings there, noting not only their structure and form but also their functional value and their relationship and balance with the surrounding landscape. When he was building his own institutions, he insisted on perfectly planned buildings that also had aesthetic appeal. He believed that the practice of science required a congenial atmosphere in every way—its practitioners needed more than just appropriate equipment to thrive and be creative.

Rudolf von Leyden observed that Bhabha was deeply conscious of the beauty of Delhi and whenever he could spare the time, he would relax at Hauz Khas, in sight of the Lodi tombs or in front of the magnificent façade of Sher Shah's Mosque at the Purana Quila.

> Once we stood together on those ramparts [...] and I remember how he spoke, in great anger, of the rape of this beautiful city by sterile PWD architects, by the tastelessness of bureaucrats and politicians, by the avarice of land speculators and by the complete lack of culture here, in this city, where a half-ruined chhattri at the end of a broken down wall has more significance than all the post-1947 Bhavans put together.[7]

According to M.G.K. Menon, Bhabha did beautiful watercolours of the Purana Quila Gate and the Mughal Gardens. He recalls being with Bhabha when he did some of these. Bhabha had nothing but contempt for the ugly matchbox buildings painted a 'government yellow'. The usual justification for such substandard structures was the paucity of funds, an explanation that Bhabha disdainfully rejected. He successfully managed to break away from such a pedestrian approach, first in his planning of TIFR, and then with Trombay, which was much larger than TIFR. At Trombay, Bhabha could dream not only of beautiful buildings but also of well manicured gardens, pools, avenues and even a whole landscaped hill.[8] The civil engineering division at Trombay had a small cell of competent architects attached to it. These were the people who helped Bhabha translate his dreams into reality.

Bhabha lavished the same attention he had given to the TIFR building on AEET's modular laboratory, considered at the time, to be the longest building in Asia. He insisted that the laboratory be placed on stilts to give the impression of a ship floating on the bay behind it to observers on Trombay Hill. Bhabha specified that each laboratory division should have its own workshops. While the heavy machinery was in the basement, the laboratories were on the higher floors alongside engineering halls with high ceilings for prototype and pilot-scale work. Bhabha had planned for inbuilt flexibility so that laboratories could be reconfigured with minimum structural changes.

Bhabha had three or four basic designs made of the modular laboratory before making his final selection. He sensitized the engineers to pay special attention to the quality of workmanship. As M.R. Srinivasan observed, 'A man with a sharp eye for detail, he could walk into a building under construction and immediately tell if the walls were plumb or if the corners were true right angles.'[9] Homi Sethna recalled how even the funnels at Trombay had to be visually aligned. 'He would harass your guts out [...] His strength was that he was damn intelligent. His weakness was he wouldn't be able to concentrate on anything for a long time. He was just too artistic. He would drive you nuts about the way institutions should come up.'[10]

Although Bhabha visited the Trombay site every other day, it was not as if the plans for it were finalized in the span of a few months. The master plan took many years to shape and the maharaja of Trombay, as scientists referred to Bhabha, was always on the lookout for new ideas on his frequent visits abroad. Upon returning, he would spend long hours sketching and trying out new ideas. Patell, who referred to Bhabha's 'intensive participation', would often get a phone call from Bhabha after dinner: 'Doing anything tonight?' This was, in effect, a polite order to visit Bhabha at his house. Upon Patell's arrival, Bhabha would unveil his latest ideas with the remark: 'Well, what do you think of it?'[11] During the construction of AEET, Bhabha would often drive up the hill and, from a vantage point (called 'Bhabha Point'), get a bird's-eye view of the entire complex.

Bhabha took Patell with him to visit nuclear establishments in Europe and the US. Patell said that it was as much a tour 'of the architecture of Versailles, of the "Gothic" and the "Baroque", as of reactors and laboratories'.[12] A believer in organic architecture that promoted harmony between human habitation and the natural world, Patell thought that he would be unable to translate Bhabha's classically oriented approach into reality. It was only later that he realized how wrong he was. Patell recalled that it would sometimes seem that architecture was Bhabha's special field of interest. However, in reality, Bhabha's contagious enthusiasm for every field made everyone feel that their subject was of special interest to him. The TIFR complex testifies to his untiring pursuit of aesthetic ideals in a vocation not really his own.

Bhabha took an interest in all aspects of the master plan for AEET. He focussed on both the macro and the micro aspects of design. Whether it was the major and minor axes of site planning or other minutiae, like calculating the cost per gallon for artificial reservoirs, Bhabha was fully involved in all matters. He drew heavily from old masters, like Le Notre, but was always willing to try something new. He had a little black diary in which he entered details he thought he would not be able to remember, including everything from a sketch of the approach to the lower gardens at the Vaux Le Vicomte or the dimensions of the Bagatelle Gardens to the level of the water in the pools at the Rashtrapati Bhavan. Patell says that one of the outcomes of such assimilation was the ingenious design for the reactor cooling ponds that were designed entirely by Bhabha.[13]

Patell commented on Bhabha's interest in the planning of AEET.

> His interest in this subject took him beyond the limits of a 'client architect' relationship on any project that he guided; it amounted to an intensive participation some (might have said interference) but he had the uncanny knack of being right more often than not in matters of conceiving and the frankness to be self-critical of his own design decisions when proved wrong. Of the powerful driving forces that made up the highly creative entity that he was, the Ego was understandably dominant: However, in frequent cases

> of disagreement on architectural ideas, formidable though he was to convince and relentless in putting through his own point of view, he disapproved of mere concurrence from others when it was without true concurrence.[14]

Rudolf von Leyden, who visited Bhabha at his home, noticed a large drawing board near his desk, which had printed plans pinned to it. They were the first layouts for the afforestation schemes and suggested gardens at AEET. By the side of the drawing board were illustrated volumes on gardens at Versailles and in countries like England, Japan and Italy. Bhabha's knowledge was detailed and he could describe the essential points of a garden design regardless of whether he had seen it in Vienna, Paris, Rome, Yorkshire or Kyoto. He instinctively knew the intentions of garden architects—what they were aiming at and what their underlying idea was. Bhabha was also an avid gardener and experimented with exotic plants and flowers, crossbreeding bougainvillea and roses on his terrace garden. He took the initiative to start growing roses at Trombay, and by 1960, the rose garden had over 750 varieties. Since Bhabha was a lover of roses, B.P. Pal, a former director general of the Indian Council of Agricultural Research, developed a variety of white roses and called them 'Dr Homi Bhabha roses'.

Bhabha was a perfectionist in everything he did. Perhaps that was what Sethna meant when he said he would harass your guts out. No detail was too small for him. If a wall had to be painted a certain colour then small patches of sample colours would be painted on a wall for Bhabha to see and approve. Such approval often took months since Bhabha was often travelling abroad, but there was no doubt in anyone's mind that the final painting could only proceed after Bhabha's approval. This attention to detail was apparent in every choice, be it the choice of paint or washbasins. Bhabha took particular interest in the modular laboratory and, given his knowledge and interest, obtaining his approval was a major task. He was a hard taskmaster and a challenge for the architect. The stilts Bhabha wanted for the modular laboratory underwent many changes. Patell recalled:

> For us it was often a challenge to our skill and endurance to keep up to the pace he would set, particularly on a scheme such as the

Modular Laboratory at Trombay, the only single building (besides the Tata Institute of Fundamental Research) to which he gave his special care and attention. In its innumerable design details that had to go through his surveillance, every season had its major architectural hurdle to cross before he could be convinced and give his final consent.[15]

LANDSCAPES AND GARDENS

Trombay's immense landscape allowed Bhabha free rein to set his reactors and laboratories according to his aesthetic sensibilities. He did not view individual buildings in isolation—to him, they formed an integrated complex, and he gave great importance to how they should be located with respect to each other. This meant laying out roads and filling the spaces between them in an artistic manner. As M.G.K. Menon aptly put it:

> He believed in a landscaping which took into account what already existed. His aim was not to bulldoze all that existed and plant science and technology as a shining imported item and to create a uniform faceless society—but to grow science indigenously as a way of life in the midst of all that was good and great from the past, a science which would bear the imprint of the traditions, the culture and natural gifts of the Indian people.[16]

During June 1959, Bhabha spent two days in Cambridge regarding an honorary degree of Doctor of Science that he was to receive. Upon returning on 2 July, he wrote to Nehru, telling the latter that he had stayed in Cambridge at the Master's Lodge in Trinity as Lord Adrian's guest. It had been a particularly good year for roses, and he said that he had never seen such profusion of beautiful roses as were to be found in the garden at the back adjoining the river. Bhabha hoped that some of the scientific laboratories and establishments they were building would have a beauty of their own and a salutary effect on those who worked there. He thought that both AEET and TIFR would be architecturally and botanically beautiful when they were completed.[17]

When Trombay was selected as the site for the project, the long hill there formed the western boundary of the site. On the other side were several industries and an oil refinery. The hill was barren, having been denuded over the years by people in search of firewood. Bhabha's first priority was to reforest the atomic energy side of the hill with more than a million trees and shrubs. As Leslie Stuart and Indira Chowdhury have observed, 'The all-Indian botanical garden he created—an intricate geometry patterned on the European classics but with subtropical flair—was maintained by a dedicated professional staff.'[18]

In Shreepad Dwarkanath Vaidya, who joined AEET as Head of Parks and Gardens, Bhabha possibly found India's first Versailles-trained landscape architect. Vaidya who had a bachelor's degree in horticulture from Poona, was responsible for the landscaping of both TIFR and AEET. Although Bhabha didn't live long enough to see what he had planned at Trombay, he worked closely with Vaidya for the garden of the new TIFR building. Vaidya's grandfather, Sitaram Khanderao Vaidya, designed the Taj Mahal Hotel and his grand uncle was the chief engineer for the same. Later, Vaidya joined the Taj Group as chief landscape designer and designed, among others, the garden of the Fort Aguada Resort in Goa.

Vaidya recalled how the trees had been planted well before the main laboratories were built. Although he faced some criticism about the obsession with aesthetics, Bhabha countered it by saying that since trees take time to grow the early planting had been done to save time.

Vaidya narrated an incident that attests to Bhabha's deep interest in trees. Once, when he was in Bangalore, Bhabha spotted a casuarina tree with foliage that appeared to differ from the usual. He checked with Vaidya if it could be a different species and when the latter said it could be possible, he wanted to bring the fallen cones with him. When Bhabha was told that the fallen cones would not have seeds in them, he couriered seeds from cones still hanging on trees to Bombay. The trees grown from such seeds can be seen in Trombay to this day.[19] Vaidya observed, that Bhabha's sensitive mind perceived the trees as living sculptures, giving a character of their own to the place where they stood ungrudgingly for generations.[20]

An office note issued by Bhabha in early 1964 gives details of the arrangements for the sale of flowers from Trombay. It was decided that a notice would be issued from time to time, indicating the type of flowers available for sale at the Trombay site, along with their prices. The note also indicated from where and at what time the flowers could be collected. 'The flowers to the extent available will be ready for collection from 4.30 pm at the South Gate, Trombay, and from 7.30 pm at the Old Yacht Club on the same day or at any time in the following day.'[21]

Vaidya is also the source for another anecdote. In order to have a straight road, the civil engineers had recommended the removal of an old mango tree. Bhabha was very unhappy about this. He was of the view that the tree, which had stood there for a hundred years, had every right to continue being in the same spot. On Bhabha's intervention the road was realigned and the tree, which had given character to the whole area for over a century, was saved.

On another occasion, Bhabha was instrumental in saving a mutilated rain tree on Pedder Road and relocating it to a neighbouring compound. When Vaidya objected to the cost of this activity, especially given the possibility that the tree would not survive, Bhabha said that artists and scientists should not bother about the monetary involvements of their projects. It was the job of financial wizards and administrators to find the funds for any good cause.[22] Bhabha purchased the tree for ₹30 from the municipal authorities and replanted it at Kenilworth.

He played a prominent role in the Citizen's Committee for Beautiful Bombay and, with the intervention of Nehru, halted the construction of a multi-storeyed building that would have obstructed the beautiful westerly view from the Hanging Gardens.[23] As with TIFR, AEET also got its collection of modern Indian art, but the predominant ethos remained western. Bhabha insisted that the place feel western and, until his death, its researchers wore western style clothing and the canteen, too, had western crockery.

Bhabha went into great detail while planning any activity, but this should not be taken to mean that he believed in micromanagement. His style was quite the opposite. He was a shrewd judge of character and ability and knew who to assign a task to, and having done so, gave them full freedom to do as they pleased.

Devendra Lal narrated how Bhabha told Sethna that he wanted the plutonium plant to be operational soon, and what exactly he had in mind. Having said this, he just went away, leaving Sethna to get on with the task.[24] Sethna himself confirmed Lal's observation. 'Dr Bhabha never micro-managed the tasks assigned. You do what you want, here is the money, if you want more money, take it.'[25] Obaid Siddiqi who set up molecular biology at TIFR had a similar recollection. 'I do not remember ever being asked by Dr Bhabha what I was doing or why. I suppose he took it for granted that the people he selected knew what they were doing.'[26]

On another occasion, Lal got some good advice when Bhabha was entertaining Patrick Blackett at Mehrangir and Lal was also present. Bhabha wanted to entrust a certain task to someone and asked Lal to give him a name. Lal gave him a name, and mentioned, in passing, that the person had little work on hand. Bhabha responded with advice that may sound counter-intuitive but is, nevertheless, true: 'Never give an important task to someone who has nothing to do. Give it to a busy person; he will find time to do it.'[27]

ADMINISTRATIVE CULTURE

Bhabha wanted the colonial administrative machinery to be adapted to suit the requirements of scientific work. In a speech he made in January 1966, shortly before he died, Bhabha asserted that while India had been fortunate to inherit an administrative machinery that was competent to deal with a static and underdeveloped economy, it did not possess the necessary skills to develop science and technology. This lack of a bureaucracy that could adapt to the needs of the newly emergent nation state was, in his opinion, a bigger obstacle to the progress of science than the paucity of scientists because even the existing scientists had been made less effective through the lack of the right type of administrative support. M.G.K. Menon, while recalling Bhabha's contribution to institution building, noted his focus on creating a culture and ambience of a scientific institution since the administrative culture was not conducive to growth.[28]

Bhabha was very clear that he would not allow the convoluted, inflexible and dilatory processes associated with a bureaucracy to find a place in his institute. The administrative mechanisms of a science institute had to be entirely different from a normal bureaucratic setup. Bhabha was highly critical of the system where a file used to be marked to a number of persons who merely signed it without making any contribution to it whatsoever. Bhabha observed that the secretariat system of noting by half-educated clerks was one of the worst features of government and should not be introduced in the institute. He believed that each person should be encouraged and taught to dispose of a paper in the most direct manner possible, and as much responsibility as possible should be delegated right down the line.[29]

In Bhabha's time, the number of administrative staff remained small, but, in later years, as AEET grew and its needs changed, its administrative structure began to resemble the bureaucracy that Bhabha had been so careful to discard. He was also highly meticulous about mundane details, like referring to people as 'Shri', and not 'Mr', and how authors of scientific papers were to refer to AEET while writing their address. He discovered that AEET was being referred to as the Atomic Energy Establishment, Bombay, in scientific papers sent for publication. He wanted it to be Trombay, in place of Bombay, since the ultimate aim was that the establishment should be known in atomic energy circles in the world as simply 'Trombay' much like 'Harwell' and 'Saclay' in England and France, respectively.

Bhabha brought many changes to the administrative culture. He abolished the system of peons sitting on stools outside the offices and carrying files from one table to the other. He abolished long notes on the files, starting from the lowest level and passing through every level before it reached him. He insisted on the notes being precise and to the point so that decisions could be taken faster. He introduced his own system of purchase and stores, where things could be done faster and discouraged scientists from performing administrative duties, like signing cheques, making their own pay bills and checking the attendance registers of subordinates. As Bhabha observed, 'To apply existing administrative and financial procedure devised for an entirely different purpose to scientific institutions is largely to defeat the purpose which the Government has in view by letting the tail wag the dog.'[30]

Anil Kakodkar recalls reading some of Bhabha's noting related to the performance appraisal of AEET personnel and related documentation. Apart from a record of the officer's work and the final assessment by their higher ups, all the individuals the officer had worked with also assessed them on loose sheets. The idea was to promote cross-hierarchical interaction and cooperative work. Over the years, Kakodkar laments, this concept has been lost. He tried to revive the old culture, but it did not find many takers.[31] In another departure from the norm, February and August were considered the months for the promotion of scientists. In the central government, the date of joining was usually considered for promotion. This system was abolished at the institute, for promotion did not mean changing the officer's desk or moving to another office. It was a recognition of the scientist's contribution.

We have already seen that Bhabha created his own norms for construction rather than relying on the CPWD. The civil engineering group brought about visible results in the high quality of their construction work at TIFR and other institutions under the DAE. No foreign consultants were needed for the laboratories since the scientists prescribed the norms, checked the designs and made sure that the laboratories were well constructed and maintained. Air conditioning was considered a necessity and not a luxury. The canteens and kitchens were planned so that the utmost cleanliness was maintained. Sufficient attention was paid to designing schools, hospitals, playgrounds, roads and shopping centres, and the colonies at Anushakti Nagar and later Kalpakkam, typified the new way of living.

From the early days of AEET, two scientific bodies, the Trombay Scientific Committee and the Trombay Council, were formed to make decisions about a variety of subjects, including administration. The Trombay Scientific Committee consisted of all the heads of divisions and was involved in policymaking and reviewing the scientific and technological programmes. Members of this committee had the freedom to travel, interact with other bodies and arrange scientific meetings and discussions.

The Trombay Council consisted of the directors of major groups including physics, electronics, instrumentation and safety, metallurgy, chemical engineering, reactor engineering and biology. The directors of these groups were highly respected scientists who had already made a name for themselves in their own scientific fields. They advised Bhabha on various issues. However, the different administrative structure for TIFR that Bhabha had envisaged remained incomplete. As Bhabha became more and more preoccupied with atomic energy, he had little time to devote to this aspect, and it remained a work in progress. In fact, Bhabha was so busy that he would visit the institute only on Wednesdays for the weekly colloquium. The day-to-day administration was left to Allardice, whose training in the bureaucracy had not prepared him for the kind of administrative mechanisms that were compatible with the needs of science.

Chapter 30

THE GENEVA CONFERENCE AND AFTER

For the full industrialization of the underdeveloped countries, for the continuation of our civilization and its further development, atomic energy is not merely an aid, it is an absolute necessity. The acquisition by man of knowledge of how to release and use atomic energy must be recognised as the third epoch of human history.

—Address at the first Geneva Conference in August 1955[1]

On 15 November 1945, three months after the Second World War had ended, the heads of the US, British and Canadian governments met in Washington and decided to adopt a policy of secrecy in the nuclear field, pending the establishment of a system for the effective international control of nuclear power. The policy was sought to be implemented by blocking the transfer of technical knowledge and uranium. In December that year, the USSR accepted a proposal to establish an atomic energy commission consisting of the 11 countries represented on the United Nations Security Council (UNSC), along with Canada. On 24 January 1946, the UN approved the establishment of such a commission.[2]

EARLY ATTEMPTS AT INTERNATIONAL DISARMAMAMENT

In March 1946, David Lilienthal, who went on to become the first chairman of the US AEC, was entrusted with the task of studying the problem of the peaceful development of nuclear energy. His report proposed that all operations concerning nuclear weapons' development be entrusted to a single international authority. An international administrative body would own, operate and develop the nuclear industry on behalf of all nations. The international authority would own nuclear ores and fuels, conduct research and operate nuclear fuel fabrication plants and power reactors. International inspectors would be responsible for discovering any clandestine activities that could take place.

The draft report was presented at the inaugural session of the United Nations Atomic Energy Commission (UNAEC) on 14 June 1946, by the US delegate Bernard Baruch, a financier turned political consultant, and came to be known as the Baruch Plan. It proposed to internationalize fission energy via an international atomic development authority, which would have a monopoly over mining fissile materials like uranium and thorium, refining the ores, owning materials and constructing and operating nuclear plants. This authority would fall under the umbrella of the UNAEC. The Baruch Plan was rejected by the USSR, and, following this rejection, the Atomic Energy Act of 1946 put in place a US policy to prevent the spread of nuclear technology through denial and secrecy. It prohibited the exchange of information even with those US allies that had cooperated in the development of the atomic bomb. If it had been implemented, it would have had implications for India as well. Bhabha resisted the US plans because they entailed international control over India's thorium and uranium deposits.

The fruitless negotiations continued and, after two years, the UNAEC informed the UNSC that it had reached an impasse and discontinued its work. The first attempts to achieve international nuclear disarmament had failed. From 1951 onwards, the negotiations on nuclear controls were linked with those on traditional disarmament. There was no more talk about the idea of international ownership and management.

'ATOMS FOR PEACE'

The policy of restriction proved to be a failure. The USSR tested a device as early as 1949, ending the US's nuclear monopoly. The Baruch Plan was given a quiet burial. The British and French were quickly catching up with the US and the USSR. Research on the peaceful uses of nuclear power, especially in commercial applications, had also progressed. At the end of 1953, the relative ineffectiveness of the policy of secrecy induced the US to change its policy quite suddenly.[3] It realized that continuing such a policy would only restrict its influence and allow others to enter the field.

The transition from a policy of secrecy to active promotion of the peaceful applications of atomic energy was first clearly articulated in President Dwight Eisenhower's famous 'Atoms for Peace' speech before the United Nations General Assembly (UNGA) in December 1953, with Vijaya Lakshmi Pandit presiding. Eisenhower acknowledged that other states would eventually acquire nuclear know-how and emphasized the need to exploit the potential of the atom for progress and development. More specifically, he proposed that the governments principally involved in nuclear research and development jointly contribute fissionable materials to an International Atomic Energy Agency (IAEA). The change in US policy reflected a belated realization that the country's monopoly over atomic energy had been irretrievably lost.

The IAEA was to be set up under the jurisdiction of the UN and would be responsible for storing and protecting the fissionable materials contributed by the member states. The idea was to share nuclear material and technology for peaceful purposes, but restrict the proliferation of nuclear weapons through an international agency that would undertake the mandated oversight of national nuclear programmes. The contributing powers would, in Eisenhower's words, be dedicating some of their strength to serve the needs rather than the fears of mankind. As far as India was concerned, there were several problems regarding this proposal. All states were not members of the UN and they would reject control over their nuclear programmes. There was also a distinct asymmetry in nuclear knowledge between countries: the recipient states would always be dominated by the big nuclear powers.

Following Eisenhower's speech, the US and the USSR worked hard on a mutually acceptable plan to implement the proposals. The US and six other western powers proposed the establishment of an international agency that would supervise and foster the growth of atomic technology for peaceful purposes. The UN agreed to place the issue on the agenda of the UNGA and the USSR too seemed willing to discuss the matter. A key factor in easing the negotiations with the USSR was the substantial revision made in the very restrictive US Atomic Energy Act of 1946. The changes were incorporated in the Atomic Energy Act of 1954 and included removing most controls on the classifications of information regarding nuclear research, approving ownership of nuclear facilities and fissionable material by private industry and authorizing the government to enter into agreements for cooperation with other nations on the peaceful uses of nuclear energy.

In December 1954, there was a call to organize an international technical conference under the auspices of the UN, no later than August 1955. The resolution stated that all members of the UN and its specialized agencies were to be invited to the conference. A scientific advisory committee, consisting of representatives from Canada, Brazil, France, Britain, India, the USSR and the US, was set up to prepare the agenda and issue invitations. The UN Secretary General, Dag Hammarskjöld, was chairman of this committee. Bhabha represented India on this committee. He had left Bombay to spend his Christmas vacation in Bangalore when he was told at the airport that Nehru wanted him to be India's representative at the conference. Bhabha cut short his holiday and arrived in New York in the middle of January 1955.[4]

Cold War considerations ensured that the global status quo was maintained from the beginning. The committee, too, reflected the relative power quotients of the international order. The US was represented by Isidor Isaac Rabi, a Nobel laureate in physics and a Columbia University physicist. The other members of the committee were W.B. Lewis of Canada, John Cockcroft of the UK, D.V. Skobeltsyn of the USSR and Joaquim da Costa Ribeiro of Brazil. Lewis and

Cockcroft were well known to Bhabha, as was the French representative Bertrand Goldschmidt. Hammarskjöld was an admirable chairman, adept at summing up the essential points to reach a consensus. It also helped that most of the members knew each other personally and scientifically. Since time was short, they had to work quickly. According to Cockcroft, a lot of work was done outside the committee. He remembers a lunch party at St. John's College, Cambridge, and at the weekend immediately following it on the Norfolk coast, where a good start was made of drafting the programme. At a subsequent meeting of the committee in New York, Bhabha was unanimously chosen as the president of the conference.[5]

In February 1955, Bhabha received a letter from Frédéric Joliot-Curie. He wanted Bhabha to organize a separate conference to round off the main conference. The main aim of this other conference would be to underline the dangers of atomic and thermonuclear weapons. He said that it would be a mistake to exclusively emphasize the peaceful uses of atomic energy at a time when the entire world was agitated by the threat of atomic weapons. Such a conference, unlike individual warnings, would be less likely to give rise to suspicion of bias or political manoeuvrings.[6] It is not known if Bhabha responded; in any case, nothing came of the suggestion.

THE CONFERENCE

The conference was to be held in Geneva from 8 to 22 August 1955 in neutral Switzerland since it was least likely to provoke objections from other countries. One of the most important issues was the choice of president of the conference since geopolitical considerations were paramount. The US initially wanted a Swiss national to be president but it was later agreed that 'an avowedly non-partisan figure' be appointed president. India was recognized as a potential nuclear state but was non-aligned. Bhabha was found to be universally acceptable (or least politically objectionable), and he had strong support from the British delegation. It is said that Nehru had also got V.K. Krishna Menon to lobby actively for this purpose.

According to Laura Fermi (Enrico Fermi's wife), the official historian of the conference, Bhabha's candidature was proposed by Goldschmidt, who had worked on the Manhattan Project. He later became an international authority on nuclear policy. He recalled Bhabha as 'an extraordinary man, a sort of a mixture of an Oriental potentate and a very refined, distinguished intellectual person'[7] who drew beautifully, spoke beautifully and adored music. He also said that if the IAEA is based in Vienna, and not in Geneva, it was largely because of Bhabha's love for music. Music is Vienna's most important product and the city's chief cultural export is the Vienna Philharmonic. Bhabha wanted to be able to visit Vienna more often, and it was due to his insistence that the agency is based in Vienna today.

Walter Whitman, a professor of chemical engineering at MIT was chosen as the secretary general of the conference. The plan of the meeting and the composition of its official organization were announced to the press by Hammarskjöld on 1 February 1955. Bhabha led the Indian delegation of 24 members and staff to Geneva. K.S. Krishnan was its senior-most member and the others included V.R. Khanolkar, director of the Indian Cancer Research Centre and the diplomat Arthur Lall. The others were young nuclear physicists who subsequently played an important role in India's nuclear programme and included A.S. Rao, Homi Sethna, K.S. Singwi, V.N. Meckoni and K.G. Vohra.

As the only other credible institute for nuclear physics research, the Institute of Nuclear Physics was asked to nominate delegates for the conference and Saha recommended a list of scientists, including B.D. Nagchaudhuri and his son Ajit Kumar Saha. Bhabha excluded the Calcutta delegation by making an unsustainable distinction between nuclear technology and nuclear physics. He claimed that the Geneva conference was concerned with nuclear technology and energy and would, therefore, be of limited interest to researchers in nuclear physics. Bhabha was very powerful in matters of nuclear research. But his power had to be justified, sustained, and amplified on a regular basis. The composition of the Indian delegation to Geneva was one such event.[8]

While this denial was unsustainable on the grounds Bhabha had posited, he perhaps had a reason for doing what he did, and he certainly must also have had Nehru's support in the matter.

In October 1954, Saha and D.S. Kothari, then the scientific advisor to the defence ministry, had been delegates of the Government of India to a conference in Moscow. Carried away by his own enthusiasm, Saha had asked for permission to negotiate a deal for a reactor on behalf of the government. On his visit to the US that year, Saha had heard rumours about the US helping Pakistan with a reactor. He had felt that since India did not count as a friendly country in the eyes of the US, it stood a better chance with the USSR. What Saha had not known at the time was that Cockcroft had offered India designs for a nuclear reactor only a month before, and the slightest hint of a deal with the USSR would have caused much embarrassment.[9] Bhabha did not want a similar predicament in Geneva and cut out the delegation from Calcutta altogether.

Saha had been unable to reconcile himself to his marginal role, along with the fact that he could no longer participate in international meetings or take part in negotiations as an individual. Gone were the days when he could attend such events on his own steam. This change was necessitated both by the change in India's status from a colony to a sovereign state as well as by the use of nuclear technology to build nuclear weapons. The Cold War also played a role. Saha was certainly aware of all this but his ambitions for the Calcutta group, as well as his own, led him to continue trying. But, by this time, Bhabha's hold over nuclear research was complete.

The conference was a big event with 1,428 delegates from 73 nations and seven specialized agencies of the UN. There were just as many observers and some 900 media persons attended the biggest scientific conference ever held. The venue was the grand Palais des Nations (the home of the UN in Geneva), overlooking Lake Geneva. More than a thousand papers were submitted for discussion. As Jahnavi Phalkey has observed, 'If the Empire Scientific Conference (1946) was a plea to maintain imperial preference in post-war international politics, the Atoms for Peace conference in Geneva (August 1955) was an invitation to establish American preference (over the USSR) in Cold War politics.'[10] The centrepiece of the American participation was a swimming-pool-type research reactor that had been flown to Geneva and assembled during the conference. Eisenhower, who was also in

Geneva at the time, took time off to visit the American exhibit and see the reactor.

A landmark event in the history of science and technology, a galaxy of scientists from all over the world (with the exception of Albert Einstein) attended the conference. For the first time, all the world's top scientists came together and decided to make scientific knowledge accessible to the people. Projects undertaken during the Second World War, which had hitherto remained shrouded in secrecy, were discussed at the conference. It was also Bhabha's hour of glory.

BHABHA AS PRESIDENT

Bhabha delivered his presidential address on 8 August 1955, the opening day of the conference. His talk, delivered with ceremonious diction but without pomposity, was titled 'The Need for a New Power Source'. He declared that the purpose of the conference was to discuss the peaceful use of atomic energy and to exchange scientific and technical knowledge on the subject.

Bhabha divided his broad view of human history into three great epochs, each with its distinctive energy source. The first and longest epoch, covering almost the entire period of 250,000 years, had been characterized by the use of muscle power, both animal and human. The second epoch, which had started 200 years ago, had been marked by the burning of fossil fuels, like coal and oil, and had led to the industrial society of today. The third was going to be the epoch of atomic energy, one which the world had entered after the Second World War. Bhabha believed that electricity from atomic reactors would be all-important in promoting economic development. This was based on the belief that economic development and energy consumption are directly related. Since energy consumption would ultimately depend on energy supply, a major increase in the supply of electricity would lead to greater economic development. Thus, Bhabha linked energy supply to the development of civilization. In societies that were dependent on the muscle power of slaves, the fruits of civilization could be enjoyed by only a privileged few. With the advent of the Industrial Revolution, a larger number of people in the highly developed countries started enjoying a better standard of living, but the vast majority in underdeveloped countries were no better off.

Bhabha, in the course of his speech, placed the figures of available energy production and the rates of energy consumption before his audience and declared that if the entire population of the world consumed energy per capita at the same rate as the US, the world's resources of fossil fuels would be exhausted in less than a century. Bhabha told his audience that for the full industrialization of the underdeveloped regions and to ensure the continuation of civilization and its further development, atomic energy was not merely an aid, it was an absolute necessity. He went on to say, 'Atomic weapons lie outside the scope of this conference, but we cannot entirely separate the applications of peace from the applications of war [...] A widespread atomic power industry in the world will necessitate an international society in which the major powers have agreed to maintain peace.'[11]

Bhabha then quoted from the Russell–Einstein Manifesto, warning against nuclear war and appealing for peace. On 9 July 1955, Bertrand Russell and Albert Einstein had issued an appeal to humanity, urging that the great peril associated with nuclear weapons be recognized by both governments and the public. The Russell–Einstein Manifesto spoke of the threat of nuclear weapons to the survival of the human species and called for major efforts to contain such dangers. As a consequence of the manifesto, the first Pugwash conference was held in 1957. The origins and goals of this conference came from the manifesto—to highlight the dangers of nuclear weapons, call for their elimination and promote the peaceful resolution of conflicts.

Eight years before the conference, in April 1947, Einstein, as chairman of the Emergency Committee of Atomic Scientists, had written to Bhabha asking for a donation to help educate people about atomic energy and its implications for society. 'In this lies our only security and our only hope—we believe that an informed citizenry will act for life and not for death.'[12] In his reply, Bhabha had promised all help and requested for literature on the subject, which he would help publicize widely. The monetary assistance, Bhabha said, would be a little problematic because of 'the very stringent dollar restrictions that the Government has been compelled to bring in.'[13]

Bertrand Russell, while recalling the preparations for drafting the manifesto, had hoped that he would have the support of Indian leaders and scientists. His hopes soared when, at the beginning of Nehru's visit to London in February 1955, Nehru seemed most sympathetic and friendly and invited him to hold the first meeting in India. For some time, it appeared that the Indian government would sponsor what eventually became the Pugwash conferences. New Delhi was chosen to host the first conference and, in June 1956, Russell even dispatched invitations for a conference in January 1957. But, Bhabha was lukewarm with regard to the Pugwash movement and showed little enthusiasm for nuclear disarmament. He was not keen to be associated with the movement, which was believed to be inspired by 'fellow-travellers'.

When Russell met Bhabha, the former received, as he put it, a 'cold douche'. Bhabha, he said, had profound doubts about any such manifesto, 'let alone any such conference as I had in mind for the future [Pugwash]. It became evident that I should receive no encouragement from Indian official scientific quarters'.[14] Not a single Indian nuclear scientist signed the Russell–Einstein manifesto. The first Pugwash conference was held in July 1957, at Pugwash, a village in Nova Scotia, Canada, the birthplace of Cyrus Eaton, an industrialist and philanthropist who offered to host and finance the conference. (It is said that the Greek shipping tycoon, Aristotle Onassis, also offered to host it at Monaco but his offer was turned down). However, Nehru, on Russell's suggestion, set up an official group to study the effects of nuclear explosions.[15]

Before concluding his speech Bhabha said:

> This Conference has already broken down many barriers, and we have come to know of the remarkable advances in atomic energy, achieved in several countries of which we were totally ignorant hitherto. If so much has been achieved through the individual and isolated efforts of a few countries, how much more could be achieved by the combined effort of all...We have the unique opportunity of giving our knowledge to others for the common good. I hope this Conference will play its part in helping the progress of mankind towards the ever-widening dawn of the atomic age, with the promise of a life, fuller and happier than anything we can visualise today.[16]

Bhabha made a case for knowledge to be shared more freely and debunked the idea that secrecy could ensure security. 'Knowledge should be freely available and shared,' he said. 'It is unreal to believe that security can be ensured by secrecy.'[17]

FUSION

Bhabha's opening speech also contained two unexpected, if not controversial, statements. The first was about China's absence from the conference because it had not been admitted to the UN. Bhabha said that it was a matter of regret that there were several areas of the world not directly represented at the Conference, whose populations constituted a quarter of the world's population. Political questions were not supposed to be raised at what was a scientific conference, and although Bhabha's statement raised a few eyebrows, it was generally ignored. Bhabha must have had Nehru's consent for making such a reference since Nehru, as the leader of the non-aligned movement, was in favour of China's admission to the UN.

However, it was Bhabha's other statement regarding fusion that caused uproar. In the concluding part of his address, he said:

> The historical period we are just entering in which atomic energy released by fission process will supply some of the power requirements of the world, may well be regarded one day as the primitive period of the atomic age. It is well known that atomic energy can be obtained by a fusion process, as in the hydrogen bomb, and there is no basic scientific knowledge in our possession today to show that it is impossible for us to obtain the energy from the fusion process in a controlled manner [...] I venture to predict that a method will be found for liberating fusion energy in a controlled manner within next two decades [...] When that happens, the energy problems of the world will truly have been solved forever, for the fuel will be as plentiful as the heavy hydrogen in the oceans.[18]

Bhabha had clearly got carried away by his optimism since the prospect of using fusion energy in a controlled manner to solve the energy problems of the world remains a distant dream even today. His concluding remarks, where he spoke of the beginning of an era of international cooperation in the sphere of atomic energy, the development of which promised a prosperous future for humanity, were of greater relevance. Bhabha said:

> We have the unique opportunity of giving our knowledge to others for the common good. I hope this Conference will play its part in helping the progress of mankind towards the ever-widening dawn of the atomic age, with the promise of a life fuller and happier than anything we can visualise today.[19]

Fusion was not on the agenda nor was it mentioned in any paper presented at the conference. It was still a top-secret matter with the US, UK and USSR engaged in serious research on the subject. Bhabha must have had some inkling of the ongoing research. It was probably this informed suspicion, coupled with his unhappiness about the associated secrecy, that prodded him to make that unexpected statement on fusion. His comments created a stir because many of the countries had been working on fusion in secret: Bhabha's public declaration struck a blow for openness.

It proved to be an unpleasant surprise for the Americans. Project Sherwood—the code name for the US programme in controlled nuclear fusion initiated by the Eisenhower government—had, until then, been very much a secret, and the Americans were not free to comment on Bhabha's speech. Raja Ramanna said that Bhabha had wanted to 'shake them out of their secrecy'.[20] It has been argued that it was, in fact, Bhabha's bold declaration that forced the first step towards the programme's declassification.[21] The unfettered presentation of scientific results made it obvious that knowledge was still rudimentary and encouraged the internationalization of the field while reducing the feeling of competition.

Patrick Blackett called the incident 'operation tin-opener', and it was generally agreed that had the conference been presided over by a less forceful personality, the world would not have known of the possibility of fusion at the time it did.[22] It is also not too extravagant to suggest that Bhabha's declaration could have been a gamble he took with British support. After all, he shared a good relationship with Blackett and Cockcroft. Bhabha's statement followed by the British admission on the subject could potentially have paved the way for an American statement and more transparency on the subject.

Bhabha's declaration, perhaps, did come at the right time, and was not without effect. Abdus Salam, while talking about it many years later, said that Bhabha's 'supreme indiscretion' proved helpful to future generations since, until then, fusion was supposed to be a very 'hush-hush subject'. He said that Bhabha made a 'wonderful speech' that led, in a sense, to a declassification of the fusion business.[23] Bhabha's stock had never been higher in the international community of scientists. After the conference, he came to be respected more highly in Indian political circles. Nehru's trust in his abilities now had the seal of international acclaim.

In a press conference, Cockcroft suggested that fusion power could become a reality within a generation. Admiral Lewis Strauss, chairman of the US AEC, in his press conference, admitted that the US had been conducting research on controlled thermonuclear reactions for many years. Bhabha also met the press and not only held his ground but also said that there would always be place for fission power even in a world of fusion power. In the aftermath, Admiral Strauss held a press conference in Washington on 3 October 1955 and disclosed basic facts about Project Sherwood.[24]

Bhabha's declaration caused apprehension in the minds of the businessmen who had come to Geneva. Those in the fission power reactor business had to be reassured that they would be in business for a long time. A number of fission power reactors were either in operation or under construction in the US, UK, USSR, France and Canada. It was the commercial exploitation of fission reactors that had, in fact, been the main motive behind the US's 'Atoms for Peace' initiative since firms in the UK and the US had been interested in the sale of reactors and power stations to other nations. They had made huge investments in all aspects of atomic power plants and, in informal meetings outside the conference hall, representatives of industries discussed reactor sales with the representatives of underdeveloped countries.[25]

John Cockcroft, who was the leader of the British delegation, in his lecture on 19 August, attempted to put Bhabha's bold declaration in the right perspective:

> Although we are working seriously on this problem in Britain, my vision is not good enough for that. I am not as bold as Dr Bhabha, our President. I am sure that he would not mind my saying that the experimental physicist must inevitably have a greater appreciation of problems and difficulties than a theoretical physicist.[26]

◆

The conference proved to be a grand success. Soon after the conference, writing in *Nature*, Cockcroft observed,

> The Conference has been more successful than we had dared to hope from such a large gathering. It has brought together East and West after a long period of separation in the physical sciences. It has been a meeting ground for innumerable friends from all parts of the world. It... has done a great deal to re-establish the normal pattern of communication in the scientific world.[27]

It became clear to the delegates gathered at the conference that basic research was proceeding along similar lines in all the nuclear nations and that the secrecy around this research had been unsuccessful in restricting such knowledge in these countries. No country had absolute superiority, and the once cosmopolitan scientific community that had been nationalized by the war, was back by default. It also became clear that no country had made much progress towards harnessing the power of the atom for civilian energy production.[28]

Bhabha reported the details of the conference to Nehru. The Indians, he said, had presented 13 papers and many of them had been well received. Bhabha told Nehru about his paper on the role of atomic energy in India, which had been quoted by Cockcroft in his evening lecture. He also mentioned a paper on a new method for surveying for uranium written by K.G. Vohra, a young scientist, which was reported in *The New York Times*. In his letter on 24 August 1955, Bhabha wrote that all signs of the Cold War seemed to have disappeared and he wanted India to take this opportunity to enlarge the areas of cooperation in this field. 'The essential thing is that any future action in this field should be taken with tact and consideration, so as not to mar in any way the excellent psychological atmosphere which now exists.'[29]

Shankar's Weekly, the Indian counterpart to *Punch*, named Bhabha as 'The Man of the Week' in its 14 August issue. It observed, 'This is the inescapable paradox of the present age: we can either exterminate ourselves or we can create a world in which there is no upward limit to quantitative material prosperity [...] And Dr Bhabha is our man of the week for having upheld, as India's foremost atomic scientist our determination to achieve the latter end.'[30]

A VISIT TO THE USSR

The leader of the Soviet delegation, Skobeltsyn, had invited Bhabha and the Indian delegation to visit the USSR immediately after the conference. Such a visit had always been on the cards and a delegation of six scientists had been approved by the Government of India even before they left for Geneva. Nehru's visit to the USSR in 1954 had laid the ground for this visit, and it marked the beginning of closer relations with the USSR. Skobeltsyn was, as noted before, a pioneer in cosmic ray research and is considered the founder of Russian cosmic ray investigations. He was the first physicist to put a Wilson Chamber in a magnetic field and show that cosmic rays are high energy particles. He initiated cosmic ray research in Leningrad and Moscow, nurturing a generation of distinguished cosmic ray physicists.

Bhabha, accompanied by K.S. Krishnan, Ramanna, N.S. Prasad, Sethna and A.S. Rao, left for Moscow on 30 August 1955. Their journey in an old Ilyushin plane was a long and noisy one, but their welcome in Moscow more than made up for the inconvenience of the journey. The Indians arrived late at night and were received by all the important members of the Academy of Sciences of the Soviet Union. They were treated, according to Ramanna, like a high-level delegation. Krishnan kept a diary of his daily activities and this provided an invaluable account of the details of the visit.

The visit started on 1 September 1955 at the Institute of Scientific

Information of the Academy. There, the delegation was shown the entire range of scientific and technical publications and the abstracting services. This was followed by a visit to the Lebedev Institute of Physics in Moscow, of which Skobeltsyn was the director. Visits to a machine tools factory and Moscow University followed the next day. On 3 September, the Indians were taken to the Institute of Nuclear Problems at Dubna, a small township near a large dam on the Volga, about 120 km from Moscow. The main attraction there was a synchrocyclotron, the biggest in the world at the time.

After some sightseeing on the next day, which was a Sunday, the delegation was taken to the site of the first Soviet-built natural uranium reactor, the details of which had been presented at the Geneva Conference. The next day, they visited the Institute of Physical Problems and met its director, the colourful Pyotr Kapitza, and two of its illustrious members, Lev Landau and Evgeny Lifshitz. Kapitza was meeting Bhabha after many years and showed him and Krishnan a turbine-operated liquid air plant, which produced 200 litres of liquid air per hour. He was commissioning a larger helium plant and told Bhabha and Krishnan that, very soon, liquid helium would be used as freely as liquid air or nitrogen. They also discussed Kapitza's work on superconductivity at the institute. That afternoon, Bhabha and Krishnan had an official meeting with Deputy Foreign Minister Vasili Kuznetzov.

The next stop for the delegation was Leningrad, which included a visit to the Physico-Technical Institute of the USSR Academy, of which the eminent Russian physicist Abraham Joffe was the first director. Krishnan discovered that the group there was already familiar with the NPL's work on thermionics in New Delhi. The delegation visited the Academy of Sciences of the Soviet Union on 10 September. The evening was given over to the pleasures of ballet, with a grand performance of Serge Lifar's *Romeo and Juliet*, with the famous ballerina Galina Ulanova in the lead role. Among those who witnessed the performance were the Russian premier Nikolai Bulganin and West German Chancellor Konrad Adenauer.

A trip to Voronezh followed, where the delegation saw the new nuclear reactors that were being constructed. Subsequently, the delegation flew to Zaporozhye, in Ukraine, a city located on the Dnieper that was famous for its dam and a hydroelectric power station constructed in 1932. The dam had been destroyed in the war and fully rebuilt thereafter. The delegation was accommodated in a guest house near the dam. Ramanna narrated a trivial but embarrassing incident that took place there. While they were provided with all comforts at the guest house, water and toilet paper seemed to be in short supply. Bhabha asked Ramanna to ask the kindly old caretaker for some toilet paper. The lady knew no English, and Ramanna's attempts at sign language met with little success. She then called someone who knew English and Ramanna communicated his requirement to him. The man nodded enthusiastically and returned with a large bunch of old copies of *Pravda*![31]

This was followed by a visit to Kiev and Sverdlovsk in the Arctic Circle. The fortnight-long visit came to an end with a farewell dinner organized by the Academy at Moscow on 16 September. The delegation, then, flew to Stockholm via Helsinki and Professor Manne Siegbahn hosted a dinner for the delegation at his residence. The next morning, Krishnan and Sethna left for London while Bhabha and the others stayed back in Sweden to complete their discussions. From Stockholm, Ramanna and Bhabha flew to Canada to discuss the Canadians' offer to set up an NRX-type reactor in India. The new helpful attitude was a direct result of the Geneva Conference since the Canadians had refused to part with details of this reactor just a year earlier.

A STATUTE AND THE SCOPE OF SAFEGUARDS

Soon after the Geneva Conference, the USSR announced its willingness to participate in the future agency, accepting, as a basis for discussion, the third draft statute prepared by the US government in March 1955. The text of the final statute was arrived at in the course of two conferences, held at the beginning and end of 1956 in Washington D.C. and New York, respectively. In 1955, the UNGA entrusted the US with organizing a conference of the 12 countries most interested in the creation of the new agency in Washington. The countries invited to participate were those that had been consulted over the drafts of the statute, along with the USSR, Czechoslovakia, Brazil and India. The conference took place in February and March 1956. The negotiations lasted four weeks and were characterized by the conciliatory attitude of the USSR. The type of organization that emerged from the negotiations was to have the role of a broker rather than a banker and possessed very broad control powers.

On 23 September 1956, the draft statute was presented to a gathering of 81 countries at the UN headquarters. The most controversial issue was that of the scope of safeguards. Bhabha, who led the Indian delegation, accepted certain safeguards on enriched uranium and plutonium, but opposed safeguards on natural uranium. He likened safeguards to neocolonialism, pointing out that, in general, countries with nuclear weapons would be exempted since owing to their advanced stage of development they would never have to request the assistance of the new agency. 'We will stand on the brink of a dangerous era sharply dividing the world into atomic "haves" and "have-nots" dominated by the Agency.'[32]

Such a division would result in dangerous tensions that would defeat the purpose of the safeguards, which was securing a peaceful world. Bhabha spearheaded the opposition to a very strict application of safeguards and France, which was represented by Goldschmidt, supported him by proposing a relaxation of safeguards on natural uranium and urging that safeguards should not be so severe as to deter future member countries from turning to the new agency for help.

Bhabha, who enjoyed great personal prestige, was opposed, above all, to a perpetuation of safeguards applied to successive generations of nuclear materials, something that was highly likely to occur in the case of India, which possessed nuclear materials but needed assistance to embark on a nuclear programme. There is no doubt that Bhabha played a decisive role in India's successful effort to weaken the scope of the safeguards. The 10 September 1956 draft statute had envisioned that plutonium and other special fissionable materials were to be deposited with the agency, except for quantities that the agency would allow members to retain for specified non-military uses under the safeguards. Bhabha fiercely rejected this broad agency authority over plutonium reprocessing and possession of the resultant plutonium.

At the deliberations on 27 September 1956, Bhabha observed: 'We consider it to be the inalienable right of States to produce and hold the fissionable material required for the peaceful power programmes.'[33] The problem for Bhabha was whether the plutonium produced using foreign reactors, such as CIRUS, would be subject to the control of the IAEA. Even though CIRUS had been negotiated before the IAEA came into existence, and only bilateral safeguards applied to it, future negotiations over foreign assistance would have come under the purview of the IAEA.

Bhabha pointed to the illusory nature of strict safeguards and emphasized that any aid in the nuclear field—be it training opportunities or nuclear materials—could potentially be military aid since it might allow a country to switch resources to a military programme. He proposed that the new agency give assistance only to those countries that did not have military programmes—defined as programmes in the field of nuclear and thermonuclear explosives and radiological weapons, but not including military nuclear propulsion. Bhabha said:

> There are many states, technically advanced, which may undertake with Agency aid, fulfilling all the present safeguards, but in addition run their own parallel programmes independently of the Agency in which they could use the experience and know-how obtained in Agency-aided projects, without being subject in any way to the system of safeguards.[34]

Bhabha's argument was that safeguards would severely curtail India's three-stage nuclear energy programme which had as its base, plutonium separated from the spent fuel produced in natural uranium reactors. The logic was that the imposition of safeguards would disallow plutonium acquisition. The point that Bhabha categorically opposed was the new agency's right, under Article XII.A.5, with respect to all facilities subjected to its safeguards, 'to decide on the use of all special fissionable materials recovered or produced as a by-product and to require that such special fissionable materials be deposited with the Agency, except for those quantities which the Agency allows to be retained for specified non-military purposes under continuing Agency safeguards'. Such power in the hands of the new agency might have given it too strong a hold on a country's economy if the latter had been based on nuclear power generation to which the new agency had contributed only in the initial stages.

After consulting the Secretary of State, the US delegation refused to modify its position to any appreciable extent. On 19 October 1956, the day the conference was to end with a vote on Article XII, the USSR, which had not yet declared its position, joined its allies, which had come out clearly on the side of India. Sensing the possibility of an impasse, Goldschmidt and August Lindt, who was the permanent observer at the UN, decided to table a compromise amendment. This amendment gave a country the right to retain (from the fissionable materials which it had produced) those quantities that it considered necessary for its research activities and for fuelling the nuclear reactors that it already possessed or was constructing.

After discussions, in which the Canadian delegation's influence worked in favour of acceptance of the compromise, the Franco–Swiss proposal was accepted. Bhabha, in recognition of Goldschmidt's help, dropped his proposal that the new agency should only assist countries that did not have a military programme. The next day, Article XII was adopted unanimously. Thus, a failure of the conference was narrowly avoided and the last obstacle to the establishment of the IAEA and its safeguards—fundamental elements in the present world policy of non-proliferation—was overcome.

In the end, the final statute declared a regime of much less stringent safeguards, where recipient countries could possess the plutonium extracted through spent fuel generated in foreign-supplied nuclear reactors and use it for peaceful purposes. As Bhabha observed, the revised statute ensured that fissionable material,

> [P]roduced in Agency-aided projects in a country should be at the disposal of that country, which would have the right to decide whether it wished to go ahead with a particular use of that fissionable material or not...in this way we ensured that the Agency would not be given powers which would enable it to interfere in the economic development and the economic life of the States concerned.[35]

Bhabha realized that the proposed safeguards of the IAEA would not be able to prevent nations from developing nuclear weapons. He accepted the final statute knowing that India could receive international assistance without sacrificing its nuclear weapon capability.

Cockcroft had an amusing anecdote to narrate in connection with the meetings of the IAEA at Vienna. Bhabha's complete disregard for time was well known. Cockcroft put it a little more politely when he said, 'One of Homi's characteristics was his apparent unawareness of the passage of time.'[36] Cockcroft says that while Bhabha was very punctual at the meetings of the UN Scientific Advisory Committee chaired by Hammarskjöld, he was often very late for the other meetings of the IAEA. As a result, the secretary began to advertise the official time of the meeting half an hour before the real time. Bhabha would then appear in time for the beginning of the discussion to everyone's amusement.[37]

Eisenhower's Atoms for Peace programme had ushered in a period of relaxed control over nuclear information, and one aspect of this was the rush by both the US and the USSR to declassify and disseminate a large volume of technical information. By 1958, the US had made it possible for any nation to access almost all the basic scientific information in the field of nuclear fission. However, the fundamental dilemma of how a policy could prevent the proliferation of nuclear weapons while promoting the benefits of nuclear energy, when the basic raw materials and technology for both are essentially the same, remained unchanged.

THE SECOND AND THIRD GENEVA CONFERENCES

The second UN International Conference on the Peaceful Uses of Atomic Energy was held at Geneva in September 1958 and was even bigger than the first. Even though a lot of technical data was presented at this conference, it lacked the dramatic impact of the first. There was also reason to be less optimistic given the fact that nuclear power from fission energy was not proving to be as cheap as predicted in the first conference. Francis Perrin of France was selected to preside over the conference. Bhabha was one of the six vice-presidents, and

Sigvard Eklund of Sweden was appointed the secretary general of the conference. Sethna was appointed as deputy director general, a signal honour, and A.S. Rao was one of the 25 secretaries appointed to prepare for the conference and to service its various technical sessions.

One of the main outcomes of Bhabha touching upon fusion in the first conference in 1955 was that it became part of the agenda at the second Geneva conference. Bhabha was selected to preside over the plenary session on the 'Possibility of Controlled Fusion'. The most notable feature of the conference was the declassification of information on fusion. Though the conference gave great prominence to controlled fusion, there was also a note of caution in the predictions. Bhabha and W.B. Lewis presented a paper on the 'Canada–India Reactor: An Exercise in International Collaboration', something that is discussed in a later chapter. Apart from the general and technical sessions, there were a series of six evening lectures by eminent scientists. Bhabha was the first to speak and his subject was 'Need for Nuclear Power in Underdeveloped Countries'. His lecture served to dispel the view that atomic energy would find immediate applications only in the industrially advanced countries.[38]

At the third conference held in September 1964, Bhabha presented a paper on 'World Energy Requirements and the Economics of Nuclear Power with Special Reference to Underdeveloped Countries'. In it, he estimated that, by the year 2000, nuclear power would generate two million MW of electricity throughout the world. In presenting his paper, Bhabha made his famous remark, 'No power is as expensive as no power.' Bhabha was far too optimistic in his prediction as future events proved, but it was this quest for cheap electricity that was the driving force of the last two decades of his life.

◆

Soviet leaders Nikita Khrushchev and Nikolai Bulganin visited India in February 1960 and Bhabha gave them a tour of TIFR. In the summer of the same year Bhabha led an Indian delegation to the USSR. In February 1961 Bhabha announced that the USSR had agreed to build a reactor for India. In a public statement, Bhabha said that the inspections required by the IAEA for reactors coming in from the US were an infringement upon Indian sovereignty. The USSR, which was not a member of the IAEA, hailed the opposition from India and other non-aligned nations to IAEA controls over their nuclear programmes.[39] An Indo–Soviet agreement was signed on 7 October 1961 but nothing much came of it.

Chapter 31

APSARA

> The control system of Apsara, the first reactor in Asia was built under the auspices of this Institute in a war time hutment on this very site, and many parts of the reactor in the Institute's workshop. It is not an exaggeration to say that this Institute was the cradle of our atomic energy programme, and if the Atomic Energy Establishment has been able to develop so fast, it is due to the assisted take off which was given to it by the Institute in the early stages of its development.
>
> —Homi Bhabha's Speech at the inauguration of the new TIFR buildings, Bombay, 15 January 1962[1]

Bhabha created the conditions for an atomic energy establishment through legislation, massive funding and bureaucratic independence. Separated from the State and civil society, it gave the scientists freedom to experiment without constantly being in the public eye. Bhabha was under immense pressure to provide tangible proof of some progress in his atomic energy programme. The best and most public way was by delivering the most obvious and long overdue nuclear artefact—a nuclear reactor. This need was made more urgent by the challenge from Saha.

NEGOTIATIONS

India's early strategy for the development of atomic energy was centred on its large reserves of rare earths, such as thorium, and strategic materials, such as beryllium and manganese. These would be offered to select foreign countries in exchange for funds, material and know-how. The Atomic Energy Act of 1948 ensured a state monopoly over such minerals, so such barter could be easily negotiated. Bhatnagar ensured there was no interference in the AEC's plans. In fact, foreign governments would treat statements from Bhabha and Bhatnagar as the equivalent of state policy.

As early as January 1948, even before AEC was formed, Bhabha started approaching his old friends and colleagues in the UK for scientific assistance. On 9 January 1948, he wrote to John Cockcroft, requesting a meeting to explore the possibility of cooperation in scientific research between India and England. The British showed some interest in Indian monazite, but Bhabha's first request for scientific collaboration was turned down, perhaps due to fear of the disapproval of the US. Bhabha was also in touch with other international scientists, many of whom were his friends. He visited Canada in 1947, and the Canadians, who had an eye on Indian thorium, agreed to supply India with a tonne of uranium oxide.

Bhabha also discussed the possibility of collaboration with Fréderic Joliot-Curie. In fact, the latter first approached India with the suggestion of a joint project in atomic energy. A French research reactor went critical in 1948 thanks to his leadership. But he was a communist and it was his Left leanings, and his opposition to military research that ultimately cost him his job. As noted before, Joliot-Curie was dismissed in 1950 and was succeeded by Francis Perrin. His visit to India, in December 1950, led to a Franco–Indian agreement under which the French were to collaborate with India in the development of a monazite processing factory and training Indian scientists.

A few months after receiving the French offer, Bhabha once again approached the UK Atomic Energy Authority, asking if they were interested in formal scientific collaboration. The construction of a small atomic pile had always been one of the immediate objectives of

AEC and Sri Prakasa, the minister for natural resources and scientific research, made an announcement to this effect in Parliament as early as 1951. Henry Tizard, who was the chief scientific advisor to the British government, was sympathetic to the idea but turned down the possibility of a formal agreement given the present state of the world. Bhabha was becoming impatient and made it known that if the British continued to drag their feet, he would have to use the alternatives available to him.

His approach made the British soften their stand somewhat. They had been keeping a watchful eye on the state of India's negotiations with France and the US and realized that it would be short-sighted on their part to not exploit opportunities in their former colony. It helped that Bhabha was dealing with former Cambridge colleagues and mentors, who were at the helm of atomic affairs in the UK. Cockcroft, who Bhabha knew from his time at the Cavendish, was in charge of the UK Atomic Energy Authority's main facility at Harwell. Tizard and Edwin Plowden, senior advisers to the British government, were also inclined to help Bhabha. Thus, by the early 1950s, Britain found itself drifting into a closer relationship with India on matters concerning atomic energy.

However, realizing the dangers of putting all his eggs in one basket, Bhabha was also in touch with the US authorities. In 1952, Bhabha wrote a letter to his US counterpart, Gordon Dean, asking him, rather bluntly, for all the declassified information on reactor theory, design and technology. He added that details on the big graphite reactor at Harwell had been almost completely declassified along with the one at Chalk River in Canada. The letter also requested for ten tonnes of heavy water over the next two years. In return, Bhabha said that India was prepared to give the US certain quantities of thorium in a suitable form—uranium ore or some compound of uranium.

Bhabha also added that India would provide the US with more beryl than had been originally agreed upon. Nothing came of this for a variety of reasons, the US's annoyance with India's non-aligned stance being one of them. Furthermore, other sources of raw materials had appeared, with Brazil and South Africa willing to supply thorium for cash. The US stockpiles, too, were close to being full. Negotiations also suffered because of the Indian desire to set up processing plants within the country rather than simply negotiating a sale of unprocessed raw material.[2]

There was a history of barter between the two countries though. When the diplomat William Pawley led a mission to India, in June 1951, seeking an end to the Indian embargo on monazite, he agreed to assist India in setting up a monazite processing factory. Pawley told Cabinet Secretary N.R. Pillai that he and his colleagues had lobbied successfully to separate these negotiations from the India Food Aid Bill, which was being presented to Congress back then. The Indian government acceded to the US's request and Pawley returned home with a promise of 500 tonnes of monazite. However, the interaction between the US and the Indian AEC was driven largely by strategic and Cold War concerns. In the end, the American need to ensure that ores and compounds of uranium did not reach Iron Curtain countries was of paramount importance.[3]

A RESEARCH REACTOR

Bhabha and Bhatnagar had a series of meetings with British officials in mid-1954, requesting help for setting up a reactor. In June 1954, Bhabha requested Cockcroft for five tonnes of heavy water. Cockcroft declined, citing their own needs, and suggested that Bhabha take his request to W.B. Lewis, the vice-president of Atomic Energy of Canada Limited (AECL) and a Cambridge colleague who had been on the same rowing team as Bhabha. A few months later, in September, Cockcroft wrote to Bhabha, asking him if he had considered the possibility of building a swimming pool reactor[4] that required enriched uranium, which he offered to provide. Bhabha immediately accepted, asking him about the quantity that would be made available, the terms and conditions and the timeline for delivery. He emphasized the time aspect, indicating that he wanted to have a reactor ready in as short a time as possible so that they had something to work with till their other plans fructified. Bhabha wanted to show some concrete results as soon as possible to silence his critics, especially Saha.

The details were negotiated over the next few months. Apart from

6 kg of enriched uranium fuel rods from the UK, Bhabha managed to get detailed engineering drawings and technical data. The British got nothing in return, except a promise that the Indians would favour buying a British reactor in the near future. In March 1955, Bhabha decided to build a swimming pool research reactor, and the plans were finalized by July 1955. It was conceived as a multipurpose reactor for experiments in neutron physics, the production of radioisotopes and studies in shielding of nuclear radiation. Bhabha was confident enough to make a bet with Cockcroft that the Indian reactor would be ready in one year.

The location chosen for the reactor was AEET. Trombay was, at the time, a remote and verdant area north of Bombay. It was far away from the urban conurbation but accessible enough for water, power and telephone services to be made available. Bhabha wanted the reactor to be built entirely from Indian materials. When the reactor achieved criticality, it was presented as a purely indigenous effort, which, despite Bhabha's attempts to maximize Indian content and expertise, was not true.

A multidisciplinary team took charge of the different aspects of building the reactor. The overall design for the reactor was entrusted to a committee consisting of Bhabha, Raja Ramanna, A.S. Rao, Homi Sethna, K.S. Singwi and N. Bhanu Prasad. Ramanna drew up the requirements for neutron experiments. Singwi worked on the theoretical physics of the reactor. A.S. Rao was in charge of the control and instrumentation work. Bhanu Prasad was responsible for the auxiliary equipment and overall design of the reactor. Sethna was in charge of the IREL plant, which extended logistic support for the construction of the reactor. V.T. Krishnan, a mechanical engineer was put in charge of the construction of the reactor pool and the building.[5]

M.R. Srinivasan, who went on to become chairman of AEC in 1987, joined the DAE in September 1955. A mechanical engineer from Bangalore, he had a doctorate from McGill University, Montreal. He was selected for the position of a senior research officer by the DAE in mid-1955, and met his senior Bhanu Prasad at the High Commission in London. Bhanu Prasad had joined in 1954 and later became the project manager when CIRUS was being built. Maheshwar Dayal also joined the institute as a junior research officer at the same time. They were deputed to attend a 10-week programme at Harwell headed by Cockcroft, which was now open to non-British participants.[6]

On his return in early 1956, Srinivasan joined the reactor engineering division headed by Bhanu Prasad. His job was to get some large aluminium components fabricated locally. In 1956, few Indian fabricators had worked with thick sections of aluminium. Only after Srinivasan spent many mornings with the firm that was doing the welding job for them that they finally developed processes for successfully welding the sections. The structural frame for supporting the reactor core was fabricated at Mazagon Dock Limited, which also prepared the base plate that had precision-bored holes with keyways for locating the fuel elements.[7]

In December 1955, Bhabha was optimistic enough about the progress of the team to write to Cockcroft that he was close to winning their bet. Six months later, the two exchanged notes, deciding the text of the official letters of congratulation and the reply. However, things were not always easy.

Ramanna recalled that the completion of the building that would house the reactor coincided with one of the heaviest monsoons in Bombay. The roof leaked as did the walls of the reactor hall. To make matters worse, the sides of the swimming pool also began to leak. These were ultimately fixed and the civil engineers gained crucial experience that allowed them to build non-leaky tanks of all dimensions in the future. Ramanna observed with a tinge of wry humour that his years with the atomic energy programme had always had something to do with leaks of one sort or the other, be it leaks of water or heavy water from pools, pumps, heat exchangers or even merely just information.[8]

GOING CRITICAL

The first attempt to make the reactor critical[9] was made on 31 July 1956. Srinivasan recalled that it was a typical monsoon day in Bombay, with heavy rain and strong winds. After a working dinner in the conference room, Bhabha and his team attempted to start the reactor. The fuel loading configuration had four control rods in the central part of the core. Since the leakage of electrons was much higher than expected, the reactor could not achieve criticality that day. Bhabha was personally loading the fuel elements—something that the safety groups disapproved of. While doing this, one of the fuel elements slipped and fell on to the grid plate and made a burr on a keyway. The team worked into the wee hours of the morning, before Bhabha suspended work.

A little more work was required before the second attempt. Some of the fuel elements had been deformed during transportation. Srinivasan and A.S. Rao undertook minor repairs before the next attempt. The physics calculations showed that the control rods were to be moved into less reactive positions. The second attempt at criticality was made on 4 August 1956. Unlike the first time, when a large number of people were present, Bhabha asked only key personnel to be present with no distractions from onlookers. All went very smoothly during the second start-up and when India's first research reactor went critical that afternoon, it became the first country in Asia to have a functioning nuclear reactor. Bhabha lost his bet with Cockcroft by a few days and offered to take him out to dinner in Paris, unless he preferred London. It catapulted the DAE from its initial private phase into the glare of public view.

Jubilation followed what may now seem to be a relatively minor achievement. As Ramanna noted, the first taste of success was unparalleled.[10] Bhabha called Nehru to give him the good news. Since it had been raining heavily on that day, given the primitive state of telecommunications at the time, it was nothing short of a miracle that Bhabha could even speak to Nehru. The successful commissioning of the reactor marked the culmination of frenetic activity and showed the Indians that they could do it. There is a one rupee note in the TIFR Archives that Bhabha inscribed with the following words for his mother: 'Apsara, first reached criticality 4 Aug,' 56. To Twit with fondest love and kisses, Homi.' Bhabha, like some Parsis, perhaps considered a one rupee note to symbolize good luck.

That same evening, Bhabha wrote to Nehru enclosing a copy of the approved press note. When he reached home, his mother told him that the Prime Minister had made an announcement on the subject.

During the last phase of the commissioning, some 50 scientists and engineers had worked continuously for a week, and Bhabha had been hands-on with them for the last two days. Trombay, at the time, was not easily accessible. It was almost an hour away from the centre of the city and transporting people at odd hours proved problematic. On 1 August 1956, Bhabha wrote to Nehru, requesting two cars to be placed at the disposal of the scientists round the clock. Additionally, he wanted lunch and dinner to be provided to his team from an appropriate restaurant. Both his requests, Bhabha said, were not permitted by the rules, and he wished to have the Prime Minister's approval for these actions since 'it is clear that in the circumstances, everything should be done which should relieve the physical strain'.[11]

Bhabha informed Nehru about the late hours they had all been keeping. He also did not hesitate to tell Nehru how unsuited government rules and regulations were to 'executive work which was to be done at speed and pressure'. He ended by saying, 'It would be unreasonable to try and apply Government rules to groups and departments which work like this. The result will only be to reduce output and enthusiasm.'[12] The note was rushed to Delhi and Nehru approved it with the words 'I agree' on the same day, yet another example of how Bhabha's relationship with the Prime Minister eased his way at every step. It is inconceivable for such a thing to happen today; in fact, even in those days, it would have been very difficult since 'through proper channel' had been the mantra even then. The bureaucracy knew of Bhabha's special relationship with Nehru but that still did not deter civil servants from trying to block him whenever they could.

The 1 MW reactor was named 'Apsara', after the water nymph of Hindu mythology. K. S. Krishnan suggested the name, which Bhabha and Nehru approved. Nehru, in his speech on 20 January 1957, while inaugurating AEET, told his audience how it got its name. 'Dr Krishnan from the fund of his knowledge of Sanskrit lore suggested various names. Dr Bhabha pondered over it; the governor who is also a Sanskrit scholar was consulted [...] The name we suggest for it is Apsara which as you know, means a celestial damsel or a water nymph.'[13] Apsara achieved its rated capacity without any mishaps and was used for the production of isotopes, basic research, shielding experiments, neutron activation analysis, neutron radiography and the testing of neutron detectors for over 50 years.

After Apsara started functioning, all important visitors were brought to Trombay to see it. The dignitaries who visited it included Chinese premier Chou En-lai and Emperor Haile Selassie of Ethiopia. Saha did not live long enough to see Apsara go critical. On 16 February of that year, he collapsed on a pavement as he was walking to the office of the Planning Commission to attend a meeting. A long-standing hypertensive, Saha had been a member of the Lok Sabha at the time. He was rushed to Willingdon Hospital (now Dr Ram Manohar Lohia Hospital) but he had already passed away. Bhabha and Nehru could rest easy since the criticism that Saha had spearheaded would now end. In the atmosphere of self-congratulation that followed Apsara reaching criticality, the reactor was presented as a purely indigenous affair. Even though it was admitted that the fuel rods and some electronic valves had been imported, it was less well known that plans and drawings crucial to the project has also been sourced from the British.

The local media represented Apsara as an Indian reactor built with local ingredients and expertise. The political interpretation of this event continued: India had stolen a march over China, the only other country that was remotely in a position to develop nuclear capability. Nehru's speech at the inauguration of the reactor, on 20 January 1957, reinforced the image of a completely indigenous effort and is a good example of the propensity to claim indigenous production of all nuclear artefacts. Nehru said: 'We are not reluctant in the slightest degree to take advice and help from other countries [...] But it is to be remembered that this swimming pool reactor in front of you is the work, almost entirely, of our young Indian scientists and builders.'[14]

In a speech in the Lok Sabha later that year, Nehru dropped the qualifier and confidently asserted, 'The putting up of the swimming pool reactor [...] was done entirely by Indian scientists and engineers.' Cockcroft must have been a little put off when he noted in an internal memo, 'Did you see the press release from Delhi [...] this seems rather ungracious in view of the advice and help we have given and are asked to give. Presumably detailed plant designs and drawings do not constitute outside help.'[15]

When the new buildings of TIFR were being inaugurated in January 1962, Bhabha referred to how the control system of Apsara had been designed there.

> [T]he control system of Apsara, the first reactor in Asia was built under the auspices of this Institute in a war time hutment on this very site, and many parts of the reactor in the Institute's workshop. It is not an exaggeration to say that this Institute was the cradle of our atomic energy programme, and if the Atomic Energy Establishment has been able to develop so fast, it is due to the assisted take -off which was given to it by the Institute in the early stages of its development.[16]

After the initial supplies of fuel from the UK were consumed, enriched uranium was obtained from France. After more than half a century of successful operation, Apsara was permanently shut down in 2010. In September 2018, an upgraded version of the reactor, with a maximum output of 2 MW was recommissioned. It is fuelled with plate-type dispersion fuel plates made of low enriched uranium. By virtue of its higher neutron flux, this reactor has increased the production of radioisotopes for medical application by about 50 per cent. It is also extensively used for research in nuclear physics, material science and radiation shielding.

PRODUCTION OF RADIOISOTOPES

The production of radioisotopes started soon after Apsara was commissioned. V.K. Iya, who returned to India with a doctoral degree from Paris in January 1957, was an early pioneer in this field. Iya set up a small laboratory in a shed in the Bombay Dyeing godown on Cadell Road in Bombay for the extraction of small quantities of iodine-131, gold-198, phosphorus-32 and sodium-24 using irradiated samples from the reactor. The isotopes were purified and tested for medical use with the help of King Edward Memorial Hospital in Bombay, first on animals, and then in human trials.[17]

The production of isotopes was a priority. Srinivasan recalled how C.B.G. Taylor, in his enthusiasm, transported the first radioactive capsule produced at Apsara in his trouser pocket while he rode his motorcycle to the TMH, Bombay's cancer hospital. Taylor, who had worked at Harwell, built neutron detectors at TIFR and, later, worked on the production of radioisotopes, heading the division at Trombay.[18]

At the World Agricultural Fair in Delhi in 1960, the US had set up a Training, Research, Isotopes, General Atomic reactor—a swimming pool reactor that could be installed without a containment building, designed for research by scientific institutions and universities. Iya, on Bhabha's advice, set up an isotope laboratory with glass walls so that the operations inside were visible to all. The isotopes produced in the American reactor were processed in the Indian pavilion. Later, they were supplied to Safdarjung Hospital in New Delhi, which became the starting point for the establishment of the Institute of Nuclear Medicine and Allied Sciences in New Delhi.[19] The isotope activities moved to a temporary facility in Trombay, and, by the early 1960s, an isotopes laboratory was set up at the South Site in Trombay. The design for the full-fledged radiological laboratories at the north site started in 1961–1962, but it was only in 1965 that work commenced. It was completed in 1970.

THE TATA MEMORIAL HOSPITAL

The TMH was yet another sterling venture in philanthropy associated with the Sir Dorabji Tata Trust. It was inaugurated by the Governor of Bombay, Sir Roger Lumley, on 1 March 1941. Lady Meherbai Tata, it may be recalled, had died of leukaemia, sparking Dorabji Tata's interest in fighting cancer. Nine years after his own demise, a hospital to fight the dreaded disease was inaugurated thanks to a grant of ₹31 lakh from the trust formed from his bequest. The trust ran the hospital entirely on its own from 1941 to 1948. In 1949, the Government of Bombay sanctioned an annual grant of ₹1 lakh. However, the trust soon discovered that it was becoming a bottomless pit into which money was disappearing with astonishing speed. By 1957, the trust had made its single highest contribution of ₹1 crore to the hospital, with 75 per cent of the beds being free. By 1957, the trustees realized that they needed to seek assistance for further expansion and such aid could only come from the government. In May 1957, the hospital was taken over by the Ministry of Health, but things only deteriorated due to bureaucratic delays and financial constraints.[20]

Bhabha was instrumental in lending a helping hand to the TMH. He was told of this sad state of affairs by Dr J.C. Paymaster and Dr D.J. Jussawalla, two of the four doctors who had been with the hospital since its inception. They urged Bhabha to take the hospital under the umbrella of the DAE, citing the example of the US AEC, which ran cancer research hospitals. When Bhabha told Nehru about it, he was initially sceptical. 'What Homi! [...] You now want to run a cancer hospital? I am not so sure you should, but go ahead if you feel it is necessary.'[21] What changed Nehru's mind was Bhabha's argument that the US AEC's concerns about radiation had made them take over the cancer hospitals.

Bhabha had already established a cell biology unit at the hospital in 1948 and two years later the Indian Cancer Research Centre (ICRC) was set up in close proximity to the hospital. The DAE, TMH and TIFR all functioned as a team. The Indian Cancer Society, a voluntary organization formed by Dr Jussawalla in 1951 to fight cancer on all fronts, was closely associated with these institutions. Bhabha wrote to Rustum Choksi, asking for the transfer of administrative responsibilities to the DAE. He was on the board of the hospital as were his successors, Vikram Sarabhai and Sethna. In 1967, the hospital merged with the ICRC, and became the Tata Memorial Centre. The chairman of AEC became the chairman of the governing council. From the beginning, the governing council of the Tata Memorial Centre has included three representatives from the Sir Dorabji Tata Trust.[22]

Bhabha had mooted a proposal in 1958 to locate the Centre for Radiation Medicine at Trombay, but later he decided to locate the centre at Parel, as four major hospitals were already located there. He arrived at this decision based on the advice of John Lawrence of Donner Laboratory in the US, who had visited India on Bhabha's invitation. John was the brother of the famous E.O. Lawrence, the inventor of the cyclotron. He suggested the type of equipment required, arranged funding, and sent one of his colleagues to assist in starting the centre. R.D. Ganatra—who had been sent to the US for training in this field under the auspices of the World Health Organisation—had, by a lucky coincidence, been assigned to the Donner Laboratory. Upon returning in 1962, he immediately commenced work on this project.[23] Bhabha declared the centre open in August 1963, putting Dr K.N. Jeejeebhoy in charge, under the overall supervision of Dr Paymaster, the director of TMH. When Dr Jeejeebhoy migrated to Canada in 1968, Ganatra took his place.

'ACCELERATING' RESEARCH

Saha and B.D. Nagchaudhuri had built a 38 in. cyclotron, modelled after the one at Berkeley in Calcutta in the 1940s. All the components of this cyclotron, except for the magnet yoke and pole pieces, were fabricated indigenously. In 1950, Bhabha had attended the international conference organized by Cockcroft at Harwell and discussed Indian requirements with him. The betatron seems to have been forgotten. In a letter, Bhabha told Cockcroft that he was thinking of ordering a linear accelerator of the type they had at Harwell, which gave a maximum energy of 15 MeV. Bhabha wanted to know the details of the firm and the contact person so that Raja Ramanna could be put directly in touch with the accelerator group at Harwell.[24] Ramanna headed the experimental physics group at AEC, then housed in TIFR.

An order was placed, in September 1951, for a cascade generator with Philips Electrical (India) Limited. The one-million-volt Cockcroft–Walton accelerator (commonly known as a cascade generator) was installed in 1953. It was originally installed in a shed at the site where TIFR is now located. The site was covered with hutments and sheds which were used for research work before the new buildings came up. It was shifted in 1960 to a high hall at the southern end of the institute's building. It was used extensively for studying a variety of nuclear reactions as well as the slowing down and diffusion of neutrons in different moderator assemblies.

Before the arrival of the cascade generator, cosmic ray research had dominated experimental work. The TIFR was now officially conducting research in nuclear physics for AEC. There was also a workshop for AEC housed within TIFR that manufactured Geiger counters, amplifiers and radiation survey meters, among other things. However, Bhabha continued to focus on cosmic ray research because the study of elementary particles needed increasingly larger accelerators that were very expensive to build and maintain. He believed that given India's limited technical and financial resources, it could not compete effectively in the field of elementary particle physics using large accelerators that are resource-heavy to build.[25]

In the mid-1950s, since accelerator energies had yet to surpass the energies obtained in cosmic rays, Bhabha was happy to continue concentrating on cosmic ray research. Accelerator building at TIFR must be understood in this context. It is also pertinent to note that AEC and TIFR were not two institutions with well-defined mandates. Their research agendas were intertwined, a result of Bhabha's dual position as a science administrator and the head of a research institution. The TIFR had been declared as the institution for nuclear research, but it did not develop its own agenda of applied and industrial research. Cosmic ray research, and later, the AEC's priority of building reactors became the focus.

The accelerator programme, thus, began on the margins of atomic energy research and was subordinate to research in cosmic rays. In September 1950, Bhabha had written a note to the Planning Commission, bemoaning the lack of coordination and waste caused by the Ministry of Education administering institutes of higher learning while national laboratories were under the Ministry of Natural Resources and Scientific Research. Bhabha felt that the correct place for institutes of higher learning in science and technology, including university teaching departments, should be under a ministry that dealt with scientific research combined with education. This way, universities would benefit from the research facilities in national laboratories and from teaching by leading researchers in the field. In February 1952, he wrote to remind Nehru about these views, opining that the separation of advanced scientific and technical teaching from research was not advisable.

In 1952, Bhabha organized three groups overseen by D.Y. Phadke, who was in charge of the electronics and instrument division that

TIFR ran for AEC. Phadke, who had taught electronics at St Xavier's College, Bombay, supervised the work of three sub-groups involved in constructing a 12-inch cyclotron, an open-air 1 MeV Van de Graaff generator and a 1 MeV linear proton accelerator. T.G. Varghese was in charge of the 12-inch cyclotron, K.A. George led the Van de Graaff group and R.V.S. Sitaram headed the 1 MeV linear proton accelerator group. Each of the groups had a few members trained in vacuum technology, physics and engineering. It is interesting to note that most parts of the machines, including the magnet, were sourced from Bombay's Chor Bazaar, which was selling war supplies left behind by the Allied forces when they withdrew after the Second World War ended.

The Philips cascade generator was finally installed in 1953 after two postponements. It was purchased by AEC but was housed at TIFR. When Nehru came to Bombay for the foundation stone-laying ceremony of the new TIFR building, he and the other guests were invited to see the new cascade generator. There is a photograph of Bhabha wearing a fine Italian suit sitting in front of the machine. Bhabha asked Mark Oliphant for advice on a 60-inch cyclotron at TIFR. Oliphant said that a 60-inch cyclotron would be useful for training purposes, but the experience gained in its design and construction would be only of limited use for work upon accelerators for higher energies. Also, such a cyclotron, built specifically for manufacturing short-lived isotopes could be justified only if the demand for such substances was sufficiently large.

By this time the ambiguity of TIFR's position as a research institution within the larger agenda of AEC was coming into focus. The cascade generator was not available for fundamental research. It was required for work on nuclear reactors, which got priority over fundamental research (after all, it was the AEC's machine). Ramanna, who was in charge of the nuclear physics group at AEC, concluded that a very small machine should not be purchased. The machines that were available were larger machines and AEC did not need another larger machine just then. A larger Van de Graaff machine was being constructed at Trombay then, which would be available for fundamental research when ready. The accelerator builders were being trained for bigger work and could not be asked to build a smaller machine for the TIFR's experimental physicists.

In August 1964, Bhabha called for a two-day meeting to discuss the need for a medium energy accelerator. By then, 400 keV Cockcroft–Walton accelerators had been built at the Saha Institute and the Bose Institute, Calcutta. The TIFR, as noted, had bought a 1 MeV Cockcroft–Walton accelerator. A large accelerator became available for the first time when a 5.5 MeV Van de Graaff accelerator purchased from the High Voltage Engineering Corporation, Burlington, US, was installed at AEET in 1962. Though the accelerator had been purchased from abroad, the switching magnet and the beam transport system were built indigenously. The need for a bigger accelerator was being felt, and Bhabha convened a meeting at IISc on 3 August 1964 to discuss this issue.[26]

This meeting was the first to consider particle accelerator research after July 1947, and certainly the first in which all important laboratories were represented. A committee had already been appointed to review this question in 1962, and Ramanna read out the recommendations. The choice was between two types of accelerators: a tandem machine and an AVF (azimuthally varying field) cyclotron. A tandem Van de Graaff machine could be bought from the High Voltage Energy Corporation and the estimates for the AVF cyclotron were taken from the University of Maryland, College Park. The cyclotron would have to be contracted out to a company abroad with the design specifications being determined by Indians. It was estimated that construction of both machines would take about four years.

Bhabha decided that they would go in for both the machines since they were complementary. The tandem could be located at TIFR or AEET, where similar work was being done. Bhabha felt that the technology of an AVF cyclotron was entirely different, and this was the field they should now develop.[27] A proposal for the establishment of an inter-university centre for particle accelerator research was already being considered, and Bhabha said the cyclotron facility could be installed in a university setting, perhaps near Calcutta.

Bhabha's untimely demise less than two years after this meeting did not affect the plans for the two proposed facilities. The AEET and TIFR acquired a tandem Van de Graaff machine and the plans to construct a variable energy cyclotron continued with assistance from the USAID and the Lawrence Radiation Laboratory, Berkeley. The permission for an inter-university centre only came in 1984, leading to the establishment of the Nuclear Science Centre, which is now known as the Inter-University Accelerator Centre in New Delhi.

Chapter 32

CIRUS

The CIR being a Canadian gift under the Colombo Plan, all parts of it came from Canada, and the erection, though a Canadian responsibility was carried out jointly with our staff.

528

—Lecture delivered at the International Council of Scientific Unions, Bombay, 7 January 1966[1]

Apsara was useful for conducting scientific experiments and gaining experience in reactor maintenance, but it was not something on which a nuclear industry could be based. For that, a larger and more complex reactor was needed, expertise for which was not available in India at the time. As it turned out, owing to a conjuncture of international events, India did not need to produce such a reactor with indigenous expertise. The assistance it needed came from a surprising quarter—Canada. In an unforeseen event, Canada decided to use the Geneva conference to enter the international atomic arena. Like the British, the Canadians had also paid the price of close alignment with US strategic desires during the Second World War and wanted to internationally assert their difference from the US.[2]

THE CANADIAN OFFER

A few months before the Geneva conference, the Canadians decided to offer India one of their national research experimental (NRX) research reactors. They hoped that the attendant publicity would help create an international market for atomic energy. Among developing countries, India was the most advanced in nuclear matters and the Canadians thought it prudent to start with India in their attempt to exploit the market. This gesture gave Canada a chance to showcase its atomic technology and set the stage for further commercial ventures in this area. While Canada's offer was motivated by its amicable diplomatic relations with India, there was little doubt that it saw India as a potential customer for its nuclear reactors.[3]

When a Canadian team visited Delhi in early 1955, they found, to their consternation, that the US embassy personnel were discussing nuclear cooperation with the Indians, and they feared that the Soviets, too, were in the fray. By March 1955, the Canadian Cabinet had given its approval for the offer; on 5 April, W.B. Lewis wrote to Bhabha informally conveying Canada's interest in nuclear cooperation with India.[4]

The Indian response to this adventitious offer was cautious. Bhabha wanted to know if the Canadian offer would result in less aid for development. In his reply to Lewis, he hinted that he was interested in a more advanced British reactor. In June, Bhabha wrote that he still preferred a British reactor but was also considering a swimming pool reactor built in the US. The Canadians agreed to supply extra funds for the reactor under the Commonwealth's Colombo Plan, ensuring that normal development aid would not be affected. Bhabha conveyed a message through Nehru that he was interested in the more advanced National Research Universal reactor, but this elicited a strong and instant rejection.

In the end, Lewis managed to convince Bhabha, when they met at the Geneva conference in August that year, that the 40MW NRX reactor on offer was suitable for India's needs.[5] M.R. Srinivasan said that, by then, the UK had also offered to assist India in building a high flux reactor using enriched uranium, but Bhabha preferred the

Canadian offer because the heavy water reactor could be fuelled with Indian uranium.[6] Bhabha had been toying with the idea of building a E 443-type British reactor but abandoned this idea in favour of the Canadian NRX, a decision that appears to have been influenced by a conversation Nehru had with the Canadian prime minister.[7] Initially, the reactor was called Canada India Reactor (CIR) but was later renamed CIRUS when the US supplied heavy water for the reactor.

The development of heavy water reactors was the result of international cooperation during the war. Canada had been assigned the task of developing heavy water in the nuclear collaboration between the US, UK and Canada. The Americans were not interested in heavy water reactor development as a part of their own programme so Canada was allowed to take the lead as a joint British–Canadian activity. The Montreal Road Laboratories, outside Ottawa, were set up under the leadership of John Cockcroft.

The Zero Energy Experimental Pile (ZEEP), the first heavy water reactor, was built at Chalk River, a nuclear research complex 180 km from Ottawa, which eventually superseded the Montreal Road Laboratories. The AECL was set up to manage all these activities. Lorne Gray, the president of AECL, and Lewis, who had immigrated to Canada during the war, led the overall development efforts. By 1947, AECL constructed an NRX reactor of 10 MW output, the most powerful heavy water reactor in the world at the time. But India signed an agreement, in September 1955, to build CIRUS—a 40 MW reactor moderated by heavy water, cooled by light water and fuelled by natural uranium—which was four times the capacity of the AECL's NRX reactor. It was to be a slightly modified version of the NRX-type reactor at Chalk River Nuclear Laboratories.

Ramanna observed that the Canadian offer suited their requirements and the most important aspect of it was the hands-on training of Indian personnel at Chalk River. Ironically, a year earlier when they had visited Chalk River, they had been subjected to strict security measures. No answers were given to their questions and the Canadians tried their level best to be secretive.'[8]After the Second World War, when Cockcroft returned to the UK to head the atomic energy programme, Lewis remained in Canada to develop the Canadian programme. Lewis had ambitious plans for a large atomic energy establishment that could not be located in a big city like Montreal. He shifted to Chalk River, where the large complex of the AECL developed. The residential area was in a small village called Deep River, a planned community that had been developed to house the employees of Chalk River, located nearby. Along with Los Alamos, New Mexico, and Oak Ridge, Tennessee, Chalk River was an offshoot of the nuclear efforts of the Allies to make the atom bomb. After the war, Canada continued its research into the peaceful uses of the atom. Deep River was situated far enough upwind and upriver of Chalk River to avoid radioactive fallout.

Half the cost of CIRUS was to be borne by India and the other half by Canada, under the Colombo Plan. Canada offered to pay all the foreign exchange costs of building the $14 million reactor (which ultimately cost $24 million). For operational reasons, Bhabha insisted that the project be kept outside the Colombo Plan. The Department of Economic Affairs (DEA), which was controlling the Indian side of the Colombo Plan activities, thought that the reactor project was under their purview. The DEA controlled the purse strings and Bhabha realized that he would have to navigate bureaucratic waters in that ministry, something he was not prepared to do. He got himself out of this potential mess in a way that had, by then, become quite common for him—he wrote directly to the Prime Minister. Bhabha said that unless it was made clear to the DEA that this project was outside the Colombo Plan, all inter-governmental correspondence would have to be routed through the Department of Economic Affairs of the Ministry of Finance. The Canadian-Indian Reactor was a highly technical project and there was clearly no advantage in routing all this correspondence through the Department of Economic Affairs. Indeed, it would only slow down the entire process.[9]

The construction of CIRUS also necessitated the import of 21 tonnes of heavy water from the US AEC. Even though the Canadians had realized that the possibility of producing plutonium for nuclear weapons existed, Bhabha stubbornly refused to accept safeguards on the spent fuel from CIRUS. The Canadians gave in because they anticipated fierce competition in the nuclear market and did not want to endanger the contract. In the end, Canada attached no strict safeguards on how the plutonium produced by the reactor would be used—it merely obtained a commitment from India in a secret annexure in the treaty that the fissile material would only be used for peaceful purposes.[10] The agreement signed between Canada and India was only a few pages long. But the clause about the peaceful use would become problematic after the Pokhran I test in May 1974.

STARTING WORK

Work started on CIRUS in 1956, and it took four years to complete the construction. G. Venkataraman recalls how he first visited the site in late September 1955, when the entire area was one huge chickoo forest. It was known as Chickoowadi and was famous for illicit liquor brewing. Homi Sethna was initially in charge of the project, and, after the early activities related to civil construction were completed, N. Bhanu Prasad took over from him. More than a thousand engineers and skilled artisans worked on the project, including 30 Canadians. The designs, equipment and specialists were provided by Canada, along with supervising the construction, installation and commissioning activities. India's responsibility was the construction of the reactor building, installation of equipment, piping and electrical works, air conditioning and ventilation. A group of 40 young engineers from different streams of engineering were sent to Chalk River for 12 to 18 months after being given introductory courses in reactor physics. The team leader of the group, V. Surya Rao, eventually headed reactor operations for CIRUS.

Most of the newly recruited members who were sent for training to Canada during 1956–1957 came from very conservative backgrounds and were strict vegetarians. They were from different regions of the country and spoke different languages. Towards the end of 1956, M.R. Srinivasan delivered a few lectures to them on nuclear reactor design. They were taken to the Taj Hotel so that they would learn to use knives and forks while eating. E.C. Allardice conducted orientation courses to prepare the team for a different style of living. Padmanabhan Krishnagopala Iyengar, former chairman of the AEC, recalled that the

Kenilworth

Group photo from the Elementary Particles Conference
at the Old Yacht Club, 1950.

Bhabha speaking at the foundation stone-laying ceremony of TIFR on 1 January 1954. Seated on the stage are Nehru, Morarji Desai, J.R.D. Tata and S.S. Bhatnagar.

Bhabha shows Nehru a model of the new buildings as S.S. Bhatnagar, Morarji Desai and J.D. Choksi look on. M.C. Chagla is in the background.

Bhabha and Nehru at the foundation stone-laying ceremony of TIFR.

Bhabha launches a balloon at Central College, Bangalore, with A.S. Rao and G.V. Vasudevachar.

Bhabha with Paul Dirac, TIFR, 1954.

From left, Patrick Blackett, H.J. Taylor and Bhabha.

From left, Albert Einstein, Hideki Yukawa, John Wheeler and Bhabha at the Institute for Advanced Study, Princeton, circa 1948.

From left to right, Vikram Sarabhai, Cecil Powell, Bhabha and Patrick Blackett in a discussion.

Bhabha discussing the design of the Apsara reactor with his colleagues, May 1955.

Niels Bohr being introduced to J.R.D. Tata while Jamshed and Homi Bhabha look on.

Bhabha showing Nehru around the Atomic Energy Establishment, Trombay, during its inauguration, 20 January 1957.

Bhabha and Nehru at Trombay.

Bhabha, N.B. Prasad and Nehru at Trombay, 1958.

An aerial view of Trombay with the dome of CIRUS.

The new TIFR buildings

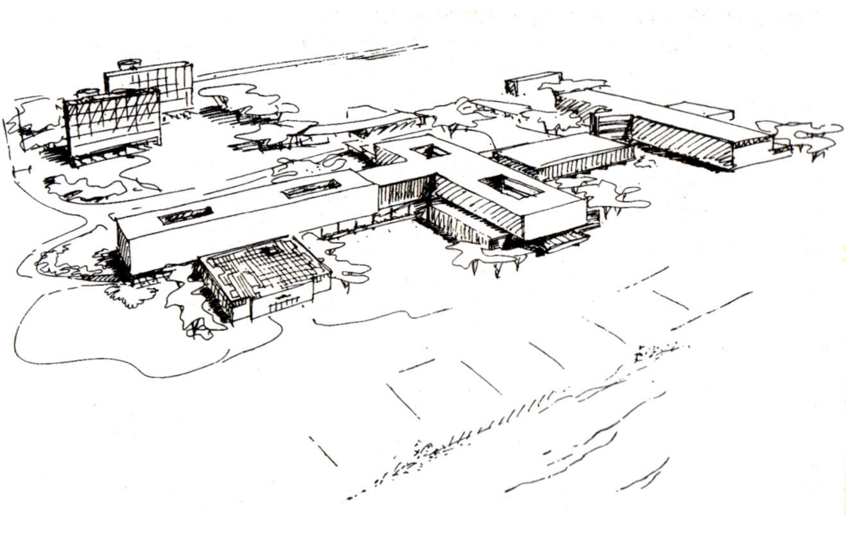

Helmut Bartsch's sketch of the TIFR buildings.

Bhabha speaking at the inauguration of the new TIFR buildings on 15 January 1962. On the dais are Nehru, Y.B. Chavan, Sri Prakasa and J.R.D. Tata.

Bhabha and Nehru at the inauguration of the new buildings, 15 January 1962.

Bhabha shows Nehru a piece of equipment at TIFR.

K. Chandrasekharan being introduced to Nehru as Raja Ramanna, E.C. Allardice and Rustum Choksi look on.

Nehru being shown the facsimile of the notebooks of S. Ramanujan. To his right is Bhabha and to his left is K. Chandrasekharan.

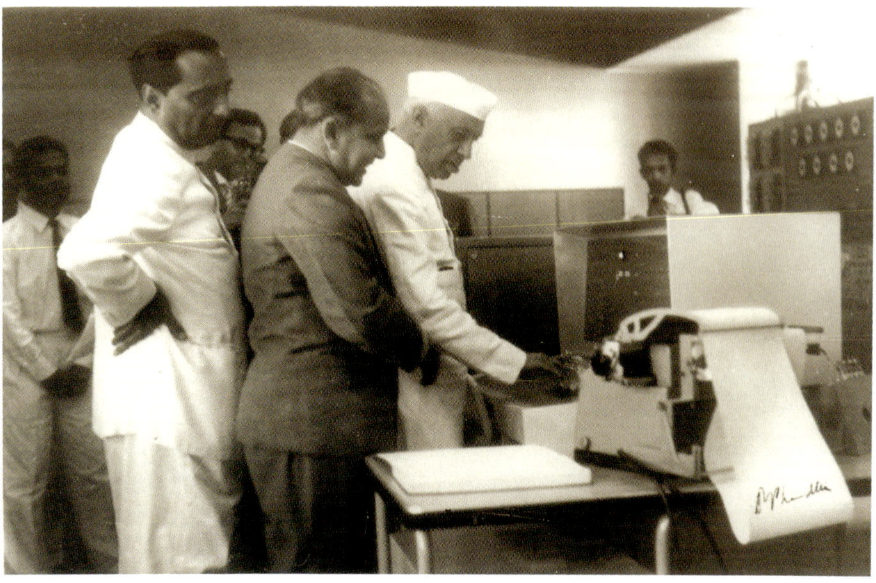

Nehru naming the TIFR computer, 'TIFRAC', in the presence of Bhabha and D.Y. Phadke.

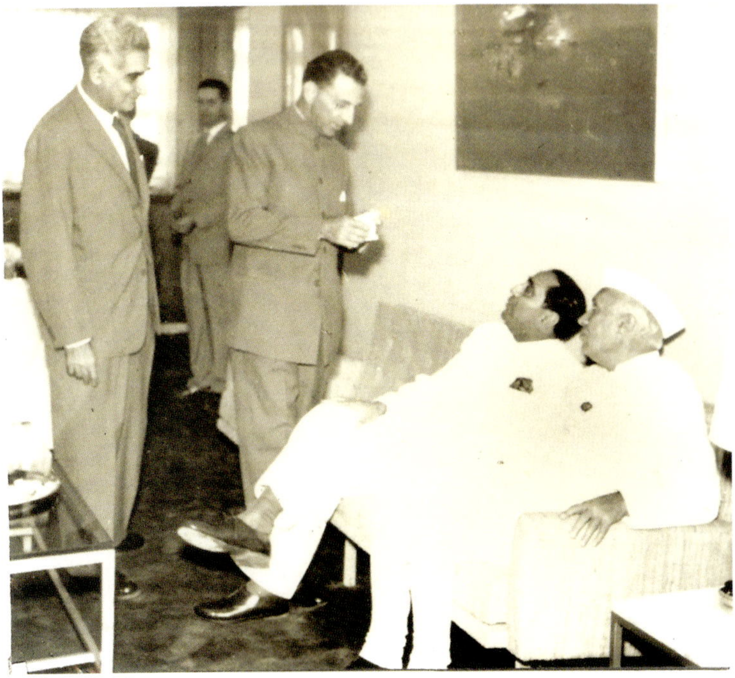

Rustum Choksi, J.R.D. Tata, Bhabha and Nehru during the inauguration of the new TIFR buildings, 15 January 1962.

Queen Elizabeth II, the Duke of Edinburgh, Bhabha, Meherbai and others during the Queen's visit to Trombay in 1961.

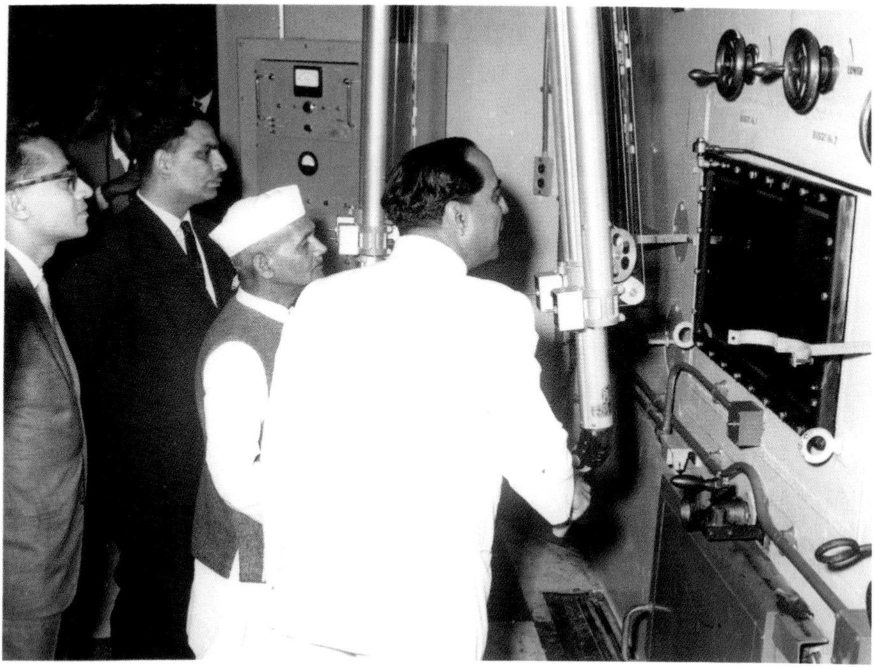

Prime Minister Lal Bahadur Shastri's visit to Trombay.

Members of the scientific advisory committee of the International Atomic Energy Agency at Vienna paying their respects to Bhabha on 25 January 1966.

Uthamna ceremony at TIFR, January 1966.

Vikram Sarabhai speaking at the condolence meeting at Trombay, January 1966.

J.R.D. Tata and Jamshed Bhabha with B. Vithal's bust of Bhabha gifted to TIFR by J.R.D. on 24 January 1967.

training school programme that Bhabha spearheaded helped national integration—it brought people from different parts of the country together.

In Deep River, the Indians were accommodated in two apartments, one by the riverside on a street called Beach and another some distance away on Algonquin Street. They came to be called Beach Indians and Algonquin Indians. Lewis stayed on Beach Street.[11] Since Canadians found the Indian names difficult to pronounce, they shortened them or gave them nicknames. Thus, Narahari became Harry, Sundaram became Sandy and Lakshminarasayya became Lucky! Two Indians for whom suitable names could not be found were called Tom and Jerry, drawing inspiration from the famous cartoon show. G. Venkataraman, who himself had been to Chalk River for training in 1957–1958, narrates that these names stuck, and even after they returned, they often used these nicknames while referring to their colleagues or addressing each other.[12]

Aesthetics played an important part in every element of the project: the emergency water storage tank of CIRUS was spherical, made out of pre-stressed concrete, and the other tanks were shaped like mushrooms. The roof of the workshop used an inverted umbrella shell as the basic element. The construction was completed in early 1960 and the operational testing took a few months. The uranium metal plant producing nuclear grade uranium and the fuel element fabrication plant were already operational and so half the fuel used in CIRUS came from Trombay's own production.

The Canadians had offered to supply the entire first fuel charge of natural uranium, mistakenly believing that the Indians would find the fuel technology too difficult to handle. Bhabha surprised Gray by offering to produce half the initial fuel charge for the new reactor. Nuclear reactor fuel was a difficult technology mastered by only a few countries and it is hardly surprising that Gray did not think Bhabha was serious about his offer. It was only later that he realized that Bhabha was in right earnest.

Bhabha turned to Brahm Prakash, AEET's senior metallurgist and an outstanding technologist, to help him manufacture the fuel. After a doctorate in chemistry from Punjab University, where he was put through his paces by Bhatnagar, Brahm Prakash went to MIT, where he received his second doctoral degree in mineral engineering and metallurgical thermodynamics. On his return to India in 1949, Bhabha invited him to join his team as a metallurgist. As the organization of the atomic energy programme was still in a nascent stage, Brahm Prakash became head of the Department of Metallurgy at IISc, where he served with distinction. At the same time, he was actively involved in the planning and organization of the metallurgy programme for atomic energy development. The six years (1951–1957) he spent at IISc were crucial for the development of the metallurgy department. During this tenure, he was selected as one of the scientific secretaries for the Geneva conference in 1955. The paper presented by him at the conference about his work on the separation of hafnium and zirconium received fulsome praise.[13]

Since the Canadians were highly sceptical about India making fuel elements, Bhabha said he would send two Indian-made fuel elements to Canada for use in their NRX reactor. If they passed muster, he wanted half the first charge of CIRUS to utilize Indian-made fuel elements. Nuclear grade natural uranium ingots were produced by IREL under Sethna's leadership. When the first fuel elements produced by Brahm Prakash[14] at the fuel fabrication facility in Trombay were sent for testing to Canada, they performed better than the ones made locally, something the Canadians were sporting enough to concede. In June 1959, Bhabha informed Nehru about the Indian effort with pardonable pride. 'I have little doubt that we will be able to make satisfactorily fuel elements. There are hardly five countries in the world which make their own fuel elements today.'[15]

The reactor attained criticality on 10 July 1960. However, there were teething problems. The reactor's cooling water flow was inadequate and the fuel temperatures were higher than planned. In their anxiety to be home for Christmas, the Canadians had not commissioned the cooling water system properly. There was algae growth in the primary water system. Plastic waste, seeds and leaves from the ever-expanding mangroves clogged the screens in the sea water inlet system and silt accumulation damaged the pumps. Also, since Bombay's municipal water was used, dissolved solids, such as silica, got deposited on the fuel cladding, reducing coolant flow and causing damage. The Canadians did not face such a problem in their reactor because they had used fresh water from the Ottawa river. The reactor had to be shut down and the entire charge of fuel had to be changed. The water was purified using ion exchangers. It took over a year to resolve these problems.

The CIRUS reactor was an important project because it served as a training ground for heavy water reactor technology. Most importantly, it marked the foundation of India's three-stage nuclear programme and a prototype of India's most successful reactor design, the Canada Deuterium Uranium (CANDU) type of reactors. It was also India's first source of plutonium. By October 1963, they could raise the power level to its rated capacity of 40 MW and, soon, CIRUS became the workhorse of the Indian atomic research programme. It provided a platform to train engineers and technologists, serving as the cradle for future programme leaders in the field. CIRUS was also a powerful neutron source for various experiments, and Bhabha realized its importance in fostering basic research. Here, too, Bhabha's aesthetic sensibility found expression. The CIRUS reactor required a cooling system that called for sprays. Inspired by the pools of the Rashtrapati Bhavan, Bhabha designed a circular pool containing fountains, which were a visual delight when in operation.

Iyengar, who had joined TIFR in 1952 as a junior research scientist under Ramanna, undertook a wide variety of research in neutron scattering. Iyengar shifted to AEET when it came into existence in 1954 and was sent to work with Bertram Brockhouse at Chalk River in 1956–1957 where he spent 18 months working on the problems of neutron spectroscopy. Iyengar narrated how he and his colleagues had once cooked for Bhabha when the latter was visiting Chalk River which only had one restaurant called Honey Dew that used to be closed on Sundays. When Allardice told Iyengar that Bhabha had no place to eat lunch because the restaurant was closed, the latter offered to cook for him. Bhabha not only accepted but also invited Lewis, Gray and others to lunch. Iyengar and his colleagues, who were all bachelors at the time, cooked pure vegetarian food in the South Indian style, a meal that the guests appreciated greatly. The Canadians even carried some back home for their wives.

Iyengar returned to India even though Brockhouse had asked him to stay on at Chalk River's expense. He felt a sense of duty to return to India since Bhabha had paid for his expenses at Chalk River and he had received a full salary for the time he was away. Bhabha had impressed upon his team members that it was they who had the responsibility to establish new science and technology at Trombay. Iyengar built and headed the team of physicists and chemists that gained international recognition for their original research contributions in this field. In the 1960s, he designed the PURNIMA reactor and headed the team that successfully commissioned the reactor on 18 May 1972 at Trombay. He retired as chairman of the AEC in 1993.

G. Venkataraman, on his return from Chalk River, built neutron spectrometers and conducted basic research in the field. At that time, there were very few countries with such facilities and Trombay's record as a pioneer in neutron scattering was acknowledged many years later, in the early 1990s, at a conference held there. It also trained scientists from Asian countries who, upon returning to their countries, ordered neutron spectrometers from India, since these were cheaper than those made in the US or Europe. It has been commented that many nuclear physicists were South Indians. *Time* magazine, after interviewing Bhabha in February 1955, observed: 'An important asset, physicist Bhabha believes, is India's tradition of learning. "Those Brahmins who sit on their bottom all day," he says, "are not just sitting. They are thinking and have been doing it for thousands of years. When the young ones turn their thinking to physics, they quickly get rather good at it."'[16]

◆

Both Apsara and CIRUS had been built with sizeable assistance from foreign countries. Subsequently, Bhabha set out to construct a

100 per cent Indian reactor. He had considerable talent and expertise available in-house. India's third research reactor, Zerlina, modelled on the French reactor Aquilon, reached criticality on 14 January 1961. This zero-energy experimental reactor was indigenously built but used US-supplied heavy water as a moderator and coolant. The tank-type reactor was the first of a series of critical assemblies built at Trombay. The unsafeguarded reactor burned natural uranium fuel rods to produce a nominal power output of 100W. The reactor was decommissioned and dismantled in 1983. As was the case with the better known Apsara and CIRUS reactors, Zerlina was to provide the experience to set up the next generation of reactors (i.e., Dhruva). The scientists applied lessons learned from Zerlina in the design, construction, and operation of future reactors.

The Nuclear Research Centre at Saclay in France also contributed immensely to the first phase of India's nuclear programme. Saclay was one of the early foreign atomic energy establishments where the first generation of Indian scientists were trained. The other establishments were Harwell, Chalk River and Argonne, Oak Ridge (US). Saclay proved to be an important training ground. Bhabha had asked Vasudev Iya to learn the production of radio isotopes at these laboratories. Upon returning, the experience was to prove invaluable in the extraction of samples irradiated in the Apsara reactor.

A PLUTONIUM REPROCESSING PLANT

In July 1958, four years after the establishment of the DAE (and even as CIRUS was being built), Bhabha decided to build a plutonium reprocessing plant at Trombay. Known as Project Phoenix, it was to process the uranium fuel discharged from the CIRUS reactor when it was operational. It had a capacity of processing 20 tonnes of fuel per year, which was expected to yield 10 kg of plutonium, which was at the heart of Bhabha's three-stage plan. Recognizing India's limited natural uranium sources, he saw it as a necessary alternative to uranium. The importance of plutonium lay in its great potential for unlocking the huge energy reserves of thorium. This potential arises from the fact that plutonium is capable of fuelling breeder reactors, thereby enabling

Bhabha's three-phase programme. Plutonium's suitability for making nuclear weapons was well known. In India, this option was kept open through various choices the nuclear establishment made even though it was never publicly acknowledged for many years. The only use that was publicly admitted was its potential for the construction of breeder reactors.

But the nuclear weapons option was never far from Bhabha's mind. In September 1960, in a note to Nehru, he said that as long as India had no plutonium in its possession, or plutonium that was under foreign or international safeguards, people would doubt India's protestations of using nuclear energy only for peaceful purposes. With plutonium, it would be possible to demonstrate India's technical ability to produce bombs to the world, if it so wished.

> To my mind it is not only because of the inescapable needs of the second stage of our nuclear power programme, but also in order to establish India's technological position and correspondingly enhance its political prestige abroad, that we should produce the required quantities of plutonium ourselves and use it thereafter entirely for peaceful purposes, but not convert it into bombs.[17]

Since plutonium can also be used to construct nuclear weapons, the DAE deliberately kept the weapon option open, even in the early decades when India's nuclear programme was ostensibly peaceful. Reprocessing is also a way of dealing with nuclear waste. Plutonium had been considered by many as necessary for the expansion of nuclear power. First produced by Glenn Seaborg at the University of California, Berkeley, in 1941, its industrial scale production was carried out on the Hanford Site in the US as part of the Manhattan Project.

It is probable that one of the main reasons for Bhabha's insistence on producing fuel rods indigenously for CIRUS's first loading in 1960 was to have a better legal and political claim to India using the resultant plutonium as it saw fit. CIRUS using metal uranium fuel meant that the reactor had to be operated in a way where fuel would be irradiated only for a short time before it was removed. This low burn up resulted in the production of weapons-grade plutonium, some of which was used in Pokhran's peaceful nuclear explosion in May 1974.[18]

During the early years, the bismuth phosphate process and then the Redox process were used to separate plutonium. When the PUREX reprocessing technique was invented, it replaced the others as the generally accepted method for irradiating spent fuel. Reprocessing of spent fuel and blanket assemblies to recover plutonium and uranium-233 for fabrication into new fuel was central to Bhabha's three-stage plan. India's quest for technical information about reprocessing was aided by the fact the US made public extensive details about reprocessing as part of its Atoms for Peace programme at the Geneva Conference.

The first four nuclear weapon states—the US, USSR, UK and France—showed off their expertise in various nuclear technologies and vied with one another to supply these technologies to other countries at the Geneva Conference. As already noted, one of the great benefits of the conference was the sharing of knowledge and techniques kept secret during the post-war period. Thus, extensive details about reprocessing technology were made public at the conference. France described the design of its Marcoule Plant, and the US made details of its chemical processing plant at Idaho public.

The US was happy to help India in this regard, and, shortly after the decision to construct the plant was made, Bhabha discussed the design of the plant with American scientists. Four scientists, three chemists and a health physicist visited the US for training. The American company, Vitro International, was involved in its design and construction, and the plant was designed by Edward Durell Stone, the architect who had designed the American embassy in New Delhi. Bhabha, who had always been interested in the minutiae of design, had a few run-ins with Stone if Robert Kupp, the chief nuclear engineer of Vitro International, who travelled to Bombay on three occasions, is to be believed. He witnessed Bhabha and Stone, 'two incredible egotists go at it'.[19]

For Trombay, Stone decorated the walls with precast concrete panels and put open courtyards at either end of the building, one with a large cooling pond and fountain. A perfectionist, Bhabha thought nothing of asking the contractor to redo work that did not pass muster with him. When he noticed that the plant's exhaust stack was not properly aligned with the CIRUS reactor, he asked the contractor to pour the foundation again to preserve AEET's strict geometry.[20] The plant in Trombay was used to reprocess the irradiated fuel from CIRUS and, later, the Dhruva research reactor. Relatively pure plutonium-239 is preferred by weapons designers and described as 'weapons-grade'. It is generally assumed that India used this plutonium primarily for manufacturing nuclear weapons. Early on in India's nuclear programme, some of it was used to fuel research reactors as well.

The Trombay reprocessing plant, which was built under the leadership of Sethna and N. Srinivasan, started operating in 1964 and the first samples of plutonium oxide and plutonium metal produced there became available later that year and in early 1965. With this, India became one of the five countries in the world (the others being the US, UK, France and the USSR) to demonstrate capabilities of plutonium recovery and nuclear fuel reprocessing.

In August 1964, Bhabha gave a talk titled 'Development of Atomic Energy in India' on All India Radio, where he spoke at length about the reprocessing plant at Trombay. He began by explaining the importance of plutonium to India's atomic energy programme. 'It is this material which will form the bridge to the use of thorium for the generation of electricity from atomic energy.' He then went on to convey to his audience the gravity of the accomplishment. 'A plutonium plant is one of the most sophisticated types of chemical plants and has several special features [...] The design and construction of this plant by Indian scientists and engineers is a considerable achievement. There are only four other countries operating plutonium plants—France, the UK, the US, the USSR; Norway also has a small pilot plant.'[21]

The official DAE history makes no mention of any foreign assistance in the construction of this plant: it is shown to be a purely indigenous affair. 'It became one of the most important landmarks in the Indian programme, that this plant entirely designed and built by Indian engineers, under the leadership of Sethna and N. Srinivasan could be completed and commissioned by mid-1964.'[22] Similarly, Sethna, said: 'In 1961, we commenced work for setting up a reprocessing plant at the Centre completely on our own and without any foreign collaboration. The plant was successfully commissioned in 1964, making India the fifth country in the world with reprocessing facilities.'[23] The atomic energy establishment has generally tended to portray all its activities as examples of 'indigenous development' and 'self-reliance', downplaying the role of external agencies in its achievements.

The Trombay plant was initially designed to reprocess spent fuel containing 20 tonnes of heavy metal per year. The plant was formally inaugurated by Prime Minister Lal Bahadur Shastri in January 1965. He even offered its facilities to other Southeast Asian countries, if required. Unsurprisingly, there were no takers because no country in Southeast Asia had any nuclear reactors. The plant, however, failed to work properly for several years. The primary problem was the separation of plutonium from the fission products. The concentration of the latter in the final product continued to be too high for it to be of much practical use. Presumably, these problems were sorted out by the end of the decade, since the plant produced enough plutonium to fuel the small pulsed fast reactor PURNIMA—commissioned in 1972—and also for Pokhran I in 1974. By this time, however, Trombay's plant equipment and piping was seriously corroded, and it had to be shut down for about a decade for decontamination and partial decommissioning before it was rebuilt. This was done over a decade from 1972 to 1983. When it was recommissioned, its capacity was increased to 50 tonnes.

A TRAINING SCHOOL

The DAE required trained personnel in various science subjects and the different engineering streams, including chemical engineering and metallurgy. The atomic energy programme was growing fast and this growth required a large pool of trained scientists. The subject specialists needed to work together in the interdisciplinary field of nuclear power. Apart from their fields of specialization, they had to have a good grasp of the overall field. Rather than depending on only on-the-job training, Bhabha decided to set up a training school through which new recruits would be trained intensively for one year before they were assigned to the various divisions of AEET.

Candidates were recruited through interviews by expert committees rather than a written exam. Later, when the applications became too numerous, a written test was introduced to filter the candidates. The training school was started in 1957 with Ramanna, who was the head of the nuclear physics division, at its helm. Other division heads and senior scientists were members of the committee, which deliberated on policy and decided the syllabus. The interview panel consisted of senior scientists and sometimes even division heads. About 150 trainees were selected annually, approximately 20 or 30 from each discipline. Like TIFR, selection for this school was based purely on merit, without any regional or provincial bias. The teaching faculty consisted mainly of scientists from AEET, TIFR and IIT, Bombay. In the early days, stalwarts like Sethna, Ramanna, Yash Pal, M.G.K. Menon, B.M. Udgaonkar and R.R. Daniel also taught here. K.K. Damodaran, who had a PhD in nuclear science from London, returned to India to join AEET in 1956, and he was appointed head of the school.

Bhabha was assisted by Colonel Otty, who was the administrative officer of the school, and the trusty Allardice, who was the controller of AEET. Western ethos predominated in the school. The students' hostel even had a dress code for dinner—trousers and shirt. During Bhabha's monthly visit to the hostel, the students were asked to wear a tie in his presence. Bhabha's recruits came from all over the country and from widely differing backgrounds. They had to be groomed suitably to face the international community of scientists and for training in foreign countries. While Bhabha's insistence on western dress has been subject to criticism, others were happy to be able to meet the westerners on their own terms. Old-timers proudly narrated how the opportunity for foreign travel and exposure to foreigners instilled a confidence in them vis-à-vis western scientists.

Bhabha tried his best to get his trainees fair pay, and wanted to ensure quick promotion opportunities for them. The training school is an example of how Bhabha nurtured and built up a scientific cadre that he then used for the benefit of his atomic energy programme. At the end of the school year, a graduation ceremony was hosted, where the toppers were given awards (now known as Homi Bhabha awards) and all the trainees were absorbed in the AEET as per the needs of the organization and their performance. One such topper from the batch of August 1964, Anil Kakodkar, went on to become the first graduate of the school to be the director of BARC, and later, the chairman of the AEC, in 2000.

Kakodkar says that the training school matched his own wavelength. He observes, 'One year at the school exposed me comprehensively to all aspects of atomic energy.'[24] Tutelage under some of the best minds in their respective disciplines and the opportunity to interact with eminent scientists of the country in a relaxed atmosphere after dinner was highly inspiring. He first interacted with Bhabha at one such dinner. Bhabha, he said, was always willing to support them in their exploration of new avenues in research. He even challenged them to win a Nobel Prize. If that happened, he said, he would consider the investment in the school to be worthwhile. He also encouraged some of them to move to universities, saying he would be happy and supportive of such migration.[25] As the topper in his batch, Kakodkar received a book on mechanical engineering personally signed by Bhabha. To his lasting regret, he misplaced it and has been unable to locate it for the last 30 years.[26]

While the school has had its share of criticism, its catalysing role in the development of technology and in building up a cadre of trained scientists can hardly be denied. Over the years, thousands of scientists and engineers have been trained there, with many of them going on to occupy high positions in government, academia and industry. The training school became the mainstay of the DAE and, in the early years, even TIFR took its students via the school. Not all remained in the fold, or even the country, but many also rose to positions of scientific eminence in India, won national awards and gained international recognition.

REDEFINING THE ATOMIC ENERGY COMMISSION

Bhabha consolidated his already considerable powers in 1958, when the AEC was refashioned by the government. The Lok Sabha passed a resolution proposed by Bhabha redefining the constitution of AEC and vesting the full executive and financial powers of the government in it. The resolution called for the establishment of an AEC with full executive and financial powers, modelled, more or less, on the lines of the Railway Board. The new AEC replaced the old one set up in 1948. It was to include no more than seven but no less than three full-time and part-time members. The secretary of the DAE would be ex-officio chairman of the commission. A full-time member for finance and administration, who functioned as an ex-officio secretary to the Government of India, was added for advice on matters of finance. The director of AEET was the third ex-officio full-time member in charge of research and development.

With the passing of this resolution, AEC became responsible for formulating the policies of the DAE and implementing all policies concerning atomic energy. The chairman of the AEC was to advise the government on all matters of atomic policy. With this, Bhabha gained formidable powers. All recommendations of the commission would be routed through him to the prime minister, and he had the power to overrule the other members of the commission. The finance member, however, had the right to ask for any financial matter on which he disagreed with the chairman to be referred to the finance minister and the prime minister.

The government promoted nuclear energy through the media. The DAE got the Films Division to make documentary films about various parts of the nuclear programme as well as its chairman, Bhabha. Until 1996 owners of commercial movie theatres were required to screen a state-approved documentary film or newsreel before starting the film. Such films constituted an excellent form of propaganda. Author Amitava Kumar, who was growing up in the 1960s, recalled: 'By the time I had learned to read, I had become aware that the smartest person in modern India was a man in Bombay called Homi Bhabha.'[27]

Prem Nath Thapar, a senior ICS officer who had earned a good reputation as administrator of the Chandigarh project, was appointed as a member (finance and administration) of the AEC. This was an appointment that Bhabha would regret. Thapar began to take an undue interest in administration and, soon, two power centres began to emerge. Since Bhabha was hardly the kind of person who would allow anyone to poach on his own turf, at the first possible opportunity, he suggested abolishing the post and arranged for it to only be part-time. At the suggestion of the then finance minister, T.T. Krishnamachari, the financial commissioner of the Railway Board, S. Jagannathan, was made finance member of the AEC. He continued in this post for several years, even after he assumed the office of secretary (economic affairs). This set a precedent, and, since then, either the finance secretary or secretary (economic affairs) has held this position. They include I.G. Patel, Manmohan Singh, Bimal Jalan and Montek Singh Ahluwalia.

After Bhabha's death, the cabinet secretary became part of the AEC. During Indira Gandhi's tenure as prime minister, her principal secretary, P.N. Haksar, was inducted into the AEC. He served as a vital link with the Prime Minister for many years. Dharam Vira, who was the cabinet secretary, held the position of chairman after Bhabha died, and before Vikram Sarabhai was appointed. Later, P.C. Alexander, Krishnaswamy Rao Sahib and B.G. Deshmukh, all had a long association as AEC members. At the level of the commission, J.R.D. Tata served for the longest period after having been inducted into the commission by Bhabha in the 1950s. When Srinivasan became the chairman of AEC, he asked J.R.D. to continue. J.R.D. was a little hesitant saying that he was getting on in years. When Srinivasan reminded him of his link with Bhabha, he readily agreed.

FURTHER ACCLAIM FOR BHABHA

In June 1959, Bhabha was conferred an honorary degree of Doctor of Science by Cambridge University. Others similarly honoured were Vivian Fuchs, Le Corbusier, Henry Moore, Benjamin Brittain and Louis Clarke. He stayed at the Master's Lodge in Trinity as the guest of the vice chancellor, Richard Adrian. There was a very large luncheon in the hall of Trinity College immediately after the conferment, and Bhabha replied to the chancellor's speech on behalf of all the recipients of honorary degrees. His address contained a number of humorous topical references to local incidents including a prank where an entire automobile had been put on the roof of the Senate House.

Bhabha wrote to Nehru on 2 July from Bangalore, enclosing a copy of his speech. Bhabha also told him that he would be spending a quiet four weeks in Bangalore, revising the last chapter of his proposed book *Algebra for the Theory of the Elementary Particles*. He reassured

Nehru that his official work would not suffer. 'A few important files will, of course, be sent on here, as on previous occasions, and there is a direct telephone connection in my room: 71422.' Nehru wrote back warmly: 'So you are back after collecting more honours! I like your speech at the Cambridge Luncheon. I am glad you are staying at Bangalore to do some quiet work.'[28] That same year Bhabha became a foreign associate of the American Academy of Arts and Sciences.

A similar honour came Bhabha's way when he was given an honorary doctorate by London University in July 1960. He also attended the Tercentenary Celebrations of the Royal Society at which Bohr was also present. Bhabha wrote to Nehru: 'I took Nan[29] with me to the opening exhibition by the British Government at Lancaster House and the final banquet at Grosvenor House, where Macmillan made an excellent speech [...] She also told me that she was present at the ceremony in the Festival Hall where honorary degrees of Doctor of Science were conferred by London University on six people including King Gustaf VI Adolf of Sweden [...] and myself.'[30] The next year, the University of Padua honoured him in a similar manner. In 1957, Bhabha was made honorary fellow of his alma mater, Gonville and Caius College. In the same year, he was also admitted to the Royal Society of Edinburgh.

Chapter 33

TARAPUR

No power is as expensive as no power.

—Homi Bhabha at the third Geneva Conference in September 1964[1]

In 1954, the DAE put out the first of its many ambitious projections for the growth of nuclear power in the country. It was based on Bhabha's three-stage plan that aimed to utilize the country's limited reserves of relatively high grade uranium ore to pave the way for exploiting India's much larger resources of thorium. After successfully commissioning Apsara, Bhabha was under pressure to deliver commercial power, and it is no surprise that making nuclear power seem cheap was important to him and the DAE. India had been independent for barely a decade and there were numerous competing demands on the scarce finances of the government. It would have been impossible for Bhabha to get the huge funding needed for running the nuclear programme if he had been completely honest in estimating the actual cost of nuclear power or was transparent about the difficulties involved in accessing it.

Thus, from the beginning, it was an article of faith that nuclear power was economical and the panacea for India's energy problems.

Bhabha's belief persuaded policymakers to invest large sums of money in expensive reactors, reprocessing facilities and a host of other facilities. There was an urgent need for an assured supply of abundant energy to meet the development needs of the economy. It was obvious from the start that the available conventional sources of energy were inadequate. In this context, Bhabha famously declared that 'No power is as expensive as no power', implying that the pursuit of the development of any form of energy is worth the cost since being without power is the costliest option of all.

Perhaps the most infamous statement in the history of nuclear power came from Lewis Strauss, the chairman of the US AEC, when he predicted, in a speech in September 1954, that the electricity produced in reactors would be 'too cheap to meter'. This statement came nine months after Eisenhower had made his Atoms for Peace speech in December 1953. Coming at the height of the Cold War, Eisenhower's primary aim was to position the US as a leader in the effort to not only contain the arms race but also to harness nuclear power for global development. Eisenhower had asked Strauss to steer global attention away from nuclear weapons and replace it with an enthusiasm for peaceful nuclear power. Although 'experts' in other countries were also prophesying cheap nuclear power, Strauss's 'too cheap to meter' provided the catchphrase that captured the imagination of the world and helped sell the merits of nuclear power.

A LITTLE 'CALCULATED' JUGGLERY

In 1958, Bhabha stated, that even at costs based on present technology, nuclear power would become cheaper in capital investment and in cost of electricity than coal power in most parts of India within ten years. 'Most parts of India' referred to regions far away from coal mines and the higher cost of transporting coal to such areas made nuclear power more competitive. Bhabha's analysis was based on the information supplied by various consortia in the UK, but he made allowances for lower civil engineering and labour costs in India and the higher cost of equipment that would have to be imported, at least in the initial stages. According to Bhabha's assumptions, a nuclear reactor would cost ₹1,700 for each KW of electricity as compared to ₹1,050 for a thermal plant. Though the nuclear power estimate was higher than the thermal one, it in no way reflected the actual cost of construction as subsequent events showed.[2]

In a paper presented as an evening lecture at the second Geneva conference in September 1958, Bhabha made a detailed study of Indian conditions and presented comparative capital cost estimates for a 150 MW nuclear power station, a hydroelectric station of similar size and some thermal power stations in India. The capital cost was 50 per cent higher for a nuclear power station but the generation costs were similar. It was thought that for regions far away from coal fields, nuclear power would be ideal since it would obviate transportation of large quantities of coal.

In a joint paper presented at the same conference with N. Bhanu Prasad, Bhabha outlined a proposal based on his three-stage process for setting up a nuclear capacity of 1,000 MW in the first generation of natural uranium fuel reactors by 1965. These reactors would produce sufficient plutonium to set up an additional capacity of 250 MW in a converter-type reactor of the second generation, which would produce uranium-233 from thorium. With this, a base would be available to launch the third stage with fast-breeder reactors using the uranium-233 and thorium cycle. In what was a very optimistic forecast, they said that it would be possible to achieve an installed capacity of 1,600 MW in the third-generation reactors by 1974, 5,000 by 1979 and 12,000 by 1984. India is yet to touch the 8,000 MW mark even in 2022. While a lot has not happened the way Bhabha envisaged, the visionary content and substance of the three-stage strategy has been accepted as logical in the years that followed.[3]

The UK was also making similar calculations at this time. Its political leadership, which wanted to weaken the powerful coal interests in the grip of Left-leaning unions, asked government economists to rework figures to make nuclear power look competitive. One of the ways of doing this was to assign a value to plutonium (the plutonium credit), the thinking being that there would be a large demand for plutonium in the future. Bhabha, apart from underestimating construction costs, made a number of assumptions that were favourable to nuclear power.

He assumed that the operating costs of a nuclear reactor would equal that of a thermal reactor and underestimated the cost of mining uranium and fabricating it into fuel. Despite all this, the cost of nuclear power was still 5 per cent more expensive than both coal- and oil-based thermal power. The plutonium credit pioneered in the UK now came to Bhabha's rescue. He assigned an arbitrary credit of ₹56,000 for each kilogram of plutonium produced and this jugglery ended up making nuclear power marginally cheaper. The costs per kilowatt-hour then became ₹0.04050, ₹0.03910 and 0.03907 for coal, oil and nuclear power respectively.[4]

Commenting on the plutonium credit, M.V. Ramana has observed that there was no justification for that arbitrary choice, even at that time. In hindsight, however, it has become clear that the extraction and storage of plutonium is so expensive that it should be considered a liability. Its appearance in the accounting books should be in the debit column, not the credit column.[5] Bhabha also ignored two important factors in his calculations—the cost of disposal of nuclear fuel and the decommissioning of nuclear reactors, both of which are expensive and add to the cost of a nuclear installation.[6]

In 1959, Bhabha co-authored a paper with Maheshwar Dayal, a Cambridge-educated electrical engineer who had joined the DAE. He estimated that the cost of generating electricity from coal at a power station situated at a distance from the coal fields would be between ₹3.49 and ₹3.60 per kilowatt-hour. This calculation assumed an efficiency of an average of 29.6 per cent, with a load factor of 80 per cent. The corresponding estimate for nuclear power was between ₹3.43 and ₹3.99, assuming an efficiency of 28 per cent and a load factor of 80 per cent. They observed, 'The immediate nuclear power programme is the seed from which can grow the tree of a large and increasing power programme based on plutonium and thorium.'[7] Bhabha and Dayal then compared the costs of hydropower with nuclear power. Citing the case of the Kundah hydroelectric project in South India, they pointed out that the capital cost of electricity produced was ₹403,000 per million kilowatt-hour per year, whereas the corresponding cost for nuclear power, assuming an 80 per cent load factor, was merely ₹243,000.

The most significant rebuttal to Bhabha's claims about the economics of nuclear power came from the British development economist, I.M.D. Little who was aware of the bias of the estimators and their need to get their estimates down to the level of conventional energy to prove their point. He pointed out the fallacy of comparing thermal and nuclear power stations, assuming that both would be working at a load factor of 80 per cent and a plant life of 20 years. According to him, the comparison favoured the nuclear station only if the unrealistically high load factor of 80 per cent was assumed. The assumption of approximately 25 per cent for thermal plants was also too low. Using cost, performance and financial parameters that were more representative of existing technology, Little concluded that nuclear power would cost 50 to 60 per cent more than thermal power.

There is no evidence to suggest that such critical assessments had any impact because Bhabha and the DAE continued in their quest for setting up nuclear plants. Economic considerations never seem to have been important in taking decisions. Commenting on Bhabha's paper on the need for atomic energy in the underdeveloped countries in an article in the *Economic and Political Weekly*, Little observed: 'Dr Bhabha says, electricity is in short supply in India. It is likely to go on being in short supply if one uses twice as much capital as is needed to get more.'[8]

A SMALL FISHING VILLAGE

Studies on the feasibility of introducing nuclear power into Indian grids began in 1956 soon after Apsara had achieved criticality. The composite state of Bombay was identified as the ideal region for India's first nuclear power station, and in 1958 Nehru gave the go-ahead to start work on a nuclear power station. M.R. Srinivasan, who was associated with the project as the principal project engineer from the very beginning gives us a detailed account of the early history of the Tarapur nuclear reactor. A former general manager of Central Railway[9], M.N. Chakravarti, was appointed project administrator in August 1959. The railways were (and still are) the largest government-run industry and it was customary at the time for dynamic senior railway officers to be hired for new and challenging jobs. Apart from Srinivasan and Chakravarti, G. Vishwanathan, a civil engineer who had worked under Chakravarti, joined as the chief engineer in charge of civil engineering works, as did Dayal. B. Maithreyan, an Indian Audits and Accounts Service officer, was brought on board as the internal finance officer.[10]

The project group convinced Bhabha that it could select the site, draw up specifications for the plant and assist in the selection of the plant supplier without the help of foreign consultants. Assisting the project group were engineers from the Central Water and Power Commission. Bombay was bifurcated into Maharashtra and Gujarat in May 1960 and the new plant was to supply power to both these states. The team visited locations on the west coast, between Bombay and Dahej. A coastal location was preferred since seawater could be used for cooling and such a location would also facilitate the transportation of large and heavy equipment to the site.

The first task of the project team was selecting the site for the power plant. The site selected was Tarapur, a small fishing village about a 100 km north of Bombay, near Boisar in the Thane district of Maharashtra. The village was located on a rocky promontory, near a creek at Chinchini, and the road that led to it was little more than a dirt track. The site of the promontory was underwater but the team was convinced that this site would satisfy most of the criteria for a nuclear power station's location including low population around the site, remoteness from a large city, and access to fresh water.[11]

Tarapur was announced as the site for the plant in August 1960. Bhabha, Krishnan and P.N. Thapar (the other commissioner) worked on the bid document and readied it by the end of 1960. Krishnan had remarked on the difficulty of the task—if the specifications were too detailed, they would constrain the freedom of the designer. If they were couched in very general terms, comparative evaluation would become a problem. After considerable persuasion from American reactor companies, they agreed to receive bids based on enriched uranium reactors but made it clear that the DAE reserved the right to reject all such bids.[12]

Of the seven bids received, two were rejected almost immediately. The Canadian offer was for joint development and lacked a firm price or schedule. The offer from General Atomic of the US was for an unproven system that was still in its conceptual stage. Besides these two, there were two bids from the UK (English Electric Company and General Electric Company), one from France (French Industrial Consortium led by Alsthom) and two others from the US (Westinghouse and General Electric [GE]). The offers from France and that of the English Electric Company were the only two serious bids for natural uranium systems. The French offered to build a graphite-gas-cooled reactor, similar to the ones being made in France. With respect to nuclear safeguards, both India and France were on the same page. While Bhabha had accepted that enriched uranium should be subject to safeguards if it was supplied by another country, he had argued that it was unreasonable to impose safeguards on natural uranium, which was a commercial material back then.

Bhabha was supported by Bertrand Goldschmidt, since at that time, both France and India had taken an identical position on international safeguards. If India had taken the French reactor, there would have been no safeguard obligations. The French also offered a fairly long term line of credit at relatively low interest rates. According to Srinivasan, the French political and atomic leadership stretched themselves to the very limit to get the order from India.[13] However, US vendors were also in the fray and American companies were not going to give up so easily.

CLINCHING THE DEAL

The nuclear industry in the US was dominated by GE and Westinghouse. Both submitted bids that were much lower in capital costs but had a higher fuelling cost because they used low enriched uranium instead of natural uranium. Westinghouse offered a pressurized water reactor design while GE was pushing its boiling-water reactor (BWR) [14], which also used low enriched uranium. To make the offer more attractive (and to upstage its rival Westinghouse), GE offered its first BWR with a turnkey contract. GE was also prepared to consider recycling plutonium from spent fuel to reduce the burden of a lifetime of foreign currency outflow. Although Westinghouse dropped its price substantially, in the final negotiations it could not match GE's offer.

The final bid from GE was cheaper than even the most optimistic expectations of the DAE. While the anticipated cost of the 300 MW reactor was ₹51 crores, GE offered two 190 MW reactors for as little as ₹48 crores. A loan of $80 million, (the foreign exchange component of the project) was routed through the USAID. The 40-year loan had a rate of interest of only 0.75 per cent with repayment to begin only after 10 years. Another long term loan of $14.4 million was provided to pay for the import of Tarapur's first fuel loading.[15] Tarapur was mostly financed by US aid: the French ultimately lost out because they could not match the financial package offered by the US. Though enriched uranium was not available in India, and Bhabha's stated policy was to construct only natural uranium-fuelled reactors, the pressure to produce electricity at the lowest possible cost took precedence in making the final decision.

In 1963, Bhabha discussed supplies of enriched uranium via an agreement of cooperation with the US AEC and the US State Department. The chairman of the US AEC from 1961 to 1971, Glenn Seaborg, led the discussions for the Americans. Seaborg was a soft-spoken man who was interested in resolving differences between the two sides amicably. India had taken a position that safeguards did not apply to equipment and materials whereas the Americans wanted to extend safeguards to the reactor plant. Bhabha was of the view that safeguards infringed on India's sovereignty. The Americans wanted to close the deal but were compelled by their commitment to the IAEA to press for safeguards in the deal involving a state receiving nuclear technology assistance. Thanks to Bhabha's strong personality, a compromise agreeable to both parties could be reached. 'He was very self-assured,' Seaborg said while recalling Bhabha's supreme self-confidence. 'He was not easy to argue with. Polite but very sure of himself, he was never at a loss for words, and was most articulate. He was a very imposing presence.'[16]

Jerome Wiesner, the scientific adviser to President Kennedy who participated in the negotiations, observed:

> We often disagreed, not so much on basic objectives or fundamental principles as on methods, interpretations and orientations. We argued endlessly during the period—about the form of controls

to be placed on the power reactor which the US was planning to help India build at Tarapur. I was enthusiastic about the American proposal to allow the IAEA, of which Bhabha was a director, to be the supervising agent. Bhabha supported the Indian proposal for a bilateral arrangement similar to the ones the US then had with many countries, particularly in Europe. We argued the issue for many months in Washington, in Geneva, in Delhi and Bombay. I could not understand why India [...] would not accept the unique leadership role that the situation offered. Homi could not accept what appeared to be a second-class position for India despite his solid commitments to international control of nuclear materials.[17]

Wiesner added that Bhabha shared Nehru's '...dreams, fears, ambitions and viewpoints, including many that were not always understood or appreciated in the US, neutralism for example—and was often regarded as anti-American when he was indeed, only being pro-Indian'.[18]

India agreed to only use enriched uranium supplied by the US and allowed the IAEA to verify that this fuel was being used only for peaceful purposes. Bhabha also agreed that the US would approve any subsequent reprocessing of the spent fuel to separate plutonium, which, in any case, would also remain under safeguards. However, the safeguards only applied to the Tarapur facility as a quid pro quo for the supply of enriched uranium and not for the supply of the reactor. This allowed India to use non-US supplied spent fuel in whatever manner it wished.

The major American gain was in getting India to accept the IAEA safeguards.[19] India agreed to only use nuclear fuel supplied by the US for the entire life of the reactors, the exception being the plutonium and residual enriched uranium that could be fed back into the plant. The agreement also foresaw the return of spent fuel to the US so that India could get credit for the unburnt uranium and plutonium. The US, too, was obliged to use these materials only for peaceful purposes. Bhabha insisted on the acceptance of reciprocal obligations and the US agreed somewhat reluctantly.[20] The agreement was signed on 8 August 1963. Srinivasan, Chakravarti, Dayal and the additional secretary of the DAE, R. Bhaktavatsalu were part of the negotiating team assisting Bhabha.

In the 1960s, the US was funding power projects under USAID and India applied to it for a long term loan for financing the Tarapur project. The US Congress agreed, partly because two refusals in the past for aid had resulted in the USSR stepping in. The first rejection was the Bokaro Steel Plant in India and the second was the Aswan Dam in Egypt, both of which had been built through Soviet collaboration. The Americans, presumably, rued these missed opportunities and thus there was no real opposition to an atomic plant in India funded by the USAID. The loan agreement was signed in December 1963 by Bhabha and Chester Bowles, the American ambassador.

In an interview in 1986, Bertrand Goldschmidt narrated details of how the French tried to court Bhabha and the Indians but ultimately lost out to the Americans. He recalled how Bhabha came to Paris to discuss the issue with them. At a lunch hosted by Pierre Guillaumat, the French minister who was in charge of atomic energy, Goldschmidt recalls how the minister tried his best to convince Bhabha to buy a French reactor. Guillaumat's arguments included the fact that theirs was a natural uranium reactor, and India, being a large country, would eventually find some natural uranium. More importantly, there would be no safeguards. Goldschmidt observed:

> Now the…the competition came from General Electric in the States, which was selling an enriched-uranium reactor, with safeguard, because that was the American law. And therefore […] Guillaumat was telling Bhabha, that if you buy the American reactor, first you'll be totally dependent of the United States for your fuel, and furthermore you'll have inspectors to see what you're doing with your reactor.[21]

At the end of the lunch, Guillaumat asked Bhabha, 'Now, Dr Bhabha, which do you prefer, the American safeguarded reactor or the French reactor with natural uranium, which will be much freer to run?' Bhabha supposedly answered, 'The reactor we prefer is the one we don't pay for.'[22] According to Goldschmidt, two months later, Bhabha signed the contract with GE because the US financial package was too good to resist.[23] Another important reason for India to opt for the US reactor had to do with the lease arrangement for fuel. The US had built large gaseous diffusion plants for producing highly enriched, weapons-grade uranium. These plants had a high capacity for producing reactor-grade low enriched uranium. The US AEC was keen on securing long term supply contracts to keep these plants running at their optimal levels. Since the enriched uranium for the initial fuel charge constituted a substantial investment, the US AEC made the initial fuel available on the basis of a lease charge so that the user did not have to finance the initial fuel inventory. The US extended a similar arrangement to India, prompting the French to observe that the Americans were practically giving the reactors away.[24]

Once the agreement to supply fuel was signed between India and the US, the former could proceed with the drafting of a contract with GE. Two issues had to be sorted out before the contract was finally signed. The first had to do with indemnification of the supplier against any claim arising out of nuclear accidents, including third-party liability. In India's case, an executive decision of the cabinet provided this indemnification, which GE accepted. The other involved tax liability of GE and its expatriate staff. The famous lawyer, Nani Palkhivala represented GE and was, in Srinivasan's words, a formidable interlocutor on behalf of his clients. Once this was sorted out, a contract was signed with Washington D.C. in May 1964—Chakravarti signed on behalf of the DAE and George White, general manager of GE, signed for his company. Srinivasan signed as the witness.

There is an amusing anecdote regarding Bhabha and his love for food connected with their visit to Washington D.C. Like all Parsis, Bhabha loved food. Bhabha, Srinivasan, Chakravarti and Dayal were all in Washington to discuss the Tarapur project. When the rest of the team went to meet Bhabha in his hotel room, he told them that he had had an uncomfortable night because of an upset stomach. The doctor attached to the Indian embassy advised Bhabha to have yoghurt and to keep light. This did not prevent Bhabha from ordering grapefruit, poached eggs, toast, coffee and two helpings of yoghurt. After the meeting in the forenoon, he ordered soup, fish and yoghurt (which the restaurant did not serve). A waiter was dispatched to the nearby supermarket to get it for him. He had conveniently interpreted the doctor's prescription to mean yoghurt in addition to other food! [25]

At about the same time, Bhabha also worked out a deal with AECL and the Montreal Engineering Company to construct a 200MW pressurized heavy water reactor (PHWR), also known as CANDU. In April 1964, an agreement was signed between the Government of India and the Export Credit Insurance Corporation of Canada for financing the project. The design was identical to the one used for the first CANDU at Douglas Point in Ontario, Canada. In hindsight, it proved to be a hasty choice since no operational feedback from this reference reactor had been available at that time and it led to many problems in some of the PHWRs that the DAE subsequently built based on this design. Part of the construction of a twin unit of the same design had been completed when India conducted its nuclear test in 1974. Canada pulled out of the project in protest and the Indians only completed construction of the reactor in April 1981.[26]

For the architecture and engineering of the Tarapur reactor, GE engaged Bechtel Corporation. This company had, at best, a spotty record, with leaks and breakdowns reported from some of the plants it had constructed. It is thought that some of the problems Tarapur faced later could well have been caused by Bechtel's construction work, which started in October 1964 but was dogged by delays caused by internal and external factors. Initially, there was a problem with the quality of concrete used at the bottom of the foundation. Due to an error in the mix, the concrete did not develop the prescribed strength. The engineer who was responsible for the job was sent back to the US, and Bechtel agreed to redo the work.[27]

The next year, external factors affected the progress of construction. The steel locks meant for the steel vessels enclosing the reactor were loaded on a ship that docked at Karachi en route to Bombay. When the Indo–Pak war broke out, the consignment was stranded at Karachi. Bechtel reordered two more locks since the fate of the consignment stuck at Karachi was uncertain.[28] Politics also reared its ugly head in union conflicts. Workers went on a strike for many months. The two unions involved were the Indian National Trade Union Congress, controlled by S.K. Patil, and the Hind Mazdoor Sabha, under George Fernandes. The two were fighting for control over trade unions in Bombay, a rivalry that included trying to influence the workers at

Tarapur. After some 50 days, the strike turned violent and seven workers were killed in police firing. This tragedy brought sanity to both the workers and the management. GE agreed to deal with the HMS and the union leaders prevented further unrest.

The Tarapur contract had stipulated the completion of the two units within 52 months of signing the contract. But some unexpected problems delayed the project by 10 months. In 1968, several US reactors of the kind that were being built in Tarapur reported cracks in the inner lining of the reactor vessel—the structure that would contain the radioactive fuel in case of an accident. The reactor vessels, large fabrications in carbon steel, were lined with stainless steel deposited by a welding process. Hairline cracks were observed in the overlay caused by stress corrosion cracking. The cracks had to be filled using special filler materials.[29]

The two reactors at Tarapur commenced operations on 28 October 1969, more than three years after Bhabha passed away. On 19 January 1970, Prime Minister Indira Gandhi dedicated the Tarapur reactor to the nation. The DAE proudly proclaimed it to be the largest operating nuclear power plant in Asia. Unfortunately, Tarapur's performance left much to be desired. There were numerous problems and failures and it had to be shut down frequently. At that time, the Tarapur reactors were the largest producers of electricity in the western region and any failures would impact the western grid, which was extremely unstable to start with. The technical problems that plagued Tarapur and subsequent developments are beyond the scope of this discussion.

While the contract stated that the US would supply enriched uranium for 30 years, it was cut short in 1974, after India conducted its first successful nuclear test in Pokhran under Operation Smiling Buddha. Pokhran I marked the first confirmed nuclear weapons test by a nation outside the five permanent members of the UNSC. This angered the US, as it was trying to expand the ambit of the Nuclear Non-Proliferation Treaty and in retaliation Washington cut off uranium supply after 1974. After the US stopped supplying uranium, countries such as China, France and Russia stepped in to supply nuclear fuel to India.

While the two BWRs of 190 MW capacity were commissioned at Tarapur in 1969, two PHWRs of 540 MW were added to it on 12 September 2005 and 18 August 2006. The four reactors combined have a power generation capacity of 1,400 MW. That makes the Tarapur reactor the second most powerful in India, after the Kudankulam Nuclear Power Plant in Tamil Nadu, which has a power generation capacity of 2,000 MW. At present, the Tarapur Atomic Power Station is operated by the Nuclear Power Corporation of India Limited. Following the Fukushima nuclear accident in 2011, in Japan, where a tsunami disabled the power supply and cooling of three Fukushima Daiichi reactors, fears have been raised about the Tarapur plant. The Fukushima plant also relied on BWRs for power generation. Local residents in Tarapur have been calling for the plant to be shut down.

BHABHA'S FLAWED ASSUMPTIONS

We have already seen that one of the most telling features of India's nuclear policymaking was the absence of any real debate around it after Saha's demise in 1956. Bhabha faced virtually no opposition to his plans, which always had Nehru's full support. Few politicians or journalists had the expertise to criticize the internationally acclaimed physicist. Nor were the scientists in a position to critique his plans. Hindsight now reveals that India's investment in nuclear power in the 1950s represented a major diversion of capital into a relatively unproductive area for economic development.

Bhabha's assumptions about nuclear power were proved wrong. He grossly overestimated the economic merits of nuclear power, mistakenly believing that a large increase of electrical power would automatically lead to economic development. Development is predicated on the growth of transportation and other infrastructure, manufacturing capabilities, markets and much else. High levels of per capita energy consumption do not, on their own, lead to national wealth. George Perkovich has argued that by channelling India's capital investment into nuclear power before the rest of the national infrastructure and skilled manpower base had developed, Bhabha unintentionally imposed excessively high opportunity costs on the economy. Part of this can be explained by the pride of a newly independent nation thinking that it could match the powerful nations through projects that may have seemed feasible at the time, but for which it was perhaps not ready.

Bhabha also believed that India, with its lower labour costs, could build nuclear plants more cheaply. This hope was not realized. Nuclear plants, to offset the huge capital costs of construction, must have the capacity to generate at least 500 to 600 MW of power. While Indian experts realized this, it was not fully understood how big a plant was really required. Indians built plants that produced around 150 to 220 MW of power, fearing the difficulties of building and running bigger facilities. This proved to be a major economic handicap to nuclear power. Bhabha also erred when he assumed that plants would be operated at 80 per cent load factors and based his cost estimates on this optimistic expectation. In reality, the performance was closer to 40 per cent, the lowest in the world, and well below the levels where nuclear power could be economically competitive with conventional fossil fuels. Bhabha also assumed that heavy water reactor technology would be cheaper. However, in fact, capital costs of such reactors are higher than those of light water reactors that use slightly enriched uranium fuel.

Chapter 34

BHABHA AND THE BOMB

> Indeed, the importance of nuclear weapons is that they enable a country possessing them in adequate measure to deter another country also possessing them from using them against it.
>
> —Address on Nuclear Disarmament on All India Radio on United Nations Day, 24 October 1964[1]

Bhabha and Nehru were adamant about protecting India's atomic sovereignty from international efforts to control nuclear technology. They were equally emphatic in wanting to preserve international peace against the fallouts of an unmitigated nuclear arms race. While there was an element of moral repugnance towards nuclear weapons, Nehru's enthusiasm for nuclear disarmament was also driven by pragmatic considerations. Emphasizing nuclear disarmament to both the US and the USSR helped position India as a genuinely non-aligned state and helped increase its stature among developing countries. It allowed India a more important role in global diplomacy than its material power warranted.

Nehru's disarmament diplomacy dictated that international bodies such as the United Nations Disarmament Commission must not exclude non-nuclear weapon states. Multilateralism in disarmament negotiations became one of the cornerstones of India's policy. Even as India sought greater representation, it held that the primary responsibility for nuclear disarmament lay with the two power blocs. It also believed that radical solutions to disarmament were impractical, given that nuclear weapons had by then become an integral part of the security of nuclear weapon states.

The Soviet nuclear test in 1949 and the testing of a hydrogen bomb in 1952 fuelled an intense arms race. In April 1954, Nehru proposed a cessation of further nuclear tests by the US and the USSR, calling for a standstill agreement. While it was largely ignored by the superpowers, it earned considerable diplomatic clout in the UN and among the third world countries and put the need for a nuclear test ban firmly on the global disarmament agenda. In 1958, the USSR announced a unilateral moratorium, and the first global disarmament treaty that came into effect in 1963 was the Limited Test Ban Treaty. Nehru was both the originator of the idea and its most public face—he was its symbol and essence, evocator and voice.

India's engagement on nuclear disarmament came at a time when it had no major stakes in nuclear disarmament. Even though its nuclear capabilities were making good progress, it was still a non-nuclear weapons state. However, this rhetoric served it well in the future; it became the pretext under which India rejected participation in any arms control measures. 'In later decades, disarmament helped India practice realism under the cloak of idealism.'[2] Nehru realized the need to not only harness the atom for generating electricity but also use it for the defence of the country. In 1948, he wrote to Defence Minister Baldev Singh that the future belonged to those who produced atomic energy and that it was going to be the chief national power of the future. He conceded that defence was intimately concerned with this and even the political consequences were worthwhile. By then, nuclear weapons had become objects of fetish—markers of national power. Access to nuclear weapons determined a nation's place in the international pecking order.

NEHRU'S AMBIVALENCE

While Mahatma Gandhi was unambiguous about his opposition to the atom bomb, Nehru, while championing a nuclear-free world, exhibited a duality in thinking that puzzled many. P.N. Haksar, one of Indira Gandhi's closest advisers, recalled Nehru telling him to save future generations from the scourge of nuclear war. At that time, Nehru was emerging as a leader of the non-aligned movement and genuinely believed he had an important role in ensuring world peace. B.K. Nehru, Nehru's cousin and a former ambassador to the US, remembers asking Bhabha why he was not planning a bomb, in 1951. 'The old man won't let me', Bhabha told him. 'He has approved of my plans for atomic energy but said under no condition was I to manufacture a weapon.'[3]

Nehru was generally taken at his word when he said that India would not make nuclear weapons. On 20 January 1957, while inaugurating AEET, he declared that India would never use atomic energy for evil purposes, whatever the circumstances. 'There is no condition attached to circumstances because once we attach any condition, nobody knows what the condition might be and the value of such an assurance does not take us very far.'[4] Never mind that he had prefaced his announcement with, 'No man can prophesy about the future'.

Some months later, Nehru again made his pious intent clear when he reiterated in the Lok Sabha that India was not interested in making atom bombs even if it could do so, and that in no event would they use atomic energy for destructive purposes. However, the end of his speech showed his ambivalence. He said that the fact remained that if one had such resources, then one could make a bomb unless the world was wise enough to come to some decision to stop the production of such bombs.[5]

When it came to choosing between securing India's strategic interests through a nuclear arsenal and securing his legacy as that of a 'peacenik' and 'internationalist', Nehru invariably chose the latter. But, his statements from late 1955 onwards betrayed not only ambivalence and vacillation but also a clear awareness of the importance of having an option of a nuclear weapon. Notwithstanding the Prime Minister's rhetoric, Bhabha knew that a time would come when a bomb would have to be made and continued to quietly prepare for it. He regularly invited foreign scientists to lecture on the physics of chain reactions and was always on the lookout for recruiting bright young men who could help him in this quest. V.K. Iya, who was a member of the 1974 Pokhran team, first met Bhabha in Paris and was told to complete his PhD thesis before joining Trombay. Iya was, at the time, completing his thesis under Bertrand Goldschmidt and was advised by Bhabha to keep his eyes and ears open.

There is enough evidence of Nehru's dual intent. In 1958, he cleared Project Phoenix to recover bomb-grade plutonium from CIRUS. This reactor was India's shortest route to developing a nuclear weapon since it generated plutonium as a by-product. There could have been no other reason to go in for plutonium production so urgently, apart from achieving Bhabha's hidden goal. Bhabha also persuaded the Canadians not to have any formal agreement on safeguards for the reactor's spent fuel rods which contained uranium merely assuring them that it would be used for peaceful purposes. To ensure that the Canadians had no claim over the plutonium produced, Bhabha, as noted before with Brahm Prakash's help, ensured that Indians mastered the complex technology of making uranium fuel rods of the required quality.

In January 1958, Nehru, while explaining how India would deal with the stationing of nuclear weapons in Pakistan or any other Asian nation, said that India had the technical know-how for manufacturing the atom bomb and could do it in three or four years. However, it would never do so since it had assured the world that it would never use nuclear science for war. Even as Nehru was delivering this warning, Bhabha was confiding in his friend, Patrick Blackett, that he hoped to develop nuclear weapons. By the end of 1955, both the US and the USSR had exploded fission and fusion bombs, and the UK had exploded a fission bomb. Bhabha and Blackett were of the opinion that these were unlikely to be used for war. They had also agreed on the fact that the bombing of Japan had been unnecessary, a view that was aired by Admiral William Leahy, who served as the chief of staff under both Roosevelt and Truman and whom Bhabha quoted approvingly in his letter to Blackett.

John Cockcroft stated that despite Bhabha's public espousal of the peaceful uses of nuclear energy, in discussions at small closed-door meetings, he appeared to be in favour of making bombs for a Plowshare programme. In late 1957, the US established a formal programme to conduct peaceful nuclear explosions (PNE), which it envisioned using for big excavation projects and releasing natural resources such as natural gas. However, Bhabha, at least at this time, was not particularly interested in such peaceful uses of nuclear explosives. George Perkovich and others have persuasively argued that Bhabha's primary aim was to acquire weapons capability under the guise of a peaceful programme.[6] As early as December 1959, Bhabha told the Parliamentary Committee on Atomic Affairs that India had progressed to such a stage where, if a political directive was received, a bomb could be made without external assistance. The time period, however, was not specified.[7]

Perkovich narrates a conversation Nehru and Bhabha had with retired Major General Kenneth D. Nichols in 1960. Nichols was an American military engineer, who had worked on the Manhattan Project and had also served as the top military official in the US nuclear establishment. When Nichols visited India in 1960, at Bhabha's request, he came as the chairman of Westinghouse International Atomic Power Company in Geneva. He managed to convince Bhabha that American light water reactors were more suitable than British gas cooled reactors.

In the sales pitch to Nehru, Nichols wanted India to open up competition to US suppliers. Nehru agreed and asked Bhabha to include US reactors. Then, Nehru, with extraordinary candidness, asked Bhabha, in Nichols' presence, if he could develop a bomb. Bhabha responded that he could, and estimated that he would need a year to do so. Subsequently, Nehru asked Nichols if he thought Bhabha could do it. Nichols said that he saw no reason why Bhabha couldn't. Nehru then turned to Bhabha and said, 'Well, don't do it until I tell you to.' Nichols observed, 'I was really astounded to be hearing these questions from the one I thought to be one of the world's most peace-loving leaders.'[8] Perkovich notes that Bhabha's claim was not based on facts. Even with the most optimistic assumptions, a bomb could not have been made before 1963.

It was widely believed that Nehru forbade Bhabha from building a nuclear weapon but S. Gopal, Nehru's biographer, told the journalist Raj Chengappa, 'It is not generally known that Nehru wrote to Bhabha that he was against outlawing of atomic weapons. His policy was never to use it but to have it because we can't completely abjure from it.' When Bhabha wanted to publicly reject nuclear weapons Nehru said, 'No! No! Don't go that far.'[9] Thus, even as Nehru advocated a nuclear bomb-free world and pushed for a nuclear test ban treaty, he sanctioned expensive plans for Bhabha to set up the giant infrastructure capable of both generating nuclear energy and building bombs.

Goldschmidt recalled Bhabha talking in 1955 about Nehru's response when he had asked Nehru to announce publicly that India would never make a bomb. Nehru had asked Bhabha if they were ready. When Bhabha had said that they were not, Nehru had said, 'But when you are ready, when you have everything, and you are ready to make a bomb, come and see me again, and we'll discuss the matter then.' So, according to Goldschmidt, Indians had always left the door open to produce a weapon if necessary. Bhabha felt that India should show the world that they could make a bomb if they wanted.[10]

In 1953, Nehru refused to take up Eisenhower's offer of helping non-nuclear nations to set up atomic power plants and set up a 'bank' from which they could draw fissile materials instead of producing their fuel. He thought that this was a ploy to prevent nascent nuclear powers from developing nuclear capability. When Bhabha put up the file to Nehru on the subject, he wrote: 'This is a political decision not to be taken by nuclear scientists.'[11]

A MISSED OPPORTUNITY

In May 1961, when the Kennedy administration was preparing to boost loans and aid to India, Vice President Lyndon B. Johnson visited India and Pakistan. While the interaction with Nehru was positive, Johnson seems to have been more impressed with the military dictator of Pakistan, President Mohammad Ayub Khan. Nevertheless, this did not prevent President Kennedy from making an extraordinary gesture to India around that time. Former foreign secretary, M.K. Rasgotra, in his autobiography, wrote about this gesture in which Kennedy offered to help India conduct a nuclear test. It is said that Kennedy personally wrote a letter to Nehru, conveying the offer. Accompanying the letter was a technical note from the chairman of the US AEC, Glenn Seaborg, setting out the assistance the US would provide to detonate an American device atop a tower in the Thar desert.[12]

Kennedy, who admired Nehru and India's democracy, had been briefed by American intelligence about China's progress. They believed China would be ready to conduct a nuclear test in 1963. Kennedy felt that India, a democracy, should have the distinction of becoming the first Asian country to detonate a nuclear device and not communist China. In his letter, while acknowledging Nehru's strong views on nuclear tests and weapons, he stressed the political and military threat of China's test to India's security. 'Nothing', Kennedy said, 'is more important than national security.'[13]

Rasgotra's source for this information was Ashok Parthasarathy, son of the diplomat G. Parthasarathy. According to Ashok, Nehru shared the letter only with Parthasarathy, who had just returned from China, and Bhabha. In fact, John Kenneth Galbraith, the US ambassador, handed over Kennedy's letter to Nehru on the very day G. Parthasarathy returned from China. Nehru summoned Bhabha urgently to Delhi to discuss the letter. Ashok said that Bhabha recommended immediate acceptance of Kennedy's offer. Bhabha told Nehru that on matters of atomic energy, the US was far more flexible and understanding than other western powers. 'It seems clear to me that our approach to them should not be made through or with the help of the United Kingdom.'[14]

Nehru did not dismiss Kennedy's offer out of hand. He asked Bhabha to work out a plan in case they accepted the offer. G. Parthasarathy requested time to mull over the implications of the offer and discussed it with Galbraith and B.N. Mullick, the director of the Intelligence Bureau. Both of them advised G. Parthasarathy to ask Nehru to accept the offer. Strangely, in the end, G. Parthasarathy ignored their advice and told Nehru to turn down the offer. It was perhaps what Nehru wanted to hear.[15] He was not keen on taking help from the US, mainly because he wanted to keep India non-aligned. It would also damage his efforts toward global disarmament, a subject on which he was arguably the world's most emphatic public face. Nehru's efforts did lead to the signing of the Partial Test Ban Treaty in 1963, which banned all atmospheric testing of nuclear weapons. Had Nehru accepted Kennedy's offer, it would have perhaps prevented the 1962 Chinese attack on India and also served as a deterrent to Ayub Khan before the 1965 Indo–Pak conflict.

Even before the offer from the US, Bhabha had sounded out the USSR to supply a power plant to India. India's relationship with the USSR had grown slowly after Independence. Stalin held little regard for India, but, after his death, there was a distinct warming in relations. In June 1955, Nehru travelled to the USSR, and a joint declaration was signed between the two countries. As we have seen in late 1955, the Soviet Premier Nikolai Bulganin and the Communist Party First Secretary Nikita Khrushchev visited India. They also went to TIFR, where Bhabha showed them around.

On a visit to Canada in November 1960, Bhabha announced that the USSR had agreed to design natural uranium and fast breeder power plants for India. This never materialized, but it helped Bhabha leverage more American and Canadian assistance with less stringent safeguards. Bhabha had also been pressing the Russian metallurgist, V.S. Emelyanov, for a natural uranium reactor. He was told that it was premature for that kind of plant, as was his idea of using such a plant to produce plutonium for recycling in breeders. The Soviet scientist was well aware that Bhabha could use the plutonium to make nuclear weapons.

REALPOLITIK

Prior to the Chinese nuclear threat, the main motivations for going nuclear had more to do with both Nehru's and Bhabha's belief that nuclear technology offered India a shortcut to modernity. Nuclear capacity, mainly in the civil domain—but, if necessary, in military applications—would increase India's international standing and boost its self-esteem. Both men wanted India to be a modern and respected member of the international community. According to the zeitgeist of those times, technological advancement and economic growth was the way to achieve such a status. More particularly, the mastering of the atom would leapfrog India into the orbit of industrialized nations.

Since India viewed success in nuclear technology as a way of competing with western nations in the field of high technology, it also became important that such success came without the help of other nations. Self-reliance and indigenous development, thus, became critical for proving India's technological parity. The failure of civilian nuclear power production meant that it would never become the advertised panacea for India's economic development. Thus, scientists focussed on the military aspects of nuclear technology as the tool for India to join the ranks of the major global powers in the world. There was an increased emphasis on promoting the values of nuclear technology outside of the development agenda, namely, a nuclear weapon and the intangible aspect of nuclear prestige that went with it. Building nuclear weapons became an answer to the failure to produce cheap electricity using nuclear energy.

The decision to collaborate with Canada for the CIRUS reactor in late 1955 was also driven by the goal of acquiring plutonium quickly. If India wished to develop a nuclear weapon, it would need either enriched uranium or plutonium. Since no country was willing to part with either without stringent safeguards, the CIRUS reactor provided India with a shortcut to weapons capability since it generated plutonium as a by-product. In fact, the real reason for Bhabha opposing safeguards at the conference in Geneva was to keep the bomb option open right from the beginning of the nuclear programme.

Since no one in the country, outside the atomic energy establishment, was familiar with nuclear technology, Bhabha's reasons were never subject to any kind of scrutiny. However, when it suited him, he also accepted safeguards. Examples of this are the reactors at Tarapur (TAPS I and II) and Rawatbhata (RAPS I and II). When CIRUS was expected to go critical in 1960, Bhabha convinced Nehru of the importance of a reprocessing plant to extract plutonium from the reactor's spent fuel rods. Thus, in July 1958, Nehru cleared Project Phoenix with a capacity to process 20 tonnes of fuel per year for recovering 10 kg of plutonium every year.

In 1960, France shifted the Fontenay Aux Roses laboratories, where much of the work on neutron triggers for atomic bombs was done, to Saclay. The French offered to permit some Indian scientists to observe how large-scale nuclear plants were built. Bhabha sent Iya to try and find out the process for manufacturing polonium used for the trigger in a bomb. Iya, who had been personally selected by Bhabha in Paris to join AEET in 1955, was deputed to Saclay, where he also studied the production and separation of radioisotopes. He joined the chemistry division of AEET in 1957. Though the French did not share any sensitive technology, they took Iya around important laboratories during the lunch break. When C.B.G. Taylor returned to England, Iya took over as the head of the isotope division. The radioisotope for the trigger device for the Pokhran explosion in 1974 was produced under his guidance.

In August 1961, when Nehru learnt that the Chinese were close to testing their first atomic device, he told Bhabha to speed up India's capability for a peaceful nuclear explosion. While he was dead against the bomb, he believed that when India called for a nuclear-free world, it must do so from a position of strength. Bhabha optimistically estimated that it would take him two years to make a bomb. In January 1962, Nehru gave Bhabha clearance to set up a small group of top physicists to understand the physics behind nuclear explosions. Bhabha asked a group of physicists, headed by R.K. Asundi, to discreetly begin work on the high pressure physics involved in nuclear explosions. This subject was the key to understanding the compression of a plutonium sphere from a controlled reaction to an explosive mode. G. Murthy, who had just returned from Stockholm, was made a member of the group. No one from this group knew why the studies were being done and they viewed it as a purely theoretical exercise.

Since Bhabha wanted this exercise to be kept under wraps, he got M.G.K. Menon to issue a circular to the group that their work was to be kept secret. They were even assigned a separate room at TIFR, where they met regularly and submitted their papers to Bhabha.[16] Bhabha was furious when he found the technicians at CIRUS displaying the temperature needed to heat the rods for generating weapons-grade plutonium. This was highly secret information that would have been a dead giveaway to a knowledgeable visitor. The display was removed immediately after.

As India prepared to develop its nuclear weapons capability, Bhabha pressed for greater secrecy. The year 1962 also marked the adoption of a revised Atomic Energy Act, which significantly tightened secrecy, along with increasing AEC's control over all atomic energy activities. Notably, neither the Act nor the debate on the Bill, mentioned, for the most part, the traditional focus on 'peaceful uses'. There was a tacit association of nuclear power with military ends.

Nehru was in a hurry to pass the Bill and tried to restrict debate. The timing of the debate on the Bill was changed at the last minute, and a time limit of three hours was set for the discussion.[17] The Act provided for the 'development, control and use of atomic energy for the welfare of the people of India and other peaceful purposes and for matters connected therewith'. No explanation was given to clarify what the matters connected with other peaceful purposes were. The last phrase 'for matters connected therewith' was omnibus enough to include India's subsequent development of nuclear weapons. In effect, the Act of 1962 created the legal and institutional basis for India's ambiguous policy of keeping the nuclear option open.

After India's disastrous war with China in October 1962 where Indian forces were routed and India severely humiliated, Nehru was a broken man. He admitted in Parliament that he had lived in an artificial world of his own creation. Bhabha realized that it would be a long time before India could claim any kind of parity with China in the field of conventional weapons and tried to capitalize on the failed campaign against China to get the go-ahead for the bomb. A month after the war with China, Bhabha wrote a secret note to Nehru in which he discussed China's imminent nuclear explosion and the Indian plan of action. He wanted the weapons programme to be given top priority and for the Indian programme to be bigger than that of the Chinese. He said that India possessing a sufficient quantity of plutonium within 18 months would conclusively demonstrate that India could make atomic weapons but was choosing not to do so.

Bhabha met Nehru a little later to discuss the note. What transpired at the meeting remains secret, but Bhabha confided in Rasgotra some details of the meeting. Bhabha told him that he tried to persuade the Prime Minister to conduct a nuclear test but 'Nehru almost threw me out of the room'.[18] He did, however, tell Bhabha to expedite the steps leading to a peaceful nuclear explosion. Bhabha, in any case, was being a little too optimistic. At the time of these discussions, CIRUS had been functioning erratically and even the reprocessing plant by which Bhabha set such store was facing technological hurdles that delayed its commissioning.

When China invaded India, J.R.D. Tata offered to help in any way he could. The offer was taken up a little later by the new defence minister, Y.B. Chavan, who succeeded V.K. Krishna Menon. Nehru had been forced to sack his favourite minister after his inept handling of the war and the public outcry that followed. Chavan requested J.R.D. for a confidential report on India's aviation needs in the long run. The Tata report covered 'Aircraft and Ancillary Electronic Equipment Required by the Defence Services' and was compiled by Tata, the defence secretary, the foreign secretary, the three service chiefs and Bhabha. It made detailed recommendations on the equipment and expansion of the Indian Air Force for the next 10 years.[19]

In July 1963, Nehru sent Bhabha an extract from American journalist and author, Edgar Snow's book *The Other Side of the River*. Snow acquainted the western world with the Communist movement in China and was, for many years, the only American writer with regular access to Chinese communist leaders. Bhabha wrote to Nehru in early August, sharing his views on the subject. Snow had pointed out that the psychological and political impact of a China-made device would be incalculably great. Bhabha agreed and said that India had to counter this impact.

> I am inclined to the view that the only way that this can be done at present is for us to show that should China explode such a device, we are in a position to do so within a few months thereafter. This implies that we should sign the Test Ban Treaty immediately to show our bona fides and our intention not to make an atomic weapon, but we should sign it while making clear that

should any neighbouring country explode a nuclear device, we should consider ourselves free to review the decision...

...If China does not desist from making a test, and proceeds to a regular production of nuclear weapons, then a new situation will arise [...] At that stage the balance of power in Asia will definitely shift in favour of China, if we have no nuclear capability by then. For while China would not dare to attack any country having a nuclear capability like the United States or USSR, it could well attack a country without such capability, say, in Asia, knowing that in such an event American nuclear capability would not be used against China for fear of retaliation [...] This can only be countered by the country liable to be so attacked, as for example India, itself possessing a minimum nuclear deterrence.[20]

Nehru died on 27 May 1964, and two weeks later Lal Bahadur Shastri was sworn in as prime minister. Bhabha, in an article he wrote after Nehru's death, said that it had been a privilege to have worked for such a remarkable man for so many years.

He indicated at the beginning the matters on which he would like to have papers put up to him. The list became shorter and shorter as time went on. He gave a few broad directives with regard to policy and his immediate decision when any policy matter was put up to him. But he never interfered with the running of the Department, though he took the keenest interest in the progress of its work.[21]

In a talk Bhabha gave on All India Radio on 1 June 1964, he emphasized the swadeshi aspect of Nehru's thrust towards industrialization, and the part played by science in it. 'It is only when India has acquired the ability to design, fabricate and erect its own plants without foreign assistance will it have become a truly advanced and industrialized country, and for this, a much greater development of science and technology is still necessary.'[22]

A resolution passed by the Science Advisory Committee, which Bhabha headed, spoke of Nehru's 'unquenched thirst for truth, his questioning mind that admitted no man-made barriers, his essential humility, his constant willingness to learn and to teach'.[23] Bhabha went on to say that although Nehru had not been a practising scientist, his personality revealed the essential attributes of a real man of science. Nehru, in turn, had been a great admirer of Bhabha. Nehru had once said, 'It is not for me to tell you of his achievements and plans, which are both great. It has been a pleasure for me to work with him in various ways and more especially in the Atomic Energy Commission.'[24]

Chapter 35

NUCLEAR PRAGMATISM

> ...a stockpile of some 50 atomic bombs would cost under ₹10 crores and a stockpile of 50 two-megaton bombs something of the order of ₹15 crores. These expenditures are small compared with the military budgets of many countries.
>
> —Address on Nuclear Disarmament on All India Radio on United Nations Day, 24 October 1964[1]

Nehru's successor, Lal Bahadur Shastri, was a Gandhian who exhibited a moral aversion to the bomb. He also had little in common with the suave and sophisticated Bhabha, who had, until then, leveraged his friendship with Nehru to get his way in virtually all matters. Shastri came from humble beginnings and was considered rustic. The atomic tsar discovered, much to his annoyance, that he could no longer walk into the prime minister's office and talk to him. He had to seek a formal appointment if he wanted to meet the Prime Minister. Shastri also found it difficult to grasp what Bhabha was telling him, and this only added to the latter's frustration.

China exploded a nuclear device on 16 October 1964 at Lop Nor, becoming the first country to violate the Partial Test Ban Treaty that Nehru had worked so hard to set up. India had never faced a direct nuclear threat till then. All that changed when, for the first time in independent India's history, a hostile neighbour (against which it had lost a war in the recent past), acquired nuclear capability. The Chinese claimed huge swathes of Indian territory along the Himalayan frontier. What was at stake was not only India's territorial integrity but also its international prestige.

China had stolen a march over India by being the first to test a nuclear device and this threatened New Delhi's purported leadership position in Asia and the Global South. To make matters worse, after Nehru's death, India was in a state of internal flux. There was not only a void in leadership but the country had also lost its most important source of international influence. The Chinese explosion brought the issue of nuclear weapons to the forefront of India's security policy for the first time. National security had become the benchmark for India's engagement with nuclear disarmament and arms control.

AN UNEASY RELATIONSHIP

After Nehru's demise, Bhabha faced problems in getting his point across to Shastri and other politicians. There are files in the Lyndon B. Johnson Library in Austin, Texas, on Indira Gandhi's first visit to Washington as prime minister in 1966, filled with secret telegrams from the US embassy in Delhi and briefs for President Lyndon B. Johnson written by his advisers. A document with the highest classification, dated December 1964, records the conversation between Bhabha and Spurgeon Keeney, scientific adviser to the US president. Apparently, among other things, Bhabha told Keeney that while making a detailed presentation to the Indian cabinet in the immediate aftermath of China's first nuclear test, he suddenly realized that most ministers 'seemed not to understand what I was talking about'.[2]

On 24 October, just eight days after the Chinese explosion, Bhabha took to the airwaves, making his now famous broadcast on nuclear disarmament on the occasion of United Nations Day on All India Radio. Speaking on the subject of nuclear deterrence, Bhabha said, 'In short, atomic weapons give a State possessing them in adequate numbers a deterrent power against attack from a much stronger State.

Indeed, the importance of nuclear weapons is that they enable a country possessing them in adequate measure to deter another country also possessing them from using them against it.'[3]

However, Bhabha's speech is famous for stating that building a nuclear bomb was not expensive, and he estimated the cost of a 10-kiloton explosion to be ₹17.5 lakh while a 2-megaton explosion would cost ₹30 lakh. Atomic explosives, Bhabha said, were much cheaper than conventional explosives. '...a stockpile of some 50 atomic bombs would cost under ₹10 crores and a stockpile of 50 two-megaton bombs something of the order of ₹15 crores. These expenditures are small compared with the military budgets of many countries.' He wanted the UN to create 'a climate favourable to countries which have the capability of making atomic weapons, but have voluntarily refrained from doing so'.[4] Bhabha's exaggerated claim about how cheap nuclear weapons were was based on the calculations published in a paper by the Lawrence Radiation Laboratory, Livermore, US.

Bhabha was pushing the government for a decision on developing nuclear weapons. By that time the reprocessing plant in Trombay had started functioning, and India had the ability to extract plutonium, and, consequently, to make nuclear weapons. In fact, even before the Chinese explosion, Bhabha had declared in London on 4 October 1964 that India could explode an atom bomb within 18 months of a decision to do so.[5] In a direct challenge to Shastri, he added, 'But I do not think such a decision will be taken.'[6] Shastri, who was in Cairo attending a meeting of non-aligned nations at the time, said that India's nuclear establishment was 'under firm orders not to make a single experiment, not to perfect a single device which is not needed for peaceful uses of nuclear energy'.[7]

Shastri was not ready to sanction a nuclear weapons programme because it would be a major change in India's traditional position. Bhabha's public sponsorship of the bomb did not sit comfortably with him. Also, as prime minister, no one knew more about the precarious state of the economy. After the 1962 conflict, the cost of conventional defence had risen sharply. Internally, the country was suffering the worst food crisis in history. Two press commentaries at the end of October 1964 illuminated the debate. In the first, Inder Malhotra, writing for *The Statesman*, triggered a national debate, when he aired his view that it would be impossible to live with the decision not to make the bomb. He cited Bhabha's cost estimates on the bomb. 'He has dispelled the belief that the cost of nuclear weapons is totally prohibitive; and he has pointed out that the only deterrent to a nuclear bomb is another nuclear bomb. Nothing else.'[8] On the same day, an article appeared in the *Hindustan Times*, which bore Bhabha's impress. It listed the 'breakthrough' achievements of AEET and said that the Indian scientists had taken the Chinese explosion in their stride—confidence that stemmed from the knowledge that they could have developed the bomb 18 months ago.[9] Like before, there was no one to challenge this tall claim.

Romesh Thapar, writing in the *Economic and Political Weekly*, held a divergent view. He questioned Bhabha's cost estimates of the bomb and wanted to know if cost escalation could be prevented once the programme was underway.[10] This questioning of Bhabha was a brave departure from the norm. Thapar observed, 'A bomb held by a beggar does not make him a king when many others have more bombs than he and the capacity to deliver them where they hurt most.'[11]

Bhabha's broadcast pressurized Shastri to sanction an overt weapons programme. Even his party, the Indian National Congress (INC), was agitated about the issue, and the bomb lobby used Bhabha's speech to claim that nuclear weapons could be produced easily and at a relatively low cost, even by a poor country like India.[12] At a Congress Committee meeting between 7 and 9 November 1964, Shastri disputed Bhabha's estimate for a bomb. His own estimate was far higher than Bhabha's, who was forced to clarify that his estimates were based on PNEs conducted by the US. Shastri said that the production of a single bomb would cost ₹40–50 crore and not ₹10 crore for 50 atom bombs, as Bhabha claimed. Shastri also said that scientists should realize that it was the responsibility of the government to defend the country and adopt appropriate measures. Clearly, he was showing Bhabha his place: Shastri, V.K. Krishna Menon and other INC leaders managed to carry the day.[13]

We have noted how Bhabha and Krishna Menon had met in England, but they never had a friendly relationship. There is a story about how, after India's first nuclear test at Pokhran in May 1974, Krishna Menon, who was then in hospital with less than five months to live, summoned Indira Gandhi and gave her a mouthful.[14] Initially, Bhabha was also a little wary of P.N. Haksar, who was India's ambassador to Austria in the 1960s. Vienna was the headquarters of the IAEA, and, in his capacity as ambassador, Haksar represented India on its board. At first, Bhabha was suspicious of him because of his loyalty to Krishna Menon, but, within a few months, he was made a member and given full powers to explain the Indian position on nuclear non-proliferation and the need for atomic energy in India.

On 17 November 1964, somewhat chastened by Shastri and Krishna Menon's attacks on his UN day speech, Bhabha, in an interview with *The Times of India*, sought to rectify the image that he favoured the acquisition of nuclear weapons. He reiterated that his cost estimates had been based on American figures for PNEs. He asserted that AEET had already accomplished much, and the essential point was not that India was incapable of making nuclear weapons or could not afford to make them, but that it was determined not to produce them due to political considerations.[15] Shastri distrusted Bhabha's estimates so much that he even requested the Harold Wilson government in the UK to offer a cost estimate of the bomb, in case India decided to develop it. The report that followed suggested that the cost would be about \$350 million with a running cost of \$50 million.[16] (These costs were highly inflated and in a confidential report not shared with the Indian government the actual costs were assumed to be significantly less).

PEACEFUL NUCLEAR EXPLOSIONS

In the Lok Sabha debates that followed in November 1964, several speakers, led by Krishna Menon, questioned Bhabha's cost estimates as well as the appropriateness of his having appeared to advocate a bomb. Others, who favoured the bomb, defended Bhabha, citing the figures he had mentioned in his speech. His talk was good news for some politicians, mostly from the Jan Sangh, who had long advocated an Indian bomb. On 27 November, the Jan Sangh member of parliament from Dewas, Hukum Chand Kachwai, introduced a motion in the

Lok Sabha calling for the manufacture of nuclear weapons. A lively debate followed. In his response, Shastri claimed that the bomb lobby had misread Bhabha's intentions and the production costs referred to those in the US. In India, the cost would still be prohibitive. In any case, the manufacture of such weapons would be a departure from Gandhian traditions and such weapons threatened the entire human race. Shastri was on the defensive in the debate, but a large INC majority ensured that the resolution was defeated.[17]

Shastri managed to win a voice vote against the resolution, but, in his speech he made important concessions, a shift that went largely unnoticed at the time. He said that while India's nuclear establishments would remain committed to peaceful work, that work would now entail preparations of PNEs for purposes such as tunnelling through mountains. 'Dr Bhabha has made it quite clear to me [...] we can reap its peaceful benefits and we can use it for the development of our nation...'[18] Much of Shastri's speech referred to Bhabha. The two had met the same morning and Shastri explained that Bhabha had been misunderstood as advocating a nuclear bomb. What Bhabha favoured, he said, was using atomic energy for peaceful purposes, such as blasting through mountains.

What is remarkable is that nobody realized at the time that Shastri had, in fact, opened the door to the bomb. Whether Shastri was cynically grasping at a political compromise or was genuinely persuaded that work towards a peaceful explosion was qualitatively different from that which manufactured weapons will never be known. Shastri was a simple man who had left school at 17 and had never travelled outside India before his appointment as prime minister; he had very little or no understanding of nuclear physics or technology. The journalist who realized what had happened was Inder Malhotra who correctly anticipated a policy shift following China's nuclear test and understood the significance of Shastri's speech to Parliament on peaceful nuclear explosions. In a series of five articles for *The Statesman*, Malhotra explained the significance of the largely hidden policy shift.

In a cartoon that he gifted Jamshed Bhabha, the cartoonist R.K. Laxman depicted Shastri as an angel with wings holding leaves in his hands with Bhabha placing a halo on his head. Below it were the words, 'Atom for Peace'. Earlier, Laxman had made another cartoon with Nehru driving the bullock cart of progress with 'atomic wheels' and the common man as a bewildered passenger. Even the look on the face of the bull shows wonderment at the new wheels. Bhabha was fond of Laxman's cartoons and had one of them enlarged and hung in his room. One of his cartoons of Bhabha looking at a model of the atom appeared in the October 1984 issue of *Science Today*.

In a note written in November 1964, K.R. Narayanan, the director of the China Division at the Ministry of External Affairs, wrote about the changed political balance of Asia and the world due to the Chinese explosion. He wanted India to manufacture a bomb. 'With the atom bomb in our possession, we should be able to play an effective part in the field of disarmament [...] It is only when we have the bomb that we can renounce it convincingly, in the manner of Ashoka, and make a dramatic impact on the world...'[19]

Since, technically, there is no meaningful distinction between a rudimentary atom bomb and a peaceful nuclear explosion, India's nuclear policy was finally where Bhabha had wanted it to be for years—on a course leading to explosives. Bhabha also realized that an overt bomb-building programme was untenable and the US and Canada could react punitively. However, a move to develop peaceful nuclear explosives would not lead to recriminations, especially since the US had its own peaceful nuclear explosives programme, named Project Plowshare. The USSR, too, claimed that they had used such explosives for development work and so Shastri's statement was received in India without much concern. But, experts realized that a peaceful nuclear explosion would also validate India's capacity to make a bomb.

Shastri's no-bomb position had not gone down well with his own party but after permitting peaceful explosions he was in a stronger position. This decision did not disrupt India's nuclear cooperation with the US and Canada while winning over Bhabha to Shastri's side. When Raja Ramanna was asked why Shastri had permitted peaceful explosions, he was quick to point out that such an idea must have emanated from Bhabha. Bhabha's scientists were imbued with the same passionate commitment to make the nuclear bomb at the earliest so that India could become a great power. According to Ramanna:

> There was never a discussion among us over whether we shouldn't make the bomb. How to do it was more important. For us it was a matter of prestige that would justify our ancient past. The question of deterrence came much later. Also, as Indian scientists we were keen to show our western counterparts, who thought little of us those days, that we too could do it.[20]

S.S. Khera, of the ICS, who was cabinet secretary from 1962–1964, observed: 'Those who knew Dr Homi Bhabha and worked with him were aware of his urge to work towards having everything ready for the bomb.'[21]

Since India could end up taking years to develop a deterrent, Shastri took to diplomacy to counter the Chinese threat. On a state visit to London in December 1964, he broached the topic with British premier, Harold Wilson, but the Indian press did not take kindly to his seeking a nuclear shield which would negate India's non-aligned policy. During a press conference, Shastri suggested that it was for the nuclear powers to discuss some kind of a guarantee that was needed not only by India but all non-nuclear countries. Shastri clarified that he was seeking protection not just for India but for all non-nuclear nations, and it was for the big nuclear powers to consider how to maintain peace in the world. Back home, Bhabha was under pressure to live up to his 'bomb in 18 months' claim. Both Homi Sethna and Ramanna thought it could not be done. Sethna thought that they could, perhaps, assemble an explosive device, but they would be unable to manufacture a portable bomb.

Ramanna thought that it could not be done at all: Bhabha had made the statement because he had been under great pressure to deliver. Part of Ramanna's realism must have stemmed from the fact that Project Phoenix, which had been set up to reprocess spent fuel from CIRUS and extract plutonium, had run into serious problems. Though they had started reprocessing plutonium in mid-1964, the plant had been functioning erratically. It was inaugurated by Shastri on 22 January 1965, and he had emphasized its uses for India's peaceful nuclear explosion programme. Bhabha wanted to send a message to China that India possessed weapons-grade plutonium and could make a bomb, if it wanted. Shastri even naively suggested that other countries could use India's facility to reprocess plutonium. It was quite obviously his own idea since there is no way Bhabha would have asked him to say something of that nature. Bhabha even had the funnel of the dull grey reprocessing plant painted a bright red. He told Ramanna, 'Let the world be warned that we are ready.'[22]

Shastri's attempts at getting a nuclear shield for India failed, and Bhabha reported that the US was unwilling to cooperate. Faced with the truth that AEET could not produce a nuclear explosive in 18 months, Bhabha secretly sought help from the US in late 1964 or early 1965. By early 1965, Bhabha had revised his estimated time for developing the bomb from the utopian 18 months to a more realistic timeline of at least five years. At the INC's annual conference in early January 1965, several delegates argued in favour of nuclear weapons: memories of the China debacle were still fresh in their minds. Other delegates thought that India should seek shelter under a US or USSR umbrella. Shastri, who considered nuclear weapons to be morally repugnant, insisted on staying with the no-bomb policy. He managed to pass a resolution declaring that, for the present, India would restrict its nuclear programme to peaceful purposes. Shastri's emphasis on speeding up the development of nuclear energy for peaceful purposes calmed the opposition in his party and also garnered the support of the press.

During his meeting with US officials at the State Department in February 1965, Bhabha claimed that India needed to make some dramatic peaceful achievement to offset the prestige gained by communist China among Africans and Asians. Referring to the plutonium reprocessing plant, Bhabha told US officials that it was large enough to process all of the plutonium from the reactors India was building and that within five years, India would be capable of producing hundred nuclear bombs per year. Therefore, unsurprisingly, in its obituary for Bhabha, the London-based *Times* remarked on India's nuclear weapons potential: 'The Indian government has therefore, provided itself with the option of producing plutonium bombs if it so desires in the future.'[23]

A year after Bhabha's death, at the memorial lecture in his honour at the Royal Institution, London, Cockcroft observed:

> In later years, Homi Bhabha became a central figure in the discussions on proliferation of nuclear weapons, and on the issue as to whether India should develop such weapons. At this time, it was a declared policy of the government of India not to develop nuclear weapons, and Homi Bhabha of course in his official pronouncements followed the policy of his government. In discussions at small closed meetings, however, he appeared to be in favour of making bombs for a Plowshare programme, i.e., to use plutonium bombs for civil engineering purposes—making storage cavities, harbours or canals.[24]

At the same meeting of the Royal Institution, M.G.K. Menon argued that India had decided to build the reprocessing plant prior to the 1962 Sino–Indian war, and it was for reprocessing fuel rods and not nuclear bombs. His role as an apologist had little impact: the mistrust between India and China was common knowledge after China's occupation of Tibet in 1950, and border tensions had existed between the two countries even before the 1962 conflict. More importantly, even before the war in October 1962, India had been aware of the advances China had made in its nuclear weapons programmes.

TALL CLAIMS BELIED

The declassified documents on the subject show that Bhabha approached the Americans for a Plowshare device or blueprints in late 1964 and early 1965. Unknown to Bhabha, the US AEC, in late November 1964, included Plowshare projects in their consideration of areas of potential cooperation with India. Bhabha pursued the Plowshare issue himself when former presidential science adviser Jerome Wiesner visited Trombay on 19 January 1965. Wiesner left a report for US AEC member John Palfrey, who was arriving in India for the inauguration of the reprocessing plant at Trombay. A cable from the US State Department to Palfrey permitted him to discuss Plowshare-type projects with Bhabha. Wiesner, in his report, had written that Bhabha had greatly underestimated the cost of making a bomb and wanted the US to give him a better cost estimate. He observed, 'Bhabha apparently wants authority and resources to move forward without final decision regarding actual explosion. No one has estimated what a real weapon system would cost or understands what will be done with it.'[25]

Wiesner recognized what many have considered to be the greatest weakness of the nuclear programme, namely, the lack of mechanisms for independently evaluating the AEC's claims and proposals. There simply was no way of checking and balancing Bhabha, and this lack of transparency and debate hindered Indian nuclear policymaking for decades. As Perkovich has observed, 'The secrecy and autonomy of the nuclear establishment and the lack of qualified scientific expertise outside that establishment badly impaired the policy-making process.'[26] Neither Nehru nor Shastri and others debating nuclear policy had the expertise to scrutinize Bhabha's claims.

Bhabha visited Washington towards the end of February 1965 and discussed nuclear issues with the undersecretary of state, George Ball. While discussing the relative capabilities of various nations, Bhabha thought that China owed its rapid progress to the blueprints left by the USSR for a nuclear device. Bhabha then told Ball—in what was, effectively, a request for an American blueprint or device—that with a US blueprint, India could produce a device in six months, whereas, on their own, they would take 18 months.[27]

In his report the scientific attaché of the American embassy revealed details about the problems faced by the reprocessing plant. The report noted that the entire operation was dependent on supplies from the US and elsewhere and had been shut for some time since it was awaiting foreign parts. The report concluded by saying that the possibility of aiding India with something dramatic to give her status comparable with the Chinese bomb looks very difficult. By April 1965, the US AEC became even more conservative regarding nuclear cooperation with India. Bhabha and Glenn Seaborg met in February to discuss nuclear cooperation and Seaborg wanted to limit this cooperation to lectures and to solving some engineering problems. The Americans were keen to maintain strict controls over technology and political considerations dictated that no Plowshare device or plans could be given to India then.[28]

In a letter to the chairman of the US Congress Joint Committee on

Atomic Energy on 30 April 1965, Seaborg listed potential cooperative ventures between India and the US but avoided any reference to cooperation regarding a Plowshare device. It also discussed Bhabha's request for a moderate amount of plutonium, which, given the recent opening of the reprocessing plant, was a strange request. It amounted to an open admission that the Trombay plant was functioning suboptimally. The omission of a reference to a Plowshare device meant that Bhabha's request was considered to have undesirable proliferation implications.[29]

Bhabha's request also reaffirmed the baseless nature of his claims about India being able to make a nuclear weapon in 18 months. Given Bhabha's autonomy in nuclear matters it is not known if he had informed Shastri of his request to the Americans. There is evidence to show that Shastri, for his part, excluded Bhabha from certain high-level discussions on nuclear policy. Bhabha's request to the US came at a time when the US non-proliferation policy was taking a concrete shape. As the administration of Lyndon B. Johnson rallied behind the non-proliferation treaty, the chances of acquiring any technology from the Americans gradually receded. At the end of it all, India received no assistance from the US and was left to its own devices.

After finally getting Shastri's nod for something he had been advocating for years, Bhabha moved quickly. Realizing that there would be no help from the Americans, Bhabha set up a small group called the Study of Nuclear Explosions for Peaceful Purposes (SNEPP) in early April 1965 and put Ramanna in charge. Ramanna recalls that Bhabha called him to the office at the OYC and gleefully waved the document that had Shastri's approval on it. Bhabha appointed two physicists to assist Ramanna. He also asked him not to tell E.C. Allardice about it and housed the group in a separate building.[30]

In the spring of 1965, during a visit to Paris, Bhabha summoned T.S. Murthy, a young scientist who was training at Saclay, to his hotel and told him about the need for mastering polonium extraction. Polonium is used in the trigger mechanism for atom bombs. Bhabha asked Murthy who was Iya's understudy and later played a major role in the Pokhran explosion in 1974, to learn as much as he could about polonium. In December 1965, Shastri asked Bhabha to speed up plans for a PNE. According to Sethna, during the 1965 war, a desperate Shastri asked Bhabha to do something. Bhabha then told him about the work on PNE. Shastri asked him to go ahead but not carry out any experiment without clearance from the Cabinet. Bhabha called Sethna and asked him to form the groups required to get the device ready. There was a group for engineering, another for measurements, and another that would produce the device.

Shastri died in Tashkent on 11 January 1966, hours after signing the Tashkent Declaration. His successor, Indira Gandhi, was in charge of nuclear affairs but was deprived of Bhabha's advice since he was killed in an air crash on 24 January 1966. With the secrecy surrounding all atomic affairs, very few knew the whole agenda. Shastri and Bhabha were both dead and no one else was in the loop, since the debates and rationale informing major decisions were generally not recorded on files. Bhabha's death left a huge void in India's nuclear policymaking.

CODA

On Sunday, 23 January 1966, Bhabha worked almost through the day at his TIFR office on the fourth floor, giving last-minute instructions on a variety of subjects. He was to travel to Vienna later in the day to attend the meeting of the advisory committee of the IAEA. M.G.K. Menon recalls how Bhabha telephoned him that morning, asking him if he had any particular plans for the day. When Menon said that he did not, Bhabha asked him to come to TIFR since he wanted to discuss some aspects of a cultural event on the West lawns. Pipsy Wadia was also going to be there. After that, Bhabha said that he would like to discuss some important matters with him in his office.*

Bhabha and Menon spoke for two hours and, in the course of their discussion, Bhabha told Menon about his conversation with Indira Gandhi, who had been elected as the prime minister of India on 19 January. Bhabha said that the prime minister-elect wanted him to assist her in all matters relating to science and technology, a responsibility which if he accepted would mean relocation to New Delhi. With such deep roots in Bombay, Bhabha was understandably not too happy about moving to New Delhi but realized the opportunity that had been given to him of growing science and using it effectively in the service of the nation. He told Menon that he had decided to accept her offer.†

Bhabha did not reveal the exact nature of the offer or the position he would occupy, but Menon thought that it would be equivalent to a cabinet minister. Although he would continue overseeing the working of the DAE from New Delhi, Bhabha had to consider how he would handle TIFR and AEET, the two big establishments in Bombay. It was then that he told Menon bluntly: 'Goku, I plan on return from Vienna, to put up a proposal to the Council for you

*Ghosh, Dipan K. and Arun K. Grover, *Tribute to a Titan: Birth Centenary of Homi Jehangir Bhabha*, Indian Physics Association, Mumbai, 2009, p. 76.
†Ibid. 77.

to be director of TIFR.'* Menon says that others who knew about this decision in different forms included Bhabha's brother Jamshed, his mother Meherbai, J.R.D. Tata, Pipsy Wadia and Rustum Choksi. Menon's expostulations that there was no need for such a step left Bhabha unmoved. Menon says that Bhabha's dentist, Dr Fali Mehta, was also privy to this information.†

At that same meeting, Menon got Bhabha's approval for new buildings in the housing colony across the road from TIFR. Bhabha wanted bigger flats for his senior scientists and asked Menon if the approval could wait. Menon told him that it was becoming difficult to keep scientists in the institute when they had no proper accommodation. He told Bhabha that they would be quite happy to stay in the smaller but modern flats planned for the B-1 block (now called Bhaskara) in the colony. Bhabha smiled and signed. It was his last official act. Menon recalls that Bhabha also spoke at length about his plans for rapid development in several fields of national endeavour—science, technology and industry.

*Ibid. 77–8.
†Ibid. 77–8.

Chapter 36

AIR INDIA FLIGHT 101

> But the span of one's life is limited. What comes after death no one knows. Nor do I care. Since, therefore, I cannot increase the content of life by increasing its duration, I will increase it by increasing its intensity.
>
> —Letter to his friend Jessie Maver, 8 December 1934[1]

THE CRASH

Bhabha took Air India flight 101 to Geneva en route to Vienna. In those days there was no direct flight to Vienna and one had to go via Geneva. Bhabha was booked to Geneva on an earlier flight the day before, but for reasons that have never been made clear he took Air India flight 101.

On 24 January, the Boeing 707, *Kanchenjunga*, crashed into the Glacier des Bossons on the southwest face of Mont Blanc at 4,807 m at around 7.02 a.m. local time. It was flying on its regular route from Bombay to London via Delhi, Beirut and Geneva. All 106 passengers, including Bhabha and 11 crew members died in the crash. The *Kanchenjunga* crashed in almost exactly the same spot as Air India's *Malabar Princess* had on 3 November 1950, killing 48 passengers and crew. Like the wreckage of *Kanchenjunga*, the bodies of the victims of the *Malabar Princess* crash were never recovered. Air India flight 245 was a Lockheed Constellation that was flying from Bombay to London, with stopovers in Cairo and Geneva. The *Malabar Princess* had flown the very first international flight of Air India on 8 June 1948 on the same route.

The black box from the 1966 crash was never recovered and rescue operations were halted the day after the crash owing to inclement weather conditions. The investigation resumed in September 1966, and the French inquiry commission completed its report in March 1967. It concluded that severe white-out conditions near the summit of the mountain, coupled with miscommunication over phraseology between the air traffic controllers in Geneva and the pilot led to the crash.[2] The report concluded that the crash was the result of a miscalculation rather than another plane intercepting it. According to the commission, the most likely hypothesis was that the commander miscalculated his position in relation to Mont Blanc. He seems to have mistakenly believed that he had passed the ridge leading to the summit, and, thinking that he was still at a level that afforded sufficient safety clearance over the top of Mont Blanc, continued his descent. One of the receivers on the plane had also not been working, and the pilot misunderstood the verbal data sent to him on the plane's position.[3] The French report was accepted by the Indian government.

Even though the largest parts of the debris were on the Italian Alps near Courmayeur, Italy did not participate in the inquiry commission but eyewitness accounts of Italian individuals were incorporated into the French report. According to the transcript of communication of the air traffic controller in Geneva included in the French report, flight 6029 was scheduled to fly over Geneva around the same time as Air India 101 was to make its descent into the city. When the latter suddenly disappeared from the radar, the controller asked flight 6029, around 7.04 am, to report on what it saw around Mont Blanc. The pilot remarked, in Italian, that he saw a black cloud at around 16,000 ft, which was different from the other clouds, and he thought that the black cloud had possibly been caused by an explosion.[4]

Bhabha's sudden death stunned India and the world; for J.R.D. Tata, it proved to be a cruel double blow. His brother-in-law,

Gianni Bertoli, who was Air India's regional director for Europe, was on the same flight. Married to Kitty, Thelma Tata's sister, he had come to India from Geneva for consultations with his brother-in-law. The two were very fond of each other. A few days before Bertoli's flight, he had been invited by the politician Minoo Masani for dinner. Since he was scheduled to depart on the day of the proposed dinner, he refused. Masani asked him to postpone his flight by a day, but he declined. He had a tryst with his destiny and stepped on the ill-fated flight. J.R.D. Tata displayed photos of both Bhabha and Bertoli in the bookshelf near his writing desk at home.[5] Apart from Bertoli, others who died in the flight were P.C.R. Coates, the chief representative of Burmah-Shell in India, who, ironically, was travelling to London for his father's funeral; G.V. Shivaswami, operating superintendent of Air India; and the commander of the flight, Captain J.T. D'Souza.

Just two days before this tragedy, on 22 January, Bhabha had presided over a condolence meeting for Shastri at Trombay. Perhaps the very last letter he signed before his demise was to the late prime minister's widow Lalita Shastri, dated 23 January, where he told her that they had observed 22 January as a full working day, even though it was a Saturday, as a fitting tribute to her husband. This was in stark contrast to the many government holidays that were declared when a dignitary passed away. '...If there is anything that I can do to be of help to you in any way, may I request you to call on me without hesitation, as I shall deem it a privilege to do what I can in memory of Shastriji.'[6]

It is said that Shastri had also asked Bhabha to join the Cabinet as a minister, but the latter had declined, preferring his scientific work over a political post. The news of Bhabha's tragic death and the immersion of Shastri's ashes appeared on the front page on the same day.

M.G.K. Menon received the news when he was in his office at TIFR. Air India had called to say that Bhabha's flight had not landed in Geneva and that they would keep him informed of further developments. He then went to the OYC, where a large number of people had already congregated. He received the news of the crash there. Soon after Vikram Sarabhai arrived, the two decided to go to Kashmir House where Sarabhai was staying to meet in quieter surroundings. On the way, they called at Mehrangir but were not admitted inside. At Kashmir House, the news of the crash and the fact that there were no survivors was confirmed. Menon says that Sarabhai had already telephoned Menon's wife to give her the sad news—kindness and consideration that were characteristic of him.[7]

There was a tragic irony to Bhabha perishing on Mont Blanc. M.V. Kamath, the veteran journalist, recalled how he had met Bhabha in Geneva at the Palais des Nations, the headquarters of the UN a few months before his demise. Bhabha had looked in the direction of Mont Blanc and enquired if Kamath had been up there. Kamath had answered that he was not the mountain climbing kind and certainly not at his age. Bhabha told him that he could do it if he wanted to, but, in case he did not feel up to it, he could use the funicular. After Bhabha died, Kamath wrote a moving obituary in *The Times of India*. In 1979, Kamath met Jamshed Bhabha, who told him how moved both he and his mother had been after reading the piece.[8] Devendra Lal recalled Bhabha not wanting TIFR scientists to travel on the same flight together, since, if the plane crashed, it would be a big loss to the institute. Sadly, Bhabha himself perished in a plane crash.

A CONSPIRACY?

In 2017, Daniel Roche, a Swiss climber, found the remains of an aircraft in the Alps, which could be from either of the two Air India crashes. Roche, an aviation enthusiast, found a few limbs as well as a jet engine at the site. He believed that Air India flight 101 had been intercepted by another plane and crashed. According to him, it had collided with an Italian aircraft, and, since there is very little oxygen at that height, there had been no combustion to cause an explosion.[9]

Conspiracy theories implicate the Central Investigative Agency (CIA) in the crash. In 2008, an alleged conversation between former CIA officer Robert T. Crowley and journalist Gregory Douglas (born Peter Stahl) was published in a book, *Conversations with the Crow*, which led people to believe that the American intelligence agency had a hand in the crash. Known as 'The Crow' within the agency, Crowley joined the CIA at its inception and spent his entire career in the Directorate of Plans, also known as the 'Department of Dirty Tricks'. Before his death in October 2000, Crowley is said to have had numerous conversations with Douglas and apparently sent him two lockers full of documents that he wanted to be opened only after his death. In the book, Crowley's take on Bhabha's death appears in conversation 22, purportedly held on 5 July 1996.[10] Crowley had been quoted as saying:

> We had trouble, you know, with India back in the 60's when they [...] started work on an atomic bomb [...] bragging how clever they were and how they too were going to be a great power in the world. The thing is, they were getting into bed with the Russians.[11]

Referring to Bhabha as the 'head expert', he said, that Bhabha was fully capable of building a bomb. He added that Bhabha had been warned several times but had been too arrogant to listen. Bhabha had made it clear that no one would stop him and India from getting nuclear parity with the big boys. He went on to add that Bhabha could be a threat. 'He had an unfortunate accident. He was flying to Vienna [...] when his Boeing 707 had a bomb go off in the cargo hold [...] There was no real evidence and the world was much safer.'[12]

Crowley also allegedly bragged that they could have blown up the plane over Vienna but decided the high mountains were safer. 'I think a possible death or two among mountain goats is much [*sic*] preferable than bringing down a huge plane right over a big city.' Crowley's real expertise within the agency was the Soviet KGB. He went on to add something even more alarming. Calling Shastri 'a political type who started the programme in the first place', he claimed that they did away with Shastri as well. Since Bhabha was a genius and could get things done, the CIA 'aced both of them'. He added, 'India was quieter after Bhabha...'[13] Bhabha's death was similar to the death of Enrico Mattei—the Italian oil magnate, who started work on Italy's first nuclear reactor and was allegedly killed by the CIA by sabotaging his private plane.

The veracity of these startling admissions has not been established, and, in all probability, the truth will never be known. Douglas is an unreliable source at best. A Nazi sympathizer, conman and

Holocaust denier who has been accused of forgery, he has also made a number of other startling revelations with no factual basis. Douglas is considered something of a pariah even among conspiracy theorists. Undoubtedly, the Americans did not want India to develop a nuclear weapon, but even persons who tend to believe in conspiracy theories would find the murder of 117 people to get rid of one man difficult to digest.[14]

DECLASSIFIED BY THE GLACIER

Due to climate change, the melting Bossons glacier in Chamonix, France, has been revealing pieces of India's nuclear history—official documents marked 'top secret' and 'secret' from the Ministry of External Affairs, long buried with the wreckage of Air India flight 101. In August 2012, a 19.8-pound bag marked 'Diplomatic Mail Ministry of External Affairs' was found by a rescue worker and returned to the Indian embassy in Paris. It contained newspaper reports from January 1966. Letters, diplomatic documents, cameras and jewellery continue to be found at the wreckage site at Mont Blanc.

New Delhi's assessments of Chinese defence and nuclear capabilities in early 1966 are included among these documents. Some are Indian analyses of Chinese foreign policy, particularly Beijing's relationship with the West and the Sino–Soviet split. Some documents originated in the East Asia Division of the Ministry of External Affairs and concerned Indian assessments of Chinese defence production, missile development and the nuclear weapons programme. Others were meant for Ambassador Khub Chand in Stockholm while still others were for Indian embassies and high commissions more generally. One such Indian document dated 20 January 1966 on China's military assessment noted with frustration the difficulty of an objective or accurate estimate because of the total absence of official statistics. Another document on Chinese nuclear delivery capability stated that by the end of 1966, the size of the bomb could be reduced so that it becomes deliverable by an IL-28 light bomber. By the middle of 1967, China could have a nuclear-armed missile with a range of 1,000 miles. Another analysis of China and the West argued that China was aware of the healthy respect and fear that it commanded in western Europe.[15]

These records from Chamonix could be a window into little-known Indian discussions on nuclear security guarantees during the Geneva-based negotiations at the Eighteen Nations Disarmament Committee (ENDC), which led to the Nuclear Non-proliferation Treaty.[16] The crash took place only three days prior to the beginning of the Ninth Session of the ENDC (27 January–10 May 1966). V.C. Trivedi, India's ambassador to Switzerland, who was close to Bhabha, was India's representative to the ENDC. In February 1966, less than a month after the crash, Trivedi delivered his famous speech calling for nuclear security guarantees for India from the superpowers outside of alliance structures as a condition for signing the NPT. The documents in Chamonix may hold clues as to what discussions led to this action.[17]

Jayita Sarkar of Boston University reviewed these documents in mid-2017 during a meeting in Geneva with Françoise Rey, a Chamonix-based novelist, who has researched the crash of flight 101 for several decades. Rey possesses only a few official Indian documents—the rest are scattered across residential homes in Chamonix and elsewhere, all discovered by residents looking for 'treasures'. The crash sites are near Nid d'Aigle (the Eagle's Nest), famous as the last station on the tramway that takes tourists to the foothills of Mont Blanc in France. It is the point where the mountaineers start their ascent to scale the heights of the Alps. In August 2019, Prime Minister Narendra Modi inaugurated a memorial at Nid d'Aigle in honour of Bhabha and the victims of the two Air India crashes.

A CRUEL BLOW

In an interview with Indira Chowdhury, Jamshed Bhabha related that his mother could not get over the fact that Bhabha had actually been booked for a later flight but took the earlier one, which crashed. It was actually the other way around. If Bertoli's death was brought about by his refusal to cancel, Bhabha met his premature end because he changed his initial travel plans and postponed his departure by a day. Bhabha was to travel on the flight leaving Bombay on 22 January but cancelled at the eleventh hour. *The Statesman* reported as follows:

'Dr Bhabha [...] had booked his passage on the plane which left Bombay on Saturday, January 22, but he cancelled at the eleventh hour to travel by the ill-fated Boeing.'[18]

Jamshed, with the passage of time, perhaps got his dates mixed up, but there is no doubt that Bhabha was not supposed to travel on that ill-fated flight. Thereby hangs an unverified tale. R.M. Lala told me that it was widely believed at the time that Pipsy Wadia had been the reason why Bhabha had changed his flight at the last moment. He was unable to give a reason for the change but was certain that Meherbai held this against Pipsy for the rest of her life.[19] Jamshed took the call that brought the tragic news. He refused to believe it and could only say, 'No! No!' His wife Betty rushed to the telephone and took it from him. It was Betty, again, who did the talking when they broke the news to Meherbai.[20] Journalist Shyam Bhatia, who thought that Jamshed was jealous of his brother, says that Jamshed's resentment extended to Pipsy Wadia as well, and this can perhaps also explain the fact that Jamshed forcibly removed many of Bhabha's personal and professional papers from her home, threatening to burn them.[21]

On the day Bhabha died, B.V. Sreekantan was at MIT. He recalled that he was scheduled to participate in a colloquium on a topic in which Bhabha had great interest—antimatter. That morning, one of his colleagues, Professor George Clark, telephoned him and asked if he had heard what had happened. Television and radio reports confirmed the news. Immediately after, Sreekantan rushed to meet Professor Phil Morrison, who was in charge of the colloquium, and told him, 'Sir I am in such a mood that I can't give this talk today, it has shocked me beyond control.' Professor Morrison said, 'Look Sreekantan, you don't give up the talk, dedicate the talk to his memory. I am sure you will do well.' Sreekantan replied, 'Well in that case I will try.'[22]

Sreekantan did give the talk, which was well received. Then, he rushed home, had a very quick dinner and went to bed with tears in his eyes. That night, he had a strange dream: he saw Bhabha dressed in a traditional Parsi *dagli* (white closed-collar coat) and cap driving away in a golden chariot drawn by four white horses high above in the sky. Sreekantan said that the dream brought him immediate relief, and he woke up his wife to tell her about it. He said that perhaps this dream had been subconsciously influenced by his reading of the Bhagavad Gita: Bhabha was his Krishna, and he was Bhabha's Arjuna.[23]

The computer scientist, S. Ramani, who joined TIFR as a research associate and was the first director of the National Centre for Software Technology, recalled encountering N.R. Puthran wearing trousers instead of the customary shorts he wore on a holiday. They were at the East Canteen and when he asked him about it, he was told the grim news about the crash. It was a huge personal loss. Ramani remembered seeing Bhabha standing just 10 ft away from Puthran a few days ago. Bhabha had been furious because orders had been given for some trees to be cut.[24]

CONDOLENCES AND TRIBUTES

It is not often that the news of a scientist's death relegates a prime minister's swearing in to the middle of the page but that is what happened in Bhabha's case. Indira Gandhi was sworn in as prime minister the day Bhabha perished on the Alps, and it was the plane crash and his demise that made the headlines the next day. Expectedly, tributes and condolences poured in from all quarters, some perhaps less eloquent than others, but there was no doubting their sincerity: all came from the heart.

Sarabhai, who succeeded Bhabha as chairman of the AEC, in his speech at the condolence meeting, held on 25 January at AEET before 9,000 employees, said:

> We meet with sorrow in our hearts to mourn the tragic death of Dr Homi Bhabha, who stood at this very place three days ago. We meet to pay homage to the creator of all that we see around us here, and to much else in the modern image of India.
>
> Most of us here who have come into contact with him in his later period have known Bhabha in a different role as a leader in science, as the man who after independence enjoyed great trust of Pandit Jawaharlal Nehru and who fulfilled this trust by creating a new confidence, a new image for modern India, an image in the most sophisticated areas of science [...] He was a strong advocate for the creation of a new culture in Indian science and

> Indian engineering—in Indian life, which did not perpetuate from a different era the norms and traditions of a bureaucracy [...] If, indeed, there was a pace-setter in modern India, Bhabha was this pace-setter...
>
> Bhabha demonstrated better than most other people I know that those who are trained in the thinking of abstract science are also those who can most successfully apply themselves to the practical tasks of the nation. And I feel that Bhabha, had he been spared this cruel fate, would have made even greater contributions to the development of modern India and to the solution of many problems in the difficult times that we are all going through today.[25]

A condolence meeting was held at TIFR on 25 January, where Rustum Choksi said that 'unremitting devotion to our work' would be the best tribute to Bhabha's memory. The offices of TIFR and AEET worked as usual because Bhabha, as noted before, had always deprecated the practice of stopping work when a high dignitary passed away. He personally considered that the best homage was to not stop work but to work even harder.[26]

Choksi, in his personal tribute on 31 January, observed:

> In the hour of his death his achievement in quantitative terms was so considerable that it had no parallel in India in the combined fields of science, technology and industry. Had he lived for even another decade it would have been colossal.
>
> There is one other touch of quality—the Homi Bhabha touch. Everything was conceived and finally executed in terms of beauty. The aesthetic sense triumphed and gave new meaning to all the monuments of his hand, whether it was an office chair, a piece of laboratory equipment, a hall of engineering, an atomic reactor and finally the whole Trombay establishment, which is a garden and landscape wonder of abiding loveliness, as much as it is the finest complex of science and technology that this country will ever know.[27]

Choksi said that it was only fitting that the ceremonial service should be held at TIFR—Bhabha's institute.

> What was more fitting than that on 26 January, the ceremonial service for him should be held in his own Institute, in that temple that he had built for other worship. There amid the utter serenity and dignity of the long colonnade and the vista of sea and landscape of green lawns, men and women of all faiths could do him their last homage [...] One loved him well but one did not know how much—till now.[28]

The grieving staff of TIFR passed a resolution in memory of Bhabha:

> His exemplary patriotism rarely found expression in words; it was in deeds that he strove, together with all his colleagues, young and old, to build with unremitting toil and exalted vision that nobler India in which skilled technology in the service of man would give the lowliest among us the beginnings of a decent life.[29]

The DAE adopted a resolution condoling his death. 'Homi Bhabha was one of the great experiences in the life of the country and in this sense, he will never die unless science itself ceases to exist.'[30] The scientific advisory committee of the IAEA, whose meeting Bhabha was to attend, observed a minute's silence as a mark of respect and tributes were paid by all its members, who knew Bhabha well. Some, like W.B. Lewis, were old friends. They included Sigvard Eklund, I.I. Rabi, V.S. Emelyanov, Sir William Penney (who wrote his profile for the Biographical Memoirs of Fellows of the Royal Society), L.C. Prado and Bertrand Goldschmidt. Rabi's remarks were particularly insightful: 'I feel that we are now mourning the loss of a very unique personality [...] More and more we find people concentrated on their specialities and perhaps never again will there be a man of such broad culture, such broad vision [...] a man who spanned the whole range between Asian and western culture.'[31]

Patrick Blackett—who was, at the time president of the Royal Society—in his tribute broadcast over BBC said that apart from his scientific achievements, Bhabha was 'the most cultivated man of his time'. Referring to their long friendship, he observed:

> To me after twenty years of close friendship, his death brings overwhelming personal grief. Just a short time ago we discussed

many things in his office and he told me of his plans to develop the electronics industry of India relying much on the cadres of trained people from the Atomic Energy Establishment. Now his plans will have to be implemented by others.[32]

The *uthamna* ceremony for Bhabha was held on 26 January at TIFR at the foot of the library stairs with the entire institute gathered outside in the colonnade watching through the glass doors. Since non-Parsis are not allowed to attend Parsi religious ceremonies, the mourners watched from the outside at Meherbai's request. There were nine priests performing the funerary service, an indication of the stature of the departed person. Normally, two suffice. It is not known how comfortable the scientists were about having a religious ceremony in a secular and scientific institution. In all probability, it was agreed to in deference to Meherbai's wishes.

Meherbai had been quite closely associated with the institute in its early days. Even later, she would converse with scientists and staff who visited the Bhabha home on Malabar Hill. They recalled her gentle interventions, and after Bhabha's demise many visited her to offer their condolences. When Suresh Sawant, who worked at the TIFR workshop, went to offer his condolences, Meherbai told him, 'Homi told me he would return soon this time, now he will never come back.' He also recalled how Bhabha's dog stopped eating and ultimately died after pining away for its master. Sawant had walked the dog as a child, since his father used to work in the Bhabha household.[33]

Meherbai was close to M.G.K. Menon and his family. She gave him three gifts, all related to Bhabha: the first was a set of white dinner plates with speckled green borders which she used when she and Bhabha would share a meal together. The second was the original sketch that Bhabha had made of Niels Bohr—he had told his mother that he had made this sketch when Menon had been talking to the senior scientist on the lawns of the OYC. He had wanted to sketch both but could complete only Bohr's sketch. The third gift was an antique bronze statuette which could be seen behind Bhabha in several photos of him in his study at home. It was inscribed in memory of 'our Homi' in her handwriting.[34]

J.R.D. Tata, who presented a bust of Bhabha to TIFR exactly a year later, said that Homi Bhabha was, undoubtedly, the most outstanding of all the distinguished men he had known. He added that Bhabha was in the mould of Jamsetji Tata, a visionary with the boldness and energy to convert his vision into reality.

> Homi was one of those who made me believe that some men in human history are born with the stamp of predestination on them which leads them to accomplishments beyond ordinary human capabilities. Some of them—and Homi, alas, was one—are also predestined to die young, an unconscious premonition which drives them to superhuman effort to complete their task in the short time allowed to them [...] Scientist engineer, master-builder and administrator, steeped in humanities, in art and music, Homi was a truly complete man.[35]

Later, he wrote in the foreword of the silver jubilee volume of TIFR:

> The Institute's task will never end, for the frontiers of knowledge will always lie ahead. A relentless pursuit of excellence has characterized and indeed made possible the Institute's achievements, progress and international recognition. May that spirit, instilled in all those who worked with Homi Bhabha, be ever maintained and inspire in the years to come even those to whom he will be but a legend.[36]

J.R.D. Tata believed that Bhabha's greatest contribution lay in training a host of young scientists and administrators who were leaders of scientific establishments in the country. M.G.K. Menon voiced a similar sentiment: 'This group of [...] trained personnel, the scarcest commodity in a developing nation, products of Homi Bhabha's inspiring and warm leadership, are his richest legacy to India.'[37]

The Guardian in its obituary message on 25 January 1966 echoed that view:

> Indeed, no other single man could walk the path which Bhabha has, not because his stature was too large but because the circumstances which made him no longer exist. Five years ago, the loss of Homi J. Bhabha would have been an irreparable disaster for India's technology. Today the loss is immeasurable, yet it can

be seen that with characteristic foresight Dr Bhabha provided for the eventuality.[38]

British civil servant and diplomat, Baron Redcliffe-Maud, said:

> Affectionate and sensitive, elegant and humorous, dynamic, one of the very few people who enhance life whatever the content of their living—fantastically talented but so fastidious about standards that he was never a dilettante. Whatever he set himself to do, he did as a professional—but one who worked for love; restlessly creative, enhancing life because he loved all forms of it. So, he became a living proof that scientific excellence can go with excellence in art, and racial differences need be no bar to friendship [...] Those qualified must judge how grievous is his death for India and for science and for civilization. His friends know that for them it is irreparable loss.[39]

As M.G.K. Menon put it, 'Homi Bhabha died at the peak of his great powers, a legendary figure in his own lifetime, working towards the end with an urgency which had to be seen to be believed, working against time when he felt he had so much to accomplish [...] He achieved so much and yet had so much more, of grand designs for India.'[40] John Cockcroft observed, 'Human progress has always depended on the achievements of a few individuals of outstanding ability and creativeness. Homi Bhabha was one of these.'[41]

In the words of Cecil Powell:

> Homi was surely, in a sense, a man out of his time. He had the wide breadth of interests, the penetrating intellect, the abounding personal confidence in the ability of the human understanding, by observation and experiment, to unravel the secrets of nature, which are characteristic of the great figures of the Renaissance. But perhaps such figures have an essential creative role to play in all ages.[42]

Mulk Raj Anand, writing in *Marg*, observed:

> ...when the history of our intelligentsia comes to be written, the world will know how many departures from orthodox positions

> you made and how many innovations you carried out. I say in anticipation of such a summing up that you were one of the few possible 'Whole Men' of our time.[43]

The Government of India published a notification with a black border on 1 February 1966 in the *Gazette of India* mourning Bhabha's death. Four days later, on 5 February, the AEC adopted a condolence resolution. President S. Radhakrishnan and Prime Minister Indira Gandhi also sent their tributes. She said:

> To lose him at this critical moment in the development of our atomic energy programme is a blow to our nation. When we take up the unfinished work he has left behind, we will realize in how many fields he served us. I shall miss his wide-ranging mind and many talents, his determination to strengthen his country's science and his enthusiastic interest in life's many facets.[44]

ENTER SARABHAI

Vikram Sarabhai succeeded Bhabha as chairman of the AEC, but he was not the first person to be offered that post. When Bhabha died, the prime minister's office had offered the post to S. Chandrasekhar, not realizing that he was an American citizen: it was an issue that her office never even considered. When Chandrasekhar came to deliver the second Nehru Memorial Lecture in November 1968, he requested for a special interview with Indira Gandhi to explain why he had declined the offer. After he had narrated the facts to her, she said, 'Well, all this was not made clear to me.'[45]

Within a span of a month, India lost both its political as well as scientific leadership. Sarabhai openly argued against a nuclear weapons programme and was unsympathetic to the idea of a PNE. In some ways, Sarabhai's choice as Bhabha's successor was a curious one because he was a peace-loving Gandhian who would be expected to strongly oppose developing a nuclear explosive device even for peaceful purposes. Indira Gandhi was beset with many problems at that time, and it is doubtful that she was briefed on the extent to which Bhabha had progressed in the 'peaceful' explosive project.

Unlike Bhabha, Sarabhai thought that a nuclear weapons programme would jeopardize India's economic development, since it would impact the utilization of national resources for productive purposes. He was also of the view that merely exploding a bomb did not constitute nuclear deterrence. It required a total defence system, a means of delivery, long range missiles and much else, which would require national resources of a huge magnitude. He put an end to the theoretical exercise of developing PNE technology at least formally.[46]

Sarabhai questioned the morality and utility of nuclear weapons and took steps to reverse the peaceful nuclear explosives project. While he did not completely reject the idea of nuclear weapons for India, he apparently did not endorse the particular PNE that Bhabha and other DAE scientists had in mind. He was also against gimmicks and wanted the emphasis to be on the reality and not a mere show. M.V. Ramana has observed that this was at variance with the importance given to performative gestures by Bhabha and Nehru. Sarabhai evaluated the bomb in concrete economic and military terms; it carried no symbolic meaning for him.[47] The PNE effort continued despite Sarabhai's attempts to shift the focus. The normal autonomy accorded to scientists in their research helped the bomb makers. It was not as if Sarabhai was unaware of what lay in store. He is believed to have told the then cabinet secretary, Dharam Vira, 'It is not going to be easy to succeed Bhabha. It is not only taking over a job, it is also taking over a thinking.'[48]

Sarabhai also had to deal with Homi Sethna, who considered himself Bhabha's rightful successor and lobbied ferociously for the position. When he failed, he adopted a bitterly adversarial position throughout Sarabhai's tenure. Sethna claimed that on assuming office Sarabhai had wound up the SNEPP and asked for all the papers to be handed over to him. Sarabhai, who usually avoided confrontation and dealt with aggressiveness through restraint, would certainly have tried to make an ally of Sethna. However, the pugnacious Parsi rebuffed him, nursing his sense of hurt at missing out, needling and opposing him whenever he got the chance.

Raja Ramanna, in a private interview, acknowledged that Sarabhai could not keep the scientists from doing their work.[49] Thus, it has been suggested that the AEC scientists continued to work on the bomb, either surreptitiously or openly. Sethna, who was heading BARC during this time, expressly asked for Indira Gandhi's permission to finish the work undertaken during the Bhabha–Shastri compromise. P.N. Haksar, who was the secretary to the Prime Minister, advised her to give a decision in this connection and she did. This, however, did not translate into a decision to conduct a nuclear test since the Prime Minister's stance was that India would not produce a nuclear weapon.[50]

Design work on the nuclear explosive tested at Pokhran began in 1968. Under the watchful eyes of Ramanna and R. Chidambaram, and with the cooperation of B.D. Nagchaudhuri, director of DRDO, about 50 to 75 scientists were directly involved in the nuclear explosives project. Sarabhai certainly must have known about this but chose to ignore it. George Perkovich was told by a former atomic energy official that Sarabhai was too much of a gentleman to stop it and did not wish to enter this particular fight.[51]

Ramanna acknowledged that the AEC had been created not merely to generate nuclear electricity; its aims were explicitly to develop 'atomic energy for all purposes'.[52] M.R. Srinivasan, who headed the DAE in the 1980s, clearly stated the commission's view that nuclear technology was to be developed by a country for its own benefit, whether for peaceful or military purposes.[53] By then, India was taking up an increasingly critical stance in the negotiations for a non-proliferation treaty.

Representing India at the Geneva discussions was V.C. Trivedi, who argued that if non-weapons states were expected to forgo developing nuclear weapons, the weapons states should freeze all work on the development of further nuclear weapons and their delivery systems. This call for universal nuclear disarmament embarrassed both the US and USSR. Negotiations on the non-proliferation treaty took place between 1965 and 1968, and it was signed on 1 July 1968. India remains a non-signatory.

Like Shastri, Sarabhai's opposition to the development of a 'peaceful' nuclear device weakened with time. In a speech at the Fourth International Conference on the Peaceful Uses of Atomic Energy in Geneva in September 1971, he announced that his scientists were giving priority to the development of PNE. Unfortunately, a few months later, on 30 December 1971, he died suddenly of a heart attack when he was just 52. His successor, Homi Sethna, did not share his predecessor's aversion to nuclear explosions. During his tenure, India exploded its first nuclear device on 18 May 1974 in Pokhran, two-and-a half years after Sarabhai's death.[54]

There is an amusing aside to Ramanna's association with the bomb, though it certainly must not have been amusing for him at the time. Four years after the Pokhran blast, the Iraqi dictator Saddam Hussein approached Ramanna, asking for help to build a nuclear weapon. The offer came when Ramanna was in Baghdad as Hussein's personal guest. Ramanna was given a tour of Baghdad and Iraq's main nuclear facility at Tuwaitha. At the end of the trip, Hussein invited him to his office and asked him to take over Iraq's nuclear programme. 'I will pay you whatever you want,' Saddam told him. A very frightened Ramanna took the next flight out of Iraq.[55]

◆

Bhabha's office was on the fourth floor of TIFR, overlooking the sea. It was from a window of this office that he had directed the creation of the famous Amoeba Garden on the west lawn of the institute. It was also here that he used to meet faculty and other visitors who had to wait for what seemed to be an interminable length of time to see him. After his death, his office remained unoccupied for more than three decades.

M.G.K. Menon, who had vivid memories of working late with Bhabha in that office was unwilling and emotionally unprepared to occupy his mentor's office. Sreekantan, who succeeded Menon, also did not occupy the office despite J.R.D. Tata's insistence that he do so. However, the room was open for council and other meetings. Jamshed Bhabha also thought that TIFR directors should sit in what had once been his brother's room. However, the leadership of TIFR chose to leave it unoccupied.[56]

After Sreekantan, Virendra Singh, too, did not occupy Bhabha's office. It was only in 1997, when S.S. Jha became director that he occupied Bhabha's office. He decided to turn Bhabha's office into a museum, relocating furniture and other objects to the museum area in the auditorium foyer. His desk, bookshelves and Eero Saarinen Tulip chair (an icon of mid-century modern furniture design) were on display and could be viewed through a glass partition. Jamshed Bhabha also thought that this was a good idea. The office was renamed as the Homi J. Bhabha Room. In 2009, when the institute celebrated Bhabha's birth centenary, the museum-like display of the office was dismantled and the furniture returned to Bhabha's office on the fourth floor.[57]

EPILOGUE

Homi Bhabha's services to the nation were commemorated in various ways. The following does not purport to be an exhaustive list and is merely indicative. In a rare gesture the Ministry of Communications released a special 15 paise stamp on 4 August 1966—the tenth anniversary of Apsara. A few months later, in January 1967, Prime Minister Indira Gandhi renamed AEET as the Bhabha Atomic Research Centre. The auditorium and museum complex being constructed at TIFR was also inaugurated by her in 1968 and named after Bhabha, who had conceived this complex and its basic design. Cecil Powell delivered the first lecture at the new auditorium. The road in Navy Nagar leading to TIFR has been named Homi Bhabha road in his memory.

The Homi Bhabha Fellowships Council was established on his first death anniversary; on his 19th death anniversary, the Homi Bhabha Fellowship programme was launched to give young talent the opportunity to develop their abilities through study, research and travel. The current council includes Anil Kakodkar and former Homi Bhabha fellow and film director, Shyam Benegal.

The TIFR has directly or indirectly played a role in the formation of six other institutions.[1] The Homi Bhabha Centre for Science Education (HBCSE) is a national centre of TIFR. Its broad goals are promoting equity and excellence in science and mathematics education from the primary school level to the undergraduate college level, and encouraging the growth of scientific literacy in the country. The HBCSE is the premier institution in the country for research and development in science, technology and mathematics education. It is India's nodal centre for Olympiad programmes in mathematics, physics, chemistry, biology, astronomy and junior science. It also runs a Graduate School in Science Education, where students work towards a PhD degree at TIFR, which is a deemed university. The Homi Bhabha National Institute is a deemed university established by the DAE, which unifies academic programmes of several of its constituent institutions.

The National Centre for Radio Astrophysics has its roots in the Radio Astronomy Group of TIFR, which was set up in the early 1960s by Govind Swarup. In the early 1980s, an ambitious plan for the GMRT telescope was proposed. Since the site chosen for this new telescope was close to Pune, a new home for the group was built in the campus of the Savitribai Phule Pune University. The TIFR Centre for Applicable Mathematics in Bangalore is a part of the School of Mathematics at TIFR, Mumbai. The centre arose and evolved since the mid-1970s to develop areas in applicable mathematics. Many renowned mathematicians from India and abroad have contributed to the development of the centre. The International Centre for Theoretical Sciences of TIFR was established in 2007 as a multidisciplinary and interdisciplinary centre. The TIFR has also started a new campus in Hyderabad, adjacent to Hyderabad University, to nurture research and higher education in the region and the TIFR Centre for Interdisciplinary Sciences is the first centre in Hyderabad.

The Dr Homi Bhabha State University, the first cluster university in Maharashtra, was established in February 2019. It comprises the four oldest and most prestigious government colleges of Mumbai, including Bhabha's old alma mater, the Royal Institute of Science, which later became the Institute of Science, Mumbai. The other three institutions are Elphinstone College, Sydenham College and the Secondary Training College. The Institute of Science is the leading institute of the university. The university is an independent degree-granting entity, with its own rules and regulations. In order to give recognition and opportunities to outstanding scientists and engineers, including those who have retired, the DAE has also established the Homi Bhabha Chair for Distinguished Scientists.

Would Bhabha have ever imagined that a bulletproof vest would be named in his memory? Probably not, but, in 2020, a BARC scientist won the Shanti Swarup Bhatnagar Prize for Science and Technology for his contributions to the large-scale synthesis of carbon nano-tubes, leading to the development of a light-weight bulletproof jacket known as the Bhabha Kavach, which was unveiled in New Delhi in July 2019.

NUCLEAR POWER: LIMITED SUCCESS

What about nuclear power generation and all the early optimism associated with it? It is beyond the scope of this brief epilogue to describe or critically analyse India's track record, but indisputably India's atomic energy programme has not delivered on its tall promises. Even 75 years after Independence, nuclear power constitutes only about 3 per cent of the country's electricity generation capacity. Also, the limited amount generated has been relatively expensive. The DAE's plans have been criticized, among others, by M.V. Ramana, who believes that the various components of the nuclear establishment's plans have serious flaws. Fast breeder reactors are unreliable and difficult to construct and operate. None of the assumptions motivating breeders have proved true.[2]

Many countries, the most striking example being France, have cut back, if not completely abandoned plans involving fast breeder reactors. France, while remaining committed to nuclear power, has had a relook at fast breeder reactors. India has, however, remained loyal to Bhabha's legacy of the three-stage programme, which has never been questioned. Ramana believes that the institutional commitments to the three-stage programme are so great that the DAE will find it impossible to abandon it.[3]

In the second stage, fast breeder reactors are designed to breed more fuel than they consume. India built a fast breeder test reactor at Kalpakkam in Tamil Nadu in 1985 and has been operating it ever since. A prototype fast breeder reactor (PFBR) with 500 MWe capacity has been constructed in Kalpakkam, and the commissioning of individual systems is underway.[4] More such reactors are in the pipeline. The plan for the third stage is to set up nuclear reactors employing thorium-based (thorium plus uranium-233) fuel. Research and technology development are underway for the third stage and commercial electricity generation is expected only in the second half of this century. As part of this programme, a 300 MWe advanced heavy water reactor fuelled by thorium and using light water as a coolant and heavy water as a moderator is being developed.[5]

There are other problems associated with nuclear power: cost overruns, unanticipated problems in construction and local opposition to the acquisition of new sites. If past experience is anything to go by, public health and environmental concerns are other problems that need to be factored into any expansion plan. The safety record, too, has been patchy at best and the secrecy maintained by the DAE makes independent studies impossible.

The extremely high capital costs and the long construction period with no returns means that investment in nuclear power imposes significant opportunity costs. According to the DAE report for 2020–2021, 6,780 MW of power is generated by 22 reactors, which are managed by the Nuclear Power Corporation of India Ltd.[6] The 700 MW Unit III at Kakrapar, Gujarat, was connected to the grid in January 2021. In May 2021, Électricité de France made an offer to build six third-generation European Pressurized Reactors in Jaitapur. On completion this will be the world's biggest nuclear plant and is projected to produce 10 GW of electricity.[7] There are currently 10 reactors (including 500 MW PFBRs) totalling 8,000 MW at varying stages of construction. The government has also approved the construction of 10 indigenous 700 MW PHWRs to be set up in fleet mode. On completion of these projects, the total nuclear capacity of the country is expected to reach 22,480 MW by 2031.

◆

Jamshed Bhabha, who inherited Mehrangir and everything in it, willed his estate to the NCPA. After his death, in 2007, all the priceless paintings and other artefacts in Mehrangir were auctioned to raise money. In 2014, the Bhabha home was bought by Smita Crishna-Godrej for ₹372 crore. The bungalow was demolished two years later, with reports of a high-rise being planned in its place. This move was met with much displeasure by the Indian scientific community. There were petitions for converting Mehrangir into a Bhabha memorial, but these could not prevent the sale.

However, Bhabha's dream can never be demolished. The epitaph of Sir Christopher Wren, the architect of St Paul's Cathedral in London, is in a great circle on the floor directly under the dome. It says, '*Lector, Si Monumentum Requiris Circumspice*', which translates as 'Reader, if you seek his memorial look around you.' If a visitor looks around, he will see not only the Wren crypt but also the glorious monument he created. The same applies to Bhabha, except that for him, TIFR, BARC and the institutions they have spawned are his monuments, all as enduring as Wren's cathedral.

ACKNOWLEDGEMENTS

A book is made possible by many institutions and people. Inevitably in a venture such as this, one is indebted to those whose work has preceded mine. This biography of Homi Bhabha is based on the work of a large number of scholars and authors apart from, of course, Bhabha's correspondence and a plethora of archival records. My bibliography illustrates the full scale of my debt, but certain books belong on a personal roll of honour, since my debt to them has been especially heavy—among whose pages I have pillaged, precised and paraphrased shamelessly. It is only appropriate that I acknowledge these with a special mention.

Indira Chowdhury and Ananya Dasgupta's excellent *A Masterful Spirit: Homi J. Bhabha, 1909–1966* was the book that, in a sense, started it all, since the seed for writing this biography was sown after I read their book. If it was involved in the seeding of this book, it has also been one of the foundations for this work. Apart from providing a glimpse into Bhabha's life and work, it has proved to be immensely useful in pointing the way to relevant correspondence and documents connected with him. It is, in fact, a treasure trove of archival material and I have used this material freely in my biography (with, I hope, no failure in citing the source). In many cases, I have resorted to taking a shortcut and quoted from it in cases where I did not have access to the original archival material.

I have known Indira Chowdhury for over a decade now and her role in the making of this book can hardly be overestimated. I am beholden to her not only for providing a wonderful foundation through her numerous articles and interviews but also through her book

Growing the Tree of Science: Homi Bhabha and the Tata Institute of Fundamental Research, a detailed and scholarly account of how Bhabha built his TIFR. I have had no compunction in precising and paraphrasing with, what I believe to be, a judicious licentiousness but not unforgiveable promiscuity and can only hope that this acknowledgement is adequate compensation for her intellectual labour. As with *The Masterful Spirit*, this book, too, has provided invaluable archival material that I have not hesitated to use in my work. Apart from this, I will treasure the many conversations I have had with her on the subject that provided both interesting trivia and also pointed the way to further material.

I owe a debt of gratitude to Jahnavi Phalkey for her book *Atomic State: Big Science in Twentieth-Century India*, which has also been extremely useful in the writing of this biography. Many other books and articles have also been used but I owe an especially heavy debt to some and they deserve a special mention.

The book owes much to Itty Abraham's *The Making of the Indian Atomic Bomb: Science, Secrecy and the Post-Colonial State*, Robert S. Anderson's *Nucleus and Nation: Scientists, International Networks, and Power in India*, M.V. Ramana's *The Power of Promise: Examining Nuclear Energy in India*, Chintamani Deshmukh's *Homi Jehangir Bhabha*, Raja Ramanna's *Years of Pilgrimage: An Autobiography*, and M.R. Srinivasan's, *From Fission to Fusion: The Story of India Atomic Energy Programme* all of which have helped me in stitching this biography together. G. Venkataraman's slim volume on Bhabha, which combines science with biographical details in a masterful fashion, was indispensable as was *Journey into Light: Life and Science of C.V. Raman*, his full-length biography of C.V. Raman, which provided useful background information.

My greatest debt, however, is to the DAE's Public Awareness Division and its head, Dr Rajesh Vatsa, who provided archival material, documents, photographs, books, magazines, reports and articles with amazing dispatch over many months. The facility of being provided material on tap is best appreciated by those who have had to encounter delay and frustration due to non-availability of material or bureaucratic obstruction. His assistance has been inestimable and this book could not have been written without him. I will also greatly value the conversations I had with him on Bhabha. I would also like to thank Satish Iyer, Dr Vatsa's junior colleague, who provided books from the DAE library and supplied soft copies of the photos and other material.

The importance of a senior bureaucrat, Sanjay Kumar, in this venture can hardly be overestimated. He not only put me in touch with Dr Vatsa but also provided crucial help in accessing the TIFR archives. This brings me to another key player, Professor Sudipta Maiti of TIFR who deserves my grateful thanks for the material provided by the TIFR archives. I cannot thank him enough for his help. I am also grateful to Professor Shiraz Minwalla of TIFR who not only spoke to me about E.C. George Sudarshan but also treated me to a fine dinner in the process.

Help often comes through a chain of interconnecting persons and institutions. The first link in this chain was my long-standing friend and colleague Reena Ranjan, who in a sense started the entire process when she put me in touch with Sanjay Kumar. Thanks, Reena, for your help and for being such a good friend all these years. Anupam Kapoor, another close friend and colleague, alarmed at my non-existent scientific credentials, not only pointed the way to interesting and useful books on popular science but also helped on occasion by downloading them from the internet. All these have greatly helped in providing the background knowledge and understanding essential to a neophyte like me for making a credible attempt at putting together a book on Bhabha's life and work. Navroz Seervai, Homi Seervai's son, generously allowed me to quote from the letters that Bhabha had written to his father. I owe him my grateful thanks.

I would be remiss if I did not mention the support and encouragement given to me by Dr K.N. Vyas, chairman of the AEC and secretary, DAE. Not only did he spare time from his busy schedule to meet me but also spoke at length about Bhabha and his contributions, sharing in the process his own anecdote about a distant family member's connection with Bhabha, which finds mention in the book. I cannot thank him enough for his kindness and courtesy.

Gratitude is also due to all those who oiled the wheels of progress.

Rupa's dynamic managing director Kapish Mehra, who commissioned this book, was enthusiastic about it from the start and brought his usual energy to the venture. My relationship with the Mehra family is an old one and one which I will always cherish. I greatly look forward to any interaction with both Kapish and his father, the generous and genial R.K. Mehra, whose wisdom still guides the fortunes of his esteemed publishing house.

Thanks are due to Rudra Sharma and Nishtha Kapil at Rupa Publications, who played an important role in making this book a reality. Sneha Bhagwat's diligent and conscientious editing has added polish, grace and clarity to many rough edges and greatly improved the book. While all efforts have been made to ensure that the facts are accurate and that no citation has been inadvertently missed out, I will be happy if errors, which I hope are not too numerous, are brought to my notice. In a letter to his friend, Benjamin Franklin said about his life, 'I shall not object to a new edition of mine; hoping however that the errata of the last may be corrected.' I feel the same about this book.

It is important for a writer to have his text checked by a discerning reader, for often what appears to be crystal clear to the author is confusing to the reader. My nephew Navroze Kotwal, proved to be an excellent sounding board. He shared my enthusiasm for the book and was invaluable in offering useful perspectives and making insightful suggestions. Thanks are also due to Ashish Mahajan and Bhushan Bandi, who played the role of assistants to perfection. I also thank my mother Rati, my sister Tushna, and brother-in-law Phiroze, who not only shared me with Homi Bhabha all these years but also provided moral support in a venture where I was often beset by self-doubt. Their encouragement and optimism have proved to be crucial.

This book is dedicated to the memory of S.S. Tarapore and Russi M. Lala. The former, one of India's finest central bankers, was a mentor to me when I wrote *Barons of Banking*. The latter, J.R.D. Tata's official biographer, generously allowed me to draw from his work when I was working on *Jeh: A Life of J.R.D. Tata*, my own book on the former Tata chairman. A warm friendship gradually developed and, after a few years, he read my profile on Homi Bhabha in *Sugar in Milk*, my book on eminent Parsis. He said it was well-written and expressed the hope that perhaps one day I should write a biography of Homi Bhabha. That idea held no appeal for me at the time, but life is a funny thing, and now, 13 years later, I find myself having delivered the manuscript of a very lengthy biography of Bhabha to the publisher. I wonder if Russi, whose writing was crisp and concise, would have approved.

NOTES

Chapter 1: Before Homi Bhabha

1 Dasgupta, Subrata, *Awakening: The Story of the Bengal Renaissance*, Penguin Random House, Gurugram, 2019, p. 405.

2 Parameswaran, Uma, *C.V. Raman: A Biography*, Penguin Books, New Delhi, 2011, p. 48.

3 Venkataraman, G., *Journey into Light: Life and Science of C.V. Raman*, Indian Academy of Sciences, Bangalore, 1988, p. 26.

4 Ibid. 38.

5 Ibid. 36.

6 Ibid. 37.

7 Parameswaran, Uma, *C.V. Raman: A Biography*, Penguin Books, New Delhi, 2011, p. 89.

8 Masters, Barry, 'C.V. Raman and the Raman Effect', *Optics and Photonics News*, Vol. 20, No. 2, 2009, pp. 26–31, https://tinyurl.com/2p9f34f7. Accessed on 22 July 2022.

9 Jayaraman, A., *Chandrasekhara Venkata Raman*, Indian Academy of Sciences, Bangalore, 2017, p. 22.

10 Ibid. 23.

11 Ramaseshan, S., 'A Conversation with Krishnan on the Story of the Discovery of the Raman Effect', *Current Science*, Vol. 75, No. 11, 1998, JSTOR, https://tinyurl.com/mr42d6t4. Accessed on 22 July 2022.

12 Mallik, D.C.V. and S. Chatterjee, *Kariamanikkam Srinavasa Krishnan: His Life and Work*, Universities Press, Hyderabad, 2012, p. 87.

13 Ibid. 87.

14 Ramaseshan, S., 'A Conversation with Krishnan on the Story of the Discovery of the Raman Effect', *Current Science*, Vol. 75, No. 11, 1998, JSTOR, https://tinyurl.com/mr42d6t4. Accessed on 22 July 2022.

15 Wali, Kameshwar C., *Chandra: A Biography of S. Chandrasekhar*, Penguin Books, New Delhi, 1992, p. 251.
16 Ibid. 251.
17 Sen, S.N. (ed.), *Professor Meghnad Saha: His Life, Work and Philosophy*, Meghnad Saha Sixtieth Birthday Committee, Calcutta, 1954.
18 Banerjee, Somaditya, *Bhadralok Physics and the Making of Modern Science in Colonial India*, 2018, The University of British Columbia, Vancouver, PhD thesis.
19 Venkataraman, G., *Saha and His Formula*, Universities Press, Hyderabad, 1995, p. 6.
20 Saha, Meghnad, 'Ionization in the Solar Chromosphere', *Philosophical Magazine Series* 6, Vol. 40, No. 238, pp. 472–88, 1920.
21 Chatterjee, Santimay, *S.N. Bose: The Man and His Work*, S.N. Bose National Centre for Basic Sciences, Calcutta, 1994, p. 312.
22 Mehra, Jagdish, 'Satyendra Nath Bose, 1 January 1894–4 February 1974', The Royal Society Publishing, https://tinyurl.com/2ne769ee. Accessed on 8 August 2021.
23 Wali, Kameshwar C., *Chandra: A Biography of S. Chandrasekhar*, Penguin Books, New Delhi, 1992, p. 251.
24 Venkataraman, G., *Journey into Light: Life and Science of C.V. Raman*, Indian Academy of Sciences, Bangalore, 1988, p. 58.
25 Mallik, D.C.V. and S. Chatterjee, *Kariamanikkam Srinavasa Krishnan: His Life and Work*, Universities Press, Hyderabad, 2012, p. 139.
26 Ibid. 150–51.
27 'An Indian Academy of Science', *Current Science*, Vol. 1, No. 11, 1933, pp. 335–37, JSTOR, https://tinyurl.com/3mdckfue. Accessed on 22 July 2022.
28 Anderson, Robert S., *Nucleus and Nation: Scientists, International Networks, and Power in India*, University of Chicago Press, 2010, p. 64.
29 Venkataraman, G., *Journey into Light: Life and Science of C.V. Raman*, Indian Academy of Sciences, Bangalore, 1988, p. 273.
30 Seshadri, T.R., 'Shanti Swarup Bhatnagar, 1894-1955', The Royal Society Publishing, https://tinyurl.com/3pd4rydm. Accessed on 28 August 2021.
31 Capri, Anton Z., *From Quanta to Quarks: More Anecdotal History of Physics*, World Scientific Publishing Company, 2007, p. 9.
32 Anderson, Robert S., *Nucleus and Nation: Scientists, International Networks, and Power in India*, University of Chicago Press, 2010, p. 50.
33 Seshadri, T.R., 'Shanti Swarup Bhatnagar, 1894-1955', The Royal Society Publishing, https://tinyurl.com/3pd4rydm. Accessed on 28 August 2021.
34 Abraham, Itty, *The Making of the Indian Atomic Bomb: Science, Secrecy and the Post-Colonial State*, Orient BlackSwan, Hyderabad, 1999, p. 62.
35 Ramana, M.V., *The Power of Promise: Examining Nuclear Energy in India*, Penguin Books, New Delhi, 2012, p. 7.

Chapter 2: The Heritage

1 Mehta, Prochy N., *Who is a Parsi?*, Niyogi Books, New Delhi, 2022, p. 64.
2 Cama, Shernaz, *Threads of Continuity: Zoroastrian Life & Culture*, Parzor Foundation, 2016, p. 126.
3 Dadabhoy, Bakhtiar K., *Sugar in Milk: Lives of Eminent Parsis*, Rupa & Co., New Delhi, 2008, p. xiii.
4 Sharafi, Mitra June, *Bella's Case: Parsi Identity and the Law in Colonial Rangoon, Bombay and London,1887-1925,* 2006, Princeton University, PhD dissertation.
5 Manuel, Thomas, *Opium Inc.*, HarperCollins, Noida, 2021, p.xii.
6 Ibid. xii.
7 Dadabhoy, Bakhtiar K., *Sugar in Milk: Lives of Eminent Parsis*, Rupa & Co., New Delhi, 2008, p. 5.
8 Farooqui, Amar, *Opium City: The Making of Early Victorian Bombay*, Three Essays Collective, Gurugram, 2006, pp. 17–8.
9 Palsetia, Jesse S., *The Parsis of India: Preservation of Identity in Bombay City*, Manohar Books, New Delhi, 2008, p. 31.
10 Fee, William Thomas, 'The Parsees and the Towers of Silence at Bombay, India', *The National Geographic Magazine*, Vol. 16, No. 12, 1905, https://tinyurl.com/5n67tycr. Accessed on 26 July 2022.
11 Palsetia, Jesse S., *The Parsis of India: Preservation of Identity in Bombay City*, Manohar Books, New Delhi, 2008, p. 195.
12 Patel, Simin, *Cultural Intermediaries in a Colonial City: The Parsis of Bombay c. 1860-1921*, 2015, University of Oxford, DPhil dissertation.
13 Palsetia, Jesse S., *The Parsis of India: Preservation of Identity in Bombay City*, Manohar Books, New Delhi, 2008, p. 64.
14 Lala, R.M., *For the Love of India: The Life and Times of Jamsetji Tata*, Penguin Viking, New Delhi, 2004, pp. 16–7.
15 Ibid. 18.
16 Manuel, Thomas, *Opium Inc.*, HarperCollins, Noida, 2021, p. xviii.
17 Palsetia, Jesse S., *The Parsis of India: Preservation of Identity in Bombay City*, Manohar Books, New Delhi, 2008, p. 277.
18 Ibid. 318.
19 Ibid. 153.
20 Ibid. 153–54.
21 Patel, Simin, *Cultural Intermediaries in a Colonial City: The Parsis of Bombay c. 1860-1921*, 2015, University of Oxford, DPhil dissertation.
22 Palsetia, Jesse S., *The Parsis of India: Preservation of Identity in Bombay City*, Manohar Books, New Delhi, 2008, p. 155.
23 Park, Robert E., 'Human Migration and the Marginal Man', *American Journal of Sociology,* Vol. 33, No. 6, 1928, pp. 881–93, JSTOR, https://tinyurl.com/2p834tru. Accessed on 26 July 2022.

24 Bhabha, Homi K., *The Location of Culture*, Routledge Classics, New York, 1994, p. 85.
25 Ibid. 85
26 Weaver, David John, *The Parsi Dilemma: A New Zealand Perspective*, 2012, Victoria University of Wellington, PhD dissertation.
27 Keshavdev, V., 'Humata, Hukhta, Hvarshta', *Outlook Business*, 19 July 2014, https://tinyurl.com/vdah2fkn. Accessed on 24 August 2022.
28 Palsetia, Jesse S., *Jamsetjee Jejeebhoy of Bombay*, Oxford University Press, New Delhi, 2015, p. 101.

Chapter 3: A Scientist in Pupa

1 Chowdhury, Indira and Ananya Dasgupta, *A Masterful Spirit: Homi J. Bhabha, 1909-1966*, Penguin Books, New Delhi, 2010, p. 17.
2 Iyer, Meera, 'Homi Bhabha's IISc Stint and Science Research in India', *Deccan Herald*, 25 July 2020, https://tinyurl.com/2p9xwhfp. Accessed on 26 July 2022.
3 'Saklat vs Bella on 22 October, 1925', Indian Kanoon, https://tinyurl.com/mvu9s9n4. Accessed on 26 July 2022.
4 This is a thread ceremony that initiates one into the Zoroastrian faith.
5 In 1903, Ratanji Dadabhoy Tata (J.R.D. Tata's father) married a Frenchwoman, Suzanne Briere, under the Parsee Marriage and Divorce Act of 1865, after she converted to Zoroastrianism. R.D. Tata then claimed that his wife, who had been renamed Sooni, was entitled to participate in all religious events, including entering the fire temple. A suit was filed in 1906 against the trustees of the Parsi Panchayat in Bombay, and the principal question raised was whether a person who had converted to Zoroastrianism was entitled to the benefits of the religious institutions and funds managed by the panchayat. A Special Bench of the Bombay High Court ruled against the plaintiff.
6 Chowdhury, Indira and Ananya Dasgupta, *A Masterful Spirit: Homi J. Bhabha, 1909-1966*, Penguin Books, New Delhi, 2010, p. 12.
7 Deshmukh, Chintamani, *Homi Jehangir Bhabha*, National Book Trust, India, New Delhi, 2014, p. 4.
8 Chowdhury, Indira and Ananya Dasgupta, *A Masterful Spirit: Homi J. Bhabha, 1909-1966*, Penguin Books, New Delhi, 2010, p. 13.
9 Deshmukh, Chintamani, *Homi Jehangir Bhabha*, National Book Trust, India, New Delhi, 2014, p. 3.
10 Chowdhury, Indira and Ananya Dasgupta, *A Masterful Spirit: Homi J. Bhabha, 1909-1966*, Penguin Books, New Delhi, 2010, p. 233.
11 Ibid. 71.
12 Cockcroft, John and M.G.K. Menon, *Homi Jehangir Bhabha 1909-1966*, Royal Institution of Great Britain, London, 1967, p. 19.
13 'Our History', Bai Ruttonbai F D Panday Girls' High School, https://tinyurl.

com/nhj8vvt7. Accessed on 24 August 2022.

14 Lala, R.M., *For the Love of India*, Penguin Viking, New Delhi, 2004, p. 119.

15 Chowdhury, Indira and Ananya Dasgupta, *A Masterful Spirit: Homi J. Bhabha, 1909-1966*, Penguin Books, New Delhi, 2010, pp. 12–3.

16 Ibid. 12.

17 Menon, M.G.K., 'Tribute to Dr Jamshed Bhabha', *Annual Report, 2006–07*, The Trustees of the Sir Dorabji Tata Trust, Tata Trust, https://tinyurl.com/7eavk6jk. Accessed on 26 July 2022.

18 Bhatia, Shyam, 'Bhabha's Rare Assets Up for Auction in Mumbai', *The Tribune*, 13 October 2011.

19 Chowdhury, Indira and Ananya Dasgupta, *A Masterful Spirit: Homi J. Bhabha, 1909-1966*, Penguin Books, New Delhi, 2010, p. 14.

20 Ibid. 14.

21 The school became co-educational in 1967. In 1922, it consisted of three schools: (1) The Cathedral Boys School; (2) The Cathedral Girls School; and (3) The John Connon School.

22 Cockcroft, John and M.G.K. Menon, *Homi Jehangir Bhabha 1909-1966*, Royal Institution of Great Britain, London, 1967, p. 6.

23 Anderson, Robert S., *Nucleus and Nation: Scientists, International Networks, and Power in India*, University of Chicago Press, 2010, pp. 125–6.

24 Chowdhury, Indira and Ananya Dasgupta, *A Masterful Spirit: Homi J. Bhabha, 1909-1966*, Penguin Books, New Delhi, 2010, p. 21

25 Ibid. 22.

26 Ibid. 21.

27 Ibid. 21.

28 This was renamed the Mechanical Engineering Tripos in 1970.

Chapter 4: Cambridge

1 Deshmukh, Chintamani, *Homi Jehangir Bhabha*, National Book Trust, India, New Delhi, 2014, p. 7.

2 Sager, Peter, *Oxford & Cambridge: An Uncommon History*, Thames and Hudson, London, 2005, p. 9.

3 Ibid. 11–4.

4 Ibid. 276–7.

5 Chowdhury, Indira and Ananya Dasgupta, *A Masterful Spirit: Homi J. Bhabha, 1909-1966*, Penguin Books, New Delhi, 2010, p. 37.

6 Ibid. 37.

7 Isaacson, Walter, *Einstein: His Life and Universe*, Simon & Schuster, 2007, p. 31.

8 Bhabha, Homi, Received by Homi Seervai, Seervai family collection, 2 January 1929.

9 Ibid.

10 Ibid.

11 A coxswain is in charge of a boat, particularly its navigation and steering. He sits in either the bow or the stern (depending on what type of boat it is) while controlling its steering, speed and timing. He is also responsible for motivating the crew and for steering as straight a course as possible to minimize the distance to the finishing line.

12 Venkataraman, G., *Bhabha and his Magnificent Obsessions*, Universities Press, 1994, p. 5.

13 'IAEA Honours Dr Bhabha's Memory', *Nuclear India*, February-March 1966, p. 11.

14 Chowdhury, Indira and Ananya Dasgupta, *A Masterful Spirit: Homi J. Bhabha, 1909-1966*, Penguin Books, 2010, p. 48.

15 Deshmukh, Chintamani, *Homi Jehangir Bhabha*, National Book Trust, India, New Delhi, 2014, p. 106.

16 Bhabha, Homi, Received by Homi Seervai, Seervai family collection, 1 February 1931.

17 Ibid.

18 Bhabha, Homi, Received by Homi Seervai, Seervai family collection, September 1929.

19 Bhabha, Homi, Received by Homi Seervai, Seervai family collection, 20 March 1928.

20 Bhabha, Jamshed (ed.), *Homi Bhabha as Artist: A Selection of His Paintings, Drawings, and Sketches*, Marg, India, 1968.

21 Leyden, R.V., 'The Drawings of Homi Bhabha', *Marg*, Vol. 6, No. 3, June 1953.

22 Chowdhury, Indira and Ananya Dasgupta, *A Masterful Spirit: Homi J. Bhabha, 1909-1966*, Penguin Books, 2010, p. 206.

23 Ibid.

24 Ibid. 207

25 Ibid.

26 Ibid.

27 From Rudyard Kipling's famous poem *If*. ('If you can fill the unforgiving minute/With sixty seconds' worth of distance run.')

28 Homi Bhabha A Scientist In Action, YouTube, https://tinyurl.com/397c63tw. Accessed on 26 July 2022.

29 Film on Dr Homi Bhabha by TIFR Marking His Birth Centenary in 2009, YouTube, https://tinyurl.com/ywwz6rkj. Accessed on 18 December 2022.

30 Cockcroft, John and M.G.K. Menon, *Homi Jehangir Bhabha 1909-1966*, Royal Institution of Great Britain, London, 1967, p. 9.

31 Bhabha, Homi, Received by Homi Seervai, Seervai family collection, n.d.

32 Ibid. 41.

33 Deshmukh, Chintamani, *Homi Jehangir Bhabha*, National Book Trust, India, New Delhi, 2014, p. 104.

34 Venkataraman, G., *Bhabha and his Magnificent Obsessions*, Universities Press, 1994, p. 190.

35 Bhabha, Homi, Received by Homi Seervai, Seervai family collection, 3 July 1930.

36 Anderson, Robert S., *Nucleus and Nation: Scientists: International Networks, and Power in India*, University of Chicago Press, 2010, p. 99.

37 Dadabhoy, Bakhtiar K., *Sugar in Milk: Lives of Eminent Parsis*, Rupa & Co. New Delhi, 2008, p. 268.

38 Deshmukh, Chintamani, *Homi Jehangir Bhabha*, National Book Trust, India, New Delhi, 2014, p. 32.

39 Ibid. 7.

40 Bhabha, Homi, Received by Homi Seervai, Seervai family collection, September 1929.

41 Ibid.

42 Venkataraman, G., *Bhabha and his Magnificent Obsessions*, Universities Press, 1994, p. 5.

43 Dimri, Keshav, *Atomic Postalgia: Antiquity, Modernity and the Origins of the Nuclear Program in India*, 2014, Stanford University, Thesis submitted to the interschool honours programme in international security studies, https://tinyurl.com/t7as72np. Accessed on 29 December 2022.

Chapter 5: The Quantum Revolution

1 Bhabha, Homi, Received by Homi Seervai, Seervai family collection, 11 September 1929.

2 Fraser, Gordon, *Cosmic Anger: Abdus Salam–The First Muslim Nobel Scientist*, Oxford University Press, Oxford, 2008, p. 65.

3 Kanigel, Robert, *The Man Who Knew Infinity: A Life of the Genius Ramanujan*, Abacus, London, 1991, p. 132.

4 Kaku, Michio, *The God Equation: The Quest for a Theory of Everything*, Penguin Random House, UK, 2021, p. 17.

5 Devons, S., 'Rutherford's Laboratory', University of Cambridge, https://tinyurl.com/296v5eea. Accessed on 3 August 2022.

6 Feldman, Burton, *The Nobel Prize: A History of Genius, Controversy, and Prestige*, Arcade Publishing, New York, 2000, p. 135.

7 Ibid. 136.

8 Watson, Peter, *A Terrible Beauty: A History of the People and Ideas that Shaped the Modern Mind*, Weidenfeld & Nicolson, London, 2000, p. 257.

9 Ibid. 257.

10 Ibid. 256.

11 Goran, Liljestrand (ed.), *Les prix Nobel en 1963*, P.A. Norstedt & Soner, Stockholm, 1964.

12 Strathern, Paul, *Bohr and Quantum Theory*, Arrow, London, 1998, pp. 37–8.
13 Ibid. 39–40.
14 Gribbin, John, *Science: A History 1543–2001*, Allen Lane, London, 2002, p. 513.
15 Strathern, Paul, *Bohr and Quantum Theory*, Arrow, London, 1998, p. 28.
16 This refers to an English–Danish dictionary.
17 Strathern, Paul, *Bohr and Quantum Theory*, Arrow, London, 1998, p. 29.
18 Kumar, Manjit, *Quantum: Einstein, Bohr, and the Great Debate about the Nature of Reality*, Hachette, New Delhi, 2009, pp. 93–4.
19 Strathern, Paul, *Bohr and Quantum Theory*, Arrow, London, 1998, pp. 56–7.
20 Ibid. 29–30.
21 Ibid. 64.
22 Ibid. 56.
23 Kumar, Manjit, *Quantum: Einstein, Bohr, and the Great Debate about the Nature of Reality*, Hachette, New Delhi, 2009, p. 106.
24 Strathern, Paul, *Bohr and Quantum Theory*, Arrow, London, 1998, pp. 45–6.
25 Kumar, Manjit, *Quantum: Einstein, Bohr, and the Great Debate about the Nature of Reality*, Hachette, New Delhi, 2009, pp. 100–1.
26 Strathern, Paul, *Bohr and Quantum Theory*, Arrow, London, 1998, pp. 47–8.
27 Gribbin, John, *Science: A History 1543–2001*, Allen Lane, London, 2002, pp. 517–18.
28 Kumar, Manjit, *Quantum: Einstein, Bohr, and the Great Debate about the Nature of Reality*, Hachette, New Delhi, 2009, p. 1.

29 Walter, Isaacson, *Einstein: His Life and Universe*, Simon & Schuster, 2007, pp. 327–8.
30 'This Month in Physics History: January 1925: Wolfgang Pauli Announces the Exclusion Principle', APS Physics, https://tinyurl.com/bdh5rkk2. Accessed on 24 August 2022.
31 Kumar, Manjit, *Quantum: Einstein, Bohr, and the Great Debate about the Nature of Reality*, Hachette, New Delhi, 2009, pp. 160–1.
32 Frisch, Otto Robert, *What Little I Remember*, Plunkett Lake Press, 2017, p. 47.
33 Ibid. 48.
34 Kumar, Manjit, *Quantum: Einstein, Bohr, and the Great Debate about the Nature of Reality*, Hachette, New Delhi, 2009, p. 171.
35 Ibid. 186–7.
36 Baggott, Jim, *The Quantum Story: A History in 40 Moments*, Oxford University Press, 2011, p. 47.
37 Ibid. 48.
38 Kumar, Manjit, *Quantum: Einstein, Bohr, and the Great Debate about the Nature of Reality*, Hachette, New Delhi, 2009, p. 194.
39 Ibid. 197.
40 Baggott, Jim, *The Quantum Story: A History in 40 Moments*, Oxford University Press, 2011, p. 60.

41 An unsalaried university lecturer or teacher in a German-speaking country remunerated directly by students' fees.
42 Kumar, Manjit, *Quantum: Einstein, Bohr, and the Great Debate about the Nature of Reality*, Hachette, New Delhi, 2009, p. 206.
43 Baggott, Jim, *The Quantum Story: A History in 40 Moments*, Oxford University Press, 2011, p. 64.
44 Ibid. 64.
45 Ibid. 65.
46 Kumar, Manjit, *Quantum: Einstein, Bohr, and the Great Debate about the Nature of Reality*, Hachette, New Delhi, 2009, p. 208.
47 Baggott, Jim, *The Quantum Story: A History in 40 Moments*, Oxford University Press, 2011, p. 66.
48 Ibid. 62.
49 Kumar, Manjit, *Quantum: Einstein, Bohr, and the Great Debate about the Nature of Reality*, Hachette, New Delhi, 2009, p. 213.
50 Ibid. 218.
51 Ibid. 218.
52 Ibid. 219.
53 Isaacson, Walter, *Einstein: His Life and Universe*, Simon & Schuster UK Ltd., 2007, p. 335.
54 Dhyani, Vanshika, 'Einstein, Stop Telling God What to Do!', University Observer, https://tinyurl.com/527p6j32. Accessed on 2 November 2022.
55 Isaacson, Walter, *Einstein: His Life and Universe*, Simon & Schuster UK Ltd., 2007, 331–2.
56 Miller, Arthur I., *Empire of the Stars: Friendship, Obsession and Betrayal in the Quest for Black Holes*, Little, Brown and Company, London, 2005, p. 39.

Chapter 6: The Romance with Physics

1 Bhabha, Homi, Received by Homi Seervai, Seervai family collection, 2 August 1928.
2 Miller, Arthur I., *Empire of the Stars: Friendship, Obsession and Betrayal in the Quest for Black Holes*, Little, Brown and Company, London, 2005, p. 72.
3 Ibid. 72.
4 Abraham, Itty, *The Making of the Indian Atomic Bomb: Science, Secrecy and the Post-Colonial State*, Orient BlackSwan, Hyderabad, 1999, p. 38.
5 Miller, Arthur I., *Empire of the Stars: Friendship, Obsession and Betrayal in the Quest for Black Holes*, Little, Brown and Company, London, 2005, p. 73.
6 Ibid. 96.
7 Wali, Kameshwar C., *Chandra: A Biography of S. Chandrasekhar*, Penguin Books, New Delhi, p. 83.

8 Graham, Farmelo, *The Strangest Man: The Hidden Life of Paul Dirac, Mystic of the Atom*, Basic Books, New York, Reprint 2009, p. 183.

9 Wali, Kameshwar C., *Chandra: A Biography of S. Chandrasekhar*, Penguin Books, New Delhi, p. 77.

10 Ibid. 82.

11 Ibid. 84.

12 Ibid. 84.

13 Singh, Virender, 'Bhabha's Contributions to Elementary Particle Physics and Cosmic Rays Research', *Resonance*, Vol. 14, 2009, pp. 430–54, https://tinyurl.com/25mtennv. Accessed on 1 August 2022.

14 Polkinghorne, John, *Quantum Theory: A Very Short Introduction*, Oxford University Press, 2002, p. 21.

15 Farmelo, Graham, *The Strangest Man: The Hidden Life of Paul Dirac, Mystic of the Atom*, Basic Books, New York, 2009, p. 179.

16 Ibid. 295.

17 Abraham, Itty, *The Making of the Indian Atomic Bomb: Science, Secrecy and the Post-Colonial State*, Orient BlackSwan, Hyderabad, 1999, p. 37.

18 Feldman, Burton, *The Nobel Prize: A History of Genius, Controversy, and Prestige*, Arcade Publishing, New York, 2000, pp. 148–9.

19 Miller, Arthur I., *Empire of the Stars: Friendship, Obsession and Betrayal in the Quest for Black Holes*, Little, Brown and Company, London, 2005, p. 113.

20 Casimir, Hendrik B.G., *Haphazard Reality: Half a Century of Science*, Harper and Row, New York, 1983, p. 139.

21 Ibid. 139

22 Ibid. 139.

23 Ibid. 140.

24 Freund, Peter G.O. et al., 'Gregor Wentzel', Cornell University, 11 September 2008, https://tinyurl.com/4xckjmss. Accessed on 1 August 2022.

25 Perkovich, George, *India's Nuclear Bomb: The Impact on Global Proliferation*, University of California Press, 2001, p. 16.

26 Abraham, Itty, *The Making of the Indian Atomic Bomb: Science, Secrecy and the Post-Colonial State*, Orient BlackSwan, Hyderabad, 1999, pp. 39–40.

27 Ibid. 41.

28 Frisch, Otto R., *What Little I Remember*, Cambridge University Press, Cambridge, 1980, p. 46.

29 Ibid. 46.

30 Ibid. 53–4.

31 Weisskopf, Victor, *The Joy of Insight: The Passions of a Physicist*, Plunkett Lake Press, 2013.

32 Abraham, Itty, *The Making of the Indian Atomic Bomb: Science, Secrecy and the Post-Colonial State*, Orient BlackSwan, Hyderabad, 1999, pp. 38–9.

33 Miller, Arthur I., *Empire of the Stars: Friendship, Obsession and Betrayal in the Quest for Black Holes*, Little, Brown and Company, London, 2005, p. 175.
34 Ibid. 176.
35 Infield, Leopold, *Why I Left Canada: Reflections on Science and Politics*, McGill-Queen's University Press, 1978, p. 161.
36 Chengappa, Raj, *Weapons of Peace: The Secret Story of India's Quest to be a Nuclear Power*, HarperCollins, New Delhi, 2000, p. 80.
37 Segrè, Gino and Bettina Hoerlin, *The Pope of Physics*, Henry Holt and Company, New York, 2016, p. 65.
38 Feldman, Burton, *The Nobel Prize: A History of Genius, Controversy, and Prestige*, Arcade Publishing, New York, 2000, p. 156.
39 Ibid. 156–7.
40 Gopal, Sarvepalli, *Jawaharlal Nehru: A Biography*, Vol. 2., Oxford University Press, New Delhi, 2015, p. 67.
41 Ramesh, Jairam, *A Chequered Brilliance: The Many Lives of V.K. Krishna Menon*, Penguin Viking, 2019, p. 177.

Chapter 7: Cosmic Ray Research

1 Bhabha, Homi, 'The Fundamental Length Introduced by the Theory of the Mesotron (Meson)', *Nature*, Vol. 143, pp. 276–7, 1939, https://tinyurl.com/ycks8jjr. Accessed on 30 July 2022.
2 Raychaudhuri, Sreerup, 'Cosmic Ray Research in India: A Historical Perspective', *Physics News*, Vol. 51, No. 1-2, pp. 24–33, https://tinyurl.com/mwzprp43. Accessed on 30 July 2022.
3 Ibid.
4 e/m ratio is the ratio of charge on an electron to its mass.
5 Close, Frank et al., *The Particle Odyssey: A Journey to the Heart of Matter*, Oxford University Press, 2002, p. 55.
6 Raychaudhuri, Sreerup, 'Cosmic Ray Research in India: A Historical Perspective', *Physics News*, Vol. 51, No. 1-2, pp. 24–33, https://tinyurl.com/mwzprp43. Accessed on 30 July 2022.
7 A singularity is a point where some property is infinite. Physics breaks down at a singularity because we cannot describe it with our known physical laws.
8 Singh, Virender, 'Bhabha's Contributions to Elementary Particle Physics and Cosmic Rays Research', *Resonance*, Vol. 14, 2009, pp. 430–54, https://tinyurl.com/25mtennv. Accessed on 29 July 2022.
9 Anderson, Carl D. and Seth Neddermeyer, 'Mesotron (Intermediate Particle) as a Name for the New Particles of Intermediate Mass', *Nature* Vol. 142, No. 878, 1938, https://tinyurl.com/y3fwnkv3. Accessed on 30 July 2022.
10 Close, Frank et al., *The Particle Odyssey: A Journey to the Heart of Matter*, Oxford University Press, 2002, p. 71.

11 Ibid. 61.
12 Bhabha, Homi, 'On the Penetrating Component of Cosmic Radiation', *Proceedings of the Royal Society of London. Series A: Mathematical and Physical Sciences*, Vol. 164, No. 917, 1938, pp. 257–94, JSTOR, https://tinyurl.com/mrvy23v9. Accessed on 28 July 2022.
13 Bhabha, Homi, 'The Fundamental Length Introduced by the Theory of the Mesotron (Meson)', *Nature*, Vol. 143, pp. 276–7, 1939, https://tinyurl.com/ycks8jjr. Accessed on 30 July 2022.
14 Singh, Rajinder, 'Homi J. Bhabha and Niels Bohr', *Current Science*, Vol. 97, No. 4, 2009, https://tinyurl.com/bdzk9r5j. Accessed on 30 July 2022.
15 Kemmer, N., 'The Impact of Yukawa's Meson Theory on Workers in Europe: A Reminiscence', *Progress of Theoretical Physics Supplement*, Vol. E65, 1965, pp. 602–8, https://tinyurl.com/5bje5rhv. Accessed on 30 July 2022.
16 Venkataraman, G., *Bhabha and his Magnificent Obsessions*, Universities Press, 1994, p. 66.
17 Singh, Virendra, 'Homi Jehangir Bhabha: Architect of Modern Science and Technology in India', Tata Institute of Fundamental Research, 2009, https://tinyurl.com/ffjz6jz4. Accessed on 30 July 2022.
18 Farmelo, Graham, *The Strangest Man: The Hidden Life of Paul Dirac, Mystic of the Atom*, Basic Books, New York, Reprint 2009, p. 64.
19 Ibid. 66.
20 Watson, Peter, *A Terrible Beauty: The People and Ideas that Shaped the Modern Mind: A History*, Weidenfeld & Nicolson, London, 2000, p. 264.
21 Farmelo, Graham, *The Strangest Man: The Hidden Life of Paul Dirac, Mystic of the Atom*, Basic Books, New York, Reprint 2009, p. 381.
22 Chowdhury, Indira and Ananya Dasgupta, *A Masterful Spirit: Homi J. Bhabha, 1909-1966*, Penguin Books, New Delhi, 2010, p. 46.
23 Ibid. 47.
24 Glazer, Anthony Michael and Patience Thomson (eds.), *Crystal Clear: The Autobiographies of Sir Lawrence and Lady Bragg*, Oxford University Press, 2015, p. 158.
25 Brown, Andrew, *The Neutron and the Bomb: A Biography of Sir James Chadwick*, Oxford University Press, 1997, p. 164.
26 Abraham, Itty, *The Making of the Indian Atomic Bomb: Science, Secrecy and the Post-Colonial State*, Orient BlackSwan, Hyderabad, 1999, p. 42.
27 Ibid. 42.
28 Mallik, D.C.V. and S. Chatterjee, *Kariamanikkam Srinavasa Krishnan: His Life and Work*, Universities Press, Hyderabad, 2012, p. 213.

Chapter 8: Return of the Native

1 Chowdhury, Indira, *Growing the Tree of Science: Homi Bhabha and the Tata Institute of Fundamental Research*, Oxford University Press, New Delhi, 2016, p. 23.

2 Dimri, Keshav, *Atomic Postalgia: Antiquity, Modernity and the Origins of the Nuclear Program in India*, 2014, Stanford University, Thesis submitted to the interschool honours programme in international security studies, https://tinyurl.com/t7as72np. Accessed on 25 August 2022.

3 This spin is expressed in units of Planck's constant $h/2\pi = 1.055 \times 10^{-14}$ joule seconds.

4 Singh, Virender, 'Bhabha's Contributions to Elementary Particle Physics and Cosmic Rays Research', *Resonance*, Vol. 14, 2009, pp. 430–54, https://tinyurl.com/25mtennv. Accessed on 29 July 2022.

5 Ibid.

6 Penney, William George, 'Homi Jehangir Bhabha, 1909-1966', Biographical Memoirs of Fellows of the Royal Society, https://tinyurl.com/2p89c828. Accessed on 29 July 2022.

7 Pauli, Wolfgang, Received by Homi Bhabha, *Pauli Letter Collection*, CERN Archives, 12 April 1940, https://tinyurl.com/bdevjnt6. Accessed on 29 July 2022.

8 Pauli, Wolfgang, Received by Homi Bhabha, *Pauli Letter Collection*, CERN Archives, 17 May 1940, https://tinyurl.com/ycxxmdm5. Accessed on 29 July 2022.

9 Pauli, Wolfgang, Received by Homi Bhabha, *Pauli Letter Collection*, CERN Archives, 15 October 1940, https://tinyurl.com/dm4bbysz. Accessed on 29 July 2022.

10 Chowdhury, Indira, *Growing the Tree of Science: Homi Bhabha and the Tata Institute of Fundamental Research*, Oxford University Press, New Delhi, 2016.

11 Ibid.

12 Pauli, Wolfgang, Received by Homi Bhabha, *Pauli Letter Collection*, CERN Archives, 13 April 1942, https://tinyurl.com/hzs9vcv4. Accessed on 29 July 2022.

13 Ibid.

14 Pauli, Wolfgang, Received by Homi Bhabha, *Pauli Letter Collection*, CERN Archives, 1 August 1942, https://tinyurl.com/2s3mxaan. Accessed on 29 July 2022.

15 Chowdhury, Indira, *Growing the Tree of Science: Homi Bhabha and the Tata Institute of Fundamental Research*, Oxford University Press, New Delhi, 2016.

16 Ibid. 22.

17 Ibid. 24.

18 Ibid. 22.

19 Abraham, Itty, *The Making of the Indian Atomic Bomb: Science, Secrecy and the Post-Colonial State*, Orient BlackSwan, Hyderabad, 1999, p. 44.

20 Sreekantan, B.V., 'Sixty Years of The Tata Institute of Fundamental Research 1945-2005: The Role of Young Men in the Creation and Development of this Institute', *Current Science*, Vol. 90, No. 7, 2006, pp. 1012–25, JSTOR, https://tinyurl.com/3u9m4tnr. Accessed on 29 July 2022.

21 Ibid.

22 Singh, Virender, 'Bhabha's Contributions to Elementary Particle Physics and Cosmic Rays Research', *Resonance*, Vol. 14, 2009, pp. 430–54, https://tinyurl.com/25mtennv. Accessed on 29 July 2022.

23 Latitude effect is the variation of any physical quantity with latitudes. More specifically, it is an increase of intensity of cosmic rays with magnetic latitude, especially at high altitudes.

24 Raj, Gopal, *Reach for the Stars: The Evolution of India's Rocket Programme*, Penguin Viking, New Delhi, 2000, pp. 5–6.

25 Shah, Amrita, *Vikram Sarabhai: A Life*, Penguin Viking, New Delhi, 2007, p. 44.

26 Ibid. 47.

27 'Homi Bhabha Introduced My Papa and Amma': Dr Mallika Sarabhai on 'Rocket Boys' | The Quint, YouTube, https://tinyurl.com/2czwhb99. Accessed on 29 July 2022.

28 Raj, Gopal, *Reach for the Stars: The Evolution of India's Rocket Programme*, Penguin Viking, 2000, p. 3.

29 Penney, William George, 'Homi Jehangir Bhabha, 1909-1966', Biographical Memoirs of Fellows of the Royal Society, https://tinyurl.com/2p89c828. Accessed on 29 July 2022.

30 Ghosh, Dipan K. and Arun K. Grover, *Tribute to a Titan: Birth Centenary of Homi Jehangir Bhabha*, Indian Physics Association, Mumbai, 2009, p. 134.

31 Ramanna, Raja, *Years of Pilgrimage: An Autobiography*, Penguin Viking, New Delhi, 1991, p. 34.

32 Ibid. 18.

33 Ibid. 35.

34 Navratna, Deepti, *The Maverick Maharaja: The Life and Times of His Highness Maharaja Jayachamarajendra Wadiyar*, HarperCollins, Gurugram, 2022, p. 69.

35 Wadiyar, Jaya Chamaraja, Received by Homi Bhabha, Tata Institute of Fundamental Research Archive, 12 July 1948, Mumbai, Maharashtra, India.

36 Navratna, Deepti, *The Maverick Maharaja: The Life and Times of His Highness Maharaja Jayachamarajendra Wadiyar*, HarperCollins, 2022, p. 70.

37 Deshmukh, Chintamani, *Homi Jehangir Bhabha*, National Book Trust, India, New Delhi, 2014, p. 106.

38 Venkataraman, G., *Journey into Light: Life and Science of C.V. Raman*, Indian Academy of Sciences, Bangalore, 1988, p. 513.

39 Kanigel, Robert, *The Man Who Knew Infinity: A Life of the Genius Ramanujan*, Abacus, London, 1991, p. 291.
40 Mallik, D.C.V. and S. Chatterjee, *Kariamanikkam Srinavasa Krishnan: His Life and Work*, Universities Press, Hyderabad, 2012, pp. 232–4.

Chapter 9: Bangalore

1 Parthasarathy, K.S., 'A Wreath of White Roses Over the Ruins of Mehrangir, Homi Bhabha's Home', *Science: The Wire*, 25 June 2016, https://tinyurl.com/j2k3hz7r. Accessed on 25 August 2022.
2 Harish-Chandra, Lily, 'Childhood Memories', *Connect with the Indian Institute of Science*, Vol. 5, No. 3, pp. 42–3, 2018, https://tinyurl.com/y8mayy3x. Accessed on 9 August 2022.
3 Howe, Robert, 'Harish Chandra 1923–1983', National Academy of Sciences, 2011, https://tinyurl.com/5eubezdy. Accessed on 9 August 2022.
4 Bhabha, Homi and Harish-Chandra, 'On the Theory of Point-Particles', *Proceedings of the Royal Society*, Vol. 183, No. 993, 1944, https://tinyurl.com/4zrkur3c. Accessed on 8 August 2022.
5 Langlands, Robert P., 'Harish-Chandra', Biographical Memoirs of the Royal Society, https://tinyurl.com/yt5rr36j. Accessed on 9 August 2022.
6 Singh, Jagjit, *Abdus Salam: A Biography*, Penguin Books, New Delhi, 1992, p. 11.
7 Kursunoglu, Behram N. and Eugene Paul Wigner (eds.), *Reminiscences About a Great Physicist: Paul Adrien Maurice Dirac*, Cambridge University Press, New York, 1987, pp. 34–5.
8 Howe, Robert, 'Harish Chandra 1923–1983', National Academy of Sciences, 2011, https://tinyurl.com/5eubezdy. Accessed on 9 August 2022.
9 Dirac, Paul, Received by Homi Bhabha, Tata Institute of Fundamental Research Archive, 31 July 1942, Mumbai, Maharashtra, India.
10 Bhabha, Homi, Received by Birbal Sahni, Tata Institute of Fundamental Research Archive, 7 January 1943, Mumbai, Maharashtra, India.
11 Ibid.
12 A Gara is an embroidered sari predominantly worn by women of the Parsi community at weddings and special occasions.
13 Chowdhury, Indira and Ananya Dasgupta, *A Masterful Spirit: Homi J. Bhabha, 1909-1966*, Penguin Books, New Delhi, 2010, p. 84.
14 Ibid. 84.
15 Ibid. 84.
16 Dadabhoy, Bakhtiar K., *Zubin Mehta: A Musical Journey*, Penguin Random House, Gurugram, 2016, p. 20.
17 Parthasarathy, K.S., 'A Wreath of White Roses Over the Ruins of Mehrangir, Homi Bhabha's Home', *Science: The Wire*, 25 June 2016, https://tinyurl.com/j2k3hz7r. Accessed on 25 August 2022.

18 Ibid.

19 Ibid.

20 Ibid.

21 Ibid.

22 Chowdhury, Indira and Ananya Dasgupta, *A Masterful Spirit: Homi J. Bhabha, 1909-1966*, Penguin Books, New Delhi, 2010, p. 76.

23 Bhabha, Homi, Received by Wolfgang Pauli, Tata Institute of Fundamental Research Archive, 2 May 1942, Mumbai, Maharashtra, India.

24 Chowdhury, Indira and Ananya Dasgupta, *A Masterful Spirit: Homi J. Bhabha, 1909-1966*, Penguin Books, New Delhi, 2010, p. 78.

25 Ibid. 79.

26 Pauli, Wolfgang, Received by Homi Bhabha, Pauli Letter Collection, CERN Archives, 23 September 1943, https://tinyurl.com/2s3ub7c9. Accessed on 9 August 2022.

27 Pryce, Maurice, Received by Homi Bhabha, Tata Institute of Fundamental Research Archive, n.d., Mumbai, Maharashtra, India.

28 Bhabha, Homi, 'Historical Note on Tata Institute of Fundamental Research', Wilson Center Digital Archive, 1 January 1954, https://tinyurl.com/43dpnxh6. Accessed on 9 August 2022.

29 Chowdhury, Indira, *Growing the Tree of Science: Homi Bhabha and the Tata Institute of Fundamental Research*, Oxford University Press, New Delhi, 2016, p. 25.

30 Anderson, Robert S., *Nucleus and Nation: Scientists, International Networks, and Power in India*, University of Chicago Press, 2010, p. 104.

31 Sreekantan, B.V. and Richa Malhotra, 'Homi Jehangir Bhabha: A Visionary', *Resonance*, Vol. 15, No. 5, May 2010, pp. 462–75, https://tinyurl.com/mrxyyz53. Accessed on 9 August 2022.

32 Pulakkat, Hari, *Space Life Matter: The Coming of Age of Indian Science*, Hachette, Gurugram, 2021, p. xvii.

33 Deshmukh, Chintamani, *Homi Jehangir Bhabha*, National Book Trust, India, New Delhi, 2014, p. 26.

34 Menon, M.G.K., 'Homi Jehangir Bhabha', *Proceedings of the Royal Institution of Great Britain*, Vol. 41, Part 4, No. 191, p. 430.

35 Phalkey, Jahnavi, *Atomic State: Big Science in Twentieth-Century India*, Permanent Black, New Delhi, 2013, p. 67.

36 Ibid. 69.

37 Ibid. 69.

38 Bhabha, Homi et al., Received by Archibald Vivian Hill, 2 December 1943, Birbal Sahni Papers, Nehru Memorial Museum and Library, New Delhi, Delhi, India.

39 Bhabha, Homi, Received by Birbal Sahni, Birbal Sahni Papers, Nehru Memorial Museum and Library, 21 December 1943, New Delhi, Delhi, India.

40 'H.E. Lord Wavell of Indian Science', *Current Science*, Vol. 13, No. 1, 1944, p. 3, JSTOR, https://tinyurl.com/2up5wxde. Accessed on 9 August 2022.
41 Phalkey, Jahnavi, *Atomic State: Big Science in Twentieth-Century India*, Permanent Black, New Delhi, 2013, p. 70.
42 Ibid. 70.
43 Ibid. 72.
44 Anderson, Robert S., *Nucleus and Nation: Scientists, International Networks, and Power in India*, University of Chicago Press, 2010, p. 115.

Chapter 10: An Idea and a Wish

1 Chowdhury, Indira and Ananya Dasgupta, *A Masterful Spirit: Homi J. Bhabha, 1909-1966*, Penguin Books, New Delhi, p. 111.
2 Bhabha, Homi, 'Historical Note on Tata Institute of Fundamental Research', Wilson Center Digital Archive, 1 January 1954, https://tinyurl.com/43dpnxh6. Accessed on 11 August 2022.
3 Ibid.
4 Ibid.
5 Ibid.
6 Chowdhury, Indira, *Growing the Tree of Science: Homi Bhabha and the Tata Institute of Fundamental Research*, Oxford University Press, New Delhi, 2016, p. 11.
7 Ibid. 11.
8 Lala, R.M., *The Heartbeat of a Trust: Fifty Years of the Sir Dorabji Tata Trust,* TataMcGraw Hill, New Delhi, 1984, pp. 22–3.
9 Ibid. 25.
10 Chowdhury, Indira and Ananya Dasgupta, *A Masterful Spirit: Homi J. Bhabha, 1909-1966*, Penguin Books, New Delhi, p. 94.
11 Ibid. 94.
12 Lala, R.M., *Beyond the Last Blue Mountain: A Life of J.R.D. Tata*, Penguin Books, New Delhi, 1993, p. 204.
13 Ibid. 205.
14 Chowdhury, Indira and Ananya Dasgupta, *A Masterful Spirit: Homi J. Bhabha, 1909-1966*, Penguin Books, New Delhi, p. 95.
15 Lala, R.M., *Beyond the Last Blue Mountain: A Life of J.R.D. Tata*, Penguin Books, New Delhi, 1993, p. 195.
16 Chowdhury, Indira, *Growing the Tree of Science: Homi Bhabha and the Tata Institute of Fundamental Research*, Oxford University Press, New Delhi, 2016, p. 18.
17 Lala, R.M., *The Heartbeat of a Trust: Fifty Years of the Sir Dorabji Tata Trust*, TataMcGraw Hill, New Delhi, 1984, p. 25.
18 Ibid. 89.

19 Chowdhury, Indira and Ananya Dasgupta, *A Masterful Spirit: Homi J. Bhabha, 1909-1966*, Penguin Books, New Delhi, p. 111.

20 Bhabha, Homi, Received by J.R.D. Tata, Tata Institute of Fundamental Research Archive, 24 March 1944, Mumbai, Maharashtra, India.

21 Ibid.

22 Tata, J.R.D., Received by Homi Bhabha, Tata Institute of Fundamental Research Archive, 5 April 1944, Mumbai, Maharashtra, India.

23 Ibid.

24 Dadabhoy, Bakhtiar K., *Jeh: A Life of J.R.D. Tata*, Rupa & Co., New Delhi, 2005, p. 9.

25 Tata, J.R.D., Received by Homi Bhabha, Tata Institute of Fundamental Research Archive, 5 April 1944, Mumbai, Maharashtra, India.

26 Lala, R.M., *The Heartbeat of a Trust: Fifty Years of the Sir Dorabji Tata Trust*, Tata McGraw Hill, New Delhi, 1984, pp. 90–1.

27 In 1998, Robert Anderson made similar queries, but Lala refused to name the industrialist because the Tatas had asked him not to. Anderson said that other sources suggested that it was possibly the Godrej family. Anderson, Robert, *Nucleus and Nation: Scientists, International Networks, and Power in India*, University of Chicago Press, 2010, p. 610.

28 Lala, R.M., Personal interview, 12 March 2011.

29 Raianu, Mircea, *Tata: The Global Corporation that Built Indian Capitalism*, Harvard University Press, 2021, p. 101.

30 Bhabha, Homi, 'Historical Note on Tata Institute of Fundamental Research', Wilson Center Digital Archive, 1 January 1954, https://tinyurl.com/43dpnxh6. Accessed on 11 August 2022.

31 Chowdhury, Indira, *Growing the Tree of Science: Homi Bhabha and the Tata Institute of Fundamental Research*, Oxford University Press, New Delhi, 2016, p. 21.

32 Lala, R.M., *Beyond the Last Blue Mountain: A Life of J.R.D. Tata*, Penguin Books, New Delhi, 1993, p. 303.

33 Tata, J.R.D., 'Foreword', *Tata Institute of Fundamental Research 1945—1970 (Silver Jubilee Commemoration Volume)*, Tata Institute of Fundamental Research, 1970.

34 Chowdhury, Indira, *Growing the Tree of Science: Homi Bhabha and the Tata Institute of Fundamental Research*, Oxford University Press, New Delhi, 2016, p. 29.

35 Bhabha, Homi, 'Historical Note on Tata Institute of Fundamental Research', Wilson Center Digital Archive, 1 January 1954, https://tinyurl.com/43dpnxh6. Accessed on 11 August 2022.

36 Pauli, Wolfgang, Received by Homi Bhabha, Pauli Letter Collection, CERN Archives, 12 July 1945, https://tinyurl.com/3p93emc3. Accessed on 11 August 2022.

Chapter 11: Cathedral of Science

1 Chowdhury, Indira, *Growing the Tree of Science: Homi Bhabha and the Tata Institute of Fundamental Research*, Oxford University Press, New Delhi, 2016, p. xv.

2 Panday, Cooverbai, Received by the Trustees of the Sir Dorabji Tata Trust, Tata Central Archives, 11 May 1945, Pune, Maharashtra, India.

3 Chowdhury, Indira, *Growing the Tree of Science: Homi Bhabha and the Tata Institute of Fundamental Research*, Oxford University Press, New Delhi, 2016, pp. 39–40.

4 'A Masterful Spirit, Homi Bhabha 1909-1966', an exhibition by the Public Awareness Division, Department of Atomic Energy, Government of India, New Delhi, India.

5 Chowdhury, Indira and Ananya Dasgupta, *A Masterful Spirit: Homi J. Bhabha, 1909-1966*, Penguin Books, 2010, p. 85.

6 Ibid. 85.

7 Chowdhury, Indira, *Growing the Tree of Science: Homi Bhabha and the Tata Institute of Fundamental Research*, Oxford University Press, New Delhi, 2016, p. 242.

8 Lala, R.M., *The Heartbeat of a Trust: Fifty Years of the Sir Dorabji Tata Trust*, Tata McGraw Hill, New Delhi, 1984, pp. 115–6.

9 Ibid. 92–3.

10 Saklatvala, Sorab, 'Speech of Welcome', Inauguration of the Tata Institute of Fundamental Research, 19 December 1945, Tata Institute of Fundamental Research Archive, Mumbai, Maharashtra, India.

11 Bhabha, Homi, 'Speech at the Inauguration of the Tata Institute of Fundamental Research', Inauguration of the Tata Institute of Fundamental Research, 19 December 1945, Tata Institute of Fundamental Research Archive, Mumbai, Maharashtra, India.

12 'Tata Institute of Research Opened in Bombay', *The Morning Standard*, 20 December 1945.

13 Chowdhury, Indira and Ananya Dasgupta, *A Masterful Spirit: Homi J. Bhabha, 1909-1966*, Penguin Books, 2010, p. 82.

14 Bhabha, Homi, Received by S. Chandrasekhar, Tata Institute of Fundamental Research Archive, 20 April 1944, Mumbai, Maharashtra, India.

15 Ibid.

16 Chandrasekhar, S., Received by Homi Bhabha, Tata Institute of Fundamental Research Archive, 22 May 1944, Mumbai, Maharashtra, India.

17 Ibid.

18 Wali, Kameshwar C., *Chandra: A Biography of S. Chandrasekhar*, Penguin Books, New Delhi, 1991, p. 154.

19 Ibid. 154.

20 Ibid. 154.
21 Ibid. 282.
22 Chandrasekhar, S., Received by Homi Bhabha, Tata Institute of Fundamental Research Archive, 2 April 1946, Mumbai, Maharashtra, India.
23 Ibid.
24 Wali, Kameshwar C., *Chandra: A Biography of S. Chandrasekhar*, Penguin Books, New Delhi, 1991, p. 283.
25 Venkataraman, G., *Bhabha and his Magnificent Obsessions*, Universities Press, Hyderabad, 1994, p. 139.
26 Bhabha, Homi, Received by Sorab Saklatvala, Tata Central Archives, 9 May 1945, Pune, Maharashtra, India.
27 Bhabha, Homi, Received by Rustum Choksi, Tata Central Archives, 30 March 1945, Pune, Maharashtra, India.
28 Ramaswamy, Ramakrishna (ed.), *D.D. Kosambi: Selected Works in Mathematics and Statistics*, Springer India, New Delhi, 2016, p. v.
29 Raju, C.K., 'Kosambi the Mathematician', *The Many Careers of D.D. Kosambi: Critical Essays*, D.N. Jha (ed.), LeftWord Books, New Delhi, 2011, p. 168.
30 Ramaswamy, Ramakrishna (ed.), *D.D. Kosambi: Selected Works in Mathematics and Statistics*, Springer India, New Delhi, 2016, p. vii.
31 Kosambi, Meera (ed.), *Unsettling the Past: Unknown Aspects and Scholarly Assessments of D.D. Kosambi*, Permanent Black, 2013.

32 Ibid. 18.
33 Ibid. 173.
34 Raju, C.K., 'Kosambi the Mathematician', *The Many Careers of D.D. Kosambi: Critical Essays*, D.N. Jha (ed.), LeftWord Books, New Delhi, 2011, p. 170.
35 Bhabha, H.J., 'Relativistic Wave Equations for the Elementary Particles', *Reviews of Modern Physics*, Vol. 17, No. 2–3, pp. 200–16, https://tinyurl.com/43ats6p3. Accessed on 16 August 2022.
36 Singh, Virender, 'Bhabha's Contributions to Elementary Particle Physics and Cosmic Rays Research', *Resonance*, Vol. 14, 2009, pp. 430–54, https://tinyurl.com/25mtennv. Accessed on 16 August 2022.
37 Singh, Jagjit, *Some Eminent Indian Scientists*, Publications Division, Ministry of Information and Broadcasting, Government of India, 1966.
38 Singh, Virender, 'Bhabha's Contributions to Elementary Particle Physics and Cosmic Rays Research', *Resonance*, Vol. 14, 2009, pp. 430–54, https://tinyurl.com/25mtennv. Accessed on 16 August 2022.
39 Chowdhury, Indira, *Growing the Tree of Science: Homi Bhabha and the Tata Institute of Fundamental Research*, Oxford University Press, New Delhi, 2016, p. 69.
40 Ibid. 124.

Chapter 12: The Atomic Imperative

1 Bhabha, Homi, 'Historical Note on Tata Institute of Fundamental Research', Wilson Center Digital Archive, 1 January 1954, https://tinyurl.com/43dpnxh6. Accessed on 19 December 2022.

2 A betatron is a type of particle accelerator that uses the electric field induced by varying magnetic field to accelerate electrons (beta particles) to high speeds in a circular orbit.

3 Bhabha, Homi, Received by Mark Oliphant, Tata Institute of Fundamental Research Archive, 5 December 1945, Mumbai, Maharashtra, India.

4 Phalkey, Jahnavi, *Atomic State: Big Science in Twentieth-Century India*, Permanent Black, New Delhi, 2013, p. 81.

5 Ibid. 95–6.

6 Ibid. 96.

7 Kosambi, Meera (ed.), *Unsettling the Past: Unknown Aspects and Scholarly Assessments of D.D. Kosambi*, Permanent Black, 2013, pp. 115–7.

8 Ibid. 115–7.

9 Bhabha, Homi, Received by D.D. Kosambi, Tata Institute of Fundamental Research Archive, 1 August 1946, Mumbai, Maharashtra, India.

10 Phalkey, Jahnavi, *Atomic State: Big Science in Twentieth-Century India*, Permanent Black, New Delhi, 2013, p. 285.

11 Kosambi, D.D., Received by Homi Bhabha, Tata Institute of Fundamental Research Archive, 23 September 1946, Mumbai, Maharashtra, India.

12 Kosambi, Meera (ed.), *Unsettling the Past: Unknown Aspects and Scholarly Assessments of D.D. Kosambi*, Permanent Black, 2013, pp. 117–20.

13 Ibid. 117–20.

14 Bhatnagar, S.S., Received by Rustum Choksi, Bhatnagar Papers, 26 July 1947, Nehru Museum Memorial Library, New Delhi, India.

15 Bhabha, Homi, Received by S.S. Bhatnagar, Bhatnagar Papers, 25 July 1947, Nehru Museum Memorial Library, New Delhi, India.

16 Taylor, H.J., P. J. Lavakare and R. R. Daniel, 'Seventy Years of Physics', *Current Science*, Vol. 59, No. 23, 10 December 1990, https://tinyurl.com/3jdcud9j. Accessed on 17 August 2022.

17 Ibid.

18 Phalkey, Jahnavi, *Atomic State: Big Science in Twentieth-Century India*, Permanent Black, New Delhi, 2013, p. 135.

19 Ibid. 140–1.

20 Born, Max, Received by Homi Bhabha, Tata Institute of Fundamental Research Archive, 29 October 1947, Mumbai, Maharashtra, India.

21 Ibid.

22 Bhattacharya, Ananyo, *The Man from the Future: The Visionary Life of John von Neumann*, Allen Lane, 2021, p. xii–iii.

23 Ibid. xiv.

24 Isaacson, Walter, *The Innovators: How a Group of Hackers, Geniuses and Geeks Created the Digital Revolution*, Simon & Schuster, 2014, p. 102.

25 Bhattacharya, Ananyo, *The Man from the Future: The Visionary Life of John von Neumann*, Allen Lane, 2021, p. xi.

26 Isaacson, Walter, *The Innovators: How a Group of Hackers, Geniuses and Geeks Created the Digital Revolution*, Simon & Schuster, 2014, p. 103.

27 Bhabha, Homi, Received by John von Neumann, Tata Institute of Fundamental Research Archive, 9 January 1948, Mumbai, Maharashtra, India.

28 von Neumann, John, Received by Homi Bhabha, Tata Institute of Fundamental Research Archive, 3 February 1948, Mumbai, Maharashtra, India.

29 In the letter, he mistakenly refers to him as Bochert. There is no doubt, however, that he is referring to Bochner.

30 Bhabha, Homi, Received by D.D. Kosambi, Tata Institute of Fundamental Research Archive, 9 January 1948, Mumbai, Maharashtra, India.

31 Raghunathan, M.S., 'Artless Innocents and Ivory-Tower Sophisticates: Some Personalities on the Indian Mathematical Scene', *Current Science*, Vol. 85, No. 4, 2003, pp. 526–36, https://tinyurl.com/55mscetu. Accessed on 17 August 2022.

32 Kosambi, D.D., Received by Homi Bhabha, Tata Institute of Fundamental Research Archive, 7 January 1948, Mumbai, Maharashtra, India.

33 Bhabha, Homi, Received by D.D. Kosambi, Tata Institute of Fundamental Research Archive, 9 January 1948, Mumbai, Maharashtra, India.

Chapter 13: Early Birds

1 Chowdhury, Indira and Ananya Dasgupta, *A Masterful Spirit: Homi J. Bhabha, 1909-1966*, Penguin Books, New Delhi, 2010, p. 24.

2 Sreekantan, B.V., 'Sixty Years of the Tata Institute of Fundamental Research 1945-2005: The Role Of Young Men in the Creation and Development of this Institute', *Current Science*, Vol. 90, No. 7, 10 April 2006, pp. 1012–25, https://tinyurl.com/3u9m4tnr. Accessed on 18 August 2022.

3 Isaacson, Walter, *The Code Breaker: Jennifer Doudna, Gene Editing, and the Future of the Human Race*, Simon & Schuster, 2021, p. 104.

4 Chowdhury, Indira and Ananya Dasgupta, *A Masterful Spirit: Homi J. Bhabha, 1909-1966*, Penguin Books, New Delhi, 2010, p. 144.

5 Ghosh, Dipan K. and Arun K. Grover, *Tribute to a Titan: Birth Centenary of Homi Jehangir Bhabha*, Indian Physics Association, Mumbai, 2009, pp. 113–4.

6 He had joined Sacramento Junior College as a first-year student and later earned a tuition scholarship to continue his studies at the University of Southern California.

7 Gill, Piara Singh, *Up Against Odds: Autobiography of an Indian Scientist*, Allied Publishers, Bombay, 1992, p. 85.

8 Virk, H.S., 'Piara Singh Gill: A Pioneer Cosmic Ray Physicist of India (1911–2002)', *Current Science*, Vol. 82, No. 11, 10 June 2002, https://tinyurl.com/5384kxkn. Accessed on 18 August 2022.

9 Gill, Piara Singh, *Up Against Odds: Autobiography of an Indian Scientist*, Allied Publishers, Bombay, 1992, p. 85.

10 Ibid 86–7.

11 Chowdhury, Indira, *Growing the Tree of Science: Homi Bhabha and the Tata Institute of Fundamental Research*, Oxford University Press, New Delhi, 2016, p. 93.

12 Choksi, Rustum, Received by Piara Singh Gill, Tata Institute of Fundamental Research Archive, 27 January 1948, Mumbai, Maharashtra, India.

13 Gill, Piara Singh, Received by Rustum Choksi, Tata Institute of Fundamental Research Archive, 28 January 1948, Mumbai, Maharashtra, India.

14 Gill, Piara Singh, *Up Against Odds: Autobiography of an Indian Scientist*, Allied Publishers, Bombay, 1992, p. 96.

15 Ibid. 95.

16 Ibid. 95.

17 Ibid. 96.

18 Alladi, Krishnaswami, 'Alladi Ramakrishnan's Theoretical Physics Seminar and the Creation of Matscience', Krishnaswami Alladi, 2 January 2013, https://tinyurl.com/4rzna2xd. Accessed on 18 August 2022.

19 Bhabha, Homi, Received by Sir Alladi Krishnaswami Ayyar, Tata Institute of Fundamental Research Archive, 27 February 1948, Mumbai, Maharashtra, India.

20 Alladi, Krishnaswami, 'IAS Visit Inspires Creation of Institute of Mathematical Sciences', Institute for Advanced Study, https://tinyurl.com/4jehhfns. Accessed on 18 August 2022.

21 Alladi, Krishnaswami, 'Alladi Ramakrishnan's Theoretical Physics Seminar and the Creation of Matscience', Krishnaswami Alladi, 2 January 2013, https://tinyurl.com/4rzna2xd. Accessed on 18 August 2022.

22 Alladi, Krishnaswami, 'IAS Visit Inspires Creation of Institute of Mathematical Sciences', Institute for Advanced Study, https://tinyurl.com/4jehhfns. Accessed on 18 August 2022.

23 Ibid.

24 Monte Carlo methods are a broad class of computational algorithms that rely on repeated random sampling to obtain numerical results.

25 Sreekantan, B.V. and Richa Malhotra, 'Homi Jehangir Bhabha: A Visionary', *Resonance*, Vol. 15, No. 5, May 2010, pp. 462–75, https://tinyurl.com/mrxyyz53. Accessed on 18 August 2022.

26 Ibid.

27 Ibid.

28 Ibid.

29 Mondal, Naba K. and Sreerup Raychaudhuri, 'M.G.K. Menon—Statesman of Indian Science', *Resonance*, Vol. 24, No. 11, November 2019, pp. 1189–233, https://tinyurl.com/24zm47a2. Accessed on 18 August 2022.

30 Pulakkat, Hari, *Space Life Matter: The Coming of Age of Indian Science*, Hachette, 2021, p. xv.

31 Bhabha, Homi, *Director's Report, 1952-53*, Tata Institute of Fundamental Research Archive, Mumbai, Maharashtra, India.

32 Chowdhury, Indira and Ananya Dasgupta, *A Masterful Spirit: Homi J. Bhabha, 1909-1966*, Penguin Books, New Delhi, 2010, p. 136.

33 His figure was 2.24 microseconds with a plus or minus variation of 0.15 microsecond; the modern value is 2.197083 with a variation of 0.000015 microsecond.

34 Raychaudhuri, Sreerup, 'Cosmic Ray Research in India: A Historical Perspective', *Physics News*, Vol. 51, No. 1–2, https://tinyurl.com/4enmz7he. Accessed on 18 August 2022.

35 Bhattacharya, Amitabha, 'A Forgotten Legend', *The Statesman*, 23 September 2018, https://tinyurl.com/4x7t977v. Accessed on 18 August 2022.

36 Ibid.

37 Roy, Suprakash C. and Rajinder Singh, 'Bibha Chowdhuri: The First Woman Scientist at the TIFR', *Physics News*, Vol. 51, No. 1–2, pp. 4–11, https://tinyurl.com/2yxxbjpj. Accessed on 18 August 2022.

38 Ibid.

39 Ibid.

40 Singh, Rajinder and Suprakash C. Roy, *A Jewel Unearthed: Bibha Chowdhuri: The Story of an Indian Woman Scientist*, Ebenezer Printing House, 2020.

41 Cockcroft, John and M.G.K. Menon, *Homi Jehangir Bhabha 1909-1966*, Royal Institution of Great Britain, London, 1967, p. 10.

42 Kakodkar, Anil and Suresh Gangotra, *Fire and Fury: Transforming India's Strategic Identity*, Rupa, New Delhi, 2019, p. 122.

Chapter 14: The Search for a Nest

1 Chowdhury, Indira, *Growing the Tree of Science: Homi Bhabha and the Tata Institute of Fundamental Research*, Oxford University Press, New Delhi, 2016, p. 42.

2 Bhabha, Homi, Received by H.M. Patel, Tata Institute of Fundamental Research Archives, 9 February 1950, Mumbai, Maharashtra, India.

3 'Bhabha and Jawaharlal Nehru Correspondence on Indian Nuclear History', Wilson Center Digital Archive, 9 December 1953, https://tinyurl.com/3e9bda2d. Accessed on 18 August 2022.

4 'Bhabha and Jawaharlal Nehru Correspondence on Indian Nuclear History', Wilson Center Digital Archive, 11 December 1953, https://tinyurl.com/3e9bda2d. Accessed on 18 December 2022.
5 'Speech by Shri J.R.D. Tata at the Foundation Stone Laying Ceremony', Nuclear History Project, Manohar Parrikar Institute for Defence Studies, 1 January 1954, https://tinyurl.com/y58net8k. Accessed on 18 August 2022.
6 Ibid.
7 'Speech by Shri Morarji R. Desai at the Foundation Stone Laying Ceremony', Nuclear History Project, Manohar Parrikar Institute for Defence Studies, 1 January 1954, https://tinyurl.com/2w48f86j. Accessed on 18 August 2022.
8 'Speech by Dr S.S. Bhatnagar at the Foundation Stone Laying Ceremony', Nuclear History Project, Manohar Parrikar Institute for Defence Studies, 1 January 1954, https://tinyurl.com/33h3frnn. Accessed on 18 August 2022.
9 Ibid.
10 Ibid.
11 'Speech by Homi Bhabha at the Foundation Stone Laying Ceremony', Nuclear History Project, Manohar Parrikar Institute for Defence Studies, 1 January 1954, https://tinyurl.com/yc74uasm. Accessed on 28 December 2022.
12 Ibid.
13 Ibid.
14 Nehru, Jawaharlal, 'Speech on 1 January 1954', Tata Institute of Fundamental Research Archives, Mumbai, Maharashtra, India.
15 Bhabha, Homi, Received by Rustum Choksi, Tata Institute of Fundamental Research Archives, 10 May 1951, Mumbai, Maharashtra, India.
16 Bhabha, Homi, 'Historical Note on Tata Institute of Fundamental Research', Nuclear History Project, Manohar Parrikar Institute for Defence Studies, 1 January 1954, https://tinyurl.com/2v36uwz4. Accessed on 18 August 2022.
17 Ibid.
18 Ibid.
19 'Memorandum dated 28 August 1954', TIFR Archives.
20 Nehru, Jawaharlal, Received by Homi Bhabha, 'Bhabha and Jawaharlal Nehru Correspondence on Indian Nuclear History', Wilson Center Digital Archive, 11 January 1954, https://tinyurl.com/3e9bda2d. Accessed on 18 December 2022.

Chapter 15: The Foreign Hand

1 Chowdhury, Indira, *Growing the Tree of Science: Homi Bhabha and the Tata Institute of Fundamental Research*, Oxford University Press, New Delhi, 2016, p. 76.
2 Bhabha, Homi, Received by S.S. Chern, Tata Institute of Fundamental Research Archive, 16 December 1948, Mumbai, Maharashtra, India.

3 Phalkey, Jahnavi, *Atomic State: Big Science in Twentieth-Century India*, Permanent Black, New Delhi, 2013, pp. 246–7.

4 Nath, Biman, *Homi J Bhabha: A Renaissance Man among Scientists*, Niyogi Books, New Delhi, 2021.

5 Singh, Rajinder, 'Indo-American Relations with Reference to Bernard Peters', *Indian Journal of History of Science*, Vol. 43, No. 3, 2008, pp. 437–54.

6 Chowdhury, Indira, *Growing the Tree of Science: Homi Bhabha and the Tata Institute of Fundamental Research*, Oxford University Press, New Delhi, 2016, p. 75.

7 Singh, Rajinder, 'Indo-American Relations with Reference to Bernard Peters', *Indian Journal of History of Science*, Vol. 43, No. 3, 2008, pp. 437–54.

8 Gopal, Sarvepalli, *Jawaharlal Nehru: A Biography*, Vol. 3, Oxford University Press, 2011, p. 287.

9 Anderson, Robert S., *Nucleus and Nation: Scientists, International Networks, and Power in India*, University of Chicago Press, 2010, pp. 178–9.

10 Peters, Bernard, 'The Decade of My Association with Research in India', *Current Science*, Vol. 61, No. 11, 10 December 1991, pp. 717–22, https://tinyurl.com/mr2kmb9x. Accessed 19 August 2022.

11 Ibid.

12 'International Conference on Elementary Particles', *Current Science*, Vol. 20, No. 1, 1951, https://tinyurl.com/2p9bceya. Accessed 19 August 2022.

13 Chowdhury, Indira and Ananya Dasgupta, *A Masterful Spirit: Homi J. Bhabha, 1909-1966*, Penguin Books, New Delhi, 2010, p. 126.

14 'Sustaining the Search for Excellence: Higher Education and the J.N. Tata Endowment', *Tata Review*, Vol. 33, No. 3, September 1998, p. 13.

15 Taylor, H.J., 'Seventy Years in Physics', *Current Science*, Vol. 59, No. 23, 1990, https://tinyurl.com/3a46ph6v. Accessed 19 August 2022.

16 Anderson, Robert S., *Nucleus and Nation: Scientists, International Networks, and Power in India,* University of Chicago Press, 2010, p. 177.

17 Chowdhury, Indira and Ananya Dasgupta, *A Masterful Spirit: Homi J. Bhabha, 1909-1966*, Penguin Books, New Delhi, 2010, pp. 191–2.

18 Khasnis, Giridhar, 'The Silent Maestro', *Deccan Herald*, 2 February 2014, https://tinyurl.com/2p82bw3h. Accessed on 22 August 2022.

19 Baggott, Jim, *Atomic: The First War of Physics and the Secret History of the Atom Bomb, 1939–1949,* Icon Books, London, 2009, p. 234.

20 Bodanis, David, *$E = MC^2$: A Biography of the World's Most Famous Equation*, Pan Books, London, 2000, p. 148.

21 Chowdhury, Indira, *Growing the Tree of Science: Homi Bhabha and the Tata Institute of Fundamental Research*, Oxford University Press, New Delhi, 2016, p. 134.

22 Ibid. 135.

Chapter 16: International Recognition and Distinguished Visitors

1 Chowdhury, Indira, *Growing the Tree of Science: Homi Bhabha and the Tata Institute of Fundamental Research*, Oxford University Press, New Delhi, 2016, p. 76.
2 Basu, Biman, *Yash Pal: A Life in Science*, Vigyan Prasar, New Delhi, 2006, pp. 11–2.
3 Ibid. 15.
4 Ibid. 15.
5 Ibid. 16–7.
6 Ibid. 18–9.
7 Cronin, James W., 'The 1953 Cosmic Ray Conference at Bagneres de Bigorre', *European Physics Journal*, Vol. 183, 2011, https://tinyurl.com/54mfrn4u. Accessed 19 August 2022.
8 Ibid.
9 Nath, Biman, *Homi J Bhabha: A Renaissance Man among Scientists*, Niyogi Books, New Delhi, 2021, pp. 109–10.
10 Fitch, Val L., 'Elementary Particle Physics: The Origins', *Reviews of Modern Physics*, Vol. 71, No. 2, 1999, https://tinyurl.com/2823wekm. Accessed 19 August 2022.
11 Cronin, James W., 'The 1953 Cosmic Ray Conference at Bagneres de Bigorre', *European Physics Journal*, Vol. 183, 2011, https://tinyurl.com/54mfrn4u. Accessed 19 August 2022.
12 *Times of India*, 19 April 1954.
13 Tata, J.R.D., Received by Rustum Choksi, Tata Institute of Fundamental Research Archives, Mumbai, Maharashtra, India.
14 Bhabha, Homi, Note, 18 May 1954, Tata Institute of Fundamental Research Archives, Mumbai, Maharashtra, India.
15 Ibid.
16 Nath, Biman, *Homi J Bhabha: A Renaissance Man among Scientists*, Niyogi Books, New Delhi, 2021, pp. 109–10.
17 Goswami, J.N. and J.D. Macdougall, 'Devendra Lal', Biographical Memoirs of Fellows of the Royal Society, The Royal Society Publishing, https://tinyurl.com/ye286v4u. Accessed 19 August 2022.
18 Raychaudhuri, Sreerup, 'Cosmic Ray Research in India: A Historical Perspective', *Physics News*, Vol. 51, No. 1–2, https://tinyurl.com/mwzprp43. Accessed 19 August 2022.
19 Peters, Bernard, 'Speech by B. Peters on 10 April 1971 on the Occasion of the Silver Jubilee Celebrations of TIFR', Tata Institute of Fundamental Research Archives, Mumbai, Maharashtra, India.

20 Peters, Bernard, 'The Decade of My Association with Research in India', *Current Science*, Vol. 61, No. 11, 10 December 1991, pp. 717–22, https://tinyurl.com/mr2kmb9x. Accessed 19 August 2022.

21 Pauli, Wolfgang, Received by Homi Bhabha, Pauli Letter Collection, CERN Document Server, 31 December 1951, https://tinyurl.com/2mhu9mrk. Accessed 19 August 2022.

22 Pauli, Wolfgang, Received by Homi Bhabha, Pauli Letter Collection, CERN Document Server, 25 June 1952, https://tinyurl.com/52a22def. Accessed 19 August 2022.

23 Pauli, Wolfgang, Received by Homi Bhabha, Pauli Letter Collection, CERN Document Server, 26 July 1952, https://tinyurl.com/2mhu9mrk. Accessed 19 August 2022.

24 Ibid.

25 Pauli, Wolfgang, Received by Homi Bhabha, Pauli Letter Collection, CERN Document Server, 1 October 1952, https://tinyurl.com/2v86kpyx. Accessed 19 August 2022.

26 Pauli, Wolfgang, Received by Homi Bhabha, Pauli Letter Collection, CERN Document Server, 2 March 1953, https://tinyurl.com/5yyf24k4. Accessed 19 August 2022.

27 Wali, Kameshwar C., *Chandra: A Biography of S. Chandrasekhar*, Penguin Books, New Delhi, 1991.

28 Ibid. 256.

29 Pauli, Wolfgang, Received by Homi Bhabha, Pauli Letter Collection, CERN Document Server, 2 March 1953, https://tinyurl.com/5yyf24k4. Accessed 19 August 2022.

30 Malhotra, Inder, 'The Uncertainty Principle', *The Indian Express*, 31 March 2007, https://tinyurl.com/23m98s3y. Accessed 19 August 2022.

31 Marshak, Robert, 'The Pain and Joy of a Major Scientific Discovery', *A Gift of Prophecy: Essays in the Celebration of the Life of Robert Eugene Marshak*, E.C.G. Sudarshan (ed.), World Scientific Publishing, New Jersey, 1994, pp. 315–6.

32 'Robert Marshak - Session III', American Institute of Physics, 4 October 1970, https://tinyurl.com/2p8zzh4w. Accessed on 19 August 2022.

33 Ibid.

34 Ibid.

35 Ibid.

36 Farmelo, Graham, *The Strangest Man: The Hidden Life of Paul Dirac, Mystic of the Atom*, Basic Books, New York, Reprint 2009, p. 352.

37 Ibid. 353.

38 Alladi, Krishnaswami, 'IAS Visit Inspires Creation of Institute of Mathematical Sciences', Institute for Advanced Study, https://tinyurl.com/4jehhfns. Accessed on 23 August 2022.

39 Singh, Rajinder, 'Homi J. Bhabha and Niels Bohr', *Current Science*, Vol. 97, No. 4, 2009, https://tinyurl.com/bdzk9r5j. Accessed on 23 August 2022.

40 Ibid.

41 Ibid.

42 Ibid.

43 Singh, Rajinder and Hardev Singh Virk, 'Homi J. Bhabha: Physics Nobel Prize Nominee and Nominator', *OmniScience: A Multi-Disciplinary Journal*, Vol. 7, No. 1, pp. 4–10, https://tinyurl.com/2p8c9bdc. Accessed on 18 August 2022.

44 Wadia, Spenta R., 'Homi Jehangir Bhabha and the Tata Institute of Fundamental Research' *Current Science*, Vol. 96, No. 5, 10 March 2009, https://tinyurl.com/yc3ka39x. Accessed on 18 August 2022.

Chapter 17: Studying the Extraterrestrials

1 Bhabha. Homi, 'Speech at the Inauguration of New Buildings', *The Tata Institute of Fundamental Research: Inauguration of New Buildings*, Bombay, 15 January 1962, S.N. Gupta, Bombay, 1962, p. 37.

2 Mondal, Naba K. and Sreerup Raychaudhuri, 'M.G.K. Menon—Statesman of Indian Science', *Resonance*, Vol. 24, No. 11, November 2019, pp. 1189–233, https://tinyurl.com/24zm47a2. Accessed on 18 August 2022.

3 The idea of parity can be understood in terms of a mirror-image. If one looks in the mirror, left becomes right and right becomes left. To put it another way, spatial coordinates are flipped. If they are flipped again, they go back to how they were, which is called a rotation. While a rotation has a parity of 1, a flip has a parity of -1.

4 A beta particle is a fast energetic electron or positron.

5 The strong force binds quarks together in clusters to make protons and neutrons.

6 Gribbin, John and Mary Gribbin, *Richard Feynman: A Life in Science*, Icon Books, London, 2018, pp. 211–2.

7 Alfred Nobel had stipulated that the prize should be given for work carried out in the previous year but this rule is almost always broken.

8 Ghosh, Dipan K. and Arun K. Grover, *Tribute to a Titan: Birth Centenary of Homi Jehangir Bhabha*, Indian Physics Association, Mumbai, 2009, p. 79.

9 Mondal, Naba K. and Sreerup Raychaudhuri, 'M.G.K. Menon—Statesman of Indian Science', *Resonance*, Vol. 24, No. 11, November 2019, pp. 1189–233, https://tinyurl.com/24zm47a2. Accessed on 18 August 2022.

10 Ghosh, Dipan K. and Arun K. Grover, *Tribute to a Titan: Birth Centenary of Homi Jehangir Bhabha*, Indian Physics Association, Mumbai, 2009, p. 79.

11 Menon, M.G.K., Interview with Indira Chowdhury, Tata Institute of Fundamental Research Archives, 15 April 2005, Mumbai, Maharashtra, India.

12 Tropopause is the fuzzy layer of the atmosphere where the denser troposphere gives way to the stratosphere.

13 Mondal, Naba K. and Sreerup Raychaudhuri, 'M.G.K. Menon—Statesman of Indian Science', *Resonance*, Vol. 24, No. 11, November 2019, pp. 1189–233, https://tinyurl.com/24zm47a2. Accessed on 18 August 2022.

14 Pulakkat, Hari, *Space Life Matter: The Coming of Age of Indian Science*, Hachette, Gurugram, 2021, pp. 103–04.

15 Mondal, Naba K. and Sreerup Raychaudhuri, 'M.G.K. Menon—Statesman of Indian Science', *Resonance*, Vol. 24, No. 11, November 2019, pp. 1189–233, https://tinyurl.com/24zm47a2. Accessed on 18 August 2022.

16 Ramachandran, R., 'Peerless Trail-Blazer', *Frontline*, 21 August 2015, https://tinyurl.com/y5n2d2ey. Accessed on 26 August 2022.

17 Raychaudhuri, Sreerup, 'Cosmic Ray Research in India: A Historical Perspective', *Physics News*, Vol. 51, No. 1–2, https://tinyurl.com/4enmz7he. Accessed on 18 August 2022.

18 Chowdhury, Indira, *Growing the Tree of Science: Homi Bhabha and the Tata Institute of Fundamental Research*, Oxford University Press, New Delhi, p. 173.

19 Raychaudhuri, Sreerup, 'Cosmic Ray Research in India: A Historical Perspective', *Physics News*, Vol. 51, No. 1–2, https://tinyurl.com/4enmz7he. Accessed on 18 August 2022.

20 Sudarshan was a Syrian Christian. The reason he changed his name to Sudarshan was that he married a Hindu girl and part of the arrangement was that his name would be changed that way.

21 The many-body problem is a general name for a large category of physical problems pertaining to the properties of microscopic systems made out of many interacting particles, which requires the use of quantum mechanics to provide an accurate description of the system.

22 Chowdhury, Indira, *Growing the Tree of Science: Homi Bhabha and the Tata Institute of Fundamental Research,* Oxford University Press, New Delhi, 2016, pp. 137–38.

23 Ibid.

24 Ibid. 138.

25 Sudarshan, E.C.G., Received by Homi Bhabha, Tata Institute of Fundamental Research Archives, 5 June 1962, Mumbai, Maharashtra, India.

26 Minwalla, Shiraz, Personal interview, 25 February 2022.

27 'Robert Marshak - Session III', American Institute of Physics, 4 October 1970, https://tinyurl.com/2p8zzh4w. Accessed on 19 August 2022.

28 Marshak, Robert E., 'The Pain and the Joy of a Major Scientific Discovery', *Current Science*, Vol. 63, No. 2, 25 July 1992.

29 Deshmukh, Chintamani, *Homi Jehangir Bhabha*, National Book Trust, India, New Delhi, 2014, p. 121.

Chapter 18: A New Home

1 *The Tata Institute of Fundamental Research: Inauguration of New Buildings, Bombay, January 15, 1962,* S.N. Gupta, Bombay, 1962.

2 Bhabha, Homi, 'Historical Note on Tata Institute of Fundamental Research', Nuclear History Project, Manohar Parrikar Institute for Defence Studies, 1 January 1954, https://tinyurl.com/2v36uwz4. Accessed on 18 August 2022.

3 Chowdhury, Indira and Ananya Dasgupta, *A Masterful Spirit: Homi J. Bhabha, 1909-1966*, Penguin Books, New Delhi, 2010, p. 92.

4 It was later renamed Kanvinde, Rai and Chowdhury.

5 Bartsch, Helmuth, 'Note by Helmuth Bartsch', *The Tata Institute of Fundamental Research: Inauguration of New Buildings, Bombay, January 15, 1962,* S.N. Gupta, Bombay, 1962, pp. 56–7.

6 Bhabha, Homi, 'Speech at the Inauguration of New Buildings', *The Tata Institute of Fundamental Research: Inauguration of New Buildings, Bombay, January 15, 1962,* S.N. Gupta, Bombay, 1962, p. 37.

7 Bartsch, Helmuth, 'Note by Helmuth Bartsch', *The Tata Institute of Fundamental Research: Inauguration of New Buildings, Bombay, January 15, 1962,* S.N. Gupta, Bombay, 1962.

8 Chowdhury, Indira, 'The Laboratory and its Double: The Making of the Scientist-Citizen at TIFR', Centre for the Advanced Study of India, University of Pennsylvania, 2012, https://tinyurl.com/hnnwv7a7. Accessed on 29 August 2022.

9 John, Joseph, 'Bhabha, The Friend of Trees', *Bharat Jyoti*, 1 May 1966.

10 Chowdhury, Indira, *Growing the Tree of Science: Homi Bhabha and the Tata Institute of Fundamental Research,* Oxford University Press, New Delhi, 2016, p. 92.

11 Cockcroft, John and M.G.K. Menon, *Homi Jehangir Bhabha 1909-1966*, Royal Institution of Great Britain, London, 1967, p. 23.

12 'Tata Institute of Fundamental Research: Report 1945–70', Wilson Center Digital Archive, p. 52, https://tinyurl.com/2p8cep6e. Accessed on 29 August 2022.

13 Prakasa, Sri, 'Speech at the Inauguration of New Buildings', *The Tata Institute of Fundamental Research: Inauguration of New Buildings, Bombay, January 15, 1962,* S.N. Gupta, Bombay, 1962.

14 Tata, J.R.D., 'Transcript of JRD Tata's Speech on 15 January 1962', *The Tata Institute of Fundamental Research: Inauguration of New Buildings, Bombay, January 15, 1962,* S.N. Gupta, Bombay, 1962.

15 Ibid.

16 Ibid.

17 Chavan, Y.B., 'Speech 15 January 1962', *The Tata Institute of Fundamental Research: Inauguration of New Buildings, Bombay, January 15, 1962,* S.N. Gupta, Bombay, 1962.

18 Bhabha, Homi, 'Speech at the Inauguration of New Buildings', *The Tata Institute of Fundamental Research: Inauguration of New Buildings, Bombay, January 15, 1962*, S.N. Gupta, Bombay, 1962.

19 Nehru, Jawaharlal, 'Speech at the Inauguration of New Buildings', *The Tata Institute of Fundamental Research: Inauguration of New Buildings, Bombay, January 15, 1962*, S.N. Gupta, Bombay, 1962.

20 Ibid.

21 Ibid.

22 Wagner, Caroline S., *The New Invisible College: Science for Development*, Brookings Institute Press, Washington D.C., 2008, p. 35.

23 Chowdhury, Indira, *Growing the Tree of Science: Homi Bhabha and the Tata Institute of Fundamental Research*, Oxford University Press, New Delhi, 2016, p. 142.

24 Ibid. 157.

25 In physics, phenomenology is the application of theoretical physics to experimental data by making quantitative predictions based on known theories. Phenomenology stands in contrast with experimentation in the scientific method, in which the goal of the experiment is to test a scientific hypothesis instead of making predictions. It is commonly applied to the field of particle physics, where it forms a bridge between the mathematical models of theoretical physics and the results of the high energy particle experiments.

26 Nath, Biman, *Homi J Bhabha: A Renaissance Man among Scientists*, Niyogi Books, New Delhi, 2021, p. 116.

27 Jayaraman, Aiyasami, *Chandrasekhara Venkata Raman*: *A Memoir*, Indian Academy of Sciences, Bangalore, 2017, p. 104.

28 Subramanian, Kapil, 'Why Nobel laureate CV Raman resented Nehru (and even took a swipe at him)', *Scroll.in*, https://tinyurl.com/2uvkr466. Accessed on 29 August 2022.

29 Ibid.

30 Ibid.

31 Kademani, B.S., Anil Sagar and Vijai Kumar, *Scientometric Portrait of Homi Jehangir Bhabha*, Bhabha Atomic Research Centre, Mumbai, 2009, pp. 10–1.

32 Bhabha had 22 co-authors in 25 of his 104 publications. The highest numbers of publications (six) were published in 1941, 1945 and 1964, respectively. Out of 104 publications, 30 received citations. These 30 publications have received a total of 1,211 citations during 1933–2008. Bhabha became a citable author in 1937. His publications have been cited by noted Nobel laureates, like V.L. Ginzberg, Wolfgang Pauli, H.A. Bethe, Max Born, W. Bothe, E. P. Wigner, H. Yukawa, P.M.S. Blackett, and C.N. Yang.

33 Bhattacharjee, S.K., Interview with Indira Chowdhury, 12 April 2006, Tata Institute of Fundamental Research Archives, Mumbai, Maharashtra, India.

34 Chowdhury, Indira and Ananya Dasgupta, *A Masterful Spirit*: *Homi J. Bhabha, 1909-1966*, Penguin Books, New Delhi, 2010, p. 149.

35 Pandit, L.K., Interview with Indira Chowdhury, Tata Institute of Fundamental Research Archives, Mumbai, Maharashtra, India, 19 April 2005.

36 Chowdhury, Indira, *Growing the Tree of Science: Homi Bhabha and the Tata Institute of Fundamental Research,* Oxford University Press, New Delhi, 2016, p. 145.

37 Deshmukh, Chintamani, *Homi Jehangir Bhabha*, National Book Trust, India, New Delhi, 2014, p. 121.

38 The eightfold way is an organizational scheme for a class of subatomic particles known as hadrons that led to the development of the quark model. American physicist Murray Gell-Mann and Israeli physicist Yuval Ne'eman both proposed the idea in 1961.

39 Physicists felt that there could not be such a large number of fundamental particles and were looking for the 'real' fundamental particles that constituted the rest. This was not the only approach though. American physicist Geoffrey Chew offered an alternative bootstrap model based on S-matrix theory in which each particle is constructed from the others. This approach is sometimes called nuclear democracy since it avoids singling out certain particles as elementary.

40 Chowdhury, Indira, *Growing the Tree of Science: Homi Bhabha and the Tata Institute of Fundamental Research,* Oxford University Press, New Delhi, 2016, p. 144.

41 Bhabha, Homi, Received by Arthur Lall, Tata Institute of Fundamental Research Archives, 21 February 1961, Mumbai, Maharashtra, India.

42 Mondal, Naba K. and Sreerup Raychaudhuri, 'M.G.K. Menon—Statesman of Indian Science', *Resonance*, Vol. 24, No. 11, November 2019, pp. 1189–233, https://tinyurl.com/24zm47a2. Accessed on 18 August 2022.

Chapter 19: Building a School of Mathematics

1 Bhabha, Homi, 'Historical Note on Tata Institute of Fundamental Research', Wilson Center Digital Archive, 1 January 1954, https://tinyurl.com/43dpnxh6. Accessed on 11 August 2022.

2 Seshadri, C.S., 'K. Chandrasekharan (1920–2017)', *Hardy-Ramanujan Journal*, Vol. 40, 2017, pp. 47–51, https://tinyurl.com/94ctc2xy. Accessed on 29 August 2022.

3 Raghunathan, M.S., 'Artless Innocents and Ivory-Tower Sophisticates: Some Personalities on the Indian Mathematical Scene', *Current Science*, Vol. 85, No. 4, 2003, pp. 526–36, https://tinyurl.com/55mscetu. Accessed on 17 August 2022.

4 Ibid. 526–36.

5 Chowdhury, Indira, *Growing the Tree of Science: Homi Bhabha and the Tata Institute of Fundamental Research*, Oxford University Press, New Delhi, 2016, p. 163.

6 Raghunathan, M.S., 'Artless Innocents and Ivory-Tower Sophisticates: Some Personalities on the Indian Mathematical Scene', *Current Science*, Vol. 85, No. 4, 2003, pp. 526–36, https://tinyurl.com/55mscetu. Accessed on 17 August 2022.

7 Seshadri, C.S., 'K. Chandrasekharan (1920–2017)', *Hardy-Ramanujan Journal*, Vol. 40, 2017, pp. 47–51, https://tinyurl.com/94ctc2xy. Accessed on 29 August 2022.

8 A Latin square is an 'n × n' array filled with 'n' different symbols, each occurring exactly once in each row and exactly once in each column. Sudokus are 9 × 9 Latin squares that also satisfy additional requirements.

9 Seshadri, C.S., 'K. Chandrasekharan (1920–2017)', *Hardy-Ramanujan Journal*, Vol. 40, 2017, pp. 47–51, https://tinyurl.com/94ctc2xy. Accessed on 29 August 2022.

10 Ibid.

11 At the airport, Choksi was seeing off his son, Jal, who was going to Cambridge for his undergraduate studies. Jal trained in mathematics and came back and joined TIFR.

12 Choksi, Rustum, Received by Homi Bhabha, Tata Institute of Fundamental Research Archives, 2 May 1951, Mumbai, Maharashtra, India.

13 Bhabha, Homi, Received by Rustum Choksi, Tata Institute of Fundamental Research Archives, 10 May 1951, Mumbai, Maharashtra, India.

14 Ibid.

15 Raianu, Mircea, *Tata: The Global Corporation that Built Indian Capitalism*, Harvard University Press, 2021, p. 103.

16 Ibid. 103.

17 Chandrasekharan, K., Received by Homi Bhabha, Tata Institute of Fundamental Research Archives, 7 July 1960, Mumbai, Maharashtra, India.

18 Ibid.

19 Chandrasekharan, K. et al, Received by the director and council of TIFR, Tata Institute of Fundamental Research Archives, 23 August 1960, Mumbai, Maharashtra, India.

20 Ibid.

21 Ibid.

22 Ibid.

23 Kosambi, D.D., Received by Homi Bhabha, Tata Institute of Fundamental Research Archives, 6 September 1960, Mumbai, Maharashtra, India.

24 'Full Text of "Atomic Energy In India - D. D. Kosambi"', Internet Archive, https://tinyurl.com/yrm72euz. Accessed on 29 August 2022.

25 Kosambi, Meera (ed.), *Unsettling the Past: Unknown Aspects and Scholarly Assessments of D.D. Kosambi*, Permanent Black, 2013. p. 89.

26 Raju, C.K., 'Kosambi the Mathematician', *The Many Careers of D.D. Kosambi: Critical Essays*, D.N. Jha (ed.), LeftWord Books, New Delhi, 2011, p. 186.

27 Ibid. 187.
28 Chowdhury, Indira, *Growing the Tree of Science: Homi Bhabha and the Tata Institute of Fundamental Research*, Oxford University Press, New Delhi, 2016, p. 219.
29 Seshadri, C.S., 'K. Chandrasekharan (1920–2017)', *Hardy-Ramanujan Journal*, Vol. 40, 2017, pp. 47–51, https://tinyurl.com/94ctc2xy. Accessed on 29 August 2022.
30 Choksi, Jal, Interview with Indira Chowdhury, Oral history interview, 29 December 2004.
31 Early Days of Particle Physics in India by Sreerup Raychaudhuri, YouTube, https://tinyurl.com/4rjpwcj2. Accessed on 29 August 2022.
32 Chowdhury, Indira, *Growing the Tree of Science: Homi Bhabha and the Tata Institute of Fundamental Research*, Oxford University Press, New Delhi, 2016, p. 219.
33 Ibid. 218.

Chapter 20: Computers and Electronics

1 *The Tata Institute of Fundamental Research: Inauguration of New Buildings, Bombay, January 15, 1962*, S.N. Gupta, Bombay, 1962.
2 Shyamasunder, R.K. and M.A. Pai (eds.), *Homi Bhabha and the Computer Revolution*, Oxford University Press, New Delhi, 2011, p. 109.
3 Rudra, Ashok, *Prasanta Chandra Mahalanobis: A Biography*, Oxford University Press, New Delhi, 1998, p. 324.
4 Ibid. 324.
5 Menon, Nikhil, *Planning Democracy: How a Professor, an Institute, and an Idea Shaped India*, Penguin Random House, Gurugram, 2022, pp. 35–6.
6 Ibid. 36.
7 Subramanian, Samanth, *A Dominant Character: The Radical Science and Restless Politics of J.B.S. Haldane*, Simon & Schuster, New Delhi, 2019, p. 306.
8 Rao, C.R., 'Prasantha Chandra Mahalanobis, 1893–1972', Biographical Memoirs of Fellows of the Royal Society, Vol. 19, 1973, pp. 454–92, https://tinyurl.com/2swsjv9x. Accessed on 1 September 2022.
9 Menon, Nikhil, *Planning Democracy: How a Professor, an Institute, and an Idea Shaped India*, Penguin Random House, Gurugram, 2022, p. 103.
10 Chowdhury, Indira and Ananya Dasgupta, *A Masterful Spirit: Homi J. Bhabha, 1909-1966*, Penguin Books, New Delhi, p. 75.
11 Shyamasunder, R.K. and M.A. Pai (eds.), *Homi Bhabha and the Computer Revolution*, Oxford University Press, New Delhi, 2011, p. 112.
12 Weiner, Norbert, *I am a Mathematician: The Later Life of a Prodigy*, MIT Press, Cambridge, Massachusetts, 1964, p. 352.
13 Ibid. 352.

14 Rudra, Ashok, *Prasanta Chandra Mahalanobis: A Biography*, Oxford University Press, New Delhi, 1998, p. 306.
15 Narasimhan, R., 'Men, Machines and Ideas: An Autobiographical Essay', *Current Science*, Vol. 76, No. 3, 10 February 1999, JSTOR, https://tinyurl.com/3yafvj25. Accessed on 1 September 2022.
16 Sharma, Dinesh C., *The Outsourcer: The Story of India's IT Revolution*, MIT Press, Cambridge, Massachusetts, 2015, p. 23.
17 Ibid. 23–4.
18 Ibid. 23–4.
19 Ibid. 25.
20 Ibid. 25.
21 Shyamasunder, R.K. and M.A. Pai (eds.), *Homi Bhabha and the Computer Revolution*, Oxford University Press, New Delhi, 2011, p. 15.
22 Ibid. 114.
23 Ibid. 114.
24 Deshmukh, Chintamani, *Homi Jehangir Bhabha*, National Book Trust, India, New Delhi, 2014, p. 88.
25 Shyamasunder, R.K. and M.A. Pai (eds.), *Homi Bhabha and the Computer Revolution*, Oxford University Press, New Delhi, 2011, p. 115.
26 Ibid. 115.
27 Sharma, Dinesh C., *The Outsourcer: The Story of India's IT Revolution*, MIT Press, Cambridge, Massachusetts, 2015, p. 44.

Chapter 21: Science, Art and Culture

1 Chowdhury, Indira and Ananya Dasgupta, *A Masterful Spirit: Homi J. Bhabha, 1909-1966*, Penguin Books, New Delhi, 2010, p. 199.
2 Arbuthnot, Mollie, 'Rudolf von Leyden: The German-born Cartoonist Who Became Bombay's Top Art Critic in the 1930s', *Scroll.in*, 16 December 2018, https://tinyurl.com/mryfz5m5. Accessed on 30 August 2022.
3 Singh, Devika, 'German-speaking Exiles and the Writing of Indian Art History', *Journal of Art Historiography*, No. 17, December 2017, https://tinyurl.com/m728amhx. Accessed on 30 August 2022.
4 Gandhy, Kekoo, 'The Beginnings of the Art Movement', *Seminar*, 2003, https://tinyurl.com/3938rtby. Accessed on 30 August 2022.
5 Lee, Rachel and Mareike Schwartz, 'Bombay', Metromod, 2021, https://tinyurl.com/5n8a24u6. Accessed on 30 August 2022.
6 Deshmukh, Chintamani, *Homi Jehangir Bhabha*, National Book Trust, India, New Delhi, 2014, p. 113.
7 Singh, Devika, 'German-Speaking Exiles and the Writing of Indian Art History', *Journal of Art Historiography*, No. 17, 2017, https://tinyurl.com/m728amhx. Accessed on 30 August 2022.

8 Lee, Rachel and Mareike Schwartz, 'Bombay', Metromod, 2021, https://tinyurl.com/5n8a24u6. Accessed on 30 August 2022.

9 Chadha, Gita and Vidya Kamat, 'A Scientist's Patronage of Art: Homi Bhabha's Art Collection', *Humanscape*, Vol. 9, No. 8, August 2002.

10 Chatterjee, Mortimer and Tara Lal, *The TIFR Art Collection*, Tata Institute of Fundamental Research, Mumbai, 2010.

11 Gandhy, Kekoo, 'The Beginnings of the Art Movement', *Seminar*, 2003, https://tinyurl.com/3938rtby. Accessed on 30 August 2022.

12 Deshmukh, Chintamani, *Homi Jehangir Bhabha*, National Book Trust, India, New Delhi, 2014, p. 61.

13 Chowdhury, Indira and Ananya Dasgupta, *A Masterful Spirit: Homi J. Bhabha, 1909-1966*, Penguin Books, New Delhi, 2010, p. 47.

14 Ibid. 223.

15 Epstein met Albert Einstein at Roughton Heath, Norfolk, in 1933 and had three sittings for a bust. He recollected his meeting with Einstein as follows: 'His glance contained a mixture of the humane, the humorous and the profound. This was a combination which delighted me. He resembled the ageing Rembrandt.' (Clark, Ronald W., *Einstein: The Life and Times*, Avon Publishing, London 2001, p. 603.)

16 Chowdhury, Indira, *Growing the Tree of Science: Homi Bhabha and the Tata Institute of Fundamental Research*, Oxford University Press, New Delhi, 2016, pp. 103–4.

17 Chowdhury, Indira, 'The Laboratory and its Double: The Making of the Scientist-Citizen at TIFR', Centre for the Advanced Study of India, University of Pennsylvania, 2012, https://tinyurl.com/hnnwv7a7. Accessed on 29 August 2022.

18 Bhabha, Homi, Received by J.D. Bernal, Tata Institute of Fundamental Research Archives, 21 November 1962, Mumbai, Maharashtra, India.

19 Chowdhury, Indira, *Growing the Tree of Science: Homi Bhabha and the Tata Institute of Fundamental Research*, Oxford University Press, New Delhi, 2016, p. 81.

20 Chadha, Gita and Vidya Kamat, 'A Scientist's Patronage of Art: Homi Bhabha's Art Collection', *Humanscape*, Vol. 9, No. 8, August 2002.

21 Bhabha, Homi, Received by Jamini Roy, Tata Institute of Fundamental Research Archives, 19 October 1962, Mumbai, Maharashtra, India.

22 Roy, Jamini, Received by Homi Bhabha, Tata Institute of Fundamental Research Archives, 16 December 1962, Mumbai, Maharashtra, India.

23 Chowdhury, Indira, *Growing the Tree of Science: Homi Bhabha and the Tata Institute of Fundamental Research*, Oxford University Press, New Delhi, 2016, p. 81.

24 A Parsi naval officer, Nanavati, had shot dead his wife's paramour, Prem Ahuja and then, surrendered to the police.

25 Chowdhury, Indira, *Growing the Tree of Science: Homi Bhabha and the Tata Institute of Fundamental Research*, Oxford University Press, New Delhi, 2016, pp. 82–3.
26 Chowdhury, Indira and Ananya Dasgupta, *A Masterful Spirit: Homi J. Bhabha, 1909-1966*, Penguin Books, New Delhi, 2010, p. 234.
27 Deshmukh, Chintamani, *Homi Jehangir Bhabha*, National Book Trust, India, New Delhi, 2014, p. 115.
28 Ibid. 232.
29 Kapoor, Coomi, Personal telephonic interview, 20 July 2020.
30 The British premiere of *Idomeneo* took place in Glasgow in 1934.
31 Chowdhury, Indira, 'Memories, Memorabilia and Memorable Interviews: Talking to Mithoo Coorlawala', The Oral Historian, 14 May 2020, https://tinyurl.com/9be76u33. Accessed on 26 December 2022.
32 Marfatia, Meher, 'From Bombay with Love' *Mid-Day*, 23 May 2022, https://tinyurl.com/57v7tp7v. Accessed on 30 August 2022.
33 Parikh, Uttara, Personal telephonic interview, 22 May 2022.
34 Chowdhury, Indira, 'Memories, Memorabilia and Memorable Interviews: Talking to Mithoo Coorlawala', The Oral Historian, 14 May 2020, https://tinyurl.com/9be76u33. Accessed on 30 August 2022.
35 Ibid.
36 Ibid.

37 Ibid.
38 Ibid.
39 Kapoor, Coomi, *The Emergency: A Personal History*, Penguin Books, New Delhi, 2015, p. 147.
40 Ghosh, Dipan K. and Arun K. Grover, *Tribute to a Titan: Birth Centenary of Homi Jehangir Bhabha*, Indian Physics Association, Mumbai, 2009, p. 125.
41 Homi Bhabha A Scientist In Action, YouTube, https://tinyurl.com/bdz7ewbr. Accessed on 30 August 2022.
42 Ibid.

Chapter 22: 'Without Him Art is Orphaned'

1 Chowdhury, Indira and Ananya Dasgupta, *A Masterful Spirit: Homi J. Bhabha, 1909-1966*, Penguin Books, New Delhi, 2010, p. 211.
2 Chowdhury, Indira, *Growing the Tree of Science: Homi Bhabha and the Tata Institute of Fundamental Research*, Oxford University Press, New Delhi, 2016, pp. 83–4.
3 Ibid. 85.
4 Chatterjee, Mortimer, 'TIFR and the Birth of the Modern Indian Art Collection', *Mumbai Mirror*, 2 November 2015, https://tinyurl.com/4nxmhvwe. Accessed on 30 August 2022.

5 Mondal, Naba K. and Sreerup Raychaudhuri, 'M.G.K. Menon—Statesman of Indian Science', *Resonance*, Vol. 24, No. 11, November 2019, pp. 1189–233, https://tinyurl.com/24zm47a2. Accessed on 18 August 2022.

6 Deshmukh, Chintamani, *Homi Jehangir Bhabha*, National Book Trust, India, New Delhi, 2014, p. 113.

7 Davierwalla, Ardeshir, 'Letter to the Editor', *The Times of India*, 27 January 1966.

8 Menon, Rashmi, 'Science in Art: Ardeshir Davierwalla's Sculpture 'Galaxy' Sold at Record Price in Auction', *The Economic Times*, 19 September 2017, https://tinyurl.com/27npf626. Accessed on 30 August 2022.

9 Chatterjee, Mortimer and Tara Lal, *The TIFR Art Collection*, Tata Institute of Fundamental Research, Mumbai, 2010, p. 157.

10 'Robert Marshak - Session III', American Institute of Physics, 4 October 1970, https://tinyurl.com/2p8zzh4w. Accessed on 19 August 2022.

11 Bernstein, Lina, 'The Great Little Lady of the Bombay Art World', Academia, https://tinyurl.com/2nykf3sy. Accessed on 30 August 2022.

12 Bhabha, Homi, Received by Abul Kalam Azad, Tata Institute of Fundamental Research Archives, 24 July 1947, Mumbai, Maharashtra, India.

13 Bernstein, Lina, 'The Great Little Lady of the Bombay Art World', Academia, https://tinyurl.com/2nykf3sy. Accessed on 30 August 2022.

14 Bhabha, Homi, Received by Abul Kalam Azad, Tata Institute of Fundamental Research Archives, 24 July 1947, Mumbai, Maharashtra, India.

15 Ibid.

16 Bernstein, Lina, 'The Great Little Lady of the Bombay Art World', Academia, https://tinyurl.com/2nykf3sy. Accessed on 30 August 2022.

17 Lee, Rachel and Kathleen James-Chakraborty, '*Marg* Magazine: A Tryst with Architectural Modernity: Modern Architecture as Seen from an Independent India', *ABE Journal*, 2012, https://tinyurl.com/3v3zjrej. Accessed on 30 August 2022.

18 Lee, Rachel and Mareike Schwartz, 'Bombay', Metromod, 2021, https://tinyurl.com/5n8a24u6. Accessed on 30 August 2022.

19 von Leyden, Rudolf, 'The Drawings of Homi Bhabha', *Marg*, Vol. 6, No. 3, June 1953.

20 Ibid.

21 Santhanam, K., 'The Da Vinci Coda', *Outlook*, 19 July 2010, https://tinyurl.com/3meys5mm. Accessed on 30 August 2022.

22 Bhabha, Homi, 'Leonardo da Vinci: 500 Years', *Marg*, Vol. 5, No. 4, July 1952.

23 Dalmia, Yashodhara, *Amrita Sher-Gil: A Life*, Penguin Viking, New Delhi, 2006, p. xiii.

24 Although Vyas could not recall the brand of the camera, it would have, in all probability, been a Leica, which was the top brand at the time.

25 Vyas, K.N., Personal interview, 12 November 2021, Mumbai.
26 Chowdhury, Indira and Ananya Dasgupta, *A Masterful Spirit: Homi J. Bhabha, 1909-1966*, Penguin Books, New Delhi, p. 230.
27 Ibid.
28 Martha Graham was an American dancer and choreographer whose dance school pioneered modern dance since its founding in 1926.

Chapter 23: An Emerging Institutional Culture

1 Bhabha, Homi, 'Science and the Problems of Development', *Science*, Vol. 151, No. 3710, https://tinyurl.com/hfmnw48s. Accessed on 31 August 2022.
2 Chowdhury, Indira and Ananya Dasgupta, *A Masterful Spirit: Homi J. Bhabha, 1909-1966*, Penguin Books, New Delhi, 2010, p. 150.
3 Raghunathan, M.S., 'K. Chandrasekharan: Forgotten Genius', *Frontline*, 23 June 2017, https://tinyurl.com/2bfkstd6. Accessed on 31 August 2022.
4 Isaacson, Walter, *The Innovators: How a Group of Hackers, Geniuses and Geeks Created the Digital Revolution*, Simon & Schuster, 2014, p. 103.
5 Raghunathan, M.S., 'K. Chandrasekharan: Forgotten Genius', *Frontline*, 23 June 2017, https://tinyurl.com/2bfkstd6. Accessed on 31 August 2022.
6 Greenstein, George, 'A Gentleman of the Old School: Homi Bhabha and the Development of Science in India', *The American Scholar*, Vol. 61, No. 3, 1992, JSTOR, https://tinyurl.com/428vjnnr. Accessed on 11 August 2022.
7 Ibid.
8 Wali, Kameshwar C., *Chandra: A Biography of S. Chandrasekhar*, Penguin Books, New Delhi, 1991, p. 256.
9 Shah, Amrita, *Vikram Sarabhai: A Life*, Penguin Viking, New Delhi, 2007, pp. 156–7.
10 Ramanna, Raja, *Years of Pilgrimage: An Autobiography*, Penguin Viking, New Delhi, 1991, p. 53.
11 Greenstein, George, 'A Gentleman of the Old School: Homi Bhabha and the Development of Science in India', *The American Scholar*, Vol. 61, No. 3, 1992, JSTOR, https://tinyurl.com/428vjnnr. Accessed on 11 August 2022.
12 Rudra, Ashok, *Prasanta Chandra Mahalanobis: A Biography*, Oxford University Press, New Delhi, 1998, p. 191.
13 Chowdhury, Indira and Ananya Dasgupta, *A Masterful Spirit: Homi J. Bhabha, 1909-1966*, Penguin Books, New Delhi, 2010, p. 47.
14 *Homi J. Bhabha: Scientist, Visionary, Artist*, Department of Atomic Energy, Mumbai, 2021, p. 54.
15 Ibid. 55.
16 Lala, R.M., *The Heartbeat of a Trust: Fifty Years of the Sir Dorabji Tata Trust,* Tata McGraw Hill, New Delhi, 1984, p. 117.
17 Ibid. 117.

18 Chowdhury, Indira, *Growing the Tree of Science: Homi Bhabha and the Tata Institute of Fundamental Research*, Oxford University Press, New Delhi, p. 106.

19 Chowdhury, Indira and Ananya Dasgupta, *A Masterful Spirit: Homi J. Bhabha, 1909-1966*, Penguin Books, New Delhi, 2010, p. 84.

20 Bhabha, Homi, Received by Homi Seervai, Seervai family collection, 20 March 1928.

21 Chowdhury, Indira, *Growing the Tree of Science: Homi Bhabha and the Tata Institute of Fundamental Research*, Oxford University Press, New Delhi, p. 106.

22 Greenstein, George, 'A Gentleman of the Old School: Homi Bhabha and the Development of Science in India', *The American Scholar*, Vol. 61, No. 3, 1992, JSTOR, https://tinyurl.com/428vjnnr. Accessed on 11 August 2022.

23 Chowdhury, Indira, 'The Laboratory and its Double: The Making of the Scientist-Citizen at TIFR', Centre for the Advanced Study of India, University of Pennsylvania, 2012, https://tinyurl.com/hnnwv7a7. Accessed on 29 August 2022.

24 Lala, R.M., *The Heartbeat of a Trust: Fifty Years of the Sir Dorabji Tata Trust,* Tata McGraw Hill, New Delhi, 1984, p. 98.

25 Ibid. 98.

26 Ibid. 98.

27 Greenstein, George, 'A Gentleman of the Old School: Homi Bhabha and the Development of Science in India', *The American Scholar*, Vol. 61, No. 3, 1992, JSTOR, https://tinyurl.com/428vjnnr. Accessed on 11 August 2022.

28 'TIFR Canteen file', Tata Institute of Fundamental Research Archives, Mumbai, Maharashtra, India.

29 Sridharan, R., 'KC and I: My Fond Remembrance', *Bhava¯na¯*, Vol. 2, No. 1, January 2018, https://tinyurl.com/yc78zphh. Accessed on 31 August 2022.

30 Wali, Kameshwar C., *Chandra: A Biography of S. Chandrasekhar*, Penguin Books, New Delhi, 1991, pp. 255–256.

31 Deshmukh, Chintamani, *Homi Jehangir Bhabha*, National Book Trust, India, New Delhi, 2014, p. 124.

32 Ibid. 123.

33 Chowdhury, Indira, 'The Laboratory and its Double: The Making of the Scientist-Citizen at TIFR', Centre for the Advanced Study of India, University of Pennsylvania, 2012, https://tinyurl.com/hnnwv7a7. Accessed on 29 August 2022.

34 Bhabha, Homi, 'Bhabha's Speech at the Sixth International Conference on Planned Parenthood', Nuclear India, Vol. 43. No. 5–6, 2008, pp. 31–34, https://tinyurl.com/4yejwuw5. Accessed on 23 August 2022.

Chapter 24: The Outliers

1 Bhabha, Homi, 'Science and the Problems of Development', *Science*, Vol. 151, No. 3710, https://tinyurl.com/hfmnw48s. Accessed on 31 August 2022.

2 Fox, Maurice S., 'Recalling Obaid', *Journal of Biosciences*, Vol. 39, No. 4, 2014, p. 545, https://tinyurl.com/3brmnt9x. Accessed on 31 August 2022.

3 Szilard, Leo, Received by Homi Bhabha, Tata Institute of Fundamental Research Archives, 3 March 1962, Mumbai, Maharashtra, India.

4 Chowdhury, Indira, *Growing the Tree of Science: Homi Bhabha and the Tata Institute of Fundamental Research*, Oxford University Press, New Delhi, 2016, p. 177.

5 Bhabha, Homi, Received by Jawaharlal Nehru, Tata Institute of Fundamental Research Archives, 2 August 1962, Mumbai, Maharashtra, India.

6 Ibid.

7 Kotwal, J.V., Interview with Indira Chowdhury, 8 December 2003.

8 Bhabha, Homi, 'Science and the Problems of Development', *Science*, Vol. 151, No. 3710, https://tinyurl.com/hfmnw48s. Accessed on 31 August 2022.

9 Ghosh, Dipan K. and Arun K. Grover, *Tribute to a Titan: Birth Centenary of Homi Jehangir Bhabha*, Indian Physics Association, Mumbai, 2009, p. 116.

10 Ibid.

11 The Wellcome Trust, a research-charity based in London, was established in 1936 with legacies from the pharmaceutical magnate Sir Henry Wellcome to fund research in human and animal health. It is one of the wealthiest charitable foundations in the world today.

12 Chowdhury, Indira, *Growing the Tree of Science: Homi Bhabha and the Tata Institute of Fundamental Research*, Oxford University Press, New Delhi, 2016, p. 179.

13 Ghosh, Dipan K. and Arun K. Grover, *Tribute to a Titan: Birth Centenary of Homi Jehangir Bhabha*, Indian Physics Association, Mumbai, 2009, p. 117.

14 Madhavan, G., Mani Ramaswami and K. S. Krishnan, 'Obaid Siddiqi (1932-2013)', *Current Science*, Vol. 105, No. 4, 2013, pp. 531–3. JSTOR, https://tinyurl.com/2p98amy5. Accessed 31 August 2022.

15 'Obaid Siddiqi (1932-2013): Man of Science', *Frontline*, 7 August 2013, https://tinyurl.com/2p92w84r. Accessed on 29 December 2022.

16 Lala, R.M., *The Heart Beat of a Trust,* Tata McGraw Hill, New Delhi, 1984, p. 123.

17 Pulakkat, Hari, *Space Life Matter: The Coming of Age of Indian Science*, Hachette, Gurugram, 2021, p. 288.

18 Swarup, Govind, 'From Potts Hill (Australia) to Pune (India): The Journey of a Radio Astronomer', *Journal of Astronomical History and Heritage*, Vol. 9, No. 1, 2006, pp. 21–33, ResearchGate, https://tinyurl.com/5d446zxx. Accessed on 31 August 2022.

19 Goss, W.M., 'Origins of Radio Astronomy at the Tata Institute of Fundamental Research and the role of J.L. Pawsey', *Bulletin of the Astronomical Society of India*, 2014, pp. 1–19, https://tinyurl.com/3ra596xw. Accessed on 31 August 2022.

20 Chowdhury, Indira, *Growing the Tree of Science: Homi Bhabha and the Tata Institute of Fundamental Research*, Oxford University Press, New Delhi, 2016.

21 Pawsey, J.L., Received by Homi Bhabha, Tata Institute of Fundamental Research Archives, 6 October 1961, Mumbai, Maharashtra, India.

22 Goss, W.M., 'Origins of Radio Astronomy at the Tata Institute of Fundamental Research and the role of J.L. Pawsey', *Bulletin of the Astronomical Society of India*, 2014, pp. 1–19, https://tinyurl.com/3ra596xw. Accessed on 31 August 2022.

23 Ibid.

24 Ibid.

25 Ibid.

26 Ibid.

27 Swarup, Govind, 'From Potts Hill (Australia) to Pune (India): The Journey of a Radio Astronomer', *Journal of Astronomical History and Heritage*, Vol. 9, No. 1, 2006, pp. 21–33, ResearchGate, https://tinyurl.com/5d446zxx. Accessed on 31 August 2022.

28 Sinha, Ramesh, 'Bhabha and the Search for a Site for the Ooty Radio Telescope', *Current Science*, Vol. 60, No. 2, 1991, https://tinyurl.com/h7uje3eu. Accessed on 31 August 2022.

29 Pulakkat, Hari, *Space Life Matter: The Coming of Age of Indian Science*, Hachette, Gurugram, 2021, p. 18.

30 A sounding rocket is an instrument-carrying rocket designed to take measurements and perform scientific experiments during its sub-orbital flight.

31 Aravamudan, R. and Gita Aravamudan, *ISRO: A Personal History*, HarperCollins, Noida, 2017, pp. 53–4.

32 Ibid. 54–6.

33 Shah, Amrita, *Vikram Sarabhai: A Life*, Penguin Viking, New Delhi, 2007, p. 123.

34 Chitnis, E.V., Personal telephonic interview, 2 April 2022.

35 Shah, Amrita, *Vikram Sarabhai: A Life*, Penguin Viking, New Delhi, 2007, p. 124.

36 Ibid. 125.

37 Chitnis, E.V., 'Scientist, Businessman, Visionary Institution Builder', *Current Science*, Vol. 118, No. 8, 25 April 2020.

38 Shah, Amrita, *Vikram Sarabhai: A Life*, Penguin Viking, New Delhi, 2007, pp. 127–8.

39 Ibid. 128.

40 Chitnis, E.V., 'Scientist, Businessman, Visionary Institution Builder', *Current Science*, Vol. 118, No. 8, 2020.

41 Ramesh, Jairam, *A Chequered Brilliance: The Many Lives of V.K. Krishna Menon*, Penguin Viking, Gurugram, 2019, p. 617.

42 Ibid. 618.

Chapter 25: The Two Faces of Janus

1 Chowdhury, Indira and Ananya Dasgupta, *A Masterful Spirit: Homi J. Bhabha, 1909-1966*, Penguin Books, New Delhi, 2010, p. 168.

2 Phalkey, Jahnavi, *Atomic State: Big Science in Twentieth-Century India*, Permanent Black, New Delhi, 2013, pp. 166–7.

3 Singh, Virendra, 'Meghnad Saha: His Science and Life', *Current Science*, Vol. 64, No. 7, 10 April 1993, JSTOR, https://tinyurl.com/yc3zcdr7. Accessed on 11 August 2022.

4 Ibid.

5 Abraham, Itty, 'The Ambivalence of Nuclear Histories', *Osiris*, Vol. 21, No. 1, 2006, pp. 49–65, JSTOR, https://tinyurl.com/2p9e74md. Accessed on 11 August 2022.

6 Singh, Virendra, 'Meghnad Saha: His Science and Life', *Current Science*, Vol. 64, No. 7, 10 April 1993, JSTOR, https://tinyurl.com/yc3zcdr7. Accessed on 11 August 2022.

7 Segrè, Gino and Bettina Hoerlin, *The Pope of Physics: Enrico Fermi and the Birth of the Atomic Age*, Henry Holt and Company, New York, 2016, pp. 248–9.

8 Menon, Nikhil, *Planning Democracy: How a Professor, an Institute, and an Idea Shaped India*, Penguin Random House, Gurugram, 2022, p. xv.

9 Ibid. xvi.

10 Ibid. xvi–ii.

11 Anderson, Robert S., *Nucleus and Nation: Scientists, International Networks, and Power in India*, University of Chicago Press, 2010, p. 87.

12 *Ardeshir Dalal: A Memoir*, Dalal Family, 1980, p. 104.

13 Phalkey, Jahnavi, *Atomic State: Big Science in Twentieth-Century India*, Permanent Black, New Delhi, 2013, p. 66.

14 Arnold, David, 'Nehruvian Science and Postcolonial India', *Isis,* Vol. 104, No. 2, 2013, p. 362, https://tinyurl.com/yup7hk6w. Accessed on 10 August 2022.

15 Ramesh, Jairam, 'Nehru's Scientific Temper Recalled', Indian Institute of Technology Guwahati, 27 May 2011, https://tinyurl.com/23neen6d. Accessed on 10 August 2022.

16 Palat, Madhavan K., 'Nehru Did Not Set the Natural, Experimental and Exact Sciences in Opposition to Human Sciences', *The Hindu*, 19 December 2021, https://tinyurl.com/mv65z2tf. Accessed on 10 August 2022.

17 'Government of India Scientific Policy Resolution', 4 March 1958, India Bioscience, https://tinyurl.com/ywb9uknc. Accessed on 18 December 2022.

18 Perkovich, George, *India's Nuclear Bomb: The Impact on Global Proliferation*, University of California Press, 2001, p. 17.

19 Pant, Harsh V. and Yogesh Joshi, *Indian Nuclear Policy*, Oxford University Press, New Delhi, 2018, p. 19.

20 Ibid. 20.

21 Kapur, Ashok, 'Nehru's Nuclear Policy', *Nehru and the Twentieth Century*, Milton Israel (eds.), University of Toronto Centre for South Asian Studies, 1991, pp. 217–32.

22 Abraham, Itty, *The Making of the Indian Atomic Bomb: Science, Secrecy and the Post-Colonial State*, Orient BlackSwan, Hyderabad, 1999, p. 47.

23 Ibid. 23.

24 Srinivasan, M.R., 'An Intermezzo', *Selected Lectures of Dr. M.R. Srinivasan,* Department of Atomic Energy for Nuclear Power Corporation of India, 1990, p. 7.

25 Srinivasan, M.R., 'Pandit Nehru's Vision of Science, Technology and Public Administration', *Selected Lectures of Dr. M.R. Srinivasan,* Department of Atomic Energy for Nuclear Power Corporation of India, p. 328.

26 Ramana, M.V., *The Power of Promise: Examining Nuclear Energy in India*, Penguin Books, New Delhi, 2012, p. 6.

27 Nehru, Jawaharlal, *An Autobiography*, Oxford University Press, New Delhi, 1980, p. 596.

28 Paz, Octavio, *Alternating Current*, Arcade Publishing, New York, 1990, p. 50.

29 Pant, Harsh V. and Yogesh Joshi, *Indian Nuclear Policy*, Oxford University Press, New Delhi, 2018, pp. 45–6.

30 Srinivasan, M.R., 'Pandit Nehru's Vision of Science, Technology and Public Administration', *Selected Lectures of Dr M.R. Srinivasan*, Department of Atomic Energy for Nuclear Power Corporation of India, p. 329.

31 Sukumar, Arun Mohan, *Midnight's Machines: A Political History of Technology in India*, Penguin Random House, Gurugram, 2019, p. 5.

32 Arnold, David, 'Nehruvian Science and Postcolonial India', *Isis*, Vol. 104, No. 2, 2013, p. 362.

33 Phalkey, Jahnavi, *Atomic State: Big Science in Twentieth-Century India*, Permanent Black, New Delhi, 2013, p. 64.

Chapter 26: An Institutional Framework

1 Bhabha, Homi, 'Note on the Organisation of Atomic Research in India', Manohar Parrikar Institute for Defence Studies and Analyses, 26 April 1948, https://tinyurl.com/mt56bzfs. Accessed on 20 August 2022.

2 Ramana, M.V., *The Power of Promise: Examining Nuclear Energy in India*, Penguin Viking, New Delhi, 2012, p. 8.

3 Bhabha, Homi, 'Note on the Organisation of Atomic Research in India',

Manohar Parrikar Institute for Defence Studies and Analyses, 26 April 1948, https://tinyurl.com/mt56bzfs. Accessed on 20 August 2022.

4 Abraham, Itty, 'Rare Earths: The Cold War in the Annals of Travancore', *Entangled Geographies: Empire and Technopolitics in the Global Cold War*, Gabrielle Hecht (ed.), MIT Press, 2011, p. 112.

5 Bhabha, Homi, Received by Jawaharlal Nehru, Tata Institute of Fundamental Research Archives, 3 October 1947, Mumbai, Maharashtra, India.

6 Bhatia, Shyam, *India's Nuclear Bomb*, Vikas Publishing House, New Delhi, 1979, p. 84.

7 Perkovich, George, *India's Nuclear Bomb: The Impact on Global Proliferation*, University of California Press, 2001, p. 18.

8 Abraham, Itty, *The Making of the Indian Atomic Bomb: Science, Secrecy and the Post-Colonial State*, Orient BlackSwan, 1999, pp. 49–50.

9 Ibid. 51.

10 Perkovich, George, *India's Nuclear Bomb: The Impact on Global Proliferation*, University of California Press, 2001, p. 20.

11 Bhabha, Homi, 'Note on the Organisation of Atomic Research in India', Manohar Parrikar Institute for Defence Studies and Analyses, 26 April 1948, https://tinyurl.com/mt56bzfs. Accessed on 20 August 2022.

12 Ibid.

13 Abraham, Itty, 'From the Commission to the Mission Model: Technology Czars and the Indian Middle Class', *The Journal of Asian Studies*, Vol. 76, No. 3, 2017, https://tinyurl.com/mu6mk3fu. Accessed on 24 August 2022.

14 Ibid.

15 Bhabha, Homi, 'Note on the Organisation of Atomic Research in India', Manohar Parrikar Institute for Defence Studies and Analyses, 26 April 1948, https://tinyurl.com/mt56bzfs. Accessed on 20 August 2022.

16 Ibid.

17 Abraham, Itty, 'From the Commission to the Mission Model: Technology Czars and the Indian Middle Class', *The Journal of Asian Studies*, Vol. 76, No. 3, 2017, https://tinyurl.com/mu6mk3fu. Accessed on 24 August 2022.

18 Bhabha, Homi, Received by S.S. Bhatnagar, Nuclear History Project, Manohar Parrikar Institute for Defence Studies, 28 April 1948, https://tinyurl.com/4js2t45n. Accessed on 24 August 2022.

19 Bhatnagar, S.S., Received by Homi Bhabha, Nuclear History Project, Manohar Parrikar Institute for Defence Studies, 3 May 1948, https://tinyurl.com/24x563ze. Accessed on 24 August 2022.

20 Ibid.

21 Bhabha, Homi, Received by John Cockcroft, Tata Institute of Fundamental Research Archives, 5 May 1948, Mumbai, Maharashtra, India.

22 Nehru, Jawaharlal, 'Note to Cabinet, 22 May 1948', *Selected Works of Jawaharlal Nehru*, Vol. 6, Jawaharlal Nehru Memorial Fund, New Delhi, 1987, pp. 349–50.

23 Anderson, Robert S., 'Growing Science in India—2: Homi J Bhabha', *Science Today*, November 1976.

24 Frey, Karsten, *India's Nuclear Bomb and National Security*, Routledge, London and New York, 2006, p. 61.

25 Phalkey, Jahnavi, *Atomic State: Big Science in Twentieth-Century India*, Permanent Black, New Delhi, 2013, p. 194.

26 Anderson, Robert S., *Nucleus and Nation: Scientists, International Networks, and Power in India*, University of Chicago Press, 2010, p. 200.

27 Ibid. 200.

28 Ibid. 195.

29 Nath, Biman, *Homi J Bhabha: A Renaissance Man among Scientists*, Niyogi Books, New Delhi, 2021, p. 121.

30 Ganguly, Jibamitra and Chitra Roy, *Meghnad Saha: His Science and Persona through Selected Letters and Writings*, Indian National Science Academy, New Delhi, 2019, p. 179.

31 Manjapra, Kris, *Age of Entanglement: German and Indian Intellectuals across Empire*, Harvard University Press, Cambridge, Massachusetts, 2014, pp. 134–5.

32 Ibid. 198–9.

33 Hore, Peter, *Patrick Blackett: Sailor Scientist, Socialist*, Frank Cass Publishers, London, 2003, p. 221.

34 Blackett, P.M.S., Interview with B.R. Nanda, Nehru Memorial Museum and Library Oral History Project, 28 July 1967, p. 22.

35 Abraham, Itty, *The Making of the Indian Atomic Bomb: Science, Secrecy and the Post-Colonial State*, Orient BlackSwan, 1999, p. 56.

36 Hore, Peter, *Patrick Blackett: Sailor Scientist, Socialist*, Frank Cass Publishers, London, 2003, p. 223.

37 Bhabha, Homi, Received by P.M.S. Blackett, Tata Institute of Fundamental Research Archives, 28 August 1948, Mumbai, Maharashtra, India.

38 Ibid.

39 Hore, Peter, *Patrick Blackett: Sailor Scientist, Socialist*, Frank Cass Publishers, London, 2003, p. 240.

40 *Nehru: The Architect of Indian Science*, Council of Scientific and Industrial Research, 1989.

Chapter 27: Fissile Materials and the French Connection

1 Bhabha, Homi, 'Note on the Organisation of Atomic Research in India', Manohar Parrikar Institute for Defence Studies and Analyses, 26 April 1948, https://tinyurl.com/mt56bzfs. Accessed on 20 August 2022.

2 Abraham, Itty, *The Making of the Indian Atomic Bomb: Science, Secrecy and the Post-Colonial State*, Orient BlackSwan, 1999, p. 57.

3 Bamzai, Sandeep, *Princestan: How Nehru, Patel & Mountbatten Made India*, Rupa, New Delhi, 2020, pp. 101–2.

4 Abraham, Itty, 'Rare Earths: The Cold War in the Annals of Travancore', *Entangled Geographies: Empire and Technopolitics in the Global Cold War*, Gabrielle Hecht (ed.), MIT Press, 2011, p. 105.

5 Ibid. 110–1.

6 'Note Dated 27 February 1947, File No. 17(4) 47-PMS', *Selected Works of Jawaharlal Nehru*, Vol. 2, Jawaharlal Nehru Memorial Fund, Teen Murti House, New Delhi, 1984, pp. 604–7.

7 Abraham, Itty, 'Rare Earths: The Cold War in the Annals of Travancore', *Entangled Geographies: Empire and Technopolitics in the Global Cold War*, Gabrielle Hecht (ed.), MIT Press, 2011, p. 112.

8 Lala, R.M., *Beyond the Last Blue Mountain: A Life of J.R.D. Tata,* Penguin Books, New Delhi, 1993, pp. 326–7.

9 Sundaram, C.V., L.V. Iyengar and T.S. Iyengar, *Atomic Energy in India: 50 Years*, Government of India, Department of Atomic Energy, August 1998, p. 82.

10 Phalkey, Jahnavi, *Atomic State: Big Science in Twentieth-Century India*, Permanent Black, New Delhi, 2013, pp. 247–8.

11 Ibid. 248.

12 Ibid. 248.

13 Sarkar, Jayita, 'From the Peaceful Atom to the Peaceful Explosion: Indo-French Nuclear Relations during the Cold War, 1950–1974', Nuclear Proliferation International History Project, Working Paper No. 3, 2013, https://tinyurl.com/yckjrz3u. Accessed on 22 August 2022.

14 Sarkar, Jayita, 'From the Peaceful Atom to the Peaceful Explosion: Indo-French Nuclear Relations during the Cold War, 1950–1974', Nuclear Proliferation International History Project, Working Paper No. 3, 2013, https://tinyurl.com/yckjrz3u. Accessed on 22 August 2022.

15 How, F.C., Received by Roger Makins, Wilson Center Digital Archive, 18 August 1951, https://tinyurl.com/26zmp2uj. Accessed on 22 August 2022.

16 Ibid.

17 Sarkar, Jayita, 'From the Peaceful Atom to the Peaceful Explosion: Indo-French Nuclear Relations during the Cold War, 1950–1974', Nuclear Proliferation International History Project, Working Paper No. 3, 2013, https://tinyurl.com/yckjrz3u. Accessed on 22 August 2022.

18 Bhabha, Homi, Received by J.R.D. Tata, Tata Institute of Fundamental Research Archives, 31 January 1951, Mumbai, Maharashtra, India.

Chapter 28: Atomic Enclave

1 Bhabha, Homi, 'Science and the Problems of Development', *Science*, Vol. 151, No. 3710, https://tinyurl.com/hfmnw48s. Accessed on 31 August 2022.
2 Sur, Abha, 'Scientism and Social Justice: Meghnad Saha's Critique of the State of Science in India', *Historical Studies in the Physical and Biological Sciences*, Vol. 33, No. 1, 2002, p. 95, https://tinyurl.com/mtp3ketk. Accessed on 22 August 2022.
3 Anderson, Robert S., *Nucleus and Nation: Scientists, International Networks, and Power in India*, University of Chicago Press, 2010, p. 230.
4 Sur, Abha, 'Scientism and Social Justice: Meghnad Saha's Critique of the State of Science in India', *Historical Studies in the Physical and Biological Sciences*, Vol. 33, No. 1, 2002, p. 87, https://tinyurl.com/mtp3ketk. Accessed on 22 August 2022.
5 Ibid. 87.
6 *Lok Sabha Debates: Sixth Session*, Vol. 5, Indian Parliament, May 1954.
7 Mukhopadhyay, Atri, 'The Untold Story of the Indian Atomic Energy Programme: Meghnad Saha and Homi Bhabha', *Journal of the Asiatic Society*, Vol. 60, No. 3, 2018, p. 30.
8 Ibid. 30.
9 Ibid. 30.
10 Ibid. 31–2.
11 Ibid. 32–3.
12 Ibid. 34.
13 Ibid. 34.
14 *Lok Sabha Debates: Sixth Session*, Vol. 5, Indian Parliament, May 1954.
15 Banerjee, B. and N. Sarma, *Nuclear Power in India: A Critical History*, Rupa & Co., New Delhi, 2008, p. 16.
16 Abraham, Itty, *The Making of the Indian Atomic Bomb: Science, Secrecy and the Post-Colonial State*, Orient BlackSwan, Hyderabad, 1999, p. 74.
17 Venkataraman, G., *Bhabha and His Magnificent Obsessions*, Universities Press, 1994, p. 148.
18 Sundaram, C.V., L.V. Krishnan and T.S. Iyengar, *Atomic Energy in India: 50 Years*, Department of Atomic Energy, Government of India, 1998, p. 1.
19 Ramanna, Raja, *Years of Pilgrimage: An Autobiography*, Penguin Viking, New Delhi, 1991, p. 51.
20 Ibid. 51.
21 Menon, Meena and T.S. Subramanian, 'Sethna–The Man Who Dared to Dream', *The Hindu*, 7 September 2010, https://tinyurl.com/yrvyt37n. Accessed on 22 August 2022.
22 Anderson, Robert S., 'Building Scientific Institutions in India: Saha and Bhabha', *Occasional Paper Series*, McGill University, Montreal, 1975, p. 50.

23 Ramana, M.V., *The Power of Promise: Examining Nuclear Energy in India*, Penguin Viking, New Delhi, 2012, p. 12.
24 Ramana, M.V., 'India and Fast Breeder Reactors', *Science and Global Security*, Vol. 17, No. 1, 2009, p. 55.
25 Ramanna, Raja, *Years of Pilgrimage: An Autobiography*, Penguin Viking, New Delhi, 1991, p. 61.
26 Ibid. 62.
27 Ganguly, Jibamitra and Chitra Roy, *Meghnad Saha: His Science and Persona through Selected Letters and Writings*, Indian National Science Academy, New Delhi, 2019, p. 196.
28 Ibid. 196.
29 Miller, Arthur I., *Empire of the Stars: Friendship, Obsession and Betrayal in the Quest for Black Holes*, Little, Brown and Company, London, 2005, p. 149.
30 Abraham, Itty, *The Making of the Indian Atomic Bomb: Science, Secrecy and the Post-Colonial State*, Orient BlackSwan, Hyderabad, 1999, p. 76.
31 Ibid. 76.

Chapter 29: Bhabha's Empire

1 Bhabha, Homi, 'Science and the Problems of Development', *Science*, Vol. 151, No. 3710, https://tinyurl.com/hfmnw48s. Accessed on 31 August 2022.
2 Venkataraman, G., *Bhabha and his Magnificent Obsessions*, Universities Press, Hyderabad, 1994, p. 149.
3 Deshmukh, Chintamani, *Homi Jehangir Bhabha*, National Book Trust, India, New Delhi, 2014, p. 116.
4 Bhabha, Homi, Received by Jawaharlal Nehru, *Nuclear India*, Vol. 43. No 5-6, Nov–Dec 2008, 18 June 1956, https://tinyurl.com/4yejwuw5. Accessed on 23 August 2022.
5 Nehru, Jawaharlal, Received by Homi Bhabha, *Nuclear India*, Vol. 43. No 5-6, Nov–Dec 2008, 19 June 1956, https://tinyurl.com/4yejwuw5. Accessed on 23 August 2022.
6 Leslie, Stuart and Indira Chowdhury, 'Homi Bhabha: Master Builder of Nuclear India', *Physics Today*, Vol. 71, No. 9, September 2018, https://tinyurl.com/5n7xjswn. Accessed on 23 August 2022.
7 Venkataraman, G., *Bhabha and His Magnificent Obsessions*, Universities Press, 1994, p. 192.
8 Ibid. 193.
9 Srinivasan, M.R., *From Fission to Fusion: The Story of India's Atomic Energy Programme*, Penguin Viking, New Delhi, 2002, p. 30.
10 Chengappa, Raj, *Weapons of Peace: The Secret Story of India's Quest to be a Nuclear Power*, HarperCollins, New Delhi, 2000, p. 79.
11 Patell, R.B.J., 'Dr Homi Bhabha and Architecture', *Homi Bhabha as Artist:*

A Selection of His Paintings, Drawings and Sketches, Jamshed Bhabha et al. (eds.), Marg, Mumbai, 1968.

12 Ibid.

13 Ibid.

14 Ibid.

15 Ibid.

16 Cockcroft, John and M.G.K. Menon, *Homi Jehangir Bhabha 1909-1966*, Royal Institution of Great Britain, London, 1967, p. 26.

17 Bhabha, Homi, Received by Jawaharlal Nehru, Tata Institute of Fundamental Research Archives, 2 July 1959, Mumbai, Maharashtra, India.

18 Leslie, Stuart and Indira Chowdhury, 'Homi Bhabha: Master Builder of Nuclear India', *Physics Today,* Vol. 71, No. 9, September 2018, https://tinyurl.com/5n7xjswn. Accessed on 23 August 2022.

19 Venkataraman, G., *Bhabha and His Magnificent Obsessions*, Universities Press, 1994, p. 199.

20 Deshmukh, Chintamani, *Homi Jehangir Bhabha*, National Book Trust, India, New Delhi, 2014, p. 109.

21 Ibid. 3.

22 Venkataraman, G., *Bhabha and His Magnificent Obsessions*, Universities Press, 1994, p. 200.

23 Kulkarni, R.P. and V. Sarma, *Homi Bhabha: Father of Nuclear Science in India*, Bombay Popular Prakashan, Mumbai, 1969, p. 45.

24 Ghosh, Dipan K. and Arun K. Grover, *Tribute to a Titan: Birth Centenary of Homi Jehangir Bhabha*, Indian Physics Association, Mumbai, 2009, p. 91.

25 *Homi J. Bhabha: Scientist, Visionary, Artist*, Department of Atomic Energy, Mumbai, 2021.

26 Ghosh, Dipan K. and Arun K. Grover, *Tribute to a Titan: Birth Centenary of Homi Jehangir Bhabha*, Indian Physics Association, Mumbai, 2009, p. 116.

27 Ibid. 92.

28 Menon, M.G.K., Interview with Indira Chowdhury, 21 April 2005, Tata Institute of Fundamental Research Archives, Mumbai, Maharashtra, India.

29 Bhabha, Homi, Received by E.C. Allardice, Tata Institute of Fundamental Research Archives, 10 April 1962, Mumbai, Maharashtra, India.

30 Bhabha, Homi, 'Science and the Problems of Development', International Council of Scientific Unions, Bombay, 7 January 1966.

31 Kakodkar, Anil and Suresh Gangotra, *Fire and Fury: Transforming India's Strategic Identity*, Rupa, New Delhi, 2019.

Chapter 30: The Geneva Conference and After

1 Bhabha, Homi, 'Address at the First Geneva Conference', Nuclear India, 1966, p. 19.

2 Goldschmidt, Bertrand, 'When the IAEA Was Born', International Atomic Energy Agency, 2007, https://tinyurl.com/2ktzypme. Accessed on 23 August 2022.

3 Ibid.

4 Deshmukh, Chintamani, *Homi Jehangir Bhabha*, National Book Trust, India, New Delhi, 2014, p. 74.

5 Cockcroft, John and M.G.K. Menon, *Homi Jehangir Bhabha 1909-1966*, Royal Institution of Great Britain, London, 1967, p. 13.

6 Joliot-Curie, Frédéric, Received by Homi Bhabha, Tata Institute of Fundamental Research Archive, 17 February 1955, Mumbai, Maharashtra, India.

7 'War and Peace in the Nuclear Age; Haves and Have-Nots; Interview with Bertrand Goldschmidt, 1986 [2]', Open Vault, https://tinyurl.com/yc8xmf2n. Accessed on 23 August 2022.

8 Phalkey, Jahnavi, *Atomic State: Big Science in Twentieth-Century India*, Permanent Black, New Delhi, 2013, p. 261.

9 Ibid. 262.

10 Ibid. 261.

11 Deshmukh, Chintamani, *Homi Jehangir Bhabha*, National Book Trust, India, New Delhi, 2014, p. 77.

12 Einstein, Albert, Received by Homi Bhabha, Tata Institute of Fundamental Research Archive, 30 April 1947, Mumbai, Maharashtra, India.

13 Bhabha, Homi, Received by Albert Einstein, Tata Institute of Fundamental Research Archive, 15 July 1947, Mumbai, Maharashtra, India.

14 Sharma, Dhirendra, 'India's Lopsided Science', *Bulletin of the Atomic Scientists*, Vol. 47, No. 4, May 1991, pp. 32–6, https://tinyurl.com/3veptprm. Accessed on 23 August 2022.

15 Ramana, M.V., 'La Trahison des Clercs: Scientists and India's Nuclear Bomb', *Prisoners of the Nuclear Dream*, M.V. Ramana and C. Rammanohar Reddy (eds.), Orient BlackSwan, Hyderabad, 2003, pp. 221–2.

16 Deshmukh, Chintamani, *Homi Jehangir Bhabha*, National Book Trust, India, New Delhi, 2014, p. 77.

17 Ibid.

18 Ibid. 78.

19 Bhabha, Homi, 'The Need for a New Power Source', *Journal of Scientific and Industrial Research*, Vol. 14A, 1955, pp. 413–8.

20 Ramanna, Raja, *Years of Pilgrimage: An Autobiography*, Penguin Viking, New Delhi, 1991, p. 64.

21 Bromberg, Joan Lisa, *Fusion, Science, Politics and the Intervention of a New Energy Source*, MIT Press, Cambridge, MA, 1985, p. 67.

22 Deshmukh, Chintamani, *Homi Jehangir Bhabha*, National Book Trust, India, New Delhi, 2014, p. 80.

23 Leading Pakistani Scientist Effusively Praising Great Indian Nuclear Physicist

Dr. Homi Bhabha, YouTube, https://youtu.be/CIgr_vnuZxc. Accessed on 9 January 2023.

24 Ibid. 79.

25 Ibid. 79.

26 Ibid. 80.

27 Mallik, D.C.V. and S. Chatterjee, *Kariamanikkam Srinavasa Krishnan: His Life and Work*, Universities Press, Hyderabad, 2012, p. 357.

28 Abraham, Itty, *The Making of the Indian Atomic Bomb: Science, Secrecy and the Post-Colonial State*, Orient BlackSwan, Hyderabad, 1999, p. 89.

29 *Nuclear India*, Vol. 43, No. 5-6, Nov–Dec 2008, https://tinyurl.com/4yejwuw5. Accessed on 23 August 2022.

30 *Shankar's Weekly*, 14 August 1955.

31 Ramanna, Raja, *Years of Pilgrimage: An Autobiography*, Penguin Viking, New Delhi, 1991, pp. 67–8.

32 *Nuclear India*, Vol. 2, Radiant Publishers, New Delhi, 1974, p. 44.

33 Ibid. 44.

34 Statement by Homi Bhabha at the Conference on the IAEA Statute, 27 September 1956.

35 Ibid. 72.

36 Cockcroft, John and M.G.K. Menon, *Homi Jehangir Bhabha 1909-1966*, Royal Institution of Great Britain, London, 1967, p. 14.

37 Ibid. 14.

38 'Report of the Indian Delegation at the Peaceful Uses of Atomic Energy: Second International Conference, Geneva, September 1-13', Atomic Energy Commission, Government of India, 1958.

39 Phalkey, Jahnavi, *Atomic State: Big Science in Twentieth-Century India*, Permanent Black, New Delhi, 2013, p. 284.

Chapter 31: Apsara

1 'Speech by Dr. H.J. Bhabha, Director, Tata Institute of Fundamental Research, at the Opening of the New Buildings of the Institute at Colaba on Monday, January 15, 1962', Nuclear History Project, Manohar Parrikar Institute for Defence Studies, 15 January 1962, https://tinyurl.com/bddp439h. Accessed on 24 August 2022.

2 Abraham, Itty, *The Making of the Indian Atomic Bomb: Science, Secrecy and the Post-Colonial State*, Orient BlackSwan, Hyderabad, 1999, pp. 79–80.

3 Ibid. 81.

4 A swimming pool reactor is so named because the core, where the uranium chain reaction takes place, is suspended in a pool of normal (or light) water. The water serves as the moderator and coolant. Some of the neutrons escaping from the core are also reflected back into the core.

5 Srinivasan, M.R., *From Fission to Fusion: The Story of India's Atomic Energy Programme*, Penguin Viking, New Delhi, 2002, pp. 19–20.
6 Ibid. 12.
7 Ibid. 20.
8 Ramanna, Raja, *Years of Pilgrimage: An Autobiography*, Penguin Viking, New Delhi, 1991, p. 62.
9 A reactor is said to be critical when each fission event releases sufficient number of neutrons to sustain a chain reaction. The reactor is controlled using neutron-absorbing control rods. Natural uranium can be used as fuel if heavy water or graphite is used as a moderating material.
10 Ramanna, Raja, *Years of Pilgrimage: An Autobiography*, Penguin Viking, New Delhi, 1991, p. 63.
11 Bhabha, Homi, Received by Jawaharlal Nehru, Tata Institute of Fundamental Research Archive, 1 August 1956, Mumbai, Maharashtra, India.
12 Ibid.
13 Nehru, Jawaharlal, *Jawaharlal Nehru on Science and Society: A Collection of His Writings and Speeches*, Nehru Memorial Museum and Library, 1988, p. 144.
14 Nehru, Jawaharlal, *Jawaharlal Nehru's Speeches: Volume 3*, Publications Division, Ministry of Information and Broadcasting, Government of India, 1958, p. 507.

15 Abraham, Itty, *The Making of the Indian Atomic Bomb: Science, Secrecy and the Post-Colonial State*, Orient BlackSwan, Hyderabad, 1999, p. 109.
16 'Speech by Dr. H.J. Bhabha, Director, Tata Institute of Fundamental Research, at the Opening of the New Buildings of the Institute at Colaba on Monday, January 15, 1962', Nuclear History Project, Manohar Parrikar Institute for Defence Studies, 15 January 1962, https://tinyurl.com/bddp439h. Accessed on 24 August 2022.
17 Sundaram, C.V., L.V. Iyengar and T.S. Iyengar, *Atomic Energy in India: 50 Years*, Government of India, Department of Atomic Energy, August 1998, p. 54.
18 Srinivasan, M.R., *From Fission to Fusion: The Story of India's Atomic Energy Programme*, Penguin Viking, New Delhi, 2002, pp. 22–3.
19 Sundaram, C.V., L.V. Iyengar and T.S. Iyengar, *Atomic Energy in India: 50 Years*, Government of India, Department of Atomic Energy, 1998, pp. 54–5.
20 Lala, R.M., *The Heartbeat of a Trust: Fifty Years of the Sir Dorabji Tata Trust*, Tata McGraw Hill, New Delhi, 1984, pp. 70–1.
21 Ibid. 71.
22 'Tata Memorial Hospital', Sands of Time, Tata Central Archives, 2016, https://tinyurl.com/y4v64b7c. Accessed on 24 August 2022.
23 Sundaram, C.V., L.V. Iyengar and T.S. Iyengar, *Atomic Energy in India: 50 Years*, Government of India, Department of Atomic Energy, August 1998, pp. 62–3.

24 Bhabha, Homi, Received by John Cockcroft, Tata Institute of Fundamental Research Archives, 24 July 1951, Mumbai, Maharashtra, India.

25 Bhabha, Homi, 'The Development of Atomic Energy in India', Manuscript, Bhabha Papers, Tata Institute of Fundamental Research Archives, Mumbai, Maharashtra, India.

26 Divatia, A.S. and C. Ambasankaran, 'Accelerator Development in India', *Pramana*, Vol. 24, No. 1 & 2, January and February 1985, pp. 227–44, https://tinyurl.com/3tbnmf6u. Accessed on 29 August 2022.

27 'Minutes of a Meeting Held on 3 August 1964 on the Need for a Medium Energy Accelerator', Tata Institute of Fundamental Research Archives, Mumbai, Maharashtra, India.

Chapter 32: CIRUS

1 Bhabha, Homi, 'Science and the Problems of Development', *Science*, Vol. 151, No. 3710, https://tinyurl.com/hfmnw48s. Accessed on 31 August 2022.

2 Abraham, Itty, *The Making of the Indian Atomic Bomb: Science, Secrecy and the Post-Colonial State*, Orient BlackSwan, Hyderabad, 1999, p. 89.

3 Ramanna, Raja, *Years of Pilgrimage: An Autobiography*, Penguin Viking, New Delhi, 1991, p. 70.

4 Abraham, Itty, *The Making of the Indian Atomic Bomb: Science, Secrecy and the Post-Colonial State*, Orient BlackSwan, Hyderabad, 1999, p. 90.

5 Ibid. 91.

6 Srinivasan, M.R., *From Fission to Fusion: The Story of India's Atomic Energy Programme*, Penguin Viking, New Delhi, 2002, p. 24.

7 Venkataraman, G., *Bhabha and His Magnificent Obsessions*, Universities Press, 1994, p. 160.

8 Ramanna, Raja, *Years of Pilgrimage: An Autobiography*, Penguin Viking, New Delhi, 1991, p. 70.

9 Venkataraman, G., *Bhabha and His Magnificent Obsessions*, Universities Press, 1994, p. 161.

10 Perkovich, George, *India's Nuclear Bomb: The Impact on Global Proliferation*, University of California Press, 2001, p. 27.

11 Venkataraman, G., *Bhabha and His Magnificent Obsessions*, Universities Press, 1994, p. 162.

12 Ibid. 162.

13 Sundaram, C.V., 'Brahm Prakash (1912–1984): Elected Fellow 1969', Indian National Science Academy, https://tinyurl.com/5n889pjp. Accessed on 26 August 2022.

14 Brahm Prakash continued to make important and diverse contributions to the atomic energy programmes as project director, Nuclear Fuels Complex, Hyderabad and chairman of the Uranium Corporation of India. He also

contributed to defence production as chairman, Mishra Dhatu Nigam Limited or MIDHANI from 1980 to 1984.

15 Venkataraman, G., *Bhabha and His Magnificent Obsessions*, Universities Press, 1994, p. 163.

16 'Atoms for India', *Time*, 7 February 1955, https://tinyurl.com/2p8tupkw. Accessed on 26 August 2022.

17 'Note, Homi Bhabha to Shri Y. D. Gundevia', Wilson Center Digital Archive, 2 September 1960, https://tinyurl.com/yz7tm6an. Accessed on 12 January 2023.

18 Perkovich, George, *India's Nuclear Bomb: The Impact on Global Proliferation*, University of California Press, 2001, p. 28.

19 Ramanna, Raja, *Years of Pilgrimage: An Autobiography*, Penguin Viking, New Delhi, 1991, p. 121.

20 Leslie, Stuart and Indira Chowdhury, 'Homi Bhabha: Master Builder of Nuclear India', *Physics Today,* Vol. 71, No. 9, September 2018, https://tinyurl.com/5n7xjswn. Accessed on 23 August 2022.

21 *Nuclear India: Volume 2*, Radiant Publishers, New Delhi, 1974, pp. 145–50.

22 Sundaram, C.V., L.V. Iyengar and T.S. Iyengar, *Atomic Energy in India: 50 Years*, Government of India, Department of Atomic Energy, August 1998, p. 90.

23 Sethna, H.N., 'India's Atomic Energy Programme - Past and Future', *IAEA Bulletin*, Vol. 21, No. 5, 1979, https://tinyurl.com/7c7y4bys. Accessed on 26 August 2022.

24 Kakodkar, Anil and Suresh Gangotra, *Fire and Fury: Transforming India's Strategic Identity*, Rupa, New Delhi, 2019, p. 9.

25 Ibid. 9–10.

26 Ibid.

27 Ramana, M.V., *The Power of Promise: Examining Nuclear Energy in India*, Penguin Viking, New Delhi, 2012, p. 19.

28 Bhabha, Homi, Received by Jawaharlal Nehru, 2 July 1959, *Nuclear India*, Vol. 43, No. 5-6, 2008, https://tinyurl.com/4yejwuw5. Accessed on 23 August 2022.

29 Nan was Vijay Lakshmi Pandit's nickname.

30 Bhabha, Homi, Received by Jawaharlal Nehru, 1 August 1960, *Nuclear India*, Vol. 43, No. 5-6, 2008, https://tinyurl.com/4yejwuw5. Accessed on 23 August 2022.

Chapter 33: Tarapur

1 Bhabha, Homi and M. Dayal, 'World Energy Requirements and the Economics of Nuclear Power with Special Reference to Underdeveloped Countries', Semantic Scholar, https://tinyurl.com/59evnaer. Accessed on 26 August 2022.

2 Ramana, M.V., *The Power of Promise: Examining Nuclear Energy in India*, Penguin Viking, New Delhi, 2012, p. 158.

3 Sundaram, C.V., L.V. Iyengar and T.S. Iyengar, *Atomic Energy in India: 50*

Years, Government of India, Department of Atomic Energy, August 1998, pp. 15–6.

4 Ramana, M.V., *The Power of Promise: Examining Nuclear Energy in India*, Penguin Viking, New Delhi, 2012, p. 159.

5 Ibid. 159.

6 Banerjee, B. and N. Sarma, *Nuclear Power in India: A Critical History*, Rupa & Co., New Delhi, 2008, p. 23.

7 Bhabha, Homi and M. Dayal, *The Role of Atomic Energy in Indian Development*, Commercial Print Press, 1960.

8 Little, I.M.D., 'Atomic Bombay? A Comment on "The Need for Atomic Energy in the Underdeveloped Countries"', *Economic and Political Weekly,* Vol. 28, No. 10, 1958, pp. 377–80.

9 M.R. Srinivasan, in his account, has said that he was general manager of western Railway. In fact, Chakravarti was general manager of Central Railway from January 1955 to June 1958.

10 Srinivasan, M.R., *From Fission to Fusion: The Story of India's Atomic Energy Programme*, Penguin Viking, New Delhi, 2002, p. 45.

11 Ibid. 46.

12 Ibid. 48–9.

13 Ibid. 51.

14 BWRs are nuclear reactors that use water as a coolant and moderator with steam being produced in the reactor that is then enriched with uranium oxide. BWRs actually boil water, which is then converted to steam and recycled back to water through a condenser to be reused in the heat-generation process.

15 Ramana, M.V., *The Power of Promise: Examining Nuclear Energy in India*, Penguin Viking, New Delhi, 2012, pp. 48–9.

16 Perkovich, George, *India's Nuclear Bomb: The Impact on Global Proliferation*, University of California Press, 2001, p. 29.

17 *Science Reporter*, October 1966.

18 Ibid.

19 Perkovich, George, *India's Nuclear Bomb: The Impact on Global Proliferation*, University of California Press, 2001, p. 56.

20 Srinivasan, M.R., *From Fission to Fusion: The Story of India's Atomic Energy Programme*, Penguin Viking, New Delhi, 2002, p. 52.

21 'War and Peace in the Nuclear Age; Haves and Have-Nots; Interview with Bertrand Goldschmidt, 1986 [2]', Open Vault, https://tinyurl.com/yc8xmf2n. Accessed on 23 August 2022.

22 Ibid.

23 Ibid.

24 Srinivasan, M.R., *From Fission to Fusion: The Story of India's Atomic Energy Programme*, Penguin Viking, New Delhi, 2002, pp. 54–5.

25 Chowdhury, Indira and Ananya Dasgupta, *A Masterful Spirit: Homi J. Bhabha, 1909-1966*, Penguin Books, New Delhi, 2010, p. 229.

26 Ramana, M.V., 'The Indian Nuclear Industry: Status and Prospects', Nuclear Energy Futures Paper No. 9, December 2009, The Centre for International Governance Innovation, https://tinyurl.com/3nu8b7fw. Accessed on 29 August 2022.

27 Srinivasan, M.R., *From Fission to Fusion: The Story of India's Atomic Energy Programme*, Penguin Viking, New Delhi, 2002, p. 57.

28 Ibid. 58.

29 Ibid. 64–5.

Chapter 34: Bhabha and the Bomb

1 Bhabha, Homi, 'Bhabha Speech on Nuclear Disarmament United Nation's Day', Wilson Center Digital Archive, 24 October 1964, https://tinyurl.com/2prc2j2r. Accessed on 18 December 2022.

2 Pant, Harsh V. and Yogesh Joshi, *Indian Nuclear Policy*, Oxford University Press, New Delhi, 2018, p. 45.

3 Chengappa, Raj, *Weapons of Peace: The Secret Story of India's Quest to be a Nuclear Power*, HarperCollins, New Delhi, 2000, p. 82.

4 Nehru, Jawaharlal, 'Significance of the Atomic Revolution', *Nuclear India*, Vol. 43, No. 5–6, 2008, pp. 26–8, https://tinyurl.com/4yejwuw5. Accessed on 1 September 2022.

5 Perkovich, George, *India's Nuclear Bomb: The Impact on Global Proliferation*, University of California Press, 2001, p. 34.

6 Ibid. 35.

7 Menon, Raja, *A Nuclear Strategy for India*, Sage Publications, New Delhi, 2000, p. 71.

8 Ibid. 36.

9 Chengappa, Raj, *Weapons of Peace: The Secret Story of India's Quest to be a Nuclear Power*, HarperCollins, New Delhi, 2000, p. 83.

10 'War and Peace in the Nuclear Age; Haves and Have-Nots; Interview with Bertrand Goldschmidt, 1986 [2]', Open Vault, https://tinyurl.com/yc8xmf2n. Accessed on 23 August 2022.

11 Chengappa, Raj, *Weapons of Peace: The Secret Story of India's Quest to be a Nuclear Power*, HarperCollins, New Delhi, 2000, p. 83.

12 Rasgotra, Maharajakrishna, *A Life in Diplomacy*, Penguin Books, Gurugram, 2016, p. 186.

13 Ibid. 187.

14 Gopal, Sarvepalli, *Jawaharlal Nehru: A Biography*, Vol. 3, Oxford University Press, 2011, p. 40.

15 Rasgotra, Maharajakrishna, *A Life in Diplomacy*, Penguin Books, Gurugram, 2016, p. 187.

16 Chengappa, Raj, *Weapons of Peace: The Secret Story of India's Quest to be

a Nuclear Power, HarperCollins, New Delhi, 2000, p. 86.

17 Abraham, Itty, *The Making of the Indian Atomic Bomb: Science, Secrecy and the Post-Colonial State*, Orient BlackSwan, 1999, pp. 114–6.

18 Chengappa, Raj, *Weapons of Peace: The Secret Story of India's Quest to be a Nuclear Power*, HarperCollins, New Delhi, 2000, p. 89.

19 Lala, R.M., *Beyond the Last Blue Mountain: A Life of J.R.D. Tata,* Penguin Books, New Delhi, 1993, pp. 267–8.

20 'Letter, Homi Bhabhi to Jawaharlal Nehru', Wilson Center Digital Archive, 2 August 1963, https://tinyurl.com/3nkpyfnu. Accessed on 9 January 2023.

21 Bhabha, Homi, 'Jawaharlal Nehru and Science', *Science Reporter*, Vol. 1, No. 7–8, July–August 1964.

22 Bhabha, Homi, 'Speech on All India Radio', Department of Atomic Energy, Government of India, 1 June 1964.

23 *Resolution Passed by the SAC on Nehru's Death*, Department of Atomic Energy, Government of India.

24 Nehru, Jawaharlal, 'Atomic Energy Commission', *Science Reporter*, Vol. 1, No. 7-8, 1964.

Chapter 35: Nuclear Pragmatism

1 'Bhabha Speech on Nuclear Disarmament United Nation's Day', Wilson Center Digital Archive, 24 October 1964, https://tinyurl.com/2prc2j2r. Accessed on 18 December 2022.

2 Malhotra, Inder, 'Because Walls Have Ears', *The Indian Express*, 28 March 2011, https://tinyurl.com/bddthxfm. Accessed on 1 September 2022.

3 'Bhabha Speech on Nuclear Disarmament United Nation's Day', Wilson Center Digital Archive, 24 October 1964, https://tinyurl.com/2prc2j2r. Accessed on 18 December 2022.

4 Ibid.

5 Perkovich, George, *India's Nuclear Bomb: The Impact on Global Proliferation*, University of California Press, 2001, p. 65.

6 Ibid.

7 Ibid.

8 Malhotra, Inder, 'Russia's New Rulers Silent on the Chinese Bomb', *The Statesman*, 30 October 1964.

9 'Indian Achievements in A-Energy Development', *Hindustan Times*, 30 October 1964.

10 Thapar, Romesh, 'In the Shadow of the Bomb', *Economic and Political Weekly*, 31 October 1964, https://tinyurl.com/5huj9yuw. Accessed on 1 September 2022.

11 Ibid.

12 Bhatia, Shyam, *India's Nuclear Bomb*, Vikas Publishing House, New Delhi, 1979, p. 113–4.

13 Perkovich, George, *India's Nuclear Bomb: The Impact on Global Proliferation*, University of California Press, 2001, p. 74.
14 Ramesh, Jairam, *A Chequered Brilliance: The Many Lives of V.K. Krishna Menon*, Penguin Viking, Gurugram, 2019, p. 177.
15 Perkovich, George, *India's Nuclear Bomb: The Impact on Global Proliferation*, University of California Press, 2001, p. 75.
16 Ibid.
17 Guha, Ramachandra, *India After Gandhi: The History of the World's Largest Democracy*, Picador India, 2017, New Delhi, p. 388–9.
18 Perkovich, George, *India's Nuclear Bomb: The Impact on Global Proliferation*, University of California Press, 2001, p. 82.
19 'K.R. Narayanan, "India and the Chinese Bomb"', 24 November 1964, Wilson Center Digital Archive, https://tinyurl.com/2wunu3kx. Accessed on 1 September 2022.
20 Creasman, Major David J., *The Evolution of India's Nuclear Program: Implications for the United States,* Fort Leavenworth, Kansas, 2008, p. 12.
21 Khera, S.S., *India's Defence Problem,* Orient Longman, New Delhi, 1968, pp. 317–8.
22 Chengappa, Raj, *Weapons of Peace: The Secret Story of India's Quest to be a Nuclear Power*, HarperCollins, New Delhi, 2000, p. 97.

23 Sarkar, Jayita, 'Sino-Indian Nuclear Rivalry: Glacially Declassified', *The Diplomat*, 2 June 2017, https://tinyurl.com/3pp4nvz9. Accessed on 29 December 2022.
24 Cockcroft, John and M.G.K. Menon, *Homi Jehangir Bhabha 1909-1966*, Royal Institution of Great Britain, London, 1967, p. 15.
25 Perkovich, George, *India's Nuclear Bomb: The Impact on Global Proliferation*, University of California Press, 2001, p. 94.
26 Ibid.
27 Ibid. 95.
28 Ibid. 96.
29 Ibid. 97.
30 Chengappa, Raj, *Weapons of Peace: The Secret Story of India's Quest to be a Nuclear Power*, pp. 98–9.

Chapter 36: Air India Flight 101

1 Deshmukh, Chintamani, *Homi Jehangir Bhabha*, National Book Trust, India, New Delhi, 2014, p. 103.
2 Sarkar, Jayita, 'Sino-Indian Nuclear Rivalry: Glacially Declassified', *The Diplomat*, 2 June 2017, https://tinyurl.com/3pp4nvz9. Accessed on 1 September 2022.
3 Majumdar, Neera, 'Sabotage or Accident? The Theories about How India Lost Nuclear Energy Pioneer Homi Bhabha', *The Print*, 24 January 2018.

4 Sarkar, Jayita, 'Sino-Indian Nuclear Rivalry: Glacially Declassified', *The Diplomat*, 2 June 2017, https://tinyurl.com/3pp4nvz9. Accessed on 1 September 2022.

5 Lala, R.M., *Beyond the Last Blue Mountain: A Life of J.R.D. Tata,* Penguin Books, New Delhi, 1993, 152–3.

6 Kulkarni, R.P. and V. Sarma, *Homi Bhabha: Father of Nuclear Science in India*, Bombay Popular Prakashan, Mumbai, 1969, p. 47.

7 Ghosh, Dipan K. and Arun K. Grover, *Tribute to a Titan: Birth Centenary of Homi Jehangir Bhabha*, Indian Physics Association, Mumbai, 2009, pp. 75–6.

8 Dadabhoy, Bakhtiar K., *Sugar in Milk: Lives of Eminent Parsis*, Rupa & Co., New Delhi, 2008, p. 295.

9 Majumdar, Neera, 'Sabotage or Accident? The Theories About How India Lost Nuclear Energy Pioneer Homi Bhabha', *The Print*, 24 January 2018.

10 Douglas, Gregory, *Conversations with the Crow: The Final Conversations of Robert Trumbull Crowley–Former Director of Clandestine Operations for the CIA*, Basilisk Press, 2013. In November 2008, a transcript of the conversation was sent to the Times of India journalist, Srinivas Laxman, who did a story on it. I am grateful to him for drawing my attention to this conspiracy theory.

11 Ibid.

12 Ibid.

13 Ibid.

14 What Really Happened to Homi J Bhabha P2, YouTube, https://tinyurl.com/394x4uea. Accessed on 1 September 2022.

15 Sarkar, Jayita, 'Sino-Indian Nuclear Rivalry: Glacially Declassified', *The Diplomat*, 2 June 2017, https://tinyurl.com/3pp4nvz9. Accessed on 1 September 2022.

16 Sarkar, Jayita, 'India's Nuclear History, Frozen in Time', *Sources and Methods*, Wilson Center, 22 June 2017, https://tinyurl.com/3canwcmb. Accessed on 1 September 2022.

17 Ibid.

18 *The Statesman*, 25 January 1966.

19 Lala, R.M., Personal interview, 12 March 2011.

20 Chowdhury, Indira and Ananya Dasgupta, *A Masterful Spirit: Homi J. Bhabha, 1909-1966*, Penguin Books, New Delhi, 2010, p. 247.

21 Bhatia, Shyam, 'Bhabha's Rare Assets up for Auction in Mumbai', *The Tribune*, 13 October 2011, https://tinyurl.com/ycks92ev. Accessed on 1 September 2022.

22 Sreekantan, B.V. and Richa Malhotra, 'Homi Jehangir Bhabha: A Visionary', *Resonance*, Vol. 15, No. 5, May 2010, pp. 462–75, https://tinyurl.com/mrxyyz53. Accessed on 9 August 2022.

23 Ibid.

24 Chowdhury, Indira and Ananya Dasgupta, *A Masterful Spirit: Homi J. Bhabha, 1909-1966*, Penguin Books, New Delhi, 2010, pp. 247–8.

25 'Excerpt from Speech by Vikram Sarabhai at the Condolence Meeting at AEET, Trombay, on 25 January 1966', *Nuclear India*, Vol. 4, No. 6-7, Feb–Mar 1966.
26 *Nuclear India*, Vol. 4, No. 6-7, Feb–Mar 1966.
27 Chowdhury, Indira and Ananya Dasgupta, *A Masterful Spirit: Homi J. Bhabha, 1909-1966*, Penguin Books, New Delhi, p. 244.
28 Ibid.
29 *Resolution at Condolence Meeting on 25 January 1966*, Department of Atomic Energy, Government of India.
30 *Nuclear India*, Vol. 4, No. 6-7, Feb–Mar 1966.
31 Ibid.
32 Ibid.
33 Chowdhury, Indira, 'When Breath Becomes Air…', The Oral Historian, https://tinyurl.com/2s3rfrea. Accessed on 1 September 2022.
34 Ghosh, Dipan K. and Arun K. Grover, *Tribute to a Titan: Birth Centenary of Homi Jehangir Bhabha*, Indian Physics Association, Mumbai, 2009, p. 82.
35 'Excerpt from J.R.D. Tata's Speech on the Occasion of the Release of the Book', *Collected Scientific Papers of Homi Jehangir Bhabha*, 27 March 1986.
36 Tata, J.R.D., 'Foreword', *Tata Institute of Fundamental Research 1945—1970 (Silver Jubilee Commemoration Volume)*, Tata Institute of Fundamental Research, 1970. p. x.
37 Lala, R.M., *The Heartbeat of a Trust: Fifty Years of the Sir Dorabji Tata Trust*, Tata McGraw Hill, New Delhi, 1984, p. 102.
38 *The Guardian*, 25 January 1966, (www.theguardian.com).
39 Penney, William George, 'Homi Jehangir Bhabha, 1909-1966', Biographical Memoirs of Fellows of the Royal Society, https://tinyurl.com/2p89c828. Accessed on 29 July 2022.
40 'The Visionary and the Vision', Permanent Exhibition on Homi Bhabha, October 2009, Tata Institute of Fundamental Research, Mumbai, Maharashtra, India.
41 Cockcroft, John and M.G.K. Menon, *Homi Jehangir Bhabha 1909-1966*, Royal Institution of Great Britain, London, 1967, p. 16.
42 *Homi: Commemorating the Birth Centenary of Homi Jehangir Bhabha 1909-1966*, Tata Institute of Fundamental Research, Mumbai, India, 2009.
43 Anand, Mulk Raj, 'In Memoriam: Homi Bhabha', *Marg*, Vol. 19, No. 2, March 1966.
44 Kulkarni, R.P. and V. Sarma, *Homi Bhabha: Father of Nuclear Science in India*, Bombay Popular Prakashan, Mumbai, 1969, p. 49.
45 Wali, Kameshwar C., *Chandra: A Biography of S. Chandrasekhar*, Penguin Books, New Delhi, 1991, p. 288.
46 Pant, Harsh V. and Yogesh Joshi, *Indian Nuclear Policy*, Oxford University Press, New Delhi, 2018, p. 67.
47 Ramana, M.V., 'La Trahison des Clercs: Scientists and India's Nuclear Bomb', *Prisoners of the Nuclear Dream*, M.V. Ramana and C. Rammanohar Reddy

(eds.), Orient BlackSwan, Hyderabad, 2003, p. 227.

48 Chengappa, Raj, *Weapons of Peace: The Secret Story of India's Quest to be a Nuclear Power*, HarperCollins, New Delhi, 2000, p. 104.

49 Perkovich, George, *India's Nuclear Bomb: The Impact on Global Proliferation*, University of California Press, 2001, p. 123.

50 Pant, Harsh V. and Yogesh Joshi, *Indian Nuclear Policy*, Oxford University Press, New Delhi, 2018, p. 68–9.

51 Perkovich, George, *India's Nuclear Bomb: The Impact on Global Proliferation*, University of California Press, 2001, p. 141.

52 Ramanna, Raja, *Years of Pilgrimage: An Autobiography*, Penguin Viking, New Delhi, 1991, p. 60.

53 Srinivasan, M.R., 'India's Atomic Adventure', *Frontline*, 15 August 1997.

54 Ramana, M.V., 'La Trahison des Clercs: Scientists and India's Nuclear Bomb', *Prisoners of the Nuclear Dream*, M.V. Ramana and C. Rammanohar Reddy (eds.), Orient BlackSwan, Hyderabad, 2003, pp. 227–8.

55 Sinha, Aditya, 'When Raja Ramanna Turned Down Saddam Hussein', *Hindustan Times*, 30 March 2003, https://tinyurl.com/3uyahrxt. Accessed on 1 September 2022.

56 Chowdhury, Indira, *Growing the Tree of Science: Homi Bhabha and the Tata Institute of Fundamental Research*, Oxford University Press, New Delhi, p. 226.

57 Ibid. 227.

Epilogue

1 The National Centre for Software Technology (Mumbai and Bangalore); the Society for Applied Microwave Engineering and Electronic Research (Mumbai and Chennai); the Inter-University Centre for Astronomy and Astrophysics (Pune); the Homi Bhabha Centre for Science Education (Mumbai); the National Centre for Radio Astrophysics (Pune); and the National Centre for Biological Sciences (Bangalore). The last three have been set up directly under the umbrella of TIFR and are governed by its Council of Management.

2 Ramana, M.V., *The Power of Promise: Examining Nuclear Energy in India*, Penguin Viking, New Delhi, 2012, p. 254.

3 Ibid. 269.

4 'Government of India Department of Atomic Energy: Annual Report 2020-21', Department of Atomic Energy, https://tinyurl.com/w576zxjw, p. xi. Accessed on 1 September 2022.

5 Ibid. xiv.

6 Ibid. iv.

7 Sethi, Manpreet (ed.), *The Global Nuclear Landscape*, KW Publishers, New Delhi, 2022, p. 23.

BIBLIOGRAPHY

Abraham, Itty, 'From the Commission to the Mission Model: Technology Czars and the Indian Middle Class', *The Journal of Asian Studies*, Vol. 76, No. 3, 2017, https://tinyurl.com/mu6mk3fu.

Abraham, Itty, 'India's Strategic Enclave: Civilian Scientists and Military Technology', *Armed Forces and Society*, Vol. 18, No. 2, 1992.

Abraham, Itty, 'The Ambivalence of Nuclear Histories', *Osiris*, Vol. 21, No. 1, 2006, pp. 49–65, JSTOR, https://tinyurl.com/2p9e74md.

Abraham, Itty, 'The Contradictory Spaces of Postcolonial Techno-Science', *Economic and Political Weekly*, Vol. 41, No. 3, 2006, pp. 210–7.

Abraham, Itty, *The Making of the Indian Atomic Bomb: Science, Secrecy and the Post-Colonial State*, Orient BlackSwan, Hyderabad, 1999.

Alladi, Krishnaswami, 'Alladi Ramakrishnan's Theoretical Physics Seminar and the Creation of Matscience', Krishnaswami Alladi, https://tinyurl.com/ykycd7te.

Alladi, Krishnaswami, 'IAS Visit Inspires Creation of Institute of Mathematical Sciences', Institute for Advanced Study, https://tinyurl.com/4jehhfns.

'An Indian Academy of Science', *Current Science*, Vol. 1, No. 11, 1933, pp. 335–7, JSTOR, https://tinyurl.com/3mdckfue.

Anderson, Carl D. and Seth Neddermeyer, 'Mesotron (Intermediate Particle) as a Name for the New Particles of Intermediate Mass', *Nature* Vol. 142, No. 878, 1938, https://tinyurl.com/y3fwnkv3.

Anderson, Robert S., 'Building Scientific Institutions in India: Saha and Bhabha', *Occasional Paper Series*, McGill University, Montreal, 1975.

Anderson, Robert S., *Nucleus and Nation: Scientists, International Networks, and Power in India*, University of Chicago Press, 2010.

Anderson, Robert S., 'Shanti Bhatnagar, Working Through/Against the Colonial System', *Indian Journal of the History of Science*, Vol. 53, No. 4, 2018.

Aravamudan, R. and Gita Aravamudan, *ISRO: A Personal History*, HarperCollins, Noida, 2017.

Arbuthnot, Mollie, 'Rudolf von Leyden: The German-born Cartoonist Who Became Bombay's Top Art Critic in the 1930s', *Scroll.in*, 16 December 2018, https://tinyurl.com/mryfz5m5.

Ardeshir Dalal: A Memoir, Dalal Family, 1980.

Arnold, David, 'Nehruvian Science and Postcolonial India', *Isis* Vol. 104, No. 2, 2013.

Athreya, K.B., 'K G Ramanathan', *Resonance*, Vol. 15, February 2010.

Baggott, Jim, *Atomic: The First War of Physics and the Secret History of the Atom Bomb, 1939–1949,* Icon Books, London, 2009.

Baggott, Jim, *The Quantum Story: A History in 40 Moments*, Oxford University Press, UK, 2013.

Bainbridge, Kenneth, '1975: All in Our Time: A Foul and Awesome Display', *Bulletin of the Atomic Scientists*, Vol. 31, No. 5, 1975.

Bamzai, Sandeep, *Princestan: How Nehru, Patel & Mountbatten Made India*, Rupa Publications, New Delhi, 2020.

Banerjee, B. and N. Sarma, *Nuclear Power in India: A Critical History*, Rupa & Company, New Delhi, 2008.

Banerjee, Somaditya, *Bhadralok Physics and the Making of Modern Science in Colonial India*, 2018, University of British Columbia, Vancouver, PhD dissertation.

Basu, Biman, *Yash Pal: A Life in Science*, Vigyan Prasar, New Delhi, 2006.

Bazilevskaya, G.A., 'Skobeltsyn and the Early Years of Cosmic Particle Physics in the USSR', *Astroparticle Physics*, Vol. 53, 2014, pp. 61–6.

Bernstein, Lina, 'The Great Little Lady of the Bombay Art World', Academia, https://tinyurl.com/2nykf3sy.

Bhabha, Homi, 'Historical Note on Tata Institute of Fundamental Research', Wilson Center Digital Archive, 1 January 1954, https://tinyurl.com/43dpnxh6.

Bhabha, Homi, 'Leonardo da Vinci: 500 Years', *Marg*, Vol. 5, No. 4, July 1952.

Bhabha, Homi, 'Note on the Organisation of Atomic Research in India', Manohar Parrikar Institute for Defence Studies and Analyses, 26 April 1948, https://tinyurl.com/mt56bzfs.

Bhabha, Homi, 'The Fundamental Length Introduced by the Theory of the Mesotron (Meson)', *Nature*, Vol. 143, pp. 276–7, 1939, https://tinyurl.com/ycks8jjr.

Bhabha, Homi, *Director's Report, 1952-53*, Tata Institute of Fundamental Research Archive, Mumbai, Maharashtra, India.

Bhatia, Shyam, 'Bhabha's Rare Assets up for Auction in Mumbai', *The Tribune*, 13 October 2011, https://tinyurl.com/ycks92ev.

Bhatia, Shyam, *India's Nuclear Bomb*, Vikas Publishing House, New Delhi, 1979.

Bhattacharya, Amitabha, 'A Forgotten Legend', *The Statesman*, 23 September 2018, https://tinyurl.com/2p8usv87.

Bidwai, Praful and Achin Vanaik, *South Asia on a Short Fuse: Nuclear Politics and the Future of Global Disarmament*, Oxford University Press, New Delhi, 1999.

Biswas, Sukumar, 'Early Years of High Energy Physics', *Current Science*, Vol. 60, No. 11, 1991.

Bodanis, David, *E = MC²: A Biography of the World's Most Famous Equation*, Pan Books, London, 2000.

Bodanis, David, *Einstein's Greatest Mistake: The Life of a Flawed Genius*, Little, Brown Book Group, London, 2016.

Bothwell, Robert, *Nucleus: The History of Atomic Energy of Canada Limited*, University of Toronto Press, 1988.

Brown, Laurie M., 'Hideki Yukawa and the Meson Theory', *Physics Today*, Vol. 39, No. 12, 1986.

Capri, Anton Z., *Quips, Quotes and Quanta: An Anecdotal History of Physics*, World Scientific Publishing Company, 2007.

Casimir, Hendrik B.G., *Haphazard Reality: Half a Century of Science*, Harper and Row Publishers, New York, 1983.

Chadha, Gita and Vidya Kamat, 'A Scientist's Patronage of Art: Homi Bhabha's Art Collection', *Humanscape*, Vol. 9, No. 8, 2002.

Chatterjee, Mortimer, 'TIFR and the Birth of the Modern Indian Art Collection', *Mumbai Mirror*, 2 November 2015, https://tinyurl.com/4nxmhvwe.

Chavan, Akshay, 'When Travancore Declared Its Independence', Live History India, 11 June 2021, https://tinyurl.com/5n7mdmpv.

Chengappa, Raj, *Weapons of Peace: The Secret Story of India's Quest to be a Nuclear Power*, HarperCollins, New Delhi, 2000.

Chidambaram, R. and C. Ganguly, 'Plutonium and Thorium in the Indian Nuclear Programme', *Current Science*, Vol. 70, No. 1, 10 January 1996, pp. 21–35.

Chowdhury, Indira and Ananya Dasgupta, *A Masterful Spirit*: *Homi J. Bhabha, 1909-1966*, Penguin Books, New Delhi, 2010.

Chowdhury, Indira, 'For 31 Years after His Death, Homi Bhabha's Office Room Was Unoccupied', *The Print*, 24 January 2018, https://tinyurl.com/3b8sb9tj.

Chowdhury, Indira, 'Fundamental Research, Self-Reliance and Internationalism: The Evolution of the Tata Institute of Fundamental Research', Obaid Siddiqi Collection, Archives at National Centre for Biological Sciences, https://tinyurl.com/5yjcv9kf.

Chowdhury, Indira, 'Local Hierarchies and International Networks: Molecular Biology at the Tata Institute of Fundamental Research, 1962–1980', Archives Internationales D'histoire des Sciences, Vol. 56, 2006, pp. 156–7.

Chowdhury, Indira, 'Memories, Memorabilia and Memorable Interviews: Talking to Mithoo Coorlawala', The Oral Historian, 14 May 2020, https://tinyurl.com/9be76u33.

Chowdhury, Indira, 'The Laboratory and its Double: The Making of the Scientist-Citizen at TIFR', Centre for the Advanced Study of India, University of Pennsylvania, 2012, https://tinyurl.com/hnnwv7a7.

Chowdhury, Indira, *Growing the Tree of Science: Homi Bhabha and the Tata Institute of Fundamental Research*, Oxford University Press, New Delhi, 2016

Close, Frank et al., *The Particle Odyssey: A Journey to the Heart of Matter*, Oxford University Press, Oxford, 2002.

Close, Frank, *Trinity: The Treachery and Pursuit of the Most Dangerous Spy in History*, Penguin Random House, UK, 2020.

Cockcroft, John and M.G.K. Menon, *Homi Jehangir Bhabha 1909-1966*, Royal Institution of Great Britain, London, 1967.

Cronin, James W., 'The 1953 Cosmic Ray Conference at Bagneres de Bigorre', *European Physics Journal*, Vol. 183, 2011, https://tinyurl.com/54mfrn4u.

Dadabhoy, Bakhtiar K., *Jeh: A Life of JRD Tata*, Rupa & Company, New Delhi, 2005.

Dadabhoy, Bakhtiar K., *Sugar in* Milk: *Lives of Eminent Parsis*, Rupa & Company, New Delhi, 2008.

Dadabhoy, Bakhtiar K., *Zubin Mehta: A Musical Journey*, Penguin Random House, Gurugram, 2016.

Dani, S.G., 'Kosambi, the Mathematician', *Resonance*, Vol. 4, No. 5, 1999, pp. 514–28.

Dasannacharya, B.A., 'Homi Jehangir Bhabha', *Current Science*, Vol. 96, No. 11, 10 June 2009, pp. 1536–8.

Deshmukh, Chintamani, *Homi Jehangir Bhabha*, National Book Trust, India, New Delhi, 2014.

Devons, S., 'Rutherford's Laboratory', University of Cambridge, https://tinyurl.com/296v5eea.

Dickson, David, *The New Politics of Science*, University of Chicago Press, 1988.

Dimri, Keshav, *Atomic Postalgia: Antiquity, Modernity and the Origins of the Nuclear Program in India*, 2014, Stanford University, Thesis submitted to the interschool honours programme in international security studies, https://tinyurl.com/t7as72np.

Divatia, A.S. and C. Ambasankaran, 'Accelerator Development in India', *Pramana*, Vol. 24, No. 1 & 2, January and February 1985, pp. 227–44, https://tinyurl.com/3tbnmf6u.

Early Days of Particle Physics in India by Sreerup Raychaudhuri, YouTube, https://tinyurl.com/4rjpwcj2.

Farmelo, Graham, *The Strangest Man: The Hidden Life of Paul Dirac, Mystic of the Atom*, Basic Books, New York, 2009.

Farooqui, Amar, *Opium City: The Making of Early Victorian Bombay*, Three Essays Collective, Gurugram, 2006.

Fawcett, Ruth, *Nuclear Pursuits: The Scientific Biography of Wilfred Bennett Lewis*, McGill's-Queen University Press, 1994.

Fee, William Thomas, 'The Parsees and the Towers of Silence at Bombay, India', *The National Geographic Magazine*, Vol. 16, No. 12, 1905, https://tinyurl.com/5n67tycr.

Feldman, Burton, *The Nobel Prize: A History of Genius, Controversy, and Prestige*, Arcade Publishing, New York, 2000.

Fitch, Val L., 'Elementary Particle Physics: The Origins', *Reviews of Modern Physics*, Vol. 71, No. 2, 1999, https://tinyurl.com/2823wekm.

Fraser, Gordon, *Cosmic Anger: Abdus Salam - The First Muslim Nobel Scientist*, Oxford University Press, 2008.

Freedman, Lawrence, *The Evolution of Nuclear Strategy*, Palgrave Macmillan, London, 1989.

Freund, Peter G.O. et al., 'Gregor Wentzel', Cornell University, 11 September 2008, https://tinyurl.com/4xckjmss.

Frey, Karsten, 'Guardians of the Nuclear Myth: Politics, Ideology, and India's Strategic Community', *South Asian Cultures of the Bomb: Atomic Publics and the State in India and Pakistan*, Itty Abraham (ed.), Indiana University Press, 2009.

Frey, Karsten, *India's Nuclear Bomb and National Security*, Routledge, London and New York, 2006.

Frisch, Otto Robert, *What Little I Remember*, Cambridge University Press, 1980.

Gandhy, Kekoo, 'The Beginnings of the Art Movement', *Seminar*, 2003, https://tinyurl.com/3938rtby.

Ganguly, Jibamitra and Chitra Roy, *Meghnad Saha: His Science and Persona through Selected Letters and Writings*, Indian National Science Academy, New Delhi, 2019.

Ghosh, Dipan K., and Arun K. Grover, *Tribute to a Titan: Birth Centenary of Homi Jehangir Bhabha*, Indian Physics Association, Mumbai, 2009.

Gill, Piara Singh, *Up Against Odds: Autobiography of an Indian Scientist*, Allied Publishers, Bombay, 1992.

Gilpin, Robert, *American Scientists and Nuclear Weapons Policy*, Princeton University Press, 1962.

Gopal, Sarvepalli, *Jawaharlal Nehru: A Biography*, Vol. 2, Oxford University Press, New Delhi, 2015.

Goswami, J.N. and J.D. Macdougall, 'Devendra Lal', Biographical Memoirs of Fellows of the Royal Society, The Royal Society Publishing, https://tinyurl.com/ye286v4u.

Greenstein, George, 'A Gentleman of the Old School: Homi Bhabha and the Development of Science in India', *The American Scholar*, Vol. 61, No. 3, 1992, JSTOR, https://tinyurl.com/428vjnnr.

Gribbin, John and Mary Gribbin, *Richard Feynman: A Life in Science*, Icon Books, London, 2018.

Gribbin, John, *Science: A History, 1543-2001*, Allen Lane, London, 2002.

Gribbin, John, *The Fellowship: The Story of a Revolution*, Penguin Books, London, 2005.

Gribbin, John, *The Scientists: A History of Science Told Through the Lives of Its Greatest Inventors*, Random House, New York, 2004.

Guha, Ramachandra, *India After Gandhi: The History of the World's Largest Democracy*, Picador India, New Delhi, 2017.

Gunderman, David, 'Nicolas Bourbaki: The Greatest Mathematician Who Never Was', *The Conversation*, 9 December 2019, https://tinyurl.com/bdersf3k.

Hecht, Gabrielle (Ed.), *Entangled Geographies: Empire and Technopolitics in the Global Cold War*, MIT Press, 2011.

Hill, A.V., *A Report to the Government of India on Scientific Research in India*, The Royal Society, London, 1944.

Homi Bhabha A Scientist in Action, YouTube, https://tinyurl.com/5eduzrt3.

Homi: Commemorating the Birth Centenary of Homi Jehangir Bhabha 1909-1966, Tata Institute of Fundamental Research, Mumbai, India, 2009.

Hore, Peter, *Patrick Blackett: Sailor Scientist, Socialist*, Frank Cass Publishers, London, 2003.

Howe, Robert, 'Harish Chandra 1923–1983', National Academy of Sciences, 2011, https://tinyurl.com/5eubezdy.

'International Conference on Elementary Particles', *Current Science*, Vol. 20, No. 1, 1951, https://tinyurl.com/2p9bceya.

Isaacson, Walter, *Einstein: His Life and Universe*, Simon & Schuster, 2007.

Isaacson, Walter, *The Code Breaker: Jennifer Doudna, Gene Editing, and the Future of the Human Race*, Simon & Schuster, 2021.

Isaacson, Walter, *The Innovators: How a Group of Hackers, Geniuses and Geeks Created the Digital Revolution*, Simon & Schuster, 2014.

Iyer, Meera, 'Homi Bhabha's IISc Stint and Science Research in India', *Deccan Herald*, 25 July 2020, https://tinyurl.com/2p9xwhfp.

Jain, J. P., *Nuclear India*, Vol. 2, Radiant Publishers, New Delhi, 1974.

Jayaraman, A., *Chandrasekhara Venkata Raman: A Memoir*, Indian Academy of Sciences, Bangalore, 2017.

Jha, D.N. (ed.), *The Many Careers of D.D. Kosambi: Critical Essays*, LeftWord Books, New Delhi, 2011.

Jorgensen, Timothy J., 'Lise Meitner—the Forgotten Woman of Nuclear Physics Who Deserved a Nobel Prize', *The Wire*, 19 February 2019, https://tinyurl.com/yn9da8wd.

Joseph, John, 'Bhabha, The Friend of Trees', *Bharat Jyoti*, 1 May 1966.

Kakodkar, Anil and Suresh Gangotra, *Fire and Fury: Transforming India's Strategic Identity*, Rupa Publications, New Delhi, 2019.

Kaku, Michio, *The God Equation: The Quest for a Theory of Everything*, Penguin Random House, 2021.

Kanigel, Robert, *The Man Who Knew Infinity: A Life of the Genius Ramanujan*, Abacus, London, 1991.

Kapur, Ashok, 'India: The Nuclear Scientists and the State, the Nehru and Post-Nehru Years', *Scientists and the State: Domestic Structures and the International Context*, Etel Solingen (ed.), University of Michigan Press, Ann Arbor, 1994.

Kemmer, N., 'The Impact of Yukawa's Meson Theory on Workers in Europe: A Reminiscence', *Progress of Theoretical Physics Supplement*, Vol. E65, 1965, pp. 602–8, https://tinyurl.com/5bje5rhv.

Kevles, Daniel J., *The Physicists: The History of a Scientific Community in Modern America*, Harvard University Press, 1995.

Kulkarni, R.P. and V. Sarma, *Homi Bhabha: Father of Nuclear Science in India*, Bombay Popular Prakashan, Bombay, 1969.

Kumar, Deepak, *Science and the Raj, 1857–1905*, Oxford University Press, 1995.

Kumar, Manjit, *Quantum: Einstein, Bohr, and the Great Debate about the Nature of Reality*, Hachette, New Delhi, 2009.

Kursunoglu, Behram N. and Eugene Paul Wigner (eds.), *Reminiscences about a Great Physicist: Paul Adrien Maurice Dirac*, Cambridge University Press, New York, 1987.

Lal, Devendra, 'The Discovery of Cosmogenic ^{10}Be in India', *Current Science,* Vol. 61, No. 11, 1991.

Lala, R.M., *Beyond the Last Blue Mountain: A Life of J.R.D Tata*, Penguin Books, New Delhi, 1993.

Lala, R.M., *For the Love of India: The Life and Times of Jamsetji Tata*, Penguin Viking, New Delhi, 2004.

Lala, R.M., *The Heartbeat of a Trust: Fifty Years of the Sir Dorabji Tata Trust,* Tata McGraw-Hill, New Delhi, 1984.

Langlands, Robert P., 'Harish-Chandra', Biographical Memoirs of the Royal Society, https://tinyurl.com/yt5rr36j.

Lee, Rachel and Mareike Schwartz, 'Bombay', Metromod, 2021, https://tinyurl.com/5n8a24u6.

Malhotra, Inder, 'Because Walls Have Ears', *The Indian Express*, 28 March 2011, https://tinyurl.com/bddthxfm.

Malhotra, Inder, 'Russia's New Rulers Silent on the Chinese Bomb, *The Statesman*, 30 October 1964.

Malhotra, Inder, 'The Uncertainty Principle', *The Indian Express*, 31 March 2007, https://tinyurl.com/23m98s3y.

Mallik, D.C.V. and S. Chatterjee, *Kariamanikkam Srinavasa Krishnan: His Life and Work*, Universities Press, Hyderabad, 2012.

Manjapra, Kris, *Age of Entanglement: German and Indian Intellectuals Across Empire,* Harvard University Press, Cambridge, Massachusetts, 2014.

Manuel, Thomas, *Opium Inc.*, HarperCollins, Noida, 2021.

Marfatia, Meher, 'From Bombay with Love', *Mid-Day*, 23 May 2022, https://tinyurl.com/57v7tp7v.

Marshak, Robert 'The Pain and Joy of a Major Scientific Discovery', *A Gift of Prophecy: Essays in the Celebration of the Life of Robert Eugene Marshak*, E.C.G. Sudarshan (ed.), World Scientific Publishing, New Jersey, 1994.

Masters, Barry, 'C.V. Raman and the Raman Effect', *Optics and Photonics News*, Vol. 20, No. 2, 2009, pp. 26–31, https://tinyurl.com/2p9f34f7.

Mehra, Jagdish, 'Satyendra Nath Bose, 1 January 1894 - 4 February 1974', The Royal Society Publishing, https://tinyurl.com/2ne769ee.

Menon, M.G.K., 'Tribute to Dr Jamshed Bhabha', *Annual Report, 2006-07*, The Trustees of the Sir Dorabji Tata Trust, Tata Trust, https://tinyurl.com/7eavk6jk.

Menon, Meena and T.S. Subramanian, 'Sethna– The Man Who Dared to Dream', *The Hindu*, 7 September 2010, https://tinyurl.com/yrvyt37n.

Mian, Zia, 'Homi Bhabha Killed a Crow', *The Nuclear Debate: Ironies and Immoralities*, Zia Mian and Ashis Nandy (eds.), Regional Centre for Strategic Studies, Colombo, 1998.

Miller, Arthur I., *Empire of the Stars: Friendship, Obsession and Betrayal in the Quest for Black Holes*, Little, Brown and Company, London, 2005.

Mondal, Naba K. and Sreerup Raychaudhuri, 'M.G.K. Menon—Statesman of Indian Science', *Resonance*, Vol. 24, No. 11, November 2019, pp. 1189–233, https://tinyurl.com/24zm47a2.

Morrison, Philip, 'Heaven and Earth One Substance: Bernard Peters and the Heavy Primaries', *Current Science*, Vol. 61, No. 11, 1991, pp. 740–4.

Mukhopadhyay, Atri, 'The Untold Story of the Indian Atomic Energy Programme: Meghnad Saha and Homi Bhabha', *Journal of the Asiatic Society*, Vol. 60, No. 3, 2018.

Mukunda, N., 'Homi Bhabha—An Appreciation and a Hope', *Current Science*, Vol. 100, No. 9, 10 May 2011.

Nair, Uma 'Progressives Mentor Walter Langhammer', *The Times of India*, 18 February 2017, https://tinyurl.com/33yj5x5k.

Narasimhan, R., 'Men, Machines and Ideas: An Autobiographical Essay', *Current Science*, Vol. 76, No. 3, 10 February 1999.

Nath, Biman, *Homi J Bhabha: A Renaissance Man Among Scientists*, Niyogi Books, New Delhi, 2021.

Palsetia, Jesse S., *Jamsetjee Jejeebhoy of Bombay*, Oxford University Press, New Delhi, 2015.

Palsetia, Jesse S., *The Parsis of India: Preservation of Identity in Bombay City*, Manohar Books, New Delhi, 2008.

Pant, Harsh V. and Yogesh Joshi, *Indian Nuclear Policy*, Oxford University Press, New Delhi, 2018.

Park, Robert E., 'Human Migration and the Marginal Man', *American Journal of Sociology*, Vol. 33, No. 6, 1928.

Patel, Simin, *Cultural Intermediaries in a Colonial City: The Parsis of Bombay c. 1860-1921*, 2015, University of Oxford, DPhil dissertation.

Penney, William George, 'Homi Jehangir Bhabha, 1909-1966', Biographical Memoirs of Fellows of the Royal Society, https://tinyurl.com/2p89c828.

Perkovich, George, *India's Nuclear Bomb: The Impact on Global Proliferation*, University of California Press, 2001.

Peters, Bernard, 'The Decade of My Association with Research in India', *Current Science*, Vol. 61, No. 11, 10 December 1991, pp. 717–22, https://tinyurl.com/mr2kmb9x.

Phalkey, Jahnavi, *Atomic State: Big Science in Twentieth-Century India*, Permanent Black, New Delhi, 2013.

Radhakrishnan, M.G., 'Rocket Boys, Chithira Thirunal and Thorium Politics', *Mathrubhumi.com*, 19 February 2022, https://tinyurl.com/45e47b5t.

Raghunathan, M.S., 'Artless Innocents and Ivory-Tower Sophisticates: Some Personalities on the Indian Mathematical Scene', *Current Science*, Vol. 85, No. 4, 2003, pp. 526–36, https://tinyurl.com/55mscetu.

Raianu, Mircea, *Tata: The Global Corporation that Built Indian Capitalism*, Harvard University Press, 2021.

Raj, Gopal, *Reach for the Stars: The Evolution of India's Rocket Programme*, Peguin Viking, New Delhi, 2000.

Ramana, M.V. and C. Rammanohar Reddy (eds.), *Prisoners of the Nuclear Dream*, Orient BlackSwan, Hyderabad, 2003.

Ramana, M.V., 'India and Fast Breeder Reactors', *Science and Global Security*, Vol. 17, No. 1, 2009.

Ramana, M.V., 'India's Nuclear Enclave and the Practice of Secrecy', *South Asian Cultures of the Bomb: Atomic Publics and the State in India and Pakistan*, Itty Abraham (ed.), Indiana University Press, 2009.

Ramana, M.V., 'La Trahison des Clercs: Scientists and India's Nuclear Bomb', *Prisoners of the Nuclear Dream*, M.V. Ramana and C. Rammanohar Reddy (eds.), Orient BlackSwan, Hyderabad, 2003.

Ramana, M.V., 'The Indian Nuclear Industry: Status and Prospects', Nuclear Energy Futures Paper No. 9, December 2009, The Centre for International Governance Innovation, https://tinyurl.com/3nu8b7fw.

Ramana, M.V., *The Power of Promise: Examining Nuclear Energy in India*, Penguin Viking, New Delhi, 2012.

Ramanna, Raja, *Years of Pilgrimage: An Autobiography*, Penguin Viking, New Delhi, 1991.

Ramaseshan, S., 'A Conversation with Krishnan on the Story of the Discovery of the Raman Effect', *Current Science*, Vol. 75, No. 11, 1998, JSTOR, https://tinyurl.com/mr42d6t4.

Ramaswamy, Ramakrishna (ed.), *D. D. Kosambi: Selected Works in Mathematics and Statistics*, Springer India, New Delhi, 2016.

Ramesh, Jairam, *A Chequered Brilliance: The Many Lives of V.K. Krishna Menon*, Penguin Viking, Gurugram, 2019.

Rao, C.R., 'Prasantha Chandra Mahalanobis, 1893–1972', Biographical Memoirs of Fellows of the Royal Society, Vol. 19, No. 19, 1973, https://tinyurl.com/4j3k3dt7.

Rasgotra, Maharajakrishna, *A Life in Diplomacy*, Penguin Books, Gurugram, 2016.

Raychaudhuri, Sreerup, 'Cosmic Ray Research in India: A Historical Perspective', *Physics News*, Vol. 51, No. 1-2, pp. 24–33, https://tinyurl.com/mwzprp43.

Remembering Dr. Homi Bhabha, the Physicist, Bhabha Atomic Research Centre, Department of Atomic Energy, Government of India, 2009.

Rhodes, Richard, *The Making of the Atomic Bomb*, Simon & Schuster, 2012.

'Robert Marshak - Session III', American Institute of Physics, 4 October 1970, https://tinyurl.com/2p8zzh4w.

Roy, S.C., and Rajinder Singh, 'Bibha Chowdhuri: An Unknown Indian Woman Scientist', *Science and Culture*, Vol. 84, No. 5–6, 2018, pp. 159–64.

Rudra, Ashok, *Prasanta Chandra Mahalanobis: A Biography*, Oxford University Press, New Delhi, 1998.

Sager, Peter, *Oxford & Cambridge: An Uncommon History*, Thames and Hudson, London, 2005.

Santhanam, K., 'The Da Vinci Coda', *Outlook*, 19 July 2010, https://tinyurl.com/3meys5mm.

Sarabhai, Vikram, 'Dr Homi J. Bhabha FRS' *Current Science*, Vol. 10, No. 3, March 1941.

Sarkar, Jayita, 'From the Peaceful Atom to the Peaceful Explosion', Wilson Centre, https://tinyurl.com/bdhb2frs.

Sarkar, Jayita, 'Sino-Indian Nuclear Rivalry: Glacially Declassified', *The Diplomat* 2 June 2017, https://tinyurl.com/4h9f2nkk.

Segrè, Gino and Bettina Hoerlin, *The Pope of Physics*, Henry Holt and Company, New York, 2016.

Seshadri, C.S., 'K. Chandrasekharan (1920–2017)', *Hardy-Ramanujan Journal*, Vol. 40, 2017, pp. 47–51, https://tinyurl.com/dkbrkcsf.

Seshadri, T.R., 'Shanti Swarup Bhatnagar, 1894-1955', Biographical Memoirs of Fellows of the Royal Society, Vol. 8, No. 8, https://tinyurl.com/3pd4rydm.

Shah, Amrita, *Vikram Sarabhai: A Life*, Penguin Viking, New Delhi, 2007.

Sharafi, Mitra June, *Bella's Case: Parsi Identity and the Law in Colonial Rangoon, Bombay and London, 1887-1925*, 2006, Princeton University, PhD dissertation.

Sharma, Dhirendra, 'India's Nuclear Estate', *Philosophy and Social Action*, Vol. 31, No. 4, 2005.

Sharma, Dhirendra, *India's Nuclear Estate*, Lancers Publishers, 1983.

Sharma, Dinesh C., 'Nehru and the Computer Revolution: Foundations and Transitions', Nehru Memorial Museum and Library, https://tinyurl.com/5eec2cvd.

Sharma, Dinesh C., *The Outsourcer: The Story of India's IT Revolution*, MIT Press, 2015.

Shyamasunder, R.K. and M.A. Pai (eds.), *Homi Bhabha and the Computer Revolution*, Oxford University Press, New Delhi, 2011.

Singh, Devika, 'German-Speaking Exiles and the Writing of Indian Art History', *Journal of Art Historiography*, No. 17, December 2017, https://tinyurl.com/m728amhx.

Singh, Jagjit, *Abdus Salam: A Biography*, Penguin Books, New Delhi, 1992.

Singh, Jagjit, *Some Eminent Indian Scientists*, Publications Division, Ministry of Information and Broadcasting, Government of India, 1966.

Singh, R. and F. Reiss, 'The Nobel Laureate Sir Chandrasekhara Venkata Raman FRS and His Contacts with the British Scientific Community in a Social and Political Context', *Notes and Records of the Royal Society*, Vol. 58, No. 1, 2004.

Singh, Rajinder, 'Homi J. Bhabha and Niels Bohr', *Current Science*, Vol. 97, No. 4, 25 August 2009.

Singh, Rajinder, 'Indo-American Relations with Reference to Bernard Peters', *Indian Journal of History of Science*, Vol. 43, No. 3, 2008, pp. 437–54.

Singh, Rajinder and Hardev Singh Virk, 'Homi J. Bhabha: Physics Nobel Prize Nominee and Nominator', *OmniScience: A Multi-Disciplinary Journal*, Vol. 7, No. 1, 2017.

Singh, Virender, 'Bhabha's Contributions to Elementary Particle Physics and Cosmic Rays Research', *Resonance*, Vol. 14, 2009.

Singh, Virendra, 'Homi Jehangir Bhabha: Architect of Modern Science and Technology in India', Tata Institute of Fundamental Research, 2009, https://tinyurl.com/ffjz6jz4.

Singh, Virendra, 'Meghnad Saha: His Science and Life', *Current Science*, Vol. 64, No. 7, 10 April 1993, JSTOR, https://tinyurl.com/yc3zcdr7.

Sinha, Aditya, 'When Raja Ramanna Turned Down Saddam Hussein', *Hindustan Times*, 30 March 2003, https://tinyurl.com/3uyahrxt.

Sinha, Jagdish N., 'Science and Culture under Colonialism: India between the World Wars', *Indian Journal of History of Science*, Vol. 39, No. 1, pp. 101–19, 2004.

Sreekantan, B.V. and Richa Malhotra, 'Homi Jehangir Bhabha: A Visionary', *Resonance*, Vol. 15, No. 5, May 2010, pp. 462–75, https://tinyurl.com/mrxyyz53.

Sreekantan, B.V. and Vinod K. Gaur, 'Devendra Lal (1929-2012)', *Current Science*, Vol. 104, No. 3, 10 February 2013, pp. 372–3.

Sreekantan, B.V., 'Homi Bhabha and Cosmic Ray Research in India', *Resonance*, Vol. 3, December 2005.

Sreekantan, B.V., 'Sixty Years of the Tata Institute of Fundamental Research 1945-2005: The Role of Younger Men in The Creation and Development of This Institute', *Current Science*, Vol. 90, No. 7, 10 April 2006.

Srinivasan, M.R., *From Fission to Fusion: The Story of India's Atomic Energy Programme*, Penguin Viking, New Delhi, 2002.

Srinivasan, M.R., 'An Intermezzo', *Selected Lectures of Dr. M.R. Srinivasan,* Department of Atomic Energy for Nuclear Power Corporation of India, 1990.

Srinivasan, M.R., 'India's Atomic Adventure', *Frontline*, 15 August 1997.

Srinivasan, M.R., 'Pandit Nehru's Vision of Science, Technology and Public Administration', *Selected Lectures of Dr. M.R. Srinivasan,* Department of Atomic Energy for Nuclear Power Corporation of India, 1990.

Strathern, Paul, *Bohr and Quantum Theory*, Arrow Books, London, 1998.

Strathern, Paul, *Oppenheimer and the Bomb*, Arrow Books, London, 1998.

Subramanian, C.V., 'Professor Birbal Sahni—the Man and His Message', *Current Science*, Vol. 61, No. 9/10, 1991, pp. 564–9.

Subramanian, Kapil, 'Why Nobel Laureate CV Raman Resented Nehru (and Even Took a Public Swipe at Him)', *Scroll.in*, 7 November 2007, https://tinyurl.com/vr6f9wkx.

Subramanian, Samanth, *A Dominant Character: The Radical Science and Restless Politics of J.B.S. Haldane*, Simon & Schuster, 2019.

Sukumar, Arun Mohan, *Midnight's Machines: A Political History of Technology in India*, Penguin Random House, Gurugram, 2019.

Sundaram, C.V., L.V. Iyengar and T.S. Iyengar, *Atomic Energy in India: 50 Years*, Government of India, Department of Atomic Energy, August 1998.

Sur, Abha, 'Scientism and Social Justice: Meghnad Saha's Critique of the State of Science in India', *Historical Studies in the Physical and Biological Sciences*, Vol. 33, No. 1, 2002, https://tinyurl.com/mtp3ketk.

'Sustaining the Search for Excellence: Higher Education and the J.N. Tata Endowment', *Tata Review*, Volume 33, No. 3, September 1998.

Swarup, Govind, 'Homi Bhabha—Personal Reminiscences', *Current Science*, Vol. 60, No. 2, 25 January 1991.

Tata Institute of Fundamental Research, *Annual Report, 1947–48*, 1948.

'Tata Memorial Hospital', Sands of Time, Tata Central Archives, 2016, https://tinyurl.com/y4v64b7c.

Taylor, H.J., P. J. Lavakare and R.R. Daniel, 'Seventy Years of Physics', *Current Science*, Vol. 59, No. 23, 10 December 1990, https://tinyurl.com/3jdcud9j.

Thakurdas, Purshotamdas, *A Plan of Economic Development for India: Parts One and Two*, Penguin Books, 1944.

Thapar, Romesh, 'In the Shadow of the Bomb', *Economic and Political Weekly*, 31 October 1964, https://tinyurl.com/5huj9yuw.

Venkataraman, G., *Bhabha and His Magnificent Obsessions*, Universities Press, Hyderabad, 1994.

Venkataraman, G., *Journey into Light: Life and Science of C.V. Raman*, Indian Academy of Sciences, Bangalore, 1988.

von Leyden, Rudolf, 'The Drawings of Homi Bhabha', *Marg*, Vol. 6, No. 3, June 1953.

Wadia, Spenta R., 'Homi Jehangir Bhabha and the Tata Institute of Fundamental Research', *Current Science*, Vol. 96, No. 5, 10 March 2009, https://tinyurl.com/yc3ka39x.

Wagner, Caroline S., *The New Invisible College: Science for Development*, Brookings Institution, Washington D.C., 2008.

Wali, Kameshwar C., *Chandra: A Biography of S. Chandrasekhar*, Penguin Books, New Delhi, 1991.

Warwick, Andrew, *Masters of Theory: Cambridge and the Rise of Mathematical Physics,* University of Chicago Press, 2003.

Watson, Peter, *A Terrible Beauty: A History of the People and Ideas that Shaped the Modern Mind*, Weidenfeld & Nicolson, London, 2000.

Weaver, David John, *The Parsi Dilemma: A New Zealand Perspective*, 2012, Victoria University of Wellington, PhD dissertation.

Weiner, Norbert, *I am a Mathematician: The Later Life of a Prodigy,* Doubleday, 1956.

Weisskopf, Victor, *The Joy of Insight: The Passions of a Physicist*, Plunkett Lake Press, 2017.
What Really Happened to Homi J Bhabha P2, YouTube, https://tinyurl.com/394x4uea.

INDEX

25-9	1.00	1.00	1.00
1616	1.26	1.86	1.73
2240	1.82	4.38	3.17
4300	2.29	8.27	4.83

Use of correct cascade theory instead of Serber's solution would not change general conclusion; though the actual numerical rates of decay to primary soft component.

Johnson & Shutt, '42 P.R. 61. 380-1.
Cloud Chamber Track of a Mesotron stopped by Gas

−ve meson track ends in gas after passing through 1 cm Pb plate. Track above plate thin; below, heavily ionizing. No ~~de~~ decay electron visible from end of track

Origin of the soft component of cosmic rays.

No: D_2 of cascade electrons due to meson decay calculated (using Serber's ~~method~~ cascade calculations) Those due to fast mesons D (penetrate 15 cm Pb, momentum $\geq 3\times10^8$ e and slow mesons D_1 separated ~~(D_1)~~.

No: of fast mesons, N, slow mesons N_1, no: of electrons due to meson collision, C.

Depth in rad. units	N_1/N	D/N	D_1/N	C/N	$\frac{N_1+D+D_1+C}{N}$	Observed
23.41	.07	.22	.010	.10	.40	.50
19.93	.09	.31	.014	"	.51	.74
16.47	.13	.39	.025	"	.65	1.20
14.33	.27	.47	.060	"	.90	1.81.

Shows that actual no: of soft component increases much more rapidly than no: ~~of~~ if produced by mesons ~~comp.~~ through decay + collision. At S.L. two numbers comparable, but at 4300 m. some half at least must be due to burst cascades produced by electrons at the top

Hence not all soft-comp: above S.L. can be

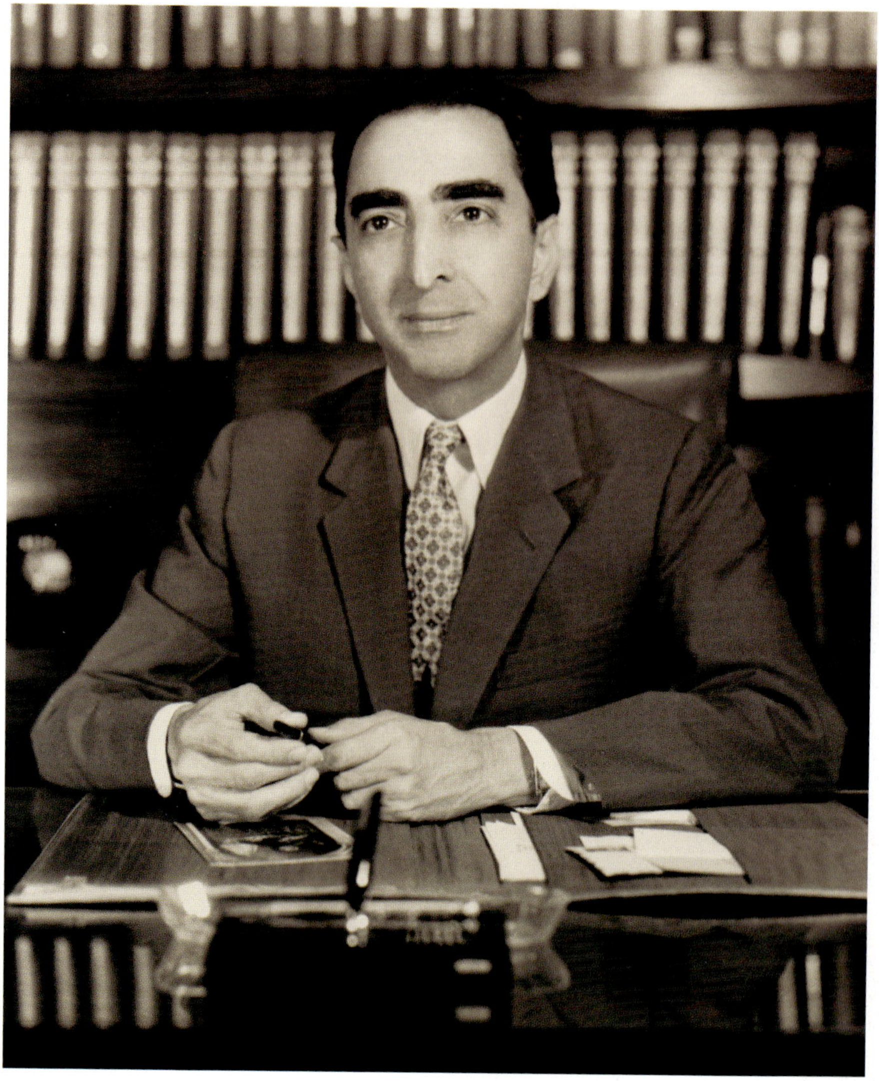

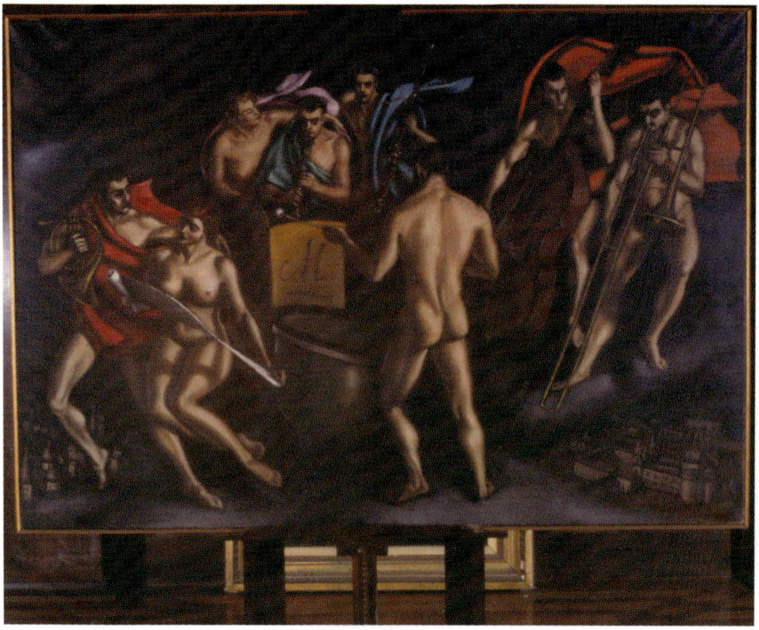

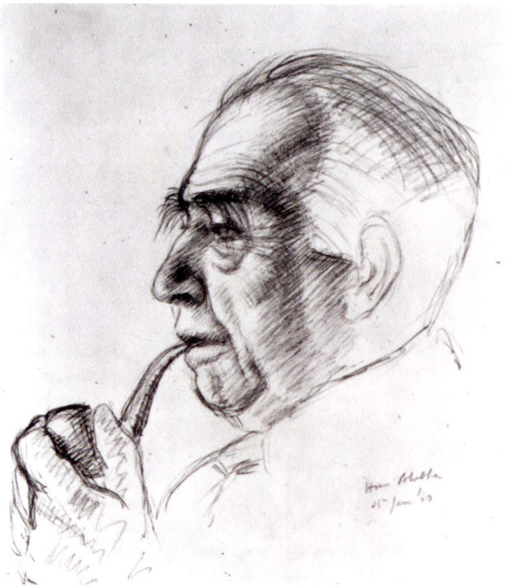

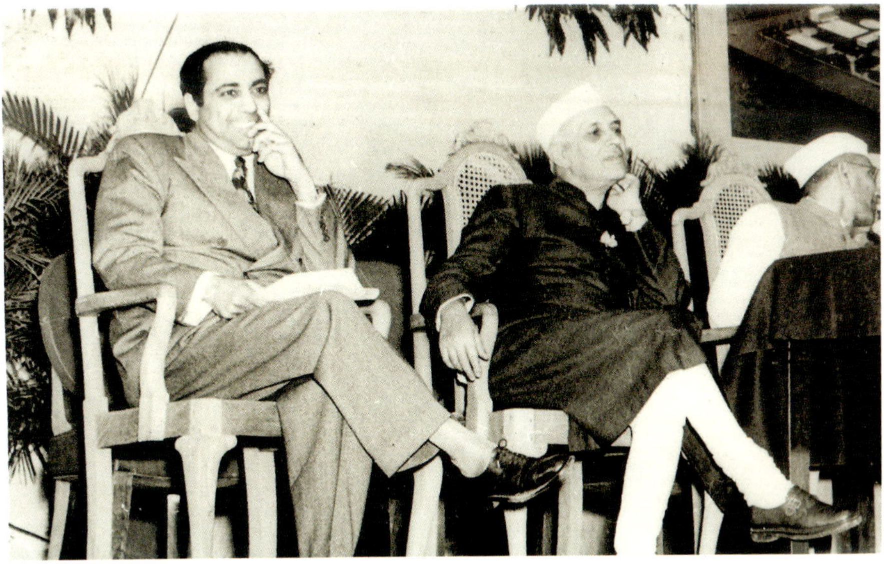

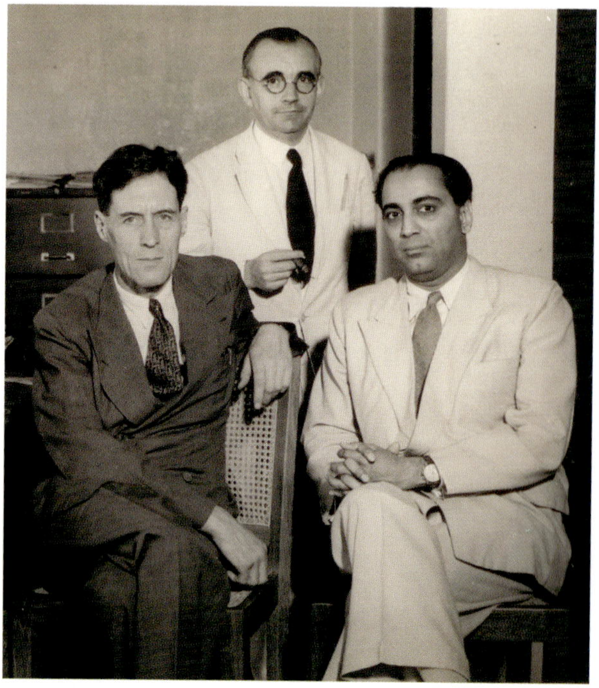

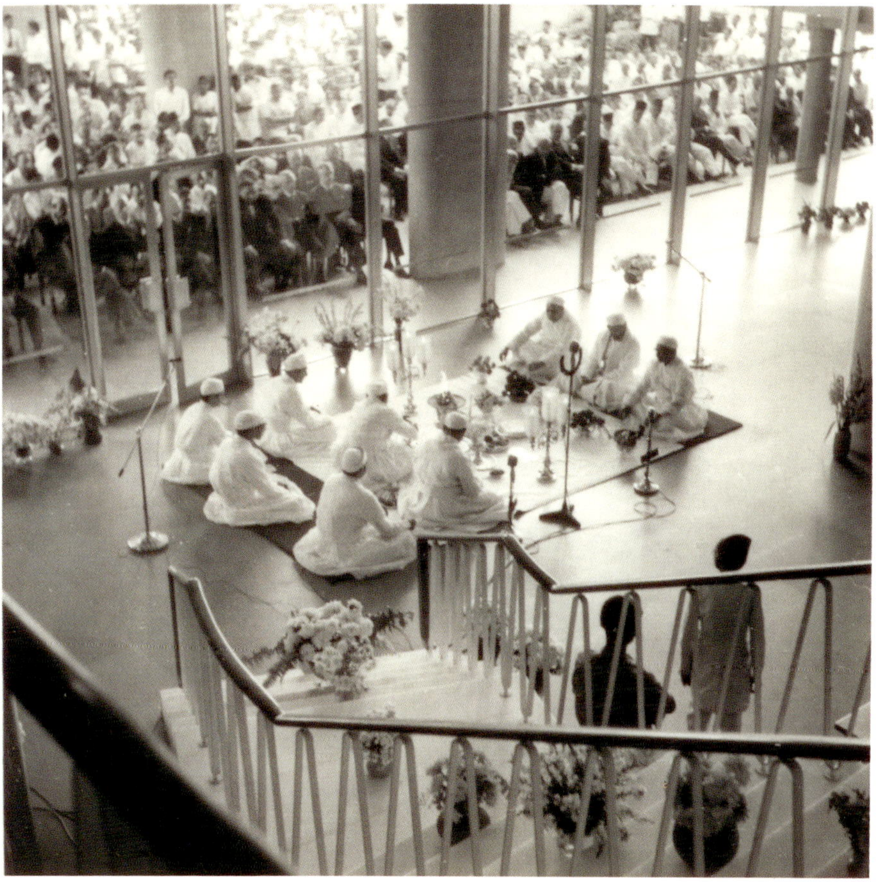

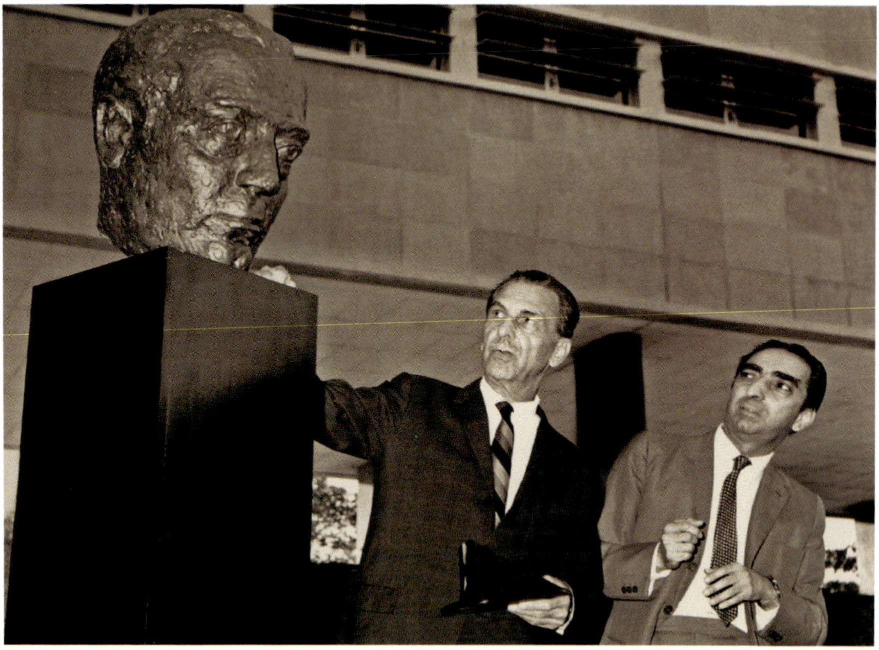